7th International Conference on Mechanical, Industrial, and Manufacturing Technologies (MIMT 2016)

MATEC Web of Conferences Volume 54 (2016)

Cape Town, South Africa
1 - 3 February 2016

Editors:

Khaled Abou-El-Hossein

ISBN: 978-1-5108-2332-7

Printed from e-media with permission by:

Curran Associates, Inc.
57 Morehouse Lane
Red Hook, NY 12571

Some format issues inherent in the e-media version may also appear in this print version.

Printed by Curran Associates, Inc. (2016)

For additional information, please contact EDP Sciences – Web of Conferences at the address below.

EDP Sciences – Web of Conferences
17, Avenue du Hoggar
Parc d'Activité de Courtabœuf
BP 112
F-91944 Les Ulis Cedex A
France

Phone: +33 (0) 1 69 18 75 75
Fax: +33 (0) 1 69 28 84 91

contact-edps@webofconferences.org

Additional copies of this publication are available from:

Curran Associates, Inc.
57 Morehouse Lane
Red Hook, NY 12571 USA
Phone: 845-758-0400
Fax: 845-758-2633
Email: curran@proceedings.com
Web: www.proceedings.com

TABLE OF CONTENTS

ISSN (electronic edition): 2261-236X

2016 7th International Conference on Mechanical, Industrial, and Manufacturing Technologies

Cape Town, South Africa
February 1-3, 2016

Editors:

MIMT was held to provide a platform for researchers, scientists and people working in the related fields to get together and share their experiences and lessons with other enthusiasts, and develop opportunities for cooperation. It arouses great interests and attentions from domestic and foreign researchers in mechanical, industrial, and manufacturing technologies field, More than 50 papers have been accepted which covers most of the current topics of the related fields, more than 70 people from scientific institutes, universities attended this conference. The conference program is rich. The participants gave excellent speeches on the topics and have lively discussions. We are sincerely grateful to the participants' interest and support as well as their contributions to the conference. We are also grateful to the great efforts of the editors, reviewers and publisher engaged in this conference proceeding.

Committee Members

Conference Co-Chairs

Prof. K. ABOU-EL-HOSSEIN, Nelson Mandela Metropolitan University, South Africa

Prof. Ramesh Kuppuswamy B.Eng. M. Tech PhD (Singapore) SM SME, Department of Mechanical Engineering Advanced Manufacturing Laboratory, University of Cape Town South Africa

Program Co-Chairs

Dr. Kumuran Kadirgama, Senior Lecturer, Faculty of Engineering, Universiti Malaysia Pahang (UMP)

Prof. Wen-Yeuan Chung, Department of Mechanical Engineering, Chinese Culture University, Taiwan

Technical Committees

Prof. Samaila Saleh, Hassan Usman Katsina Polytechnic, Nigeria

Prof. Tatsuhito AIHARA, Hosei University, Japan

Dr. Subaschandar Narayana Rao, BIUST, Botswana
Prof. Zenghui Wang, University of South Africa

Prof. Sangmo Kang, Department of Mechanical Engineering, Dong-A University, Korea

Asst. Prof. Rahul Davis, Department of Mechanical Engineering, SHIATS -Demeed-to-be-University, Allahabad, India

Asst. Prof. Sung Jae Kim, Seoul National University, Korea

Asst. Prof. Vishal Naranje, Mechanical Engineering Department BITS Pilani, Dubai campus, India

Prof. Masoud Tahani, Ferdowsi University of Mashhad, Iran

Prof. Tahir I. Khan,Department of Mechanical & Industrial Engineering, Qatar University, Qatar

Assoc. Prof. Ching-Yu Hsu, Department of Marine Mechanical Engineering, R.O.C. Naval Academy, Taiwan

Assoc. Prof. Sadok Sassi, Qatar University, Qatar

Assoc. Prof. Dinesh Seth, Qatar University, Qatar

Assoc. Prof. Samer F. Ahmed, Qatar University, Qatar

Prof. Yanxia Sun, department of Electrical and Electronic Engineering Science, University of Johannesburg, South Africa

Dr. MELLAL Mohamed Arezki, Faculty of Engineering Sciences (FSI), M'Hamed Bougara University,Boumerdes, Algeria

Prof. Xingbo WANG, Department of Mechatronics, Foshan University,Foshan City, China

PREFACE

2016 7th International Conference on Mechanical, Industrial, and Manufacturing Technologies (MIMT) was organized by IACT and was held in Cape Town, South Africa during 1-3 February 2016 at the Cape Milner Boutique Hotel. It aims to put forward the newest development and research achievement in mechanical, industrial, and manufacturing technologies and foster communication among researchers and practitioners who works in a wide variety of scientific areas with a common interest in improving mechanical, industrial, and manufacturing technologies.

The proceedings of MIMT 2016 consist of 57 selected papers which are presented orally at the conference or via poster sessions. The topics of papers include 12 branches, including material science, robot design and development technology, mechanical control and manufacturing system, automation and control engineering, automation and control engineering, automation and control engineering, automation and control engineering, image processing and visualization, satellite and communication engineering, signal analysis and processing, applied mechanics and fluid dynamics, electronic application technology. The papers are accepted and being peer-reviewed by the conference committee, international reviewers based on the topic of each paper. All the authors of accepted papers gave excellent presentations in this meeting. Participants have exchanged in detail around mechanical, industrial, and manufacturing technologies. On the basis of the conference achievement, we collect and compile these research papers into this conference proceeding, in which the content represents the latest international academic progress of mechanical, industrial, and manufacturing technologies. We hope the presented papers will be interesting for readers and a good base for inspiration for future developments.

On behalf of the conference organizing committee, we would like to express our sincere appreciation to the conference chairs, Prof. K. Abou-El-Hossein and Prof. Ramesh Kuppuswamy, program chairs, Dr. Kumuran Kadirgama and Prof. Wen-Yeuan Chung, sponsors and every technical committee member, as well as the editors and authors for their kind support to us so that we could pull off the conference and their efforts to put together the proceedings.

MIMT 2016 Conferences Organizing Committees

February 1, 2016.

Statement of Peer review

In submitting conference proceedings to MATEC Web of Conferences, I certify to the Publisher that I adhere to the **Policy on Publishing Integrity** of the journal in order to safeguard good scientific practice in publishing.

1. All articles have been subjected to peer review administered by the proceedings editors.
2. Reviews have been conducted by expert referees, who have been requested to provide unbiased and constructive comments aimed, whenever possible, at improving the work.
3. Proceedings editors have taken all reasonable steps to ensure the quality of the materials they publish and their decision to accept or reject a paper for publication has been based only on the merits of the work and the relevance to the journal.

Title, date and place of the conference

2016 7th International Conference on Mechanical, Industrial, and Manufacturing Technologies (MIMT 2016)

Cape Town, South Africa

1-3 February 2016

Proceedings editor(s): Prof. Khaled Abou-El-Hossein

Date and editor's signature

Analysis and Evaluation of Interface Mechanics for Thin Films using the Method of Image

Chongyang NIE, Le GU, Dezhi ZHENG , Xiaoli ZHAO, and Liqin WANG

Harbin Institute of Technology, Research Lab of Space & Aerospace Tribology, Hei Longjiang Province, Harbin 150001, China.

Abstract. The evaluation of interface mechanics was very useful for the design of thin solid film. According to the method of image, interface stresses and deformations of coating-substrate system were gained. Using this mathematical calculation model, the indentation test data was carried out and the results were analysed. Based on the mechanical model and test, the critical state of interface stresses from elastic deformation to plastic yield was evaluated. From experimental and calculating results, interface shear stress and normal stress had different affecting level for the coating failure under the same surface load. And a fitting relationship equation for the two stresses was also gained to determining the failure value in designing. The method of this paper could be used for thin film interface mechanics design and evaluation.

1 Introduction

Thin solid films or coatings are widely used as combining material in civil and industry applications, such as rolling bearings, gears, and many other mechanical components. With surface modified layers, these components behaved distinguished load carrying capacity, contact friction performance, and environmental adaptability.

Today many scholars researched and designed varying coatings for different applications and environments [1-3]. In high-speed, high-temperature, and heavy-load [4], as well as other extreme environments [5, 6], thin solid films were good solutions. And varies researches and designs were carried out for different coatings in different mechanical components [7], followed the interest of tribo-pairs with layered structure for their service life and reliability[8-10].

Nano indention and scratch were common tools for material mechanics tests. But these experiments were totally reliant on the test samples, instruments, environments, operations, and manipulators, so the results and criterions in these experiments were significantly diverse and impossible to compare with each other. Sometimes a conclusion from one test would even contradict the other ones found in different tests [11-13]. So a better understanding of indention data needed a mathematical model to evaluate interface stresses and deformations.

The analysis and evaluation of thin films interface could also be used for performance improvement and optimization with the aim of improving stress level in the substrates and enhancing bonding strength along the interface. Based on the analysis of interface mechanics [14, 15] in this paper, a coating-substrate system model was built using mathematical formulations, while the interface mechanics characteristics of diamond-like carbon (DLC) coating deposited on Si_3N_4 ceramics were analysed and evaluated theoretically and experimentally.

2 Theoretical analyses

For a coating-substrate system, interface was a vital layer to connect the bi-materials, and its stresses and deformations were investigated in the following sections.

2.1 Analytical model of coating-substrate system

Thin solid film was deposited on an elastic half-space body as shown in Figure 1. The elastic modulus of coating *I* and substrate *II* were defined as E_1 and E_2, Poisson's ratio v_1 and v_2, respectively. h was the thickness of the coating and, as it was extremely thin in regards to the substrate, the latter was therefore considered as semi-infinite. From Figure 1, a pressure distribution was applied on the thin film surface which meant a Hertz's normal contact state $p(\eta)$ with tangential friction traction force $q(\eta)$. The total force was the integration of normal force and tangential traction along the surface coordinate.

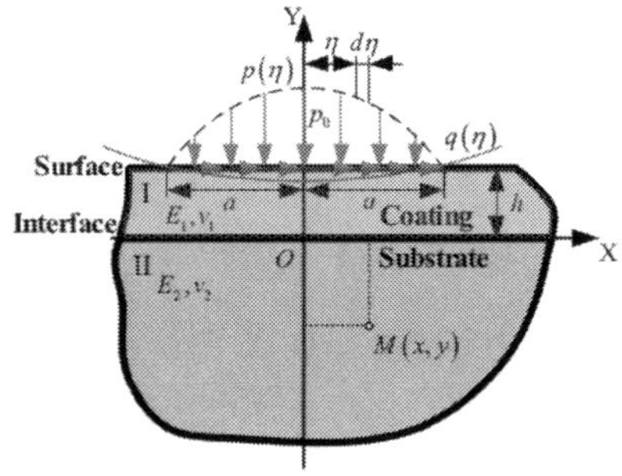

Figure 1. The model of coating-substrate system

From Figure 1, coating *I* was used to cover the surface of substrate *II*, connected only by the interface. Above the coating there was a free surface, and *X*-axis (real axis) was along the interface, and *Y*-axis (imaginary axis) was perpendicular with *X*-axis and passed through O_1. The two axes intersected at *O*, namely the global origin point and the reference frame *XOY*.

2.2 The method of image

The interface stresses shown in Figure 2 could be calculated by the method of images from the complex variable function. The boundary conditions and the constraints of the model were equivalent; furthermore, equations were composed using the relationship between stress and strain. As the method of images, the interface and the surface were imagined as mirrors that reflected point *O* or O_1 and that would generate infinite points of mirror images. These images influenced the interface stresses as a superposition form in order to maintain the boundary conditions and constraints. Namely, the conditions of interface continuum and free surface were satisfied and interface stresses were gained by superposing the stress solutions of these image points.

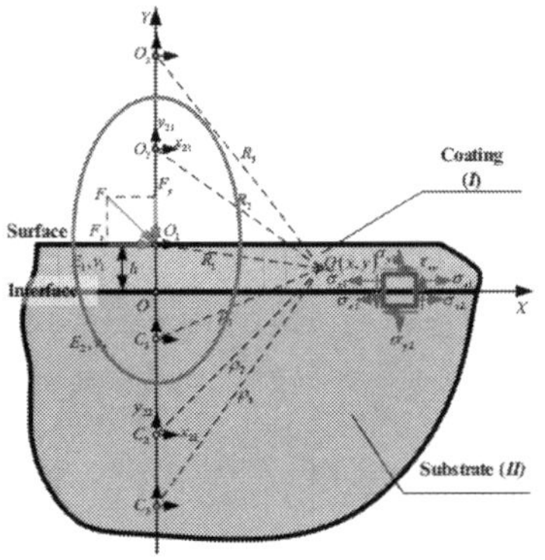

Figure 2. The scheme of the method of image

The complex interface and surface boundary condition were equation (1) and (2)

$$\sigma_{yI}+i\tau_{xyI}=\sigma_{yII}+i\tau_{xyII} \text{ and } u_I+iv_I=u_{II}+iv_{II}, \text{ for } y=0 \quad (1)$$

$$\sigma_{yI}+i\tau_{xyI}=0, \text{ for } y=h \quad (2)$$

Here subscript I and II denoted material I and II, respectively. σ_x and σ_y were stresses along *x* and *y* direction, τ_{xy} was the shear stress. *u* and *v* represented deformations along *x* and *y* direction.

The analysis procedure was rather cumbersome, and the readers who have interest could referred literature [14] and [15] for details. Here the Green's functions for the relationship between stress and deformation were gave as following equations (3)-(8) .

$$\begin{aligned} d(\sigma_{yI}+i\tau_{xyI}) &= G_1(z_k,s_k,\eta)\,d\eta \\ G_1(z_k,s_k,\eta) &= \sum_{k=1}^{\infty}\left[A_k'(z_k-\eta)+\Phi_k'(s_k-\eta)\right] \\ &+\sum_{k=1}^{\infty}\left[\overline{A_k'}(z_k-\eta)+\overline{\Phi'}_k(s_k-\eta)\right] \\ &+(\bar{z}-\eta)\sum_{k=1}^{\infty}\left[A_k''(z_k-\eta)+\Phi_k''(s_k-\eta)\right] \\ &+\sum_{k=1}^{\infty}\left[B_k'(z_k-\eta)+\Psi_k'(s_k-\eta)\right] \end{aligned} \quad (3)$$

$$\begin{aligned} d(\sigma_{xI}+\sigma_{yI}) &= G_2(z_k,s_k,\eta)\,d\eta \\ G_2(z_k,s_k,\eta) &= 4\mathrm{Re}\left(\sum_{k=1}^{\infty}\left[A_k'(z_k-\eta)+\Phi_k'(s_k-\eta)\right]\right) \end{aligned} \quad (4)$$

$$\begin{aligned} 2\mu_I d(u_I+iv_I) &= G_3(z_k,s_k,\eta)\,d\eta \\ G_3(z_k,s_k,\eta) &= \kappa_I\sum_{k=1}^{\infty}\left[A_k(z_k-\eta)+\Phi_k(s_k-\eta)\right] \\ &-(z-\eta)\sum_{k=1}^{\infty}\left[\overline{A_k'}(z_k-\eta)+\overline{\Phi_k'}(s_k-\eta)\right] \\ &-\sum_{k=1}^{\infty}\left[\overline{B_k}(z_k-\eta)+\overline{\Psi_k}(s_k-\eta)\right] \end{aligned} \quad (5)$$

$$\begin{aligned} d(\sigma_{yII}+i\tau_{xyII}) &= G_4(z_k,s_k,\eta)\,d\eta \\ G_4(z_k,s_k,\eta) &= \sum_{k=1}^{\infty}a_k'(z_k-\eta) \\ &+\sum_{k=1}^{\infty}\overline{a_k'}(z_k-\eta) \\ &+(\bar{z}-\eta)\sum_{k=1}^{\infty}a_k''(z_k-\eta)+\sum_{k=1}^{\infty}b_k'(z_k-\eta) \end{aligned} \quad (6)$$

$$\begin{aligned} d(\sigma_{xII}+\sigma_{yII}) &= G_5(z_k,s_k,\eta)\,d\eta \\ G_5(z_k,s_k,\eta) &= 4\mathrm{Re}\left[\sum_{k=1}^{\infty}a_k'(z_k-\eta)\right] \end{aligned} \quad (7)$$

$$\begin{aligned} 2\mu_{II} d(u_{II}+iv_{II}) &= G_6(z_k,s_k,\eta)\,d\eta \\ G_6(z_k,s_k,\eta) &= \kappa_{II}\sum_{k=1}^{\infty}a_k(z_k-\eta) \\ &-(z-\eta)\sum_{k=1}^{\infty}\overline{a_k'}(z_k-\eta) \\ &-\sum_{k=1}^{\infty}\overline{b_k}(z_k-\eta) \end{aligned} \quad (8)$$

3 Experimental details

The DLC coating was deposited on the surfaces of Si_3N_4 ceramics by plasma based ion implantation (PBII) [13]. The disk samples were Φ22 mm×3 mm with a surface roughness Ra 0.02-0.04 μm and had been previously ultrasonic cleaned for 5 min by acetone and anhydrous ethanol before thin film deposited onto the surfaces. The oxide and adsorbent on the samples' surface were removed by cathode deflection Ar+ ion under 2-3 mPa vacuum for 5 minutes. Then, the chamber pressure was set to 1.5 Pa and filled with reactant $V(C_2H_2)$:$V(H_2)$=10: 1. The period of deposition was for 3 hours using an 20 mA current and a voltage of 20 kV.

DOI: 10.1051/matecconf/20165401001

Tested by the spectral photometer ETA-SST, the thickness of the DLC coating was 300 nm. The load-carrying ability of the DLC coating was measured using Nano-Indenter XP so that its curvature radius of the conical indenter was 0.99 μm.

Six experiments [14] were implemented on every sample. The load was applied at the increasing rate of 0.5 mN/μm when the indenter made contact with the surface of the sample. The forces were recorded at the abrupt point of the penetration curves.

4 Results and discussions

Here based on the test data, interface stresses and deformations were calculated and discussed.

4.1. Test results

The penetration curves with the load increasing were shown in Figure 3 for different film thickness.

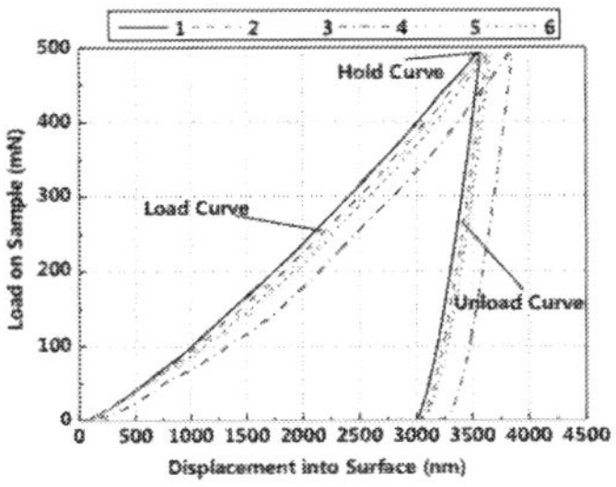

(a) The coating thickness of 110.6 nm

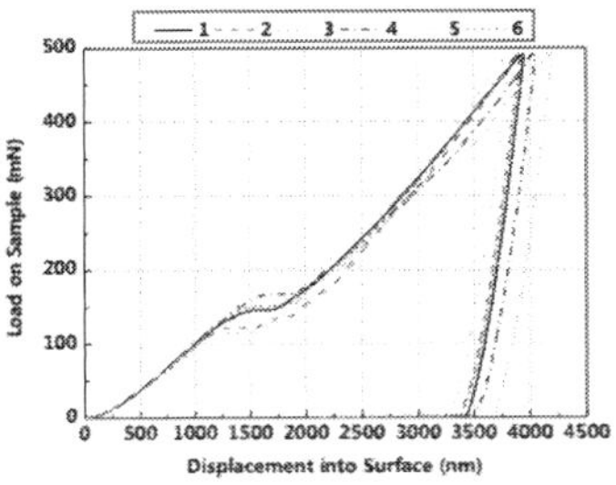

(b) The coating thickness of 381.8 nm

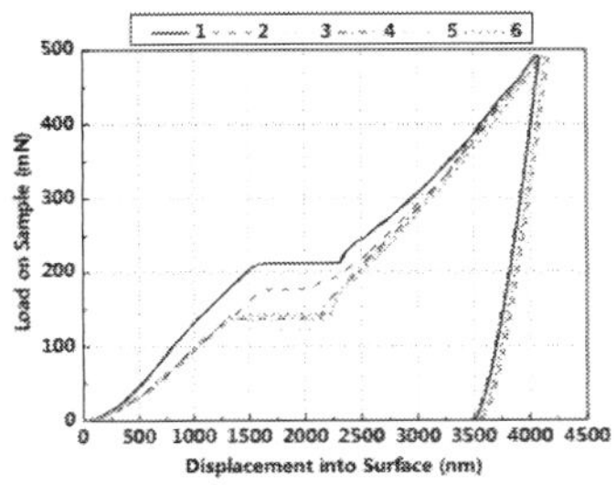

(c) The coating thickness of 763.8 nm

Figure 3. The penetration curves with the load increasing in indention tests

From the curve trends Figure 3 (a) – (c), the thicker the coating was, the more distinct of the smooth curvature was.

After the indention tests, the penetration depth by the indenter normal movement and the contact load applied on it were recorded, the data was listed in Table 1 when the curvature of test load lines emerged a step, which meant the coating-substrate system underwent from elastic deformation to elastic-plastic deformation.

Table 1. Indentation test data

h(nm) / No.	110.6	381.8	763.8	110.6	381.8	763.8
	Depth (nm)			Load (mN)		
1	764	1288	1562	72	132	210
2	682	1233	1638	59	121	175
3	751	1274	1241	65	128	123
4	769	1637	1312	52	167	137
5	745	1141	1314	60	113	135
6	815	1281	1319	72	134	139
Ave.	754	1309	1398	63	132	153
Dev.	16	63	60	3	7	12

4.2 Analysis and evaluation

Here using the mathematical model of section 2, interface stresses and deformations were calculated for different coating thicknesses. As shown in Figure 4, stresses and deformations were increased with the contact load, the thicker coating could carry a heavier load or a larger interface stress level before plastic deformation occurred.

From Figure 4 (a) to (c), if the thickness of coating was rather thin as 110.6 nm, it plastically deformed easily than thicker ones. Though the critical loads and stresses level in thinner solid film were a little lower, when the thickness increasing, the limiting value of the stress was presence and the stress divergence between thickness 381.8 nm and 763.8 nm was already very approaching.

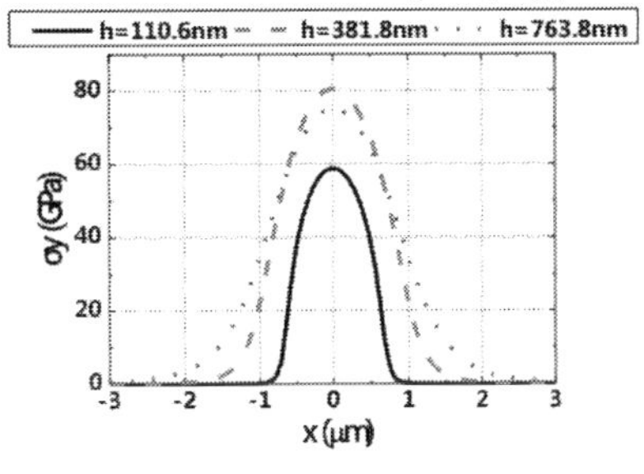

(a) Normal stresses along the interface

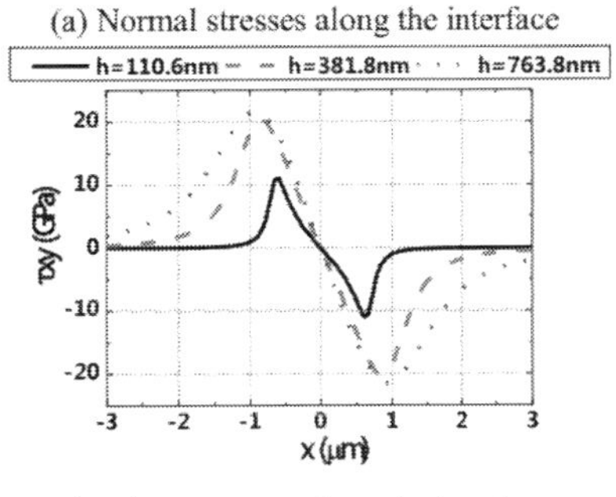

(b) Shear stresses along the interface

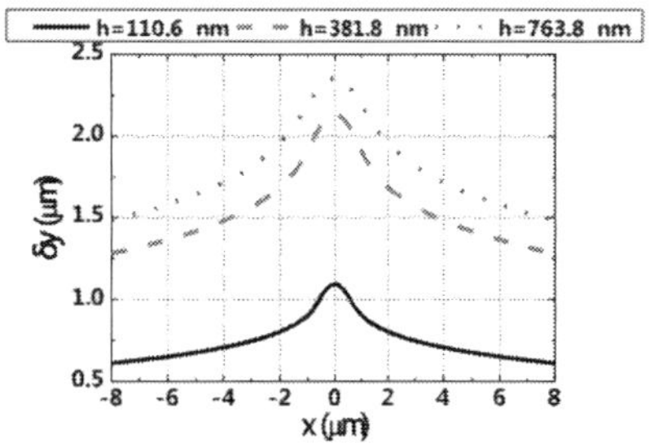

(c) Normal deformations along the interface

Figure 4. Calculating results of stresses and deformations for different coating thicknesses

About the failure mode of DLC coating, interface peeled off was similar with brittle material such as rocks. So Coulomb-Mohr criterion [15, 16] was used here for failure evaluation. The calculated maximum interface shear stresses and normal stresses were chosen from Figure 4 at the critical point of indention test. A fitting equation $\tau_{max} = 0.34857\sigma_y - 15.9873$ was gained by linear relationship as shown in Figure 5.

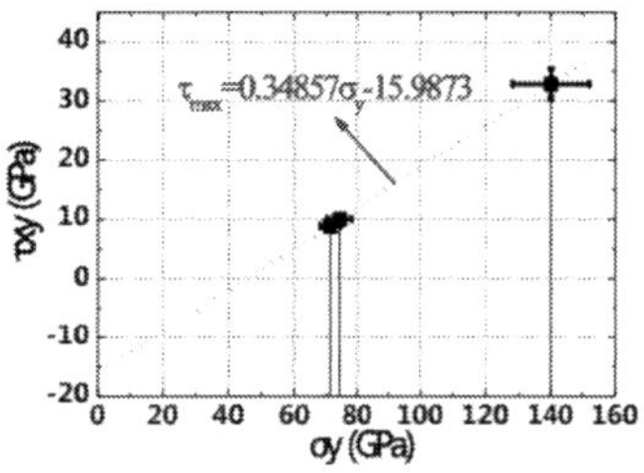

Figure 5. Fitting relationship between maximum shear stress and normal stress

The intercept of fitting line was 15.9873 GPa with *Y*-axis, and the shear bond strength of the interface could represent by this value [15, 16]. This critical value could be the plastic failure stress of coating-substrate system for coating design. From the fitting equation we found that the influence level of interface carrying capacity by tensile stress and compression stress was different.

5 Conclusions

Interface stresses and deformations of coating-substrate system were calculated by the method of image. According to the indentation test, the experimental data was analysed using this mathematical calculation model, and the interface critical state from elastic deformation to plastic deformation was evaluated.

From experimental and calculating results, interface shear stress and normal stress were in different level for interface failure under the same surface load. The analytical and evaluative method here could be used for thin film interface mechanics design and evaluation.

Acknowledgements

This research is supported by Major State Basic Research Development Program (973) (2013CB632305) and the National Natural Science Foundation of China (51475452 and 51275125). Thanks for the testing works from Cuini Jing.

References

1. K. Holmberg, A. Laukkanen, E. Turunen, and T. Laitinen, Wear resistance optimisation of composite coatings by computational microstructural modelling, Surface and Coatings Technology, **247**, pp. 1-13, 5/25/ 2014.
2. A. Ghabchi, S. Sampath, K. Holmberg, and T. Varis, Damage mechanisms and cracking behavior of thermal sprayed WC–CoCr coating under scratch testing, Wear, **313**, 97-105, 5/15/ 2014.
3. J. C. Avelar-Batista Wilson, S. Banfield, J. Eichler, A. Leyland, A. Matthews, and J. Housden, An investigation into the tribological performance of Physical Vapour Deposition (PVD) coatings on high thermal conductivity Cu-alloy substrates and the effect of an intermediate electroless Ni–P layer prior to PVD treatment, Thin Solid Films, **520**, 2922-2931, 1/31/ 2012.
4. Y. Alinia, M. A. Guler, and S. Adibnazari, On the contact mechanics of a rolling cylinder on a graded coating. Part 1: Analytical formulation, Mechanics of Materials, **68**, 207-216, 1// 2014.
5. S. B. Weber, H. L. Lein, T. Grande, and M. A. Einarsrud, Influence of the precursor solution chemistry on the deposition of thick coatings by spray pyrolysis, Surface and Coatings Technology, **221**, 53-58, 2013.
6. M. A. Guler, Y. Alinia, and S. Adibnazari, On the contact mechanics of a rolling cylinder on a graded coating. Part 2: Numerical results, Mechanics of Materials, **66**, 134-159, 11// 2013.
7. X. Zhang and B. L. Weeks, Effects on the surface structure of organic energetic materials using spin coating, Thin Solid Films, **550**, 135-139, 2014.
8. J. Scheibert, A. Prevost, G. Debrégeas, E. Katzav, and M. Adda-Bedia, Stress field at a sliding frictional contact: Experiments and calculations, Journal of the Mechanics and Physics of Solids, **57**, 1921-1933, 12// 2009.
9. R. Hauert, K. Thorwarth, and G. Thorwarth, An overview on diamond-like carbon coatings in medical applications, Surface and Coatings Technology, **233**, 119-130, 10/25/ 2013.
10. K. Holmberg, A. Laukkanen, H. Ronkainen, and K. Wallin, Surface stresses in coated steel surfaces—influence of a bond layer on surface fracture, Tribology International, **42**, 137-148, 1// 2009.
11. T. P. Nowak, L. J. Jankowski, and J. JasieŃKo, Application of photoelastic coating technique in tests of solid wooden beams reinforced with CFRP strips, Archives of Civil and Mechanical Engineering, **10**, 53-66, // 2010.

12. H. S. Park and D. Kwon, An energy approach to quantification of adhesion strength from critical loads in scratch tests, Thin Solid Films, **307**, 156-162, 10/10/ 1997.
13. C. N. Jing, L. Gu, Ch. W. Zhang, G. Z. Tang, L. Q. Wang, Preparation and tribological properties research of DLC film on the surface of Si_3N_4 ceramics, **35**, 9, Lubr. Eng. Sept. (2010) 21-25.
14. C. NIE, D. ZHENG, L. GU, X. ZHAO, L. WANG, Comparison of interface mechanics characteristics of DLC coating deposited on bearing steel and ceramics, Applied Surface Science, **317**, 188–197, (2014).
15. J. Q. Xu, Interface mechanics, Science Press, Beijing, 2006.
16. Z. Y. ZHANG & J. Q. XU, The interface failure criterion of coating materials, Journal of Shanghai Jiaotong University. pp. 983-987, 2007.

Phase Separation and Domain Formation in Multi-Component Membranes: Finsler Geometry Modeling and Monte Carlo Simulations

Satoshi Usui[1] and Hiroshi Koibuchi[2]

[1]*Advanced course of Mechanical Engineering, National Institute of Technology, Ibaraki College, Japan*
[2]*Department of Mechanical and Systems, National Institute of Technology, Ibaraki College, Japan*

Abstract. In this paper, we study a surface model for membranes of three components such as DPPC, DOPC, and Cholesterol. This membrane is experimentally well known to undergo the phase separation and to form the domain structure such as the liquid ordered (L_o) phase and the liquid disordered phase (L_d). It is also well known that this multicomponent membrane has a lot of domain pattern transitions between the circular domains and the striped domain etc. Using the new surface model constructed on the basis of Finsler geometry, we study why those morphological changes appear on the spherical vesicles. In our model, we introduce a new variable $\sigma (\in Z_2)$ to represent the domains L_o and L_d, and using the value of σ we define a metric function on the surface. As a consequence, the origin of the line tension energy, which has been used to explain the domain pattern transition in the multicomponent membranes, is naturally understood in our model.

1 Introduction

T Artificial membranes such as vesicles are used to study the mechanism of biological cell membranes [1]. The vesicles, which are composed of saturated phospholipids, unsaturated phospholipids, and cholesterols, undergo a phase separation and a domain pattern transition [2]. After the phase separation, two types of domains appear: one is the liquid ordered (L_o) phase, which is rich in the saturated phospholipids and cholesterols, and the other is the liquid disordered (L_d) phase, which is rich in the unsaturated phospholipids.

These two domains cause the domain pattern transition, which accompanies the shape deformation [2]. For example, the oblate shape membranes with two Circular L_o Domains (2CD) such as **Fig.1** (a), and the prolate shape membranes with one Striped L_o Domain (1SD) such as **Fig.1** (b) are well known. It is also well known that the L_o phase forms many round domains, and this domain structure is very similar to the so called lipid raft.

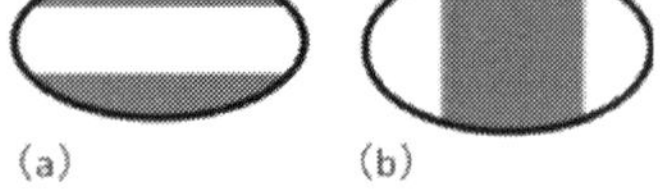

Figure 1. General form of (a) 1SD, and (b) 2CD. The blue (white) region corresponds to L_o (L_d) phase.

However, the mechanism of shape deformation is still unclear, although a lot of theoretical studies have been conducted. For example, the mechanism of shape deformation to 2CD and 1SD can be explained by using the models with the line tension on the boundary between L_o and L_d phases [3, 4]. However, the origin of the line tension is unknown. The problem that should be asked is the microscopic origin of the line tension.

The purpose of this study is to find the origin of the line tension and clarify the mechanism of shape deformation, which is produced by the phase separation. For this purpose, we define a Finsler geometry surface model [5] by extending the conventional Helfrich-Polyakov model for membranes [6, 7]. In the new model, we introduce a new degree of freedom $\sigma (\in Z_2)$, which corresponds to the two phases L_o and L_d. The interaction between this σ and the membrane shape is naturally introduced by a Finsler metric, which is defined by using the value of σ.

2 Model

2.1 Continuous model

The Hamiltonian (=energy function) of the continuous mode is defined by a mapping $\mathbf{r}$ from a two-dimensional surface M to the three-dimensional Euclidean space $\mathbf{R}^3$. The continuous description of the Hamiltonian is as follows:

$$S = \lambda S_0 + S_1 + \kappa S_2,$$
$$S_0 = \frac{1}{2}\int \sqrt{g} dx^2 g^{ab} \frac{\partial \sigma}{\partial x_a} \cdot \frac{\partial \sigma}{\partial x_b},$$
$$S_1 = \int \sqrt{g} dx^2 g^{ab} \frac{\partial \mathbf{r}}{\partial x_a} \cdot \frac{\partial \mathbf{r}}{\partial x_b}, \quad (1)$$
$$S_2 = \frac{1}{2}\int \sqrt{g} dx^2 g^{ab} \frac{\partial \mathbf{n}}{\partial x_a} \cdot \frac{\partial \mathbf{n}}{\partial x_b}.$$

The metric tensor is given by

$$g_{ab} = \begin{pmatrix} \rho^{-1} & 0 \\ 0 & \rho \end{pmatrix}, \quad \rho > 0, \quad (2)$$

where ρ is a function on the surface. The inverse metric and the determinant are $g^{ab} = (g_{ab})^{-1}, g = \det(g_{ab})$.

The metric g_{ab} in Eq.(2) comes from the most general metric such as $g_{ab} = \begin{pmatrix} E & F \\ F & G \end{pmatrix}$, with the conditions $E > 0, F > 0, EG - F^2 > 0$.

2.2 Discrete model

The discrete Hamiltonian is obtained by a discretization of the continuous Hamiltonian in (1) on the triangulated lattices such as **Fig.2.** The discrete model is considered as a Finsler geometry model [5]. We should note that the Finsler function $L = \sqrt{\sum_\Delta L_\Delta^2}$ is given by this metric such that $L_\Delta^2 = \rho^{-1} y_1^2 + \rho y_2^2 = \sum_\Delta g_{ab} y_a y_b$, where $x_a (a = 1,2)$ is a local coordinate along the bond, and $y_a = \dot{x}_a$ is a tangential vector along the local coordinate or bond.

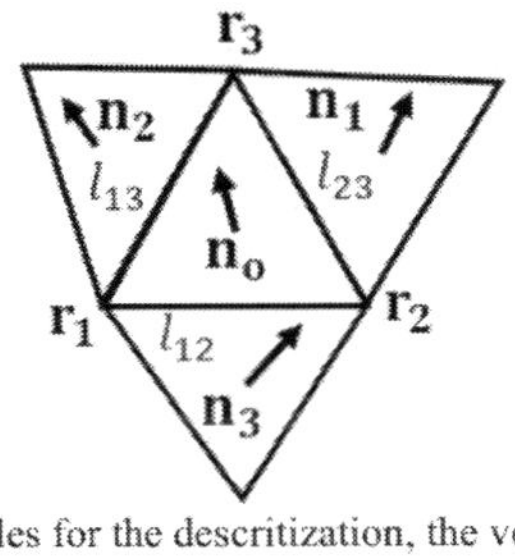

Figure 2. Triangles for the descritization, the vertex positions $\mathbf{r}_1, \mathbf{r}_2, \mathbf{r}_3$, the unit normal vectors $\mathbf{n}_0, \mathbf{n}_1, \mathbf{n}_1, \mathbf{n}_1$.

Assuming that the vertex position $\mathbf{r}_1$ is the local coordinate origin, we replace the symbols in (1) as follows :

$$\int \sqrt{g} dx^2 \to \sum_\Delta,$$
$$\frac{\partial \mathbf{r}}{\partial x_1} \to \mathbf{r}_2 - \mathbf{r}_1 = l_{12}, \frac{\partial \mathbf{r}}{\partial x_1} \to \mathbf{r}_3 - \mathbf{r}_1 = l_{13}, \quad (3)$$
$$\frac{\partial \mathbf{n}}{\partial x_1} \to \mathbf{n}_0 - \mathbf{n}_2, \frac{\partial \mathbf{n}}{\partial x_1} \to \mathbf{n}_0 - \mathbf{n}_3.$$

From (1) and (3), we get

$$S_1 = \sum_\Delta \left(\rho l_{12}^2 + \frac{1}{\rho} l_{13}^2 \right),$$
$$S_2 = \sum_\Delta \left(\rho (1 - \mathrm{n}_0 \cdot \mathrm{n}_2) + \frac{1}{\rho} (1 - \mathrm{n}_0 \cdot \mathrm{n}_3) \right) \quad (4)$$

Not only $\mathbf{r}_1$, but also $\mathbf{r}_2$ and $\mathbf{r}_3$ can be chosen as the local coordinate origin. Therefore, by the cyclic permutations $1 \to 2,\ 2 \to 3,\ 3 \to 1$ for the expressions in Eq. (4) and with the factor 1/3, we have

$$S_1 = \frac{1}{3}\sum_\Delta \left(\rho + \frac{1}{\rho} \right) (l_{12}^2 + l_{23}^2 + l_{31}^2),$$
$$S_2 = \frac{1}{3}\sum_\Delta \left(\rho + \frac{1}{\rho} \right) \quad (5)$$
$$(1 - \mathbf{n}_0 \cdot \mathbf{n}_1 + 1 - \mathbf{n}_0 \cdot \mathbf{n}_2 + 1 - \mathbf{n}_0 \cdot \mathbf{n}_3).$$

In the expressions of Eq. (5), we assume that the function ρ is independent of the chioce the local coordinate, where ρ is originallly dependent on the local coordinate.

By replacing the sum over triangles with the sum over bonds, and replacing the numerical factor 2/3 with 1/4, we get

$$S_1 = \frac{1}{4}\sum_{ij} \left(\rho_i + \frac{1}{\rho_i} + \rho_j + \frac{1}{\rho_j} \right) l_{ij}^2,$$
$$S_2 = \frac{1}{4}\sum_{ij} \left(\rho_i + \frac{1}{\rho_i} + \rho_j + \frac{1}{\rho_j} \right) (1 - \mathbf{n}_i \cdot \mathbf{n}_j). \quad (6)$$

The triangles are separated into two groups L_o and L_d, which are labeled by σ (see **Fig.3**). (Don't confuse the letters for domains and the one for Finsler function.)

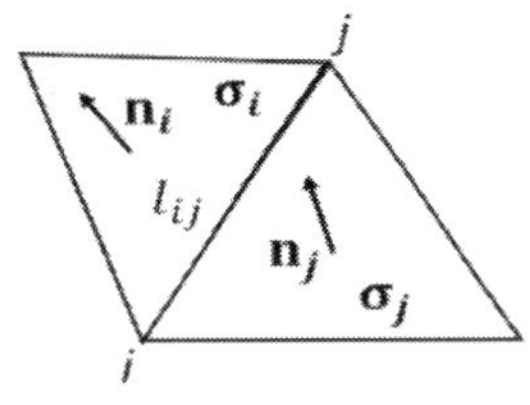

Figure 3. Geometrical quantities on the two neighboring triangles of the bond ij (on a triangulated spherical surface).

The discrete Hamiltonian can be written as

DOI: 10.1051/matecconf/20165401002

$$
\begin{aligned}
&S = \lambda S_0 + S_1 + \kappa S_2,\\
&S_0 = \sum_{(i,j)} (1 - \boldsymbol{\sigma}_i \cdot \boldsymbol{\sigma}_j),\\
&S_1 = \sum_{(i,j)} \gamma_{ij} l_{ij}^2, \quad S_2 = \sum_{(i,j)} \kappa_{ij} (1 - \mathbf{n}_i \cdot \mathbf{n}_j),\\
&\gamma_{ij} = \kappa_{ij} = \frac{c_i + c_j}{4}, \quad c_i = \rho_i + \frac{1}{\rho_i},\\
&\rho(\Delta) = \begin{cases} c\,(\Delta \in L_o) \\ 1\,(\Delta \in L_d) \end{cases},\\
&\boldsymbol{\sigma}(\Delta) = \begin{cases} 1\,(\Delta \in L_o) \\ -1\,(\Delta \in L_d) \end{cases}
\end{aligned} \tag{7}
$$

where S_0 is the aggregation energy, and $\boldsymbol{\sigma}$ (**Fig.3**) is the function on the triangle such that $\boldsymbol{\sigma} = 1$ ($\boldsymbol{\sigma} = -1$) if the triangle Δ is included in L_o (L_d).This S_0 corresponds to the line tension on the boundary between L_o and L_d, because the value of S_0 is proportional to the length of the boundary. S_1 is the Gaussian bond potential, where the symbol l_{ij} denotes the length of the bond ij (**Fig.2**). S_2 is the bending energy, where the symbol $\mathbf{n}_i$ denotes the unit normal vector of the triangle i (**Fig.3**). The quantities γ_{ij} and $\kappa\kappa_{ij}$ are respectively called effective surface tension and effective bending rigidity. Both quantities are defined by using a function ρ on the triangle, which is defined such that $\rho = c\,(\rho = 1)$ if the triangle Δ is included in $L_o (L_d)$.

The effective surface tension γ_{ij} is defined by ρ of two neighboring triangles, as mentioned above. For this reason γ_{ij} has three different values such that $\gamma_{ij} = (c + 1/c)/2$ for L_o and L_o (**Fig.4**(a)), $\gamma_{ij} = (2 + c + 1/c)/4$ for L_o and L_d (**Fig.4**(b)), and $\gamma_{ij} = 1$ for L_d and L_d (**Fig.4**(c)). For κ_{ij}, we also have three different values, which are exactly the same as those for γ_{ij}. We should note that these γ_{ij} and κ_{ij} depend on the position and the direction on the surface, because these are defined on the bonds by using ρ 's on the two neighboring triangles. Moreover, γ_{ij} and κ_{ij} are symmetric under the exchange of i and j ($\gamma_{ij} = \gamma_{ji}, \kappa_{ij} = \kappa_{ji}$).

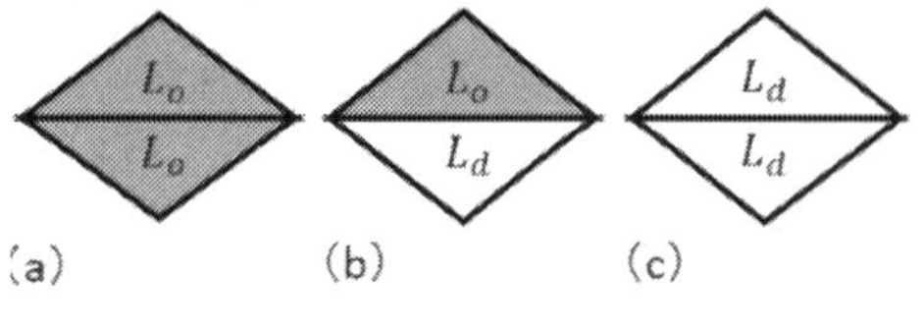

Figure 4. Three types of combinations of the triangles (a) L_o and L_o, (b) L_o and L_d, (c) L_d and L_d.

3 Simulation technique

We use the Monte Carlo (MC) technique for the membrane simulation [8, 9]. In this MC technique, the vertex position $\mathbf{r}$ is randomly moved to $\mathbf{r}' = \mathbf{r} + \delta\mathbf{r}$ (**Fig.5**) with the probability

$$
\mathrm{Min}[1, \exp(-\delta S)], \tag{8}
$$

where $\delta S = S_{\rm new} - S_{\rm old}$. This implies that $\mathbf{r}$ is updated to $\mathbf{r}'$ with the probability 1 ($\exp(-\delta S)$) if $\delta S < 0$ ($\delta S > 0$).

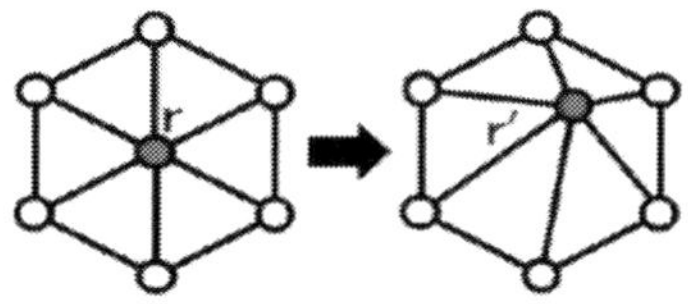

Figure 5. MC update of the vertex position $\mathbf{r}$ to $\mathbf{r}' = \mathbf{r} + \delta\mathbf{r}$

In this study, the lattice structure is changed by bond flip technique such as in **Fig.6** This bond flip makes the vertices diffuse freely over the surface, and hence the surface becomes a fluid surface. After the bond flip, $\boldsymbol{\sigma}_i$ and $\boldsymbol{\sigma}_j$, which are the functions on the triangle i and j, change to $\boldsymbol{\sigma}_i'$ and $\boldsymbol{\sigma}_j'$, which are the functions on the triangle i' and j'. In this process, we assume that the total number of triangles in L_o (and L_d) phase remains unchanged such that $\boldsymbol{\sigma}_i' = \boldsymbol{\sigma}_j'$ if $\boldsymbol{\sigma}_i = \boldsymbol{\sigma}_j$, and $\boldsymbol{\sigma}_i' \neq \boldsymbol{\sigma}_j'$ if $\boldsymbol{\sigma}_i \neq \boldsymbol{\sigma}_j$. In the latter case, the new values are randomly chosen from $\{1, -1\}$ (**Fig.7**). Thus, the free diffusion of vertices due to the dynamical triagulation formes the domain boundary between L_o and L_d on the surface. This bond flip process is essential for the domain formation.

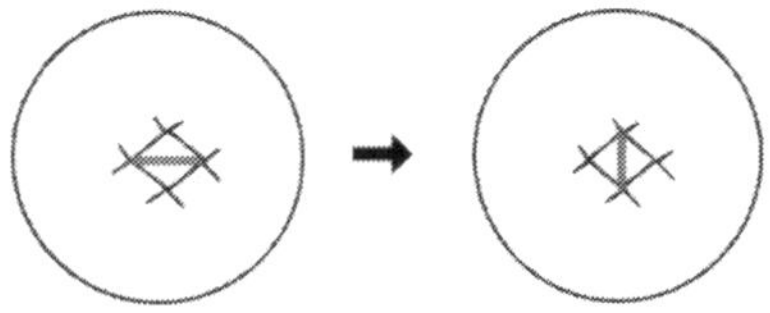

Figure 6. A bond flip as a MC process, which makes the surface fluid.

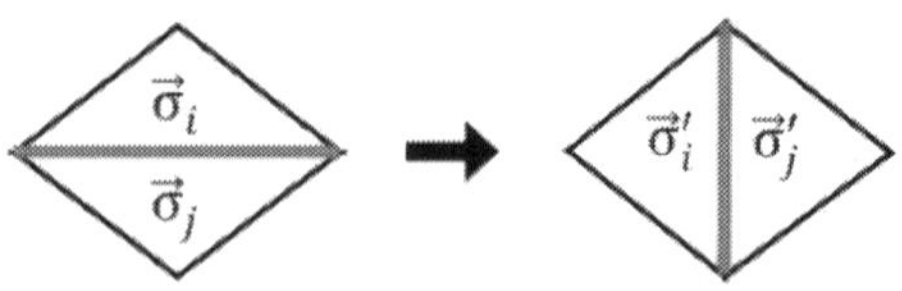

Figure 7. The variable $\boldsymbol{\sigma}$ is changed by the bond flips over the surface.

4 Simulation results

In this study, we perform the simulations for two types of models: model 1and model 2.

Model 1 is defined by $\gamma_{ij}=1$, where the Finsler metric is introduced only in S_2. S_1 in model 1 is just the same as that of the ordinary model. In model 2, the Finsler metric is introduced in both S_1 and S_2.

4.1 Model 1

The parameters used in the simulations for model 1 are shown in **Tables 1, 2**. The input parameters by hand are λ, which is the coefficient of S_0 in (4), $\phi_o(=N_{L_o}/N_T)$ (N_{L_o}, N_T are the total number of L_o triangles and the total number of triangles), which is the area fraction of L_o phase, the bending rigidity κ. These are listed in **Tables 1**. C is the parameter for the function ρ in (4). The value of C here, we assumed is $c=0.05$.

Table 1. The parameters for the simulations for model 1. λ, ϕ_o, and κ are fixed by hand as the inputs.

	λ	ϕ_o	κ
(a)	0.25	0.7	7
(b)	0.2	0.5	7
(c)	0.08	0.4	7
(d)	3.0	0.65	7

The parameters κ_{ij} is automatically determined according to the combination of the two neighboring triangles as listed in **Tables 2**.

Table 2. The parameters for the simulations for model 1. The bending rigidity κ_{ij} are automatically fixed.

	$\kappa_{ij}(L_o,L_o)$	$\kappa_{ij}(L_o,L_d)$	$\kappa_{ij}(L_d,L_d)$
(a)	10	5.5	1
(b)	10	5.5	1
(c)	10	5.5	1
(d)	10	5.5	1

Snapshots of the surface of the lattice size $N=5762$ are shown in **Fig.8.** The L_o domain is 10 times more rigid than the L_d domain, because $\kappa_{ij}(L_o,L_o)=10$ and $\kappa_{ij}(L_d,L_d)=1$. At the domain boundary, the surface stiffness is fixed to the intermediate value of these two such that $\kappa_{ij}(L_o,L_d)=5.5$. We see from **Fig.8**(a) that 1SD appears if λ and ϕ_o are large, also from **Fig.8**(b) that 2CD appears if λ and ϕ_o are small. Therefore, the membrane shape changes 1SD to 2CD, if λ and ϕ_o are reduced. We also find from **Fig.8**(c) that a random pattern appears if λ is very small, where the domains are not separated. Moreover, we find from **Fig.8**(d) that the budding appears. The budding is a phase with the domains which grow inside or outside the surface.

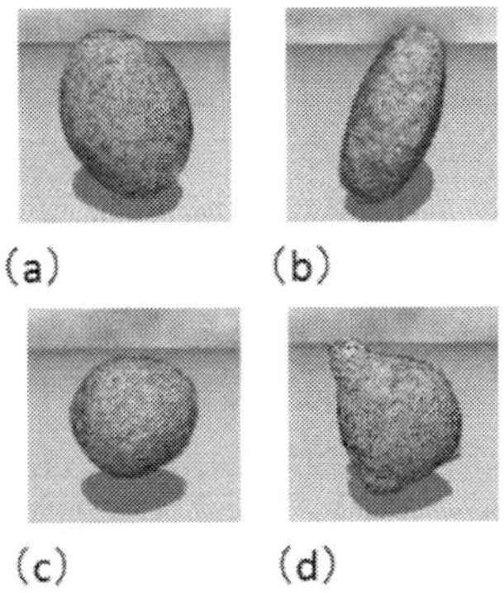

Figure 8. Snapshots of the surfaces corresponding to the domains (a) 1SD, (b) 2CD, (c) random, (d) budding, which are obtained from the simulations for model 1

4.2 Model 2

The parameters used in the simulations for model 1 are shown in **Tables 3, 4**. The parameters γ_{ij} and κ_{ij} are automatically determined from the combination of the neighboring triangles and the value of c. This is the same as that of model 1 in the previous subsection. Here, we assume $c=0.084$ and $c=0.05$ for (a), (b) and (c), (d), respectively.

Table 3. The parameters for the simulations for model 2. λ, ϕ_o, and κ are fixed by hand as the inputs.

	λ	ϕ_o	κ
(a)	0.5	0.9	10
(b)	0.2	0.7	10
(c)	1.5	0.65	7
(d)	1.5	0.8	7

Table 4. The parameters for the simulations of model 2. the bending rigidity κ_{ij}, and the surface tension γ_{ij} are automatically fixed.

	$\kappa_{ij}(L_o,L_o)$, $\gamma_{ij}(L_o,L_o)$	$\kappa_{ij}(L_o,L_d)$, $\gamma_{ij}(L_o,L_d)$	$\kappa_{ij}(L_d,L_d)$, $\gamma_{ij}(L_d,L_d)$
(a)	6	3.5	1
(b)	6	3.5	1
(c)	10	5.5	1
(d)	10	5.5	1

Snapshots of the surface of the lattice size $N=5762$ are shown in **Fig.9.**

Figure 9. Snapshots of the surfaces corresponding to the domains (a) 1SD, (b) 2CD, (c) raft, (d) budding, which are obtained from the simulations for model 2.

In this simulations, we find that the surface shape changes 1SD to 2CD, if λ and κ are reduced, just as in the cases of model 1. Moreover, we find from **Fig.9**(c) that the raft appears. The raft is a phase of the multi circular domains. The budding also appears, if λ is very large. These raft and budding can be seen in the real multi-component membranes.

5 Conclusions

In this paper, we study a coarse grained Finsler geometric surface model for three components membrane. The purpose is to understand the origin of the line tension on the boundary between L_o and L_d phases in those membranes. We have peformed Monte Carlo simulations on the dynamically triangulatted spherical surfaces.

The results are summurised as follows.

1. The domain 1SD (2CD) appears if both λ and ϕ_o are large (small).
2. The domain structure changes from 1SD to 2CD (from 2CD to 1SD) if both λ and ϕ_o are decreased (increased).
3. The bending rigidity for L_o domain ($\kappa_{ij}(L_o,L_o)$) is relatively large, therefore L_o domain is smoother than the L_d domain.

These results obtained in both models 1 and 2 are consistent with the experimental results. Recalling that the aggregation energy S_0 is proportional to the length of the boundary between L_o and L_d phases, we find that the interaction of $\boldsymbol{\sigma}$ with the surface is considered as the origin of the line tension on the boundary. This interaction between $\boldsymbol{\sigma}$ and the surface is introduced by the Finsler metric, and therefore, the origin of the line tension energy is understood in the context of Finsler geometry modeling of membranes.

In the model of this paper, we assume that the function ρ used in the Finsler metric is independent of the local coordinate on the trianle. For this reason, the effective surface tension and bending rigidity become symmetric $\gamma_{ij} = \gamma_{ji}, \kappa_{ij} = \kappa_{ji}$. Because of this property, the Hamiltonian of the surface is identical with that of the inverted surface. This implies that the model introduced in this paper is able to describe the area difference and the spontaneous curvature, both of which separates the spherical membrane from the inverted membranes, if the assumption for ρ is neglected. Therefore, it is interesting to study the origin of spontaneous curvature and area difference in bilayer membranes in the context of Finsler geometry modeling. The problem that will be asked is the microscopic origin of spontaneous curvature and area difference in bilayer membranes. This will be the future study.

References

1. D. Nelson, in *Statistical Mechanics of Membranes and Surfaces, Second Edition*, edited by D. Nelson, T. Piran, and S. Weinberg, (World Scientific, Singapore, 2004) p.1.
2. M. Yanagisawa, M. Imai, and T. Taniguchi, Physical Review Letters, **100**, 148102, (2008).
3. E. Gutlederer, T. Gruhn, and R. Lipowsky, **5**, pp. 3303-3311, (2009).
4. M. Yanagisawa, M. Imai, and T. Taniguchi, Physical Review E **82**, 051928, (2010).
5. H. Koibuchi and H. Sekino, Physica A **393**, pp. 37-50, (2014).
6. W. Helfrich, Z. Naturforsch, 28c (1973) 693.
7. A.M. Polyakov, Nucl. Phys. B **268** (1986) 406.
8. N. Metropolis, A. W. Rosenbluth, M. N. Rosenbluth and A. H. Teller, J. Chem. Phys. **21**, 1087 (1953).
9. D.P. Landau, Phys. Rev. B **13**, 2997 (1976).

MATEC Web of Conferences **54**, 02001 (2016)
MIMT 2016
DOI: 10.1051/matecconf/20165402001

Wireless Control System for Six-Legged Autonomous Insect Robot

Sim Kok Swee, Amro Al-Qudah

MMU University, Faculty of Engineering & Technology, 75450 Melaka, Malaysia

Abstract. Insect robots are a special type of robots that designed to imitate the behavior of insects. Insect robots have many advantages such as the ability to move over uneven terrain, less power consumption and smaller in size. This paper shows the progress made during the development of a six-legged robot system inspired by ants and crickets. The resulted robot is able to mimic insects in terms of gait pattern and physical size. The robot is controlled wirelessly by using a Bluetooth xBee module and remote devices including a mobile phone with android application, a personal computer with windows software, and a Bluetooth wireless controller made the Arduino development platform.

1 Introduction

Improvements made in robotics technology have spurred our economy and changed our lives [1]. Robots are perfect for tasks that need repetitive precise movements or considered dangerous for humans [2]. In manufacturing, robots have improved the quality of the product, increased production output rates and reduced operating costs [3]. Mobile robot is a device that can move autonomously from place to place to achieve a set of goals [4]. Mobile robots are used in a wide range of applications including military operations, manufacturing, search and rescue, security, health care (e.g. pharmaceutical delivery), and home applications (e.g. robotic vacuum cleaner) [4]. Insect robots are that special type of mobile robots that imitate the insects in terms of gait pattern and physical size [5]. Designs of legged and mobile robots have been influenced by insects because of their great mobility and ability to move over uneven terrain. [6]. Insect robot based manufacturing systems have less power consumption and less vibration due to its small size [7]. Insect robots can access places where human cannot such as hazardous places or small spaces. It has an advantage over wheeled robots in moving over uneven terrain [8] [9]. Wirelessly controlled robots eliminate the need of wiring and they have a longer range, which eliminate the need of people in hazardous places [10].

For insect robots, mechanical energy consumption is mainly divided into three portions; energy to support robot weight, energy to produce traction and energy to oscillate the legs [11]. In order to achieve an optimum usage of energy, the insect robot should travel with reduced duty factor, the height of the body should be maximum, step length should be as short as possible, and deviation of each leg to slip should be as low as possible [11].

The hexapod robot has inspired by insects such as ants and crickets. Having six legs, it gives the ability to move flexibly uneven various terrains and does not require any balancing mechanisms to stand upright [12].

2 Problem formulation

Autonomous walking robot is a potential research subject [13]. At the present, insect robots still face many limitations such as slow moving speed, complex mechanism and complicated control system [14]. Insect robots are a great tool to access places which are dangerous for human such as nuclear reactors, outside the atmosphere and even behind enemy lines in the war. Wireless commutations add several advantages to insect robots for instance, eliminate wiring and make it easier to replace people in risky areas [15].

3 Hardware and Software Design

In this experiment, the insect inspired robot is controlled using a PIC microcontroller unit which drives three micro-servomotors. The microcontroller receives orders from a Bluetooth module. On the other hand the Bluetooth module communicates with software installed on a Bluetooth-enabled mobile phone. Furthermore, operator commands can also be sent from PC computer or Bluetooth remote controller. An adjustable infrared sensor is used to discover obstacles. The robot is battery powered to eliminate the need of wiring.

3.1 Hardware Design

To make the insect robot lightweight and considerably strong, 3 mm acrylic sheet is used to form the main body of the robot. Body dimensions are (75x45x3) mm. The legs are constructed using a stainless steel with 1mm thickness. To increase stability, the end of the legs is made into a shape of compression coil spring; this construction is able to absorb shocks and vibrations. Nylon cable is used to make fasteners. The front legs and back legs of each side are connected by a stainless steel link.

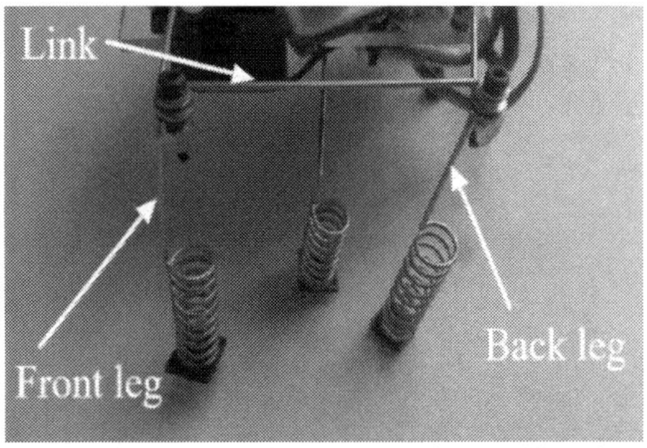

Figure 1. Stainless steel links connect the front and back legs.

PIC18F252 microcontroller has been selected for the insect robot. PIC18F252 is a powerful small 28 pin microcontroller that has enough I/O pins to take over the servomotors, the IR switch and the Bluetooth module. The microcontroller is clocked at 20 MHz using a crystal oscillator with two 330pF ceramic capacitors.

An IR switch is used to detect and avoid obstacles that are in the front of the insect robot. The range of the IR switch is adjustable from 3 cm to 80 cm using a switch in the rear of the sensor. The range is also affected by the color and surface of the obstacles. Usually white color obstacles are detected from longer range than dark obstacles. The sensor operates at 5 VDC and consumes up to 100 mA. The IR sensor internally cancels interferences caused by sunlight or bulbs. The output of the IR sensor will be low when an obstacle is detected.

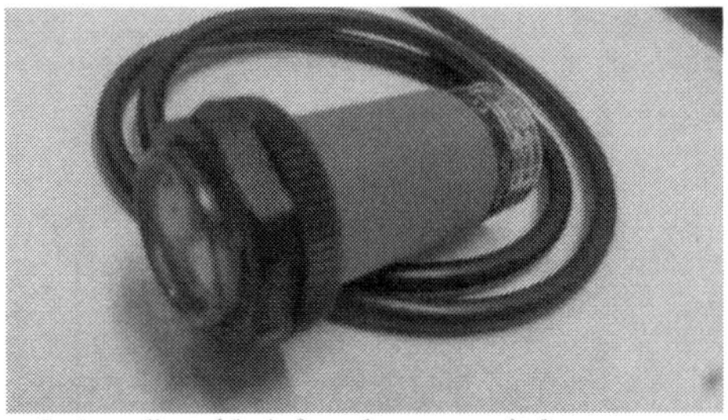

Figure 2. An adjustable infrared sensor switch.

Three servomotors (Power HD-1550A) are used to drive the robot. The dimension of this servomotor is (22x11x20.1) mm which makes it suitable for a small insect robot. The stall torque of the servomotor is 0.8 kg-cm at 4.8V, at the same voltage with no load the speed is 0.12 Sec/60 degree.

Figure 3. Power HD-1550A servomotor.

A rechargeable dual cell Lithium Polymer (LiPo) battery is chosen to power up the robot. The output voltage of the battery is 7.4 VDC and the capacity is about 900mAH. Voltage regulator L7805 is used to provide constant 5V for the microcontroller. The dimensions of the battery are (55 x 29 x 14) mm.

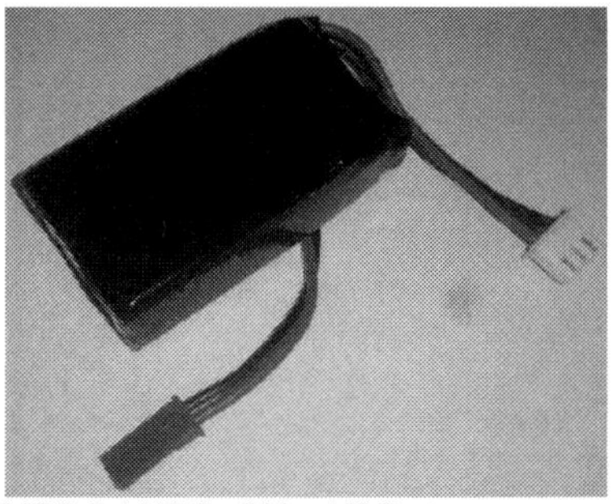

Figure 4. LIPO 7.4 VDC - 900mAH Battery.

Bluetooth module XBee is chosen as robot receiver. Also, another XBee module is used in the remote controller as the transmitter. XBee is compact in size and have a 5V power out. The XBee module has a USB mini b-type interface used to download the XBee firmware and configurations. The Bluetooth module board has 5 pins which are: 5V power source voltage, ground pin, R_X, T_X and reset pin. Reset pin is optional as there is already a reset button on XBee board itself. Pins 5V and ground are connected directly to the ground pin and 5V pin of the microcontroller while R_X and T_X of XBee and R_X and T_X of PIC18F252 are cross connected. Bluetooth module in the insect robot acts as a slave, while in the remote controller acts as the master. XBee module is connected to the mainboard through SKXBee. This board is basically a USB to UART board for the XBee module board.

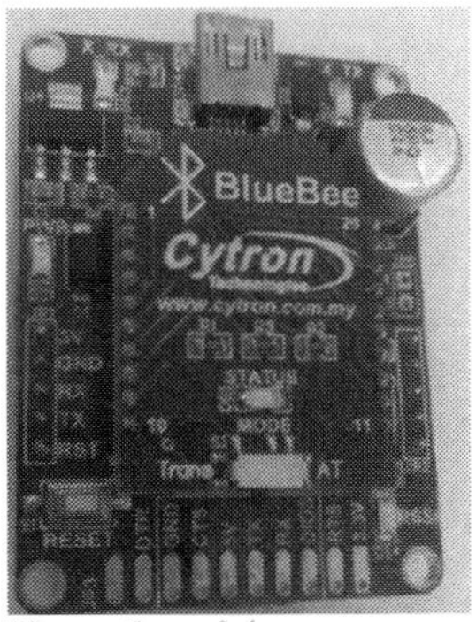

Figure 5. XBee Bluetooth module.

The Bluetooth remote controller has been constructed using Arduino UNO board plus an input shield with an XBee module mounted on the top of it. The controller is powered by a 9V battery.

Figure 6. Bluetooth remote controller.

The main board is designed to be compacted in size; the dimensions are 55mm (L) x 61 mm (W).

3.2 Software Design

For schematic diagram design the application ExpressSCH is used. ExpressSCH is a free CAD software that is easy to use [16]. However, ExcpressSCH does not have PCB layout design features, so the PCB has been designed using software called Traxmaker [17].

MikroC PRO for PIC is a full-featured integrated user environment [18] used to develop firmware for the insect robot. The firmware is coded in ANSI C programming language.

After compiling and building the firmware for the robot, PicKit 2 programmer software has been used to download the binaries to the PIC microcontroller. PicKit software is free, but requires an ICSP interface [18].

The Arduino Software [19] is the integrated development environment that was used to program the Bluetooth controller. Arduino IDE provides code editor, debugger, compiler, downloader and also a real-time monitor which is useful to check the data sent by the controller.

XBee modules are not easy to configure. The software XCTU is a Windows application developed to enable interactions with XBee modules through a simple graphical interface. XCTU includes a set of tools that made it easy to set-up, configure and test XBee models. [20]

Microsoft Visual C++ [21] is a free IDE that has enabled us to develop a windows software to control the robot. Control signals are sent from the application to a logical COM port that represents the Bluetooth adapter inside the PC.

3.3 Methodology

The main advantage of hexapod robots is the ability to move faster with a relatively high stability. That stability can be maintained by keeping three legs on the ground at the same time [6]. Table 1 shows a sample sequence of leg movements that form the gait in our robot. White color indicates that the leg is in the air while the Gray color indicates the leg in touch with the ground.

Table 1. Robot gait pattern

Stage	Moving forward	Moving Right
1	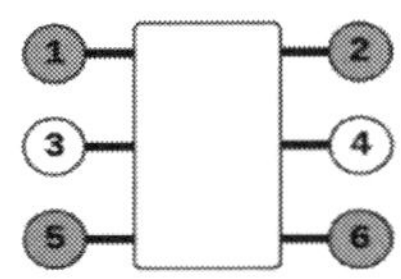 Rest Position	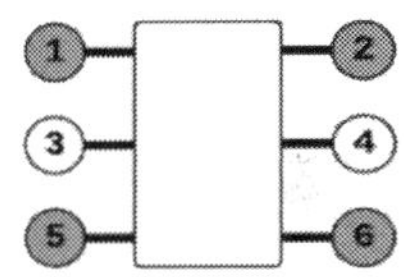 Rest Position
2	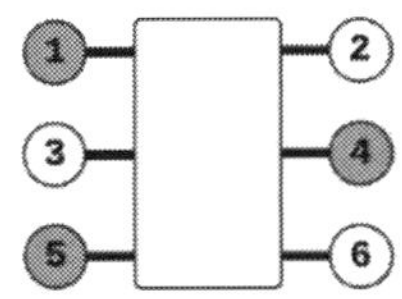	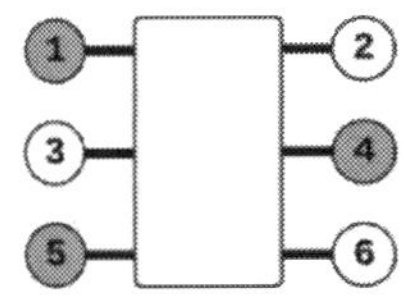
3	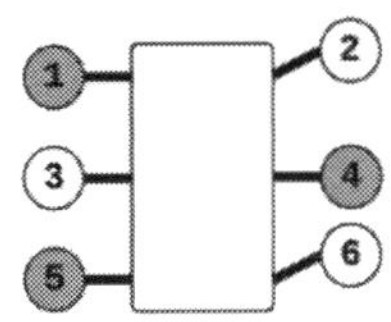	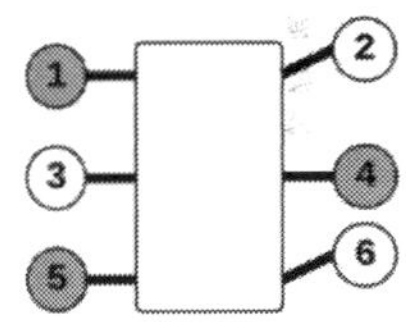
4	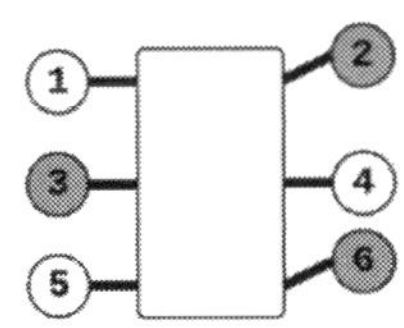	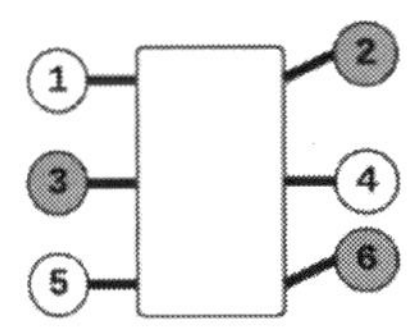
5	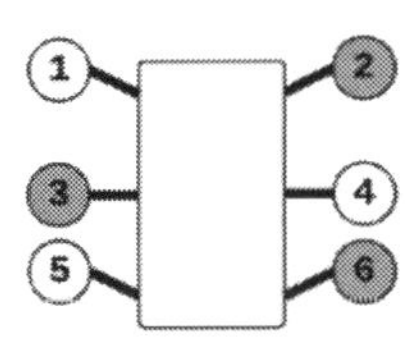	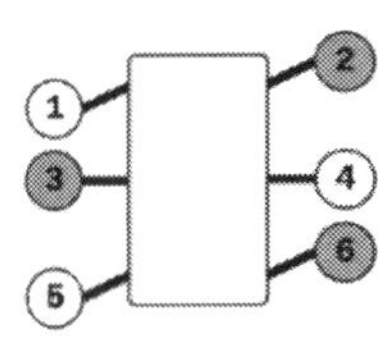

6	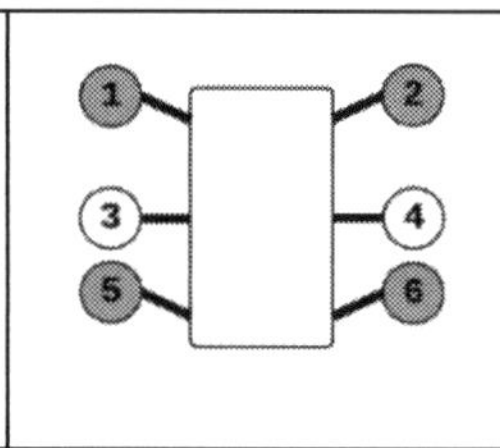	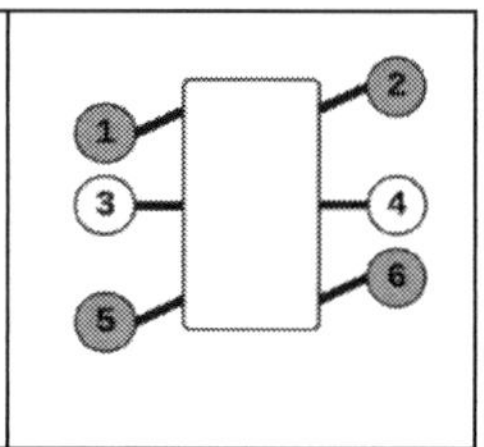
7	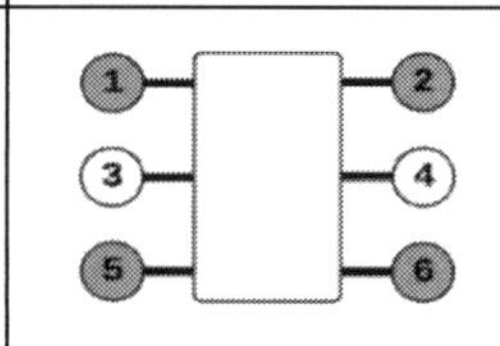 Rest Position	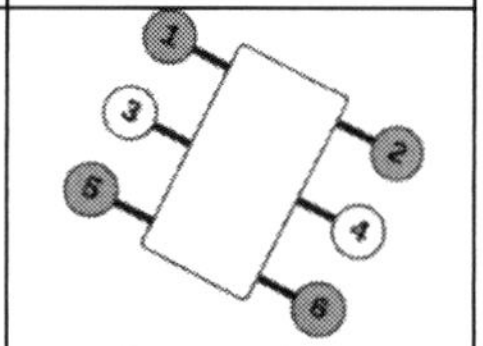 Rest Position

The microcontroller firmware receives inputs from IR sensor as well as the XBee Bluetooth module. The outputs are three control signals for the servomotors. The next figure shows the main flow diagram of the program.

The XBee module in the Bluetooth remote controller is converted to master mode in order to able to start communications and control the robot. The remote controller has pushbuttons each represents a direction of movement in the robot.

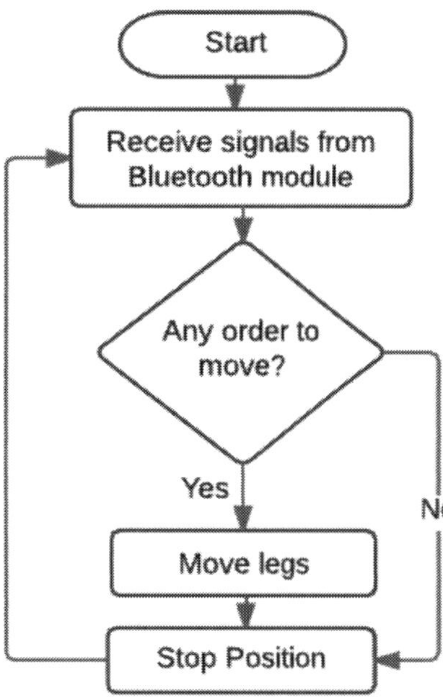

Figure 7. Main program flow diagram.

4 Results

The developed insect robot shown in figure 8 is able to mimic the behavior of insects. Infrared sensor helps the robot to avoid obstacles. The robot is lightweight and the final size of the robot is about (13x9x7) cm.

Figure 8. Six-legged autonomous insect robot.

References

1. J. F. Engelberger, *Robotics in Practice: Management and Applications of Industrial Robots*. (1980)
2. A. Tayebi, "Adaptive iterative learning control for robot manipulators," *Automatica*, **40**, 1195–1203, (2004)
3. Ifr, "History of Industrial Robots From the first installation until today," *International Federation of Robotics*, (2012)
4. S. Tzafestas, *Introduction to Mobile Robot Control*. Elsevier, (2013)
5. A. Piore, "Rise Of The Insect Drones," *Popular Science*, (2014)
6. W. a. Lewinger and R. D. Quinn, "A hexapod walks over irregular terrain using a controller adapted from an insect's nervous system," *IEEE/RSJ 2010 Int. Conf. Intell. Robot. Syst. IROS 2010 - Conf. Proc.*, 3386–3391, (2010)
7. O. Fuchiwaki, "Insect-sized holonomic robots for precise, omnidirectional, and flexible microscopic processing: Identification, design, development, and basic experiments," *Precis. Eng.*, **37**, 88–106, (2013)
8. B. G. a Lambrecht, A. D. Horchler, and R. D. Quinn, "A small, insect-inspired robot that runs and jumps," *Proc. - IEEE Int. Conf. Robot. Autom.*, 1240–1245, (2005)
9. R. Quinn, G. Nelson, R. Bachmann, D. Kingsley, J. Offi, and R. Ritzmann, "Insect designs for improved robot mobility," *Int Conf Climbing Walk. Robot.*, 69–76, (2001)
10. I. Firmansyah, B. Hermanto, and L. T. Handoko, "Control and Monitoring System for Modular Wireless Robot.", (2007)
11. M. F. Silva and J. T. Machado, "A literature review on the optimization of legged robots," *J. Vib. Control*, **18**, 1753–1767, (2012)
12. K. Aparna and S. Geeta, "Insect Inspired Hexapod Robot for Terrain Navigation," 63–69, (2013)

13. B. Klaassen, R. Linnemann, D. Spenneberg, and F. Kirchner, "Biomimetic walking robot SCORPION: Control and modeling," *Rob. Auton. Syst.*, **41**, 69–76, (2002)
14. R. D. Beer, R. D. Quinn, H. J. Chiel, and R. E. Ritzmann, "Biologically inspired approaches to robotics: what can we learn from insects?," *Commun. ACM*, **40**, 30–38, (1997)
15. R. Calcagno, F. Rusin, F. Deregibus, A. S. Vincentelli, and A. Bonivento, "Application of wireless technologies in automotive production systems," *Vdi Berichte*, 57, (2006)
16. ExpressPCB LLC, "ExpressSCH Software." [Online].Available: http://www.expresspcb.com. [Accessed: 27-Aug-2015].
17. MicroCode, "Traxmaker Software." [Online]. Available: http://microcode.com. [Accessed: 27-Aug-2015].
18. mikroElektronika, "mikroC for PIC." [Online]. Available: http://www.mikroe.com. [Accessed: 27-Aug-2015].
19. Arduino LLC, "Arduino Software IDE." [Online]. Available: http://www.arduino.cc. [Accessed: 27-Aug-2015].
20. Digi International, "XCTU software." [Online]. Available: http://www.digi.com. [Accessed: 27-Aug-2015].
21. "Microsoft Visual C ++," *Microsoft*, 1989. [Online]. Available: http://msdn.microsoft.com [Accessed: 27-Aug-2015].

MATEC Web of Conferences 54, 02002 (2016) DOI: 10.1051/matecconf/20165402002
MIMT 2016

A decentralized localization scheme for swarm robotics based on coordinate geometry and distributed gradient descent

Vy-Long Dang[1], Binh-Son Le[1], Trong-Tu Bui[1], Huu-Thuan Huynh[1] and Cong-Kha Pham[2]

[1]*University of Science - Ho Chi Minh City, 227 Nguyen Van Cu, Dist. 5, Vietnam*
[2]*The University of Electro-Communications, 1-5-1 Chofugaoka, Chofu, Tokyo 182–8585, Japan*

Abstract. In this paper, a decentralized localization scheme using coordinate geometry and distributed gradient descent (DGD) algorithm is presented. Coordinate geometry is proposed to provide a rough estimation of robots' location instead of the traditional trigonometry approach, which suffers from flip and discontinuous flex ambiguity. Then, these estimations will be used as initial values for DGD algorithm to determine robots' real position. Evaluated results on real mobile robots show an average mean error of 2.56 cm, which is closed to the minimum achievable accuracy of the testing platform (2 cm). For a team of eight robots, the total average run time of the proposed scheme is 66.7 seconds. Finally, its application in swarm robotics is verified by experimenting with a self-assembly algorithm named DASH.

1 Introduction

The studies related to swarm robotics practically focus on how to coordinate large groups of relatively simple robots to carry out complicated tasks such as area exploration, map construction, or harvesting. To achieve highest performance, it is usually necessary for all robots to agree on a global coordinate system as well as their current position in this system.

Conventionally, the work in [1] has proposed a GPS free localization method for indoor applications by using trilateration and trigonometry. However, as pointed out by authors in [2], this method suffers flip and discontinuous flex ambiguity under noisy range measurements. Therefore, they proposed the use of robust quadrilaterals to avoid these errors by excluding nodes that do not meet the noise-threshold requirement. Another approach, which has become the major trend in wireless sensor networks (WSNs) in recent years, is to use global optimization techniques to find the most corrected position of sensor nodes [3, 4]. The limitation of this method is that it requires the use of anchor nodes which usually employ GPS to have prior knowledge of their actual position.

Another problem in determining a localization scheme for both WSNs and swarm robotics is that it has to be scalable, which means the algorithm can be applied in a system with hundreds or thousands of nodes. Therefore, centralized approaches, which use a center node to distribute network information to all nodes [5], is not favorable as they puts too much strain on the communication channel in large networks. Thus, decentralized approaches, where each node only needs to communicate with its neighbors, is usually applied [1–4]. These approaches are further divided into synchronous and asynchronous methods. Synchronous algorithms mean the state of all nodes in the system are synced. In other words, a node can only go to the next state when all nodes in the system have finished the current state. On the other hand, asynchronous algorithms mean each node can freely change its state without the need to wait for others, thus, it is faster but also more complicated than synchronous ones.

In this work, a synchronous decentralized localizations scheme for swarm robotic platforms is proposed. Instead of using trigonometry as in [1, 2], a coordinate geometry method was developed to avoid flip and discontinuous flex ambiguity. To further improve the accuracy of localization results, distributed gradient descent (DGD) algorithm [4] is also integrated into this scheme. In order to verify effects of real noises in range measurements to the proposed approach, it has been implemented on eight mobile robots. Similar to other swarm-robotic platforms, such as Kilobot [6] and E-puck [7], the designed robots can measure their distance to their neighbors, with a mean error of 2 cm, and communicate through wireless signals. Evaluated results show an average mean error of 2.56 cm, which is closed to the minimum achievable accuracy of the testing platform (2 cm). For a team of eight robots, the total average run time is 66.7 seconds. To verify that this localization scheme can be used to provide robots' position for more complicated swarm robotic applications, the self-assembly algorithm DASH [8] is also implemented and shows good results.

The rest of the paper is organized as follows. Section 2 presents the proposed localization scheme. Section 3 discusses the testing platforms. The evaluation results are shown in section 4. Finally, section 5 draws the conclusion.

2 The proposed localization scheme

2.1 State 1: Distance measurement

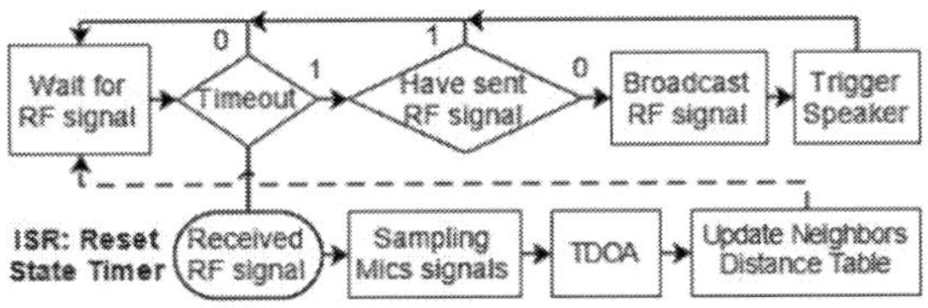

Figure 1. Distance measurement flow chart.

The flow chart of this state is shown in Fig. 1. At the start of the localization scheme, each robot will wait for a random period of time before broadcasting a message to its neighbours. The robots which receive this message will trigger a distance sensing algorithm. The measured results will be stored in a table and sent back to the broadcasting robot. After receiving all measured results, this robot will continue to listen for the same broadcasting message. If none is received until the timer stops, it will jump to the next state. At the end of this state, every robot will have one table storing its distance to its neighbours and their ID.

2.2 State 2: Exchange neighbours distance table

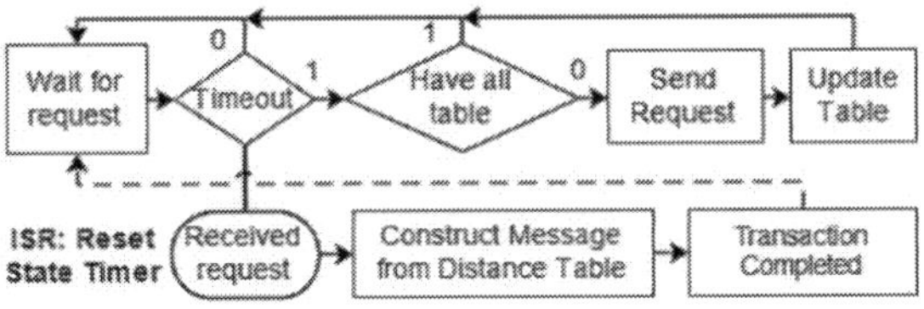

Figure 2. Exchanging measured results flow chart.

Using the ID from the table of state 1, each robot will request the table of its neighbours as shown in Fig. 2. Thus, every robot will now have a second table which stores the first table of its neighbours.

2.3 State 3: Trilateration

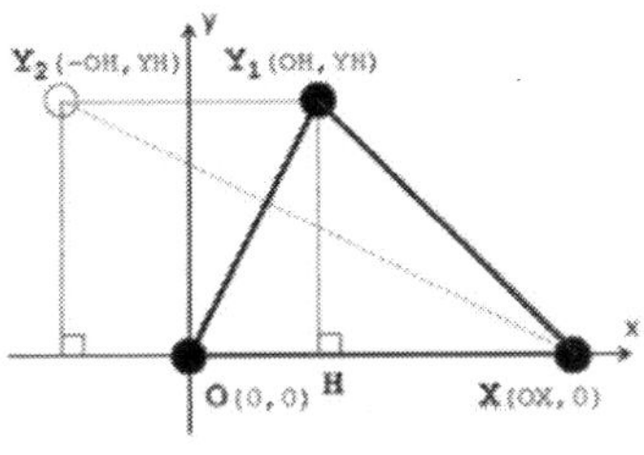

Figure 3. The coordinate of three robots in phrase 1.

Similar to [1, 2], this state has two phrases. Each robot will now establish its own local coordinate system based on the tables from state 1 and 2. It should be noted that no information is exchanged during this state.

2.3.1 Phase 1

At this state, every robot sees it as the origin, $robot_O$ or point O. Two neighbours which do not lie on the same line will be chosen. One is assumed to have the position $(d_{OX}, 0)$, $robot_X$ or point X, and the other is assumed to have a positive y coordinate, $robot_Y$ or point Y. The location of $robot_Y$ is calculated by using the triangle ΔOXY. The length of YH edge is calculated by Herons formula. The length of OH edge is determined by using Pythagoras theorem. Because $robot_Y$ has positive y coordinate, it only has two possible locations: Y_1 and Y_2. However, the correct position will have the minimum error between the distance vector and the measured distance as described in step 8 of Algorithm 1. Positions of these three robots will be stored in a third table.

Algorithm 1 Establish a local coordinate system

1: $\overrightarrow{O} \leftarrow (0,0)$
2: $\overrightarrow{X} \leftarrow (d_{OX}, 0)$
3: $s \leftarrow \frac{d_{OX}+d_{OY}+d_{XY}}{2}$
4: $d_{YH} \leftarrow \frac{2\sqrt{s(s-d_{OX})(s-d_{OY})(s-d_{XY})}}{d_{OX}}$
5: $d_{OH} \leftarrow \sqrt{d_{OY}^2 - d_{YH}^2}$
6: $\overrightarrow{Y_1} \leftarrow (d_{OH}, d_{YH})$
7: $\overrightarrow{Y_2} \leftarrow (-d_{OH}, d_{YH})$
8: **if** $|d_{XY} - \|\overrightarrow{XY_1}\|| < |d_{XY} - \|\overrightarrow{XY_2}\||$ **then**
9: $\quad \overrightarrow{Y} \leftarrow \overrightarrow{Y_1}$
10: **else**
11: $\quad \overrightarrow{Y} \leftarrow \overrightarrow{Y_2}$
12: **end if**

2.3.2 Phase 2

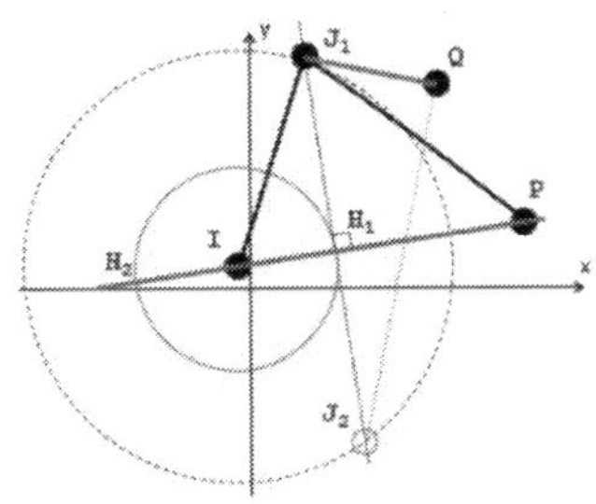

Figure 4. Finding the position of point J from three known points I, Q, and P.

Traditionally, by using trigonometry [1], we can locate the position of an unknown point J from three known points I, P, and Q, as long as they do not lie on the same line. But this approach is not accurate in noisy range measurements and can result in errors such as flip ambiguity or discontinuous flex ambiguity, which are well explained in [2]. To overcome this issue, we propose another approach as explained in Algorithm 2 and below:

- **Step 1 to 3**: As shown in Fig. 4, the linear equation $D_{(IP)}$ in general form can be constructed from any two already located robots.
- **Step 4 to 13**: Calculating the location of point H, with JH is the semiperimeter of the triangle ΔIPJ. From the circle equation, we find two possible locations of H, H_1 and H_2. The length of IH is obtained by applying Pythagoras theorem to the right triangle ΔIHJ. Similar to phrase 1, the correct location of point H has the minimum error between the distance and the measured distance.
- **Step 14 to 23**: Since JH and IP are two perpendicular lines, we can construct the linear equation in general form. Again, get two possible locations of $robot_J$. Finally, by comparing the error between distance vectors and the measured distances, we get the correct location.

Algorithm 2 Calculate position of a neighbor robot

1: $\overrightarrow{IP_x} \leftarrow P_x - I_x$
2: $\overrightarrow{IP_y} \leftarrow P_y - I_y$
3: Line equation $D_{(IP)} : (-IP_y)x + (IP_x)y + (IP_y \times I_x - IP_x \times I_y) = 0$
4: $s \leftarrow \frac{d_{IP}+d_{PJ}+d_{IJ}}{2}$
5: $d_{JH} \leftarrow \frac{2\sqrt{s(s-d_{IP})(s-d_{PJ})(s-d_{IJ})}}{d_{IP}}$
6: Circle equation $C_{(I,d_{IH})} : (x-I_x)^2+(y-I_y)^2 = d_{IH}^2$
7: $\overrightarrow{H_{1,2}} \leftarrow$ solve $D_{(IP)} = C_{(I,d_{IH})}$
8: $d_{PH} \leftarrow \sqrt{d_{PJ}^2 - d_{JH}^2}$
9: **if** $|d_{PH} - \|\overrightarrow{PH_1}\|| < |d_{PH} - \|\overrightarrow{PH_2}\||$ **then**
10: $\quad \overrightarrow{H} \leftarrow \overrightarrow{H_1}$
11: **else**
12: $\quad \overrightarrow{H} \leftarrow \overrightarrow{H_2}$
13: **end if**
14: $D_{(JH)} : (IP_x)x + (IP_y)y + (-IP_y \times H_y - IP_x \times H_x) = 0$
15: $C_{(I,d_{IJ})} : (x - I_x)^2 + (y - I_y)^2 = d_{IJ}^2$
16: $\overrightarrow{J_{1,2}} \leftarrow$ solve $D_{(JH)} = C_{(I,d_{IJ})}$
17: **if** $|d_{QJ} - \|\overrightarrow{QJ_1}\|| < |d_{QJ} - \|\overrightarrow{QJ_2}\||$ **then**
18: $\quad \overrightarrow{Robot_j} \leftarrow \overrightarrow{J_1}$
19: **else**
20: $\quad \overrightarrow{Robot_j} \leftarrow \overrightarrow{J_2}$
21: **end if**

As can be seen, the coordinate of other robots can now be calculated as long as they are the neighbor of at least three already located robots. In other words, their ID appears in the first table of $robot_O$, $robot_X$, and $robot_Y$. Algorithm 2 will go through all neighbors in the first table of $robot_O$ and try to find their position. If new robots are successfully localized, their position and ID will be stored in the third table. Obviously, they can now also be used to locate others. Therefore, whenever a new position is added, the algorithm will go through all remaining neighbors of $robot_O$ again.

To further reduce localization errors before going to the next state, each robot will use the gradient descent algorithm to optimize its calculated results as described in Algorithm 3, with $\nabla_i F(\overrightarrow{Robot_i})$ is obtained by (1), where N_i is the group of neighbors of $robot_i$ which are also belong to the local coordinate system of $robot_O$. At the end of this state, tables from state 1 and 2 are no longer needed and can be removed to free memory. The table obtained from this state is called the local coordinate table.

$$\nabla_i F(\overrightarrow{Robot_i}) = \sum_{j \in N_i} \frac{(\overrightarrow{Robot_{ij}})}{\|\overrightarrow{Robot_{ij}}\|}(\|\overrightarrow{Robot_{ij}}\| - d_{ij}) \quad (1)$$

Algorithm 3 Gradient Descent

while true **do**
 for $\overrightarrow{robot_i} \in$ `thirdTable` **do**
  if $|\overrightarrow{Robot_i} - \overrightarrow{Robot_{iold}}| >$ `allowedError` **then**
   $\overrightarrow{Robot_{iold}} \leftarrow \overrightarrow{Robot_i}$
   $\overrightarrow{Robot_i} \leftarrow \overrightarrow{Robot_{iold}} -$ `stepSize` $\times \nabla_i F(\overrightarrow{Robot_{iold}})$
  end if
 end for
 if no new update in `thirdTable` **then**
  End algorithm
 end if
end while

There are two reasons why our proposed algorithm is more accurate than trigonometry approach. Firstly, in the conventional method, angles are used to determine the sign of y coordinate of a robot. As distance measurements are not accurate, this comparison has to tolerate a noise margin. If this margin is too small or too large, the calculated y coordinate is undefined and will cause a flip ambiguity. This error can happen even with robots having high y value. On the contrary, our method uses the error between calculated and measured distances. Therefore, a large error can only happen when $robot_Q$ lies near x axis. Obviously, this issue can be solved easily by chosen the robot farthest from x axis as robot Q. Secondly, the conventional method only uses a subset of distance measurements to localize a point while in our proposed scheme, all distance measurements will be used in the gradient descent algorithm to minimize calculated errors.

2.4 State 4: Create a unique global coordinate system

After their local coordinate is established, a robot will be chosen as the global origin based on two simple rules:

- The higher number of neighbours, the higher priority.
- If the number of neighbours is equal then the smaller ID, the higher priority.

Each robot will broadcast a vote message containing its ID and its number of located neighbours from state 3. When a robot receives the higher priority vote, it will replace its current vote and broadcast this vote to its neighbours. Otherwise, it will continue to broadcast its current vote. After the origin robot has been determined, its local coordinate system will be considered as the global coordinate and no further changes are necessary. Then, this robot will request its neighbour to rotate their coordinate table to the global one. After these neighbours have rotated, they will continue to request their neighbours to rotate and so on. In short, this process starts at the origin robot and spreads out to the whole system. The rotation algorithm is the same as the one proposed in [1].

As a result of randomly choosing $robot_X$, $robot_Y$, and the global origin, the network direction will differ each time the localization scheme is run. If this is undesirable, these 3 robots can be predetermined in state 3 and the origin voting phrase can be skipped.

2.5 State 5: Errors correction & position updating

Because of the use of angle rotation in state 4, the localization results will have some inaccuracies such as reflection error [2]. For example, $robot_I$ and $robot_J$ successfully locate $robot_K$ position. If the coordinate table of these robots rotates a correct angle, then the position of $robot_K$ will be the same in both tables. However, if an error occurs, the position of $robot_K$ stored in $robot_I$ will differ with the one stored in $robot_J$. Furthermore, these errors will propagate and become larger at farther nodes. To correct this, Distributed Gradient Descent (DGD) algorithm, proposed in [4], is modified to apply to swarm robotics as described briefly in Algorithm 4. It should be noted that $\nabla_i F(\overline{Robot_i})$ is also calculated similarly as in (1).

Algorithm 4 Distributed Gradient Descent

1: **if** receive run DGD request **then**
2:   **if** Do not have a global location b yet **then**
3:     Gets its position from the coordinate table of its neighbors.
4:     b is the average value of these positions
5:   **end if**
6:   $l \leftarrow 0$
7:   $\underline{b}(l) \leftarrow b$
8:   **while** $|\underline{b}(l) - \underline{b}(l-1)| >$ `allowedError` **do**
9:     **for** $robot_i \in Neighbors$ **do**
10:       Ask $\underline{b}(l)$ of $robot_i$
11:       Update $robot_i$ position in this robot coordinate table with $\underline{b}(l)$
12:     **end for**
13:     $\underline{b}(l+1) \leftarrow \underline{b}(l) -$ `stepSize` $\times \nabla_i F(\underline{b}(l))$
14:     $l \leftarrow l + 1$
15:   **end while**
16:   Global position: $b = \underline{b}(l)$
17: **end if**

Another important feature in robotic localization is that when a robot moves, it has to quickly update its new location. To do this, after new distance measurements are updated, Algorithm 4 is run again. However the position of neighbors (step 9 to 12) will only need to be asked one time. Obviously, it still has to have at least three already located robots as its neighbors. This updating method not only offers fast response speed but also high accuracy.

3. The designed testing platform.

To verify the proposed localization scheme, the following requirements are expected:

- Each robot can measure the distance from itself to its neighbours.
- Each robot can communicate with its neighbours using wireless signals.

3.1 The robotic platform

Figure 5. The designed robot.

The testing platform includes 8 mobile robots. Each has the same main board which consists of a micro-controller (ARM Cortex-M4F TM4C123GH6PM), an RF module (nRF24L01), two electret microphones with two two-stage amplifiers, a speaker, three status LEDs, a 6 degrees of freedom IMU, and a battery management circuit as shown in Fig. 5. The total part cost of each robot is $20.68.

3.2 The Control Board and Monitoring Software

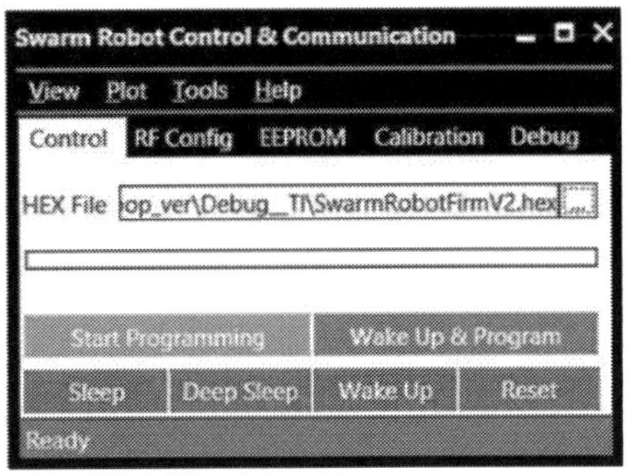

Figure 6. The monitor software.

Without a tool to command and communicate with robots, it is hard to develop and debug collective algorithms. To get over this situation, a wireless control board and a monitoring software (Fig. 6) using C# have been developed. The control board is connected to a PC using USB connection and will send user's commands to all or selective robots. In addition, a wireless bootloader protocol is also implemented to update all robots' firmware at the same time.

3.3 Distance Sensing

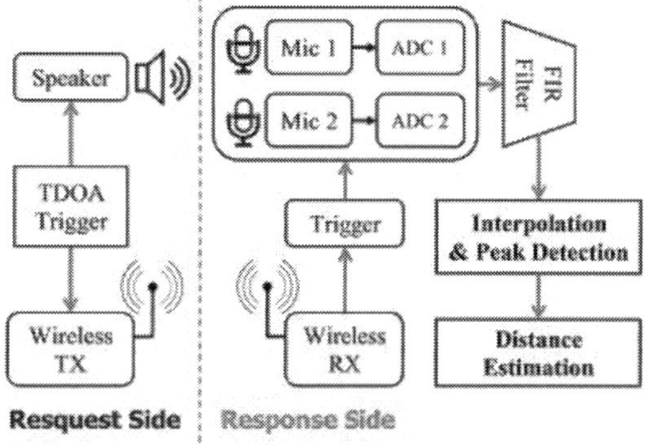

Figure 7. The distance measurement scheme proposed in [9].

For distance sensing purpose, the system and TDOA (time difference of arrival) algorithm proposed in [9] (Fig. 7) are implemented. First, the request side will broadcast an RF signal to other robots, then wait for a small amount of time before emitting an 8 KHz signal. Upon receiving the RF signal, the response side will turn on their microphones to sample the audible signal. The difference between the arrival time of the RF signal and the audible signal will be used to calculate the distance between the response and the request side. Similar to [9], each of our robot also stores two calibration parameters: S_0 (intercept) and M_s (slope), which are extracted using least square optimization. The maximum measurable distance of our system is 47.5 cm and the mean error is 2 cm.

4. Evaluation results

The proposed scheme is evaluated based on the error between measured results and known ground truth as expressed in (2), where, bx_i and by_i are the true position of $robot_i$, and x_i and y_i are its localization results.

$$\sigma_p = \sum_{N}^{i=1} \frac{(bx_i - x_i)^2 + (by_i - y_i^2)}{N} \tag{2}$$

Table 1. Evaluation results from 10 test cases.

Case	Min	Average	Max
σ_p (cm)	2.21	2.56	3.52
Run Time (seconds)	54.5	66.7	96.7

Testing with 10 different cases (by varying the number of robots and their formation), as shown in Fig. 8, the observed average mean error is 2.56 cm. For an ideal localization algorithm, its σ_p cannot be lower than the mean error of distance measurements (2 cm in our system) [10]. Therefore, this result shows that the proposed localization method achieves high accuracy. Table I summaries the evaluated results and the processing time. Since there are several states require random wait time, the run time is affected heavily and fluctuates from 54.5 to 96.7 seconds, with an average value of 66.7 seconds. It should be noted that because our approach is a synchronous decentralized method, its scalability is better than centralized methods. However, because each robot has to wait for others to finish their state, an asynchronous decentralized approach may be employed in further researches to improve runtime.

Since localization results are the foundation for more complicated swarm-robotic applications, the proposed method is continued to be verified by combining it with the self-assembly algorithm DASH [8]. The localization scheme will provide robots' initial position as well as update their location after moving as described in the final state of section 2. The result is shown in Fig. 9, which successfully changes robots' formation from a star shape to a rectangular shape.

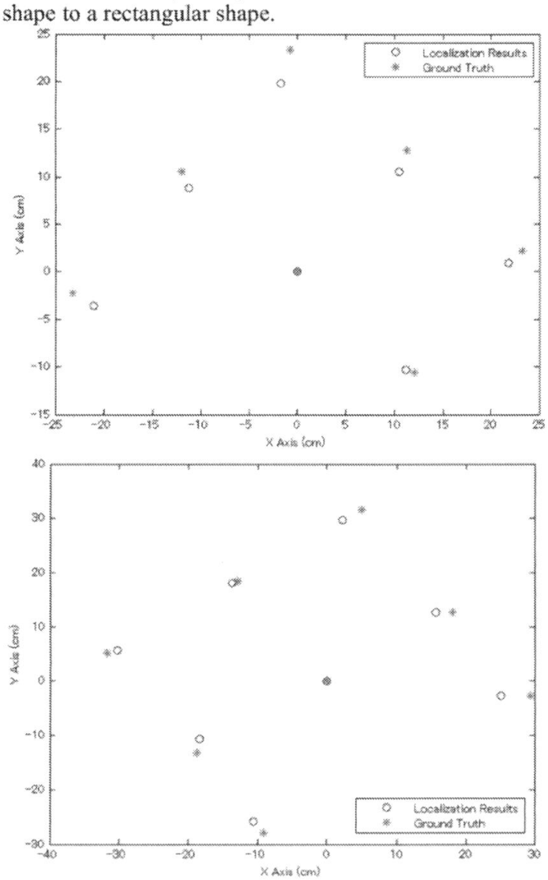

Figure 8. Localization results of (a) 7 (b) 8 robots.

Figure 9. DASH algorithm testing result.

5. Conclusion

In this work, we have presented a new localization scheme using coordinate geometry to avoid flip and discontinuous flex ambiguity errors in the conventional method. After initial positions are produced, a global optimization using DGD is carried out to improve the accuracy of the whole system. The evaluation results on eight mobile robots, which have limited sensing capabilities, show high accuracy and good processing time. However, better run time can be achieved by employing an asynchronous decentralized approach. The proposed approach can also be applied in WSNs when GPS signals are not available. Finally, the proposed method is used to provide robots' position for a self-assembly algorithm named DASH and shows good results.

References

1. S. Capkun, M. Hamdi, J. P. Hubaux, *Hawaii Int. Conf. System Sciences*, 3 (2001)
2. D. Moore, J. Leonard, D. Rus, S. Teller, *Int. Conf. Embedded Networked Sensor Systems*, 50 (2004)
3. Q. Shi, C. He, H. Chen, L. Jiang, IEEE Trans. Signal Processing, **58**, 3328 (2010)
4. G.C. Calafiore, L. Carlone, W. Mingzhu, IEEE Trans. Systems, Man and Cybernetics, **42,** 1065 (2012)
5. T.G. Karimpanal, M. Chamambaz, W.Z. Li, T. Jeruzalski, A. Gupta, E. Wilhelm, *FinE-R*, 20 (2015)
6. M. Rubenstein, C. Ahler, R. Nagpal, *IEEE Conf. Robotics and Automation*, 3293 (2012)
7. A. Gutierrez, et al., *IEEE Conf. Robotics and Automation*, 3111 (2009)
8. M. Rubenstein and W. Shen, *IEEE Conf. Intelligent Robots and Systems*, 1484 (2009)
9. G. Sineriz, M. J. Kuhlman, P. A. Abshire, *IEEE Sym. Circuits and Systems*, 717 (2012).
10. A. Savvides, W. Garber, S. Adlakha, R. Moses, M. B. Strivastava, *ISPN*, 317 (2003)

MATEC Web of Conferences **54**, 03001 (2016) DOI: 10.1051/matecconf/20165403001
MIMT 2016

Development of Critical-Siphon Theory to Fastest Deadlock Controller for Flexible Manufacturing Systems and Computer-Integrated Manufacturing

Johannes K. Chiang and Kiekang Chao

National Chengchi University, Management Information System Department, 11605 Taipei City, Taiwan

Abstract. Since decades, Flexible Manufacturing System (FMS) is a significant part of automated production and manufacturing. In the development of FMS, deadlock prevention becomes a crucial point. This paper present a critical-siphon theory to demonstrate exactly one monitor for quality FMS is required for the set of siphons in the family of a 2-compound siphons and how to assign its initial markings. The theory is aiming to avoid redundant monitors in FMS and the unnecessary associated computational burden so that the quality of a class of Flexible Manufacturing Systems can be assured latest in the run-time. Neither reachability graph nor minimal siphon needs to be computed achieving polynomial complexity- essential for large systems. This paper redevelops the theory more formally and further applies this approach to two well-known S^3PR to obtain a controller full or near maximally permissive in the context of deadlock resolution and Quality Assurance. This paper further categorizes mixture siphons into partial and full ones and the sequence among them to add monitors associated with one or different 2-compound siphons. As a result, there is no need to enumerate all siphons and the time complexity involved is polynomial. This is the first of its kind of works among all current results on the benchmark.

Introduction

After the concept of Build-to-Order, FMS becomes the most important component of automated production and manufacturing. Deadlock-freeness is essential for the automation of a FMS as well as a CIM. Various deadlock resolution approaches [1, 2] have been proposed to tackle deadlocks. Deadlock prevention has been popular to avoid deadlocks in FMS and CIM since it runs fast and statically to avoid run-time detection and computation. Classical approaches either suffer from adding too many monitors or reaching too few states [1,2,3]. Recently, maximally permissive process control policies [4,5,6] with little redundancy have emerged. They however suffer from either complete state enumeration—based on region theory [4,5,6] or the concept of selective-siphons [6]. Both are NP-hard and take exponential amount of time.

In obtaining PN based monitors for deadlock prevention (or liveness enforcing) there are three main issues tackled within the literature: behavioral permissiveness, computational complexity, and structural complexity. Behavioral complexity is related to the performance in terms of the reachable good states. In the context of FMS the number of good states in a Petri net model of an FMS, which can be provided under the deadlock prevention or liveness-enforcing policies, has been regarded as a "quality measure" . In terms of the practical implementation of these policies, this quality measure implies high efficiency, throughput, and flexibility [3]. The highest quality can be provided by the maximally permissive (optimal) control policies. Computational complexity is related to the computational cost paid in order to obtain a liveness-enforcing supervisor (LES) for a given deadlock prevention FMS problem. In this case it is desirable to obtain an answer for a given problem in the least time possible. Structural complexity means extra cost in system verification, validation, and implementation. It is related with two aspects of a LES: the number of monitors and the type of monitors. In the former, it is desirable to obtain the least number of monitors possible. In the latter, there are two types of monitors: ordinary and general. Ordinary monitors are the ones with no weighted arcs. General monitors are the ones with weighted arcs. It is obvious that ordinary monitors are preferable to general monitors due to verification, validation, and implementation issues.

Classical approaches either suffer from adding too many monitors [4] (problematic siphons growing quickly with the size of the system) or reaching too few states [5, 6]. Recently, maximally permissive control policies [7-11] with little redundancy have emerged. They however suffer from either complete state enumeration—based on region theory [7-11] or the concept of selective-siphons [10, 11]. Both are NP-hard and take exponential amount of time.

We propose in [12] to optimize the number of monitors (good states as well) if one adds monitors in the normal sequence of basic, compound, control, and other types of siphons. It is shown that among all 2-dependent siphons (depending on two component siphons), only one (called critical) siphon needs to be controlled by adding a monitor.

This greatly simplifies the synthesis as well as minimizes the number of monitors required while making the controlled net near maximally permissive. Furthermore, the computational burden is much less since there is no need to enumerate minimal siphons, nor to build the reachability graph. It requires neither iterations nor the removal of redundant monitors. In addition, no control arcs are weighted. It scales well with the initial markings and the size of the nets. However, theorems in [12] are stated and proved based on non-formal concepts and symbols.

This paper redevelops the theorems more formally by defining some new symbols so as to be able to show that for all the emptiable siphons derived from a 2-compound siphon, only one monitor is required. This paper further categorizes mixture siphons into partial and full ones and the sequence among them to add monitors associated with a or different 2-compound siphons. As a result, there is no need to enumerate all siphons and the time complexity involved is polynomial. We apply the above theory of sequence-control or critical-siphon to the control of 2 well-known benchmarks with much less computational burden than other approaches, while achieving near or full maximal permissiveness. This is the first paper that is able to identify that each monitored siphon of the benchmark is either a basic one, or a compound one, or a mixture one. This is the first of its kind among all current results on the benchmark.

The rest of the paper is organized as follows. Section II presents the preliminaries about types of siphons. Section II reviews the dependence relationship as well as the sequence of control to reduce the number of redundant monitors. It also develops new theories. Section III applies the theory to 2 well-known benchmarks. Finally, Section IV concludes the paper.

Theory

Due to the space limitation, we refer the reader to [12] for some basic terminologies in this paper. A core subnet can be obtained from an elementary circuit, called core circuit, by repeatedly adding handles. Details of handles and how an SMS is synthesized from a core subnet are in [8, 12, 13] and omitted here to shorten the paper.

Theorem 1 (Theorem 1 in [20]): Let (N_0, M_0) be a marked S^3PR and $S_1, S_2, \ldots, S_n$ a set of SMS such that $\forall i \in \{1,2, \ldots, n\}$, $Si \cap Sj \neq \varnothing$ if $|i-j|=1$ and Si is controlled $(M(Si)>0)$. Each Si is a basic siphon synthesized from basic circuit ci. S_0 is an SMS synthesized from $c_0 = c_1 \text{ o } c_2 \text{ o} \ldots \text{o} c_n$ and $R(S_0) = R(S_1) \cup R(S_2) \cup \ldots \cup R(S_n)$ ($R(S)$ is the set of resource places in S). c_0 and S_0 are called an n-compound circuit and siphon respectively. Let $Si \cap [S_0] = A_i$ $\forall i \in \{1,2, \ldots, n\text{-}1\}$. Then, (1) if $\exists M \in R(N, M_0)$, such that $M(S_0)=0$, then $M(A_1)+ M(A_2)+\ldots+ M(A_n) \geq n$. (2) $\forall M' \in R(N, M_0)$, $M'(S_0)>0$, if $M_0(S_i \cap S_{i+1})=1$, $\forall i \in \{1,2, \ldots, n\text{-}1\}$.

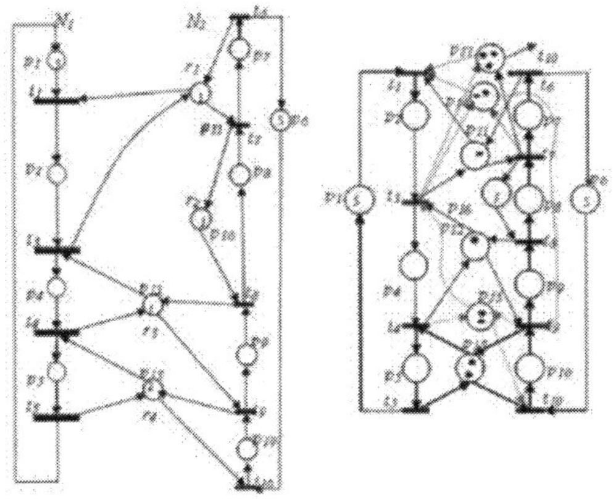

Figure 1(a). Example partial-mixture. **Figure. 1(b)** Controlled model siphon.

In Fig. 1(a), there are three SMS: $S_1=\{p_3, p_4, p_7, p_{11}, p_{12}, p_{16}\}$, $[S_1]=[VS_1]=\{p_2, p_8, p_9\}$, $S_2 =\}p_5, p_9, p_{12}, p_{13}\}$, $[S_2]=[V_{S2}]=\{p_3, p_4, p_{10}\}$, and $S_0= S_3= S^{1,2}{}_m = \{p_9, p_{10}, p_6, p_{11}, p_4\}$. $R(S_3)= \{p_9, p_{10}, p_{11}\} \Rightarrow R(S_3)= R(S_1) \cup R(S_2)$. S_1 and S_2 (resp. S_0) are basic (resp. compound) siphons; both are controlled since $M_0(V_{S1})= M_0(p_{14})=2$ and $M_0(V_{S2})= M_0(p_{15})=2$. $M_0(S_0)=3$ and $A_1= S_1 \cap [S_0]=\{p_4\}$, $A_2=S_2 \cap [S_0]=\{p_9\}$, $A_1 \cup A_2=\{p_4, p_9\}=H(p_{12})=H(S_1 \cap S_2)$. Thus, S_0 is controlled (no need for control elements) if $b=M_0(S_1 \cap S_2)= M_0(p_{12})=1$. On the other hand, if $b>1$, we need to add control elements for S_0 to be controlled.

In Fig. 1(b) (controlled model of that in Fig. 1(a)), $S_1=\{p_3, p_4, p_7, p_{11}, p_{12}, p_{16}\}$, $[S_1]=[V_{S1}]=\{p_2, p_8, p_9\}$, $S_2=\{p_5, p_9, p_{12}, p_{13}\}$, $[S_2]=[V_{S2}]=\{p_3, p_4, p_{10}\}$, and $S_3= S^{1,2}{}_m = \{p_3, p_5, p_7, p_{11}, p_{12}, p_{13}, p_{16}\}$, $S^{1,2}{}_c=\{p_3, p_4, p_8, p_9, p_{15}, p_{14}\}$, $[S^{1,2}{}_c]=\{p_2, p_{10}\}$, $b=M_0(p_{12})=1$. $R(S^{1,2}{}_m)= R(S^{1,2}{}_c)=\{p_{11}, p_{12}, p_{13}, p_{16}\}$, $R^1{}_2=\{p16\}$, $R^2{}_2=\varnothing$, $R^1{}_3=\{p_{11}\}$, and $R^2{}_3=\{p_{13}\}$.

Furthermore, $[S^{1,2}{}_p]=\{p_2, p_8, p_9, p_{10}\}=[S^{1,2}{}_c] \cup [V_{S1}]$, $[S^{2,1}{}_p]=\{p_2, p_3, p_4, p_{10}\}=[S^{1,2}{}_c] \cup [VS2]$, $[S^{1,2}{}_m]= [S^{1,2}{}_f]=\{p_2, p_4, p_8, p_9, p_{10}\}$, $[S^{1,2}{}_p] \subset [S^{1,2}{}_m]$, $[S^{1,2}{}_f]=[S^{1,2}{}_m]$, and $R([S^{1,2}{}_c])= R_3=R^1{}_3 \cup R^2{}_3$ as in the following lemma.

In Fig. 2, $R^1{}_2=\{p_{12}\}$, $R^2{}_2=\{p_{10}\}$, $R^1{}_3=R([S^{1,2}{}_c] \cap [S_1])=\{p_7, p_8\}$ and $R^2{}_3=R([S^{1,2}{}_c] \cap [S_2])=\{p_{11}, p_{13}\}$.

$$M_{max}([S^{1,2}{}_c])=M_0(R_3)=M_0(p_7)+M_0(p_8)+M_0(p_{11})+M_0(p_{13}) =a+c+d+g. \quad (1)$$

$$M_0(V_{S1})=[M_0(R^1{}_2 \cup R^1{}_3)+M_0(R_1)-1]=a+d+e+b-1. \quad (2)$$

$$M_0(V_{S2})=[M_0(R^2{}_2 \cup R^2{}_3)+M_0(R_1)-1]=c+g+f+b-1. \quad (3)$$

$$M_0(V_{S1})+M_0(V_{S2})=a+c+d+e+f+g+2(b-1)>M_{max}([S]). \quad (4)$$

Thus, S cannot become unmarked. Based on this theorem, one adds a monitor for a control siphon only if there are no non-sharing resource places ($R^1{}_2=R^2{}_2=\varnothing$) in the two basic siphons and the initial marking of $S_1 \cap S_2$ equals one. For the example in Fig. 1(b), even if $b=1$, no monitor is needed for the control siphon due to the presence of non-sharing resource place p_{16}. For the example in Fig. 1(b), $b=1$, $R^1{}_2=\{p_{16}\} \neq \varnothing$ (see the example after Def. 4), $S^{1,2}{}_c = \{p_3, p_4, p_8, p_9, p_{14}, p_{15}\}$, $M(p_{14})+ M(p_8) >0$ $\forall M \in R(N, M_0)$. Hence, the control siphon can never become unmarked since $p_{14} \in S^{1,2}_c$ and $p_8 \in S^{1,2}_c$.

Theorem 2: If $S=S^{1,2}_c$ can become unmarked, and is controlled by adding a monitor, then any mixture siphon, partial or full, corresponding to compound siphon $S^{1,2}_m$, is already controlled and needs no monitor.

This theorem states that all mixture siphons are already controlled and need no monitors after I add a monitor upon the control siphon. If one follows [14] to add a monitor with WC arcs for the mixture siphon, the monitor is redundant. In Fig. 2, the control siphon S can become unmarked if b=1. For $S'=S^{1,2}_p$, $A=H(R^1_3) \cap [S']=\{p_2\}$, $\rho(R^1_3)=\{p_6, r_1=p_9\}$, $\rho(R^1_3)\backslash A=\{p_6, r_1=p_9\}$. After adding a monitor for S so that $M([S]) < M_0(R_3)=a+c$, S' may become unmarked only if $M(p_2)=a-1$. This sets $M(\rho(R^1_3)\backslash A)=1$ or $M(S')=1$. Hence, S' is already controlled. Similar conclusion applies to $S'=S^{1,2}_p$. For $S'=S^{1,2}_f$, either $M_{max}(A_1)=M_0(R^1_3)-1$, or $M_{max}(A_2)=M_0(R^2_3)-1$. In either case, it is easy to see that S' is already controlled.

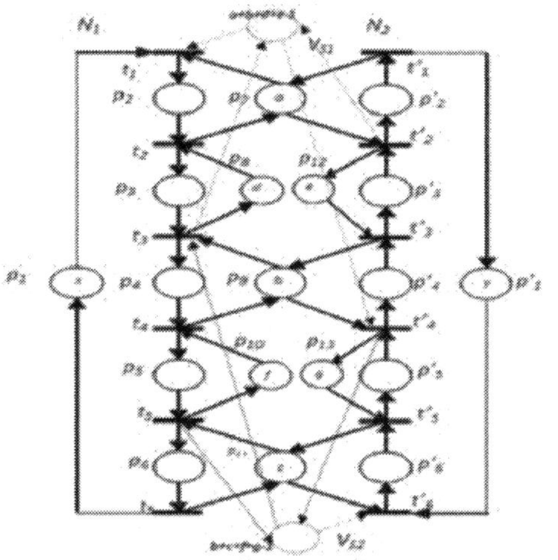

Figure 2. Example full-mixture siphon and non-redundant WC.

Theorem 3: Let $S=S^{1,2}_p$ be a partial mixture siphon as defined in Def. 10, and no monitor is added for $S^{1,2}_c$. 1) $M_{max}([S])=M_0(R^1_2 \cup R_3)+b-1$. 2) S can become unmarked, if i) $R^2_2=\emptyset$. ii) $b=M_0(S_1 \cap S_2)=1$. 3) S becomes unmarked when $M_{max}([S])=M_0(R^1_2 \cup R_3)+b-1=M_0(V_{S1})+M_0(V_{S2})=[M_0(R^1_2 \cup R^1_3)+M_0(R_1)-1]+[M_0(R^2_2 \cup R^2_3)+M_0(R_1)-1]=M_0(R_1 \cup R_2 \cup R_3)+b-2$, Simplifying the algebra, we have $M_0(R^2_2)+b-1=0$, which implies $R^2_2=\emptyset$, and $b=M_0(S_1 \cap S_2)=1$. Thus, S can become unmarked if $R^2_2=\emptyset$, and $b=M_0(S_1 \cap S_2)=1$.

This theorem states that if no monitor is added for a control siphon, then a monitor is needed for partial mixture siphon $S^{1,2}_p$ if $b=M_0(S_1 \cap S_2)=1$ and there are no non-sharing resource places (in S_2) whose holder places are in $[S_2]$. For the example in Fig. 1(b), b=1, $R^2_2=\emptyset$ (see the example after Def. 4), $S^{1,2}_p=\{p_3, p_4, p_7, p_{11}, p_{14}, p_{15}\}$ and $[S^{1,2}_p]=\{p_2, p_8, p_9, p_{10}\}$. $S^{1,2}_p$ is unmarked when $M(p_8)=M(p_9)=1$ and $M(p_{10})=M(p_2)=2$ even though there exists non-sharing resource place p_{16}. If one follows [14] to add a monitor with WC arcs for the partial mixture siphon, the monitor is not redundant. We now extend the theory to full mixture siphons.

Theorem 4: Let $S=S^{1,2}_f$ be a full mixture siphon corresponding to compound siphon $S^{1,2}_m$. Then 1) $M_{max}([S])=M_0(R_1 \cup R_2 \cup R_3)+b-2$. 2) S can never become unmarked if b>2. 3) S can become unmarked in the normal sequence, if i) $R^1_2 \neq \emptyset$ and $R^2_2 \neq \emptyset$; and ii) $b=M_0(S_1 \cap S_2)=1$.

From Part 1 of this theorem, if S becomes unmarked, then $M_{max}([S])=M_0(R_1 \cup R_2 \cup R_3)+b-2=M_0(R_1 \cup R_2 \cup R_3)=M_0(R([S^{1,2}_m]))$ since b=2 against the fact that $M_{max}([S])=M_0(R([S^{1,2}_m]))-1$ derived earlier. Thus, S can never become unmarked if b=2. Hence it must be that b=1. Assume that it does not hold that $R^1_2 \neq \emptyset$ and $R^2_2 \neq \emptyset$. There are two cases:

i) $R^1_2=R^2_2=\emptyset$. $S^{1,2}_c$ is not controlled since Condition 1 ($R^1_2=R^2_2=\emptyset$) in Theorem 2 is satisfied and b=1. Adding a monitor to $S^{1,2}_c$ leads to the inequality $M([S^{1,2}_c]) < M_0(R_3)$, which implies that $M(H(R_3) \cap S)+M(R_3)>0 \Rightarrow M(S)>0$.

ii) Exactly one of R^1_2 and R^2_2 is a nonempty set. By Theorem 2, a monitor is added for a partial mixture S' siphon such that $M([S'])$ cannot reach its maximum for S' to become unmarked, neither can $[S]$ since $[S'] \supset [S]$ by Parts 2 and 3 of Lemma 1. Hence S cannot become unmarked. Both Cases i) and ii) contradict the assumption that S can become unmarked. Thus, if S can become unmarked, then b=1, $R^1_2 \neq \emptyset$ and $R^2_2 \neq \emptyset$. If b=1. From Part 1 of this theorem, if S becomes unmarked, then $M_{max}([S])=M_0(R_1 \cup R_2 \cup R_3)+b-2=M_0(R_1 \cup R_2 \cup R_3)-1$ which is feasible since $M_{max}([S]) < M_0(R[S^{1,2}_m])$ (no monitor for $S^{1,2}_m$ by Theorem 1). Furthermore, by Theorem 2, $S^{1,2}_c$ is controlled (with no monitor) since $R^1_2 \neq \emptyset$ and $R^2_2 \neq \emptyset$ implies that Condition 1 ($R^1_2=R^2_2=\emptyset$) in Theorem 2 does not hold. Thus, it is possible that $M([S^{1,2}c])=M_0(R_3)$ (since no monitor for $S^{1,2}_c$) and $M_{max}([S])=M_0(V_{S1})+M_0(V_{S2})$ (i.e., S can become unmarked), which is not possible if a monitor was added for $S^{1,2}_c$. This theorem states that a full mixture siphon S can become unmarked if and only if b=1 and there are non-sharing resources in both R^1_2 and R^2_2. Note that if one of R^1_2 and R^2_2 is empty, then by Theorem 2, a monitor is added for a partial mixture siphon, which causes the complementary set of S not to reach its maximum.

Hence, S cannot become unmarked. For the example in Fig. 2, $R^1_2=\{p_{12}\} \neq \emptyset$ and $R^2_2=\{p_{10}\} \neq \emptyset$. $S_1=\{p_7, p_8, p_9, p_{12}, p_4, p'_2\}$, $[S_1]=\{p_2, p_3, p'_3, p'_4\}$; $S_2=\{p_9, p_{10}, p_{11}, p_{13}, p_6, p'_4\}$, $[S_2]=\{p_4, p_5, p'_5, p'_6\}$; $S_3=S^{1,2}_m=\{p_7, p_8, p_9, p_{10}, p_{11}, p_{12}, p_{13}, p_6, p'_2\}$, $[S_3]=\{p_2, p_3, p_4, p_5, p'_3, p'_4, p'_5, p'_6\}$. S_1 and S_2 (resp. S_3) are basic (resp. compound) siphons. $S^{1,2}_f=\{V_{S1}, V_{S2}, p_7, p_8, p_{11}, p_{13}, p_6, p'_2\}$ can become unmarked when b=1 and monitors are added only for basic siphons. If one follows [14] to add a monitor with WC arcs for the full mixture siphon, the monitor is not redundant. No monitors are needed for the compound, control and partial mixture siphons. The following theorem guides us on how to add a monitor to a control or a mixture siphon.

Theorem 5: Let V_{Si} be the monitor added upon S_i, i=1 or 2, such that S_i is controlled and S be a relevant control, or mixture, or full, or partial, unmarked siphon. A monitor V is added upon S such that $[V]=[S]$ and $M_0(V)=M_0(V_{S1})+M_0(V_{S2})-1$. Then S can never become unmarked.

For the example in Fig. 3, partial mixture siphon $S=S^{1,2}{}_p=\{p_3, p_4, p_7, p_{11}, p_{14}, p_{15}\}$ may become unmarked when b=1 in the normal sequence. S becomes controlled by adding monitor V such that $[V]=[S]=\{p_2, p_8, p_9, p_{10}\}$ (i.e., $V^\bullet=\{t_1, t_{10}\}$, ${}^\bullet V=\{t_2, t_7\}$) and $M_0(V)=M_0(V_{S1})+M_0(V_{S2})-1=M_0(p_{14})+M_0(p_{15})-1=3+2-1=4$. As mentioned earlier, the compound, and the control siphon, and $S=S^{2,1}{}_p$ are all controlled and need no monitor. Only three (two for basic siphons and one for $S^{1,2}{}_p$) monitors are needed to control the net to make it live.

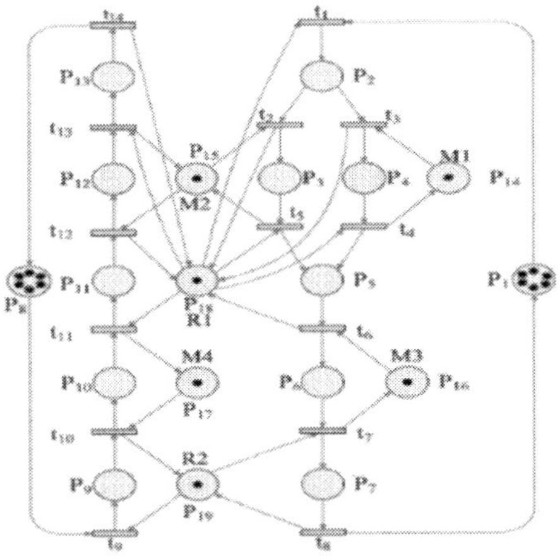

Figure 3. First benchmark example.

For the example in Fig. 2, $S=S^{1,2}{}_f$ can become unmarked when b=1 and monitors are added only for two basic siphons. S becomes controlled by adding monitor V such that $[V]=[S]=\{p_2, p_3, p_4\ p_5, p'_3, p'_4, p'_5, p'_6\}$ (i.e., $V^\bullet=\{t_1, t'_6\}$, ${}^\bullet V=\{t'_2, t_5\}$) and $M_0(V)=M_0(V_{S1})+M_0(V_{S2})-1=(a+b+d+e-1)+(b+c+f+g-1)-1$. Only three monitors are needed to control the net to make it live. For both examples, there is no need to add WC places

Theorem 6: Let $S_3=S_1$ o S_2 in a marked S^3PR (N_0, M_0), S_i (i=1 or 2) is controlled by adding monitor V_{Si}, $S_1 \cap S_2=\{r\}$. Then exactly one siphon in the set of the compound siphon, control siphon and all mixture siphons is emptiable.

Remarks: The resulting controlled model may not be maximally permissive since some live states are forbidden. For instance, in Fig. 1(b), $M=2p_{10}+p_9+p_8+p_{11}$ or $2p_{10}+p_9+p_8+p_7$ (as common practice, only operation places are included in M) is a live marking, yet forbidden by Monitor p_{17}. However, if a monitor is added for each M to control smaller region of operation places (hence less disturbance to the original model), the controlled model becomes maximally permissive. We call $\{p_8, p_9\}$ a refinement region. If a single monitor is added to cover a refinement region, some live states are lost. If for each place in a refinement region, a monitor is added accordingly (called refine operation), then no live states are lost. For the two examples in the next section, refine operations will be performed to reach more live states.

Applications

This section applies the developed theory to two well-known S^3PR. For example in Fig. 4, however, needs WC arcs to be maximally permissive. We add a monitor for each basic siphon. Among 12 compound siphons, We need only add one monitor for S_{15} (b=2) (Theorem 1) significantly reducing the total number of monitors required. For the rest of compound siphons, we have b=1. Now consider 3-compound siphon $S_{ijk}=S_i$ o S_j o $S_k=S_{ij}$ o S_k. Based on Theorem 1, all 3-compound siphons are not controlled. However, this ignores the fact that all 2-compound siphons have been controlled. Note that $S_{ij} \cap S_k=\{r\}$ and $b=M_0(r)=1$ for all 3-compound siphons. Thus, by Theorem 1, they are all controlled and needs no monitor.

In summary, there are only two 2-control siphons S'_{20} and S'_{22} with no monitors for both the associated compound and control siphons. This implies that one of the associated mixture siphons (Theorem 4) is emptiable and needs a monitor.

Now consider adding monitors for mixture siphons. By Theorem 4, no monitors need to be added for mixture siphons built from control siphon S_{19} since a control monitor has been added for S_{19}. By Theorem 2, no monitors need to be added for any partial mixture siphon S built from control siphon S_{20} since $b=M_0(S_1 \cap S_{16})>1$ and S can never become unmarked. By Theorem 4, no monitors need to be added for the full mixture siphon S built from control siphon S_{20} since monitor V_{S7} has been added for compound siphon S_{15} where $[S]=[S_{15}]$.

However, monitor V_{11} is assigned for the full mixture siphon obtained by adding TP-handles $[t_3\, p_{23}\, t_2\, p_{20}\, t_{19}\, V_5]$, $[t_{17}\, p_{26}\, t_{16}\, p_{22}\, t_5\, V_2]$, and $[t_8\, p_{25}\, t_7\, p_{20}]$ upon the core circuit of control siphon S_{22} based on Theorem 3. Also monitor V_9 is assigned for the partial mixture siphon obtained by adding TP-handle $[t_{17}\, V_8\, t_{16}\, p_{22}\, t_{10}\, V_7]$ upon the core circuit of control siphon S_{21} based on Theorem 2.

To reach more states, I consider possible refinement regions 1) for S_{21}: $\{p_{12}, p_{13}\}$ and 2) for S_{22}: $\{p_8, p_9\}$, $\{p_{12}, p_{13}\}$, and $\{p_{16}, p_{17}\}$. There are two unmarked sets of operations for S_{21}: $\Psi^1=\{p_{11}, p_{12}, p_{18}, p_{19}\}$ and $\Psi^2=\{p_{11}, p_{13}, p_{18}, p_{19}\}$; a monitor is added for each. Similarly, monitors are added for S_{22}. The final controlled monitor is shown in [10], where a theorem is developed to show how a monitor becomes redundant after adding a monitor for the refinement region $\{p_8, p_9\}$ for S_{22}.

Conclusion

For the FMS and CIM, I have redeveloped some theorem in the field of critical-siphon theory in a more formal fashion, further categorized mixture siphons, and analyzed their controllability (even among different 2-compound siphons). I have applied the theory of sequence-of-control or critical-siphon to 2 benchmarks without using weighted control arcs. The first one is maximally permissive. Although it may reach fewer states than the maximally permissive one for the 2nd benchmark, it suffers less computational burden since no reachability analysis is required and the enumeration of problematic siphons is much more efficient. The controllability among different types of siphons implies that some problematic siphons can be skipped; thus relieving the need for complete siphon enumeration. The theory is able to avoid redundant monitors in FMS and the unnecessary associated computational burden so that the quality of a class of FMSs can be assured latest in the run-time. This is the first approach that is able to identify that each monitored siphon of the benchmark is either a basic one, or a compound one, or a mixture one.

The results can only be applied to a class of S^3PR, not any S^3PR. The applied S^3PR has 2-compound siphons. That is to say, if an S^3PR has 3- or 4-compound siphons, the results may not be applicable. Future work should extend the theory to siphons calculated from an n-compound (n>2) resource circuit, to remove the assumption that any two core circuits are interconnected at a single resource place, and to extend the controllability theory to FMS with generalized arcs.

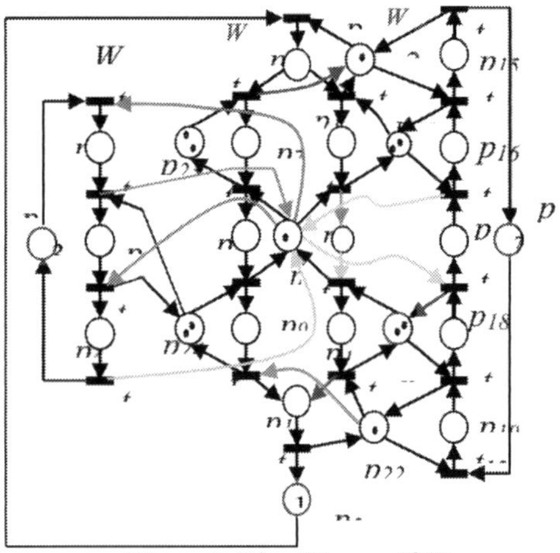

Figure 4. 2nd well-known S^3PR.

References

1. N. Viswanadham, Y. Narahari, T.L. Johnson, IEEE Trans Robot Autom Mag, **6**, 713-723 (1990)
2. H.S. Hu, M.C. Zhou, Z.W. Li, IEEE T Multimedia, **11**, 1457-1465 (2009)
3. M. Uzam, M.C. Zhou, IEEE Trans Syst Man Cybern A Syst Hum, **37**, 362-371 (2007)
4. J. Ezpeleta, J.M. Colom, J. Martinez, IEEE Trans Robot Autom **11**, 173-184 (1995)
5. Z.W. Li, M.C. Zhou, IEEE Trans Syst Man Cybern A Syst Hum, **34**, 38-51 (2004)
6. Z. Li, D. Liu, Int J Prod Res, **45**, 2161-2165 (2007)
7. M. Uzam, M. Zhou, Int J Prod Res, **44**, 1987-2030 (2006)
8. Y.S. Huang, Y.L. Pan, Int J Adv Manuf Technol **48**, 725-737 (2010)
9. L. Piroddi, R. Cordone, I. Fumagalli, IEEE Trans Syst Man Cybern A Syst Hum, **38**, 1337-1348 (2008)
10. L. Piroddi, R. Cordone, I. Fumagalli, IEEE Trans Syst Man Cybern A Syst Hum, **39**, 650-661 (2009)
11. Z. Li, M.C. Zhou, M.D. Jeng, IEEE T Autom Sci Eng, **5**, 182-188 (2008)
12. Y.Y. Shih, D. Chao, Comput J, **53**, 1691-1703 (2010)
13. D.Y. Chao, Int J Prod Res, **48**, 1217-1220 (2010)
14. Y.S. Huang, M. Jeng, D.H. Chung, IEEE Trans Syst Man [9] Cybern A Syst Hum, **36**, 1248-1256 (2006)
15. D. Chao, G.J. Liu, Asian J Control, **14**, 163-172 (2012)

Degradation of Gas Turbine Blades for a Thermal Power Plant

Ahmad Afiq Pauzi, Norlia Berahim, and Shuib Husin

Materials Engineering Group, TNB Research Sdn Bhd, Kawasan Institusi Penyelidikan dan Latihan Bangi, 43000 Kajang, Selangor, Malaysia

Abstract. There are three stages of gas turbine blade in hot gas path (turbine section) of power generation gas turbine. Occurrences of wear and crack were observed on gas turbine blades in this section. The blades are made of nickel base superalloy (GTD 111). Crack generally occurred at airfoil area of the blades while wear generally occurred at root area of the blades. The purpose of this study is to determine the degradation of two stages blades in the turbine section after being exposed to 24,000 running hours. The affected areas of the mating surfaces were analyzed through visual and microstructure testings and the type of degradation was discussed.

1 Introduction

Nickel base superalloys are widely used in applications exposed to high temperature and vibration such as hot gas path components in power generation gas turbine. Turbine blade is one of critical components in the turbine system. This component experienced temperature of 1000 °C in the gas turbine [1]. Wear and crack are the main problems of gas turbine blade [2]. The component experienced surface and materials degradation after 24,000 hours in service. Based on previous investigation of wear and crack on gas turbine blades, most blades suffer severe operation conditions characterized by the following factors: operation environment (high temperature) and mechanical action (relative movement between and forces between blade root and its supporters) [2].

In normal practice, the turbine blades are subjected to heavy repair processes thus increasing the cost of the refurbishment [3]. Detail analyses of wear and cracks on the components related to hardness, microstructure characteristics and wear mechanisms, and wear measurement have been considered to predict the wear characteristics of the components. Naeem et al had revealed through visual and metallurgical analysis, the surface of blades were very rough and experience crack and diversely colored (erosion wear) [4].

Detailed analysis is very crucial as it cannot be relied only to visual observation and measurement, but deep understanding in damage mechanisms, and wear microstructures in order to evaluate the severity of wear and crack. To date, there are many study conducted on gas turbine blades based on their design life of 72,000 hours. Bernstein had revealed that for gas turbine hot section components, wear and crack are most important for turbine hardware at their design life [5]. Karganejad et al had revealed the root of the blade is not exposed to hot combustion gasses; hence temperature in this section is much lower than the blade airfoil so the blade root is considered as cold section (reference zone) [6]. This study focuses on the degradation of gas turbine blades, which have been taken out at the earlier than their design life, which is at 24,000 hours for visual and microstructure evaluation.

2 Starting materials

The turbine blades that were under evaluation were taken from a 97MW gas turbine. Figure 1 and 2 show the location of airfoil and root on gas turbine first stage and second stage. These areas were selected due to occurrence of wear and crack based on findings by Karganejad et al. [6].

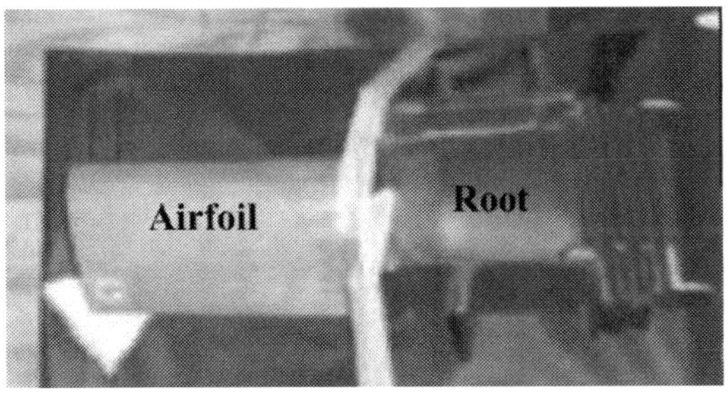

Figure 1. First Stage Blade

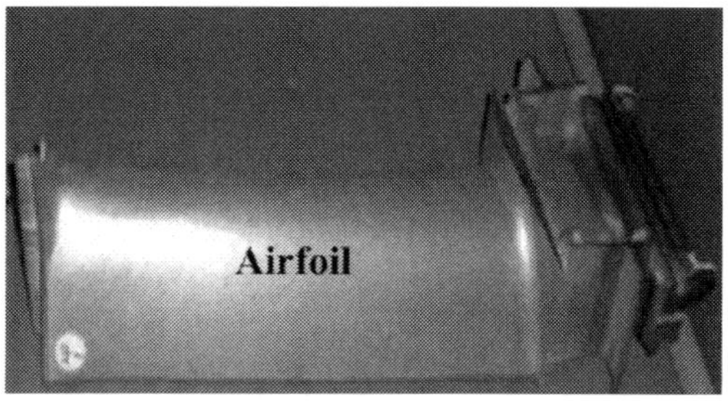

Figure 2. Second Stage Blade

3 Visual observation and microstructure

The visual observation was carried out after the blades had been operating 24,000 hours of operation. The turbine blades are made of a nickel base GTD 111 [11]. The chemical composition result for the turbine blades is shown in Table 1. These results are the average from 3 points in turbine blades. The table also includes the standard chemical-composition data for GTD 111.

Table 1. Chemical composition of GTD 111 turbine blade

Element	Cr	Ni	Co	Fe	W	Ta	Ti	Mo	Zr	Al
GTD 111	14.22	61.05	9.68	0.85	4.42	3.12	4.32	1.72	0.10	0.00

The microstructures of the turbine blades were recorded before and after its removal from the gas turbine.

4 Result and discussion

4.1 Visual observation

Visual observation was performed on first stage nozzle and second stage nozzle. Table 2 shows the visual and microstructure of gas turbine blades after 24,000 hours operation.

Table 2. Visual observation and microstructures of gas turbine blades

Components	Visual observation	Microstructure
First stage blade Operating time: ~ 24,000 hours Temperature : 1000°C	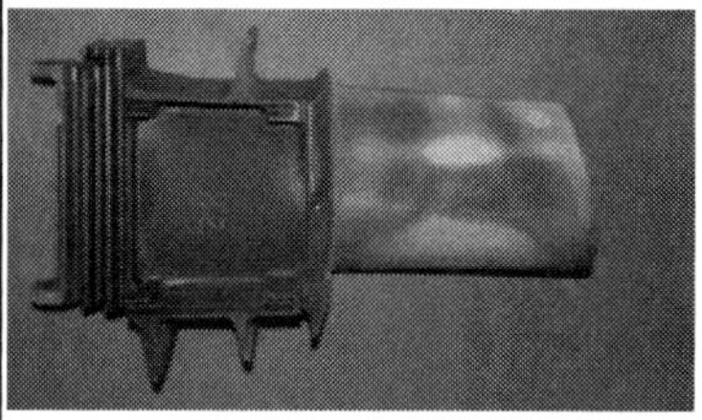 10 cm Coating diverse colour was observed at blade airfoil surface. Occurrence of wear was observed at root area. No evidence of wear was found on the airfoil surface.	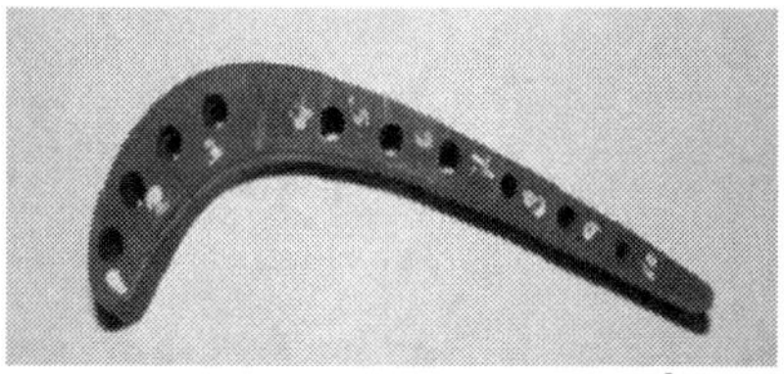 3 cm The first stage bucket was cut into small pieces for detail analysis. 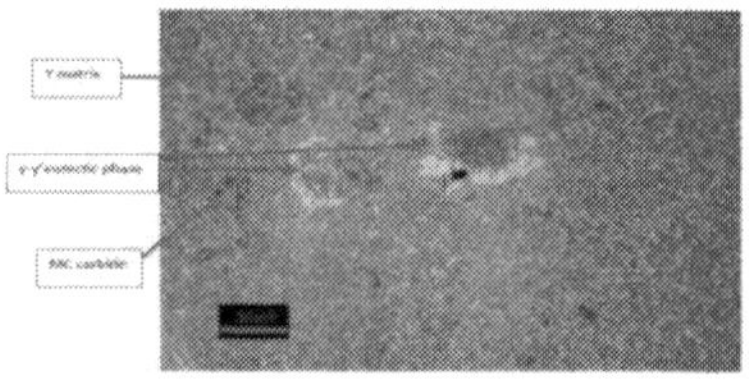

Components	Visual observation	Microstructure
Second stage blade Operating time: ~ 24,000 Temperature : 1000°C	 The airfoil surface shows erosive mode. The occurrence of brownish colour was observed at certain areas at the airfoil surface. Occurrence of wear was observed at root area.	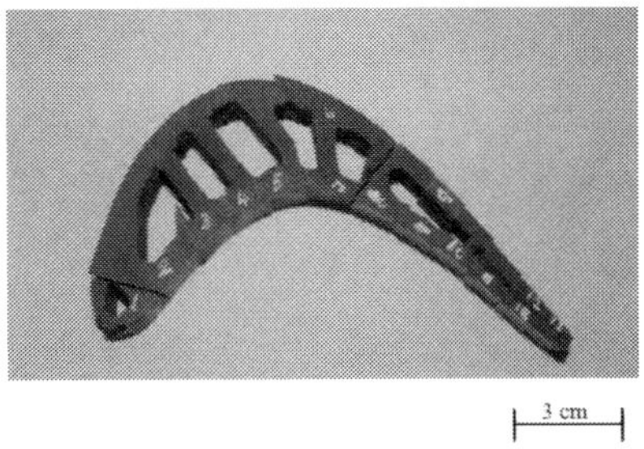 The second stage bucket was cut into small pieces for detail analysis. 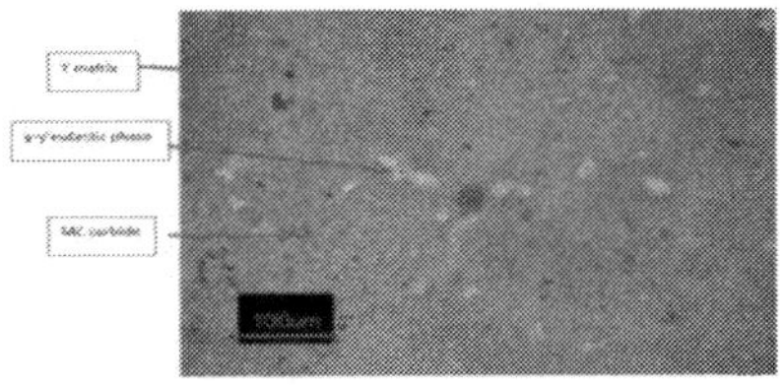

Generally, the blades are physically in good condition with no evidence of crack, oxidation or foreign object damage at the airfoil area. Vishwanathan revealed that the exposure to high temperature produced significant symptom that can initiate crack [2]. The occurrence of wear was observed at root area. Materials loss with the evidence of discolored arising was observed on these areas. Only little wear debris was retained on the surfaces [8, 9]. At these stages of damage, re-application of aluminum seal strip coating on the root area and application of thermal barrier coating (TBC) on airfoil area can protect the blades by affecting wear and crack mitigation [10].

4.2 Microstructure examination

Hyun In Kim et al revealed that when the gas turbine blades is used at high temperature for a long time, base material become damage and heat treatment has been applied of the elimanitation of cavities and cracks. [11]. Similar result had been revealed by Vardar et al, the crack initiated from the grain boundaries and propagated to the critical length. [12]. From this study, the morphology of the microstructures of gas turbine first stage and second stage blades are similar for nickel based alloy. The microstructure show fine γ and γ- γ' eutectic form in the matrix and coarse random cubic or script morphology MC carbide in mainly distribute in the matrix. No creep void and degradation were observed. Current practices of maintaining the microstructure to meet blade's design life of 72,000 hours, thermal barrier coating on was applied on the blades to make them last longer when operating at high temperature [13].

5 Conclusion

Based on microstructure analysis that has been performed, no observation of crack and degraded microstructure on the samples. If the turbine blades were decided to be re-installing into gas turbine, based on microstructure analysis, the blades are recommended to be in service for next 24,000 hours operating time.

Based on the wear mode that has been observed on the components, the main wear mechanism on the turbine blades is fretting wear.

References

1. Cyrus B. Meher Homji and George Gabriles, Gas Turbine Blade Failures – Causes, Avoidance, and Throubleshooting, Bechtel Corporation, Texas
2. R Vishwanathan, Damage Mechanisms and Life Assessment of High Temperature Components, Electric Power Research Institute (EPRI)
3. R Rajendran, MD Ganeshanchar, Jivankumar, T. Mohana Rao, Condition Assessment of Gas Turbine Blades and Coatings, Engineering Failure Analysis, 2011
4. Mehdi Tofighi Naeem, Seyed Sli Jazyeri, Nesa Rezamahdi, KN Toosi, Failure Analysis of Gas Turbine Blades, Univerisity of Technology.
5. Henry L. Bernstein, Materials Issue for Users of Gas Turbines.
6. S. Kargarnejad, F. Djvanroodi, Failure assessment of Nimonic 80A turbine blade, Engineering Failure Analysis, 2012.
7. TNBR Qats Sdn Bhd, Technical Report on F-Class Materials Identification, 2013
8. J Blau, Friction and Wear Transitions of Materials, Noyes Publication, NJ 1989
9. R.G Bayer, Wear Analysis for Engineers, HNB Publishing, New York, 2002

10. GE Energy Part and Services, Gas Turbine Condition Report, 2012
11. Hyun Ick Kim, Hong Sun Park, Jae Maen Koo, Chang Sung Seok, Sung Ho Yang, Moon Young Kim, Microstructural Investigation of GTD 111 DS Materials in The Heat Treatment Conditions
12. N Vardar, A Ekirim, Failure Analysis of Gas Turbine Blade In a Thermal Power Plant, Engineering Failure Analysis, 2007.
13. David R Clarke, Matthias Oechsner, and Nitin P Padture, Thermal Barrier Coating for More Efficient Gas Turbine Engines, Materials Research Society, 2012.

MATEC Web of Conferences **54**, 03003 (2016)
MIMT 2016
DOI: 10.1051/matecconf/20165403003

Reliability of Isotropic Model in Finite Element Simulation to Predict Wrinkling in Deep Drawing Process

M. Jasri, M.N. Iryana

Faculty of Mechanical Engineering, Universiti Malaysia Pahang, Pekan, Pahang, Malaysia

Abstract—Deep drawing is one of important sheet metal processes used in manufacturing industry. Efforts have been conducted for many years to use advance computing software in understanding the metal forming process. Finite Element (FE) simulation is considered one of the best methods. The accuracy of FE application however depends on the type of hardening model being used as part of the material input.The aim of this studyis to investigate the reliability of isotropic hardening model in predicting wrinkling in deep drawing process. The finite element resultand the experimental result are compared visually for similarity as an indicator of the reliability of the hardening model.

1 Introduction

Deep drawing is one of the sheet metal forming processes which is widely used in manufacturing industry. In the process, the blank is deformed by an action of a punch forcing the metal into the die cavity. The products usually have a complicated shape, thus undergoing several successive operations is vital to obtain a final desired shape. One of the major defects that occur in deep drawing operations is wrinkling. It is also known as surface distortions resulted from compressive stresses [1]. There are two critical places that tend to generate wrinkles which are at the flange and at the cup wall of the product. The wrinkling occurred because of radial drawing stress and tangential compressive stress during the stamping process[2, 3].

Until recently, the design of deep drawing tools were mainly gained through empirical and experimental learning and expensive trial- and- error processes. However, today the forming process are making use of finite element methods even in the early stages of die forming and process design.

Finite element simulation is an effective way to know whether it is possible or not to form a part without any defects. From a practical perspective, finite element simulation includes an input to simulate the model, a solution and an output which can be considered as a product prediction.To improve the simulation, it is necessary to have accurate information on the geometry of the tooling, the sheet metal properties, the friction coefficient of the tools and the sheet interface. In this present paper, the reliability of isotropic hardening model is investigated. This is being done by using the hardening model as material input to simulate deep drawing process. The software used for the simulation is Abaqus. The material used for the investigation is 1mm stainless steel. The blank size is 100mm x 100mm. The die size is 50mm x 50mm with edges radius 2mm which is actually the desired square hole. The punch size is 44mm x 44mm with 2mm edges radius.

2 Finite element model

The finite element model consists of tools and blank. The tools are punch and die and the blank is the sheet metal. The tools are designed as rigid whereas blank is deformable part. The element used for the tools and blank is shell type. The coefficient of friction between punch and blank and between die and blank is 0.125. The value is taken from study conducted by Ayari and Bayraktar[4]. One quarter sector of the tools and blank been constructed for the simulation's model. The meshing size for the modeling is 1.5mm. The data for input material properties is acquired from tensile testing as shown in **Table 1**.

Table1.Stainless steel properties

Young Modulus (Mpa)	210000
Poisson ratio	0.33

Figure 1 shows the FE model at the initial stage, loading stage and unloading stage.

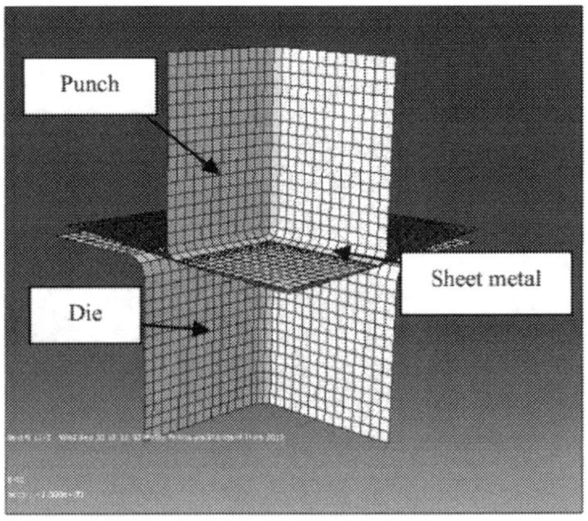

(a)

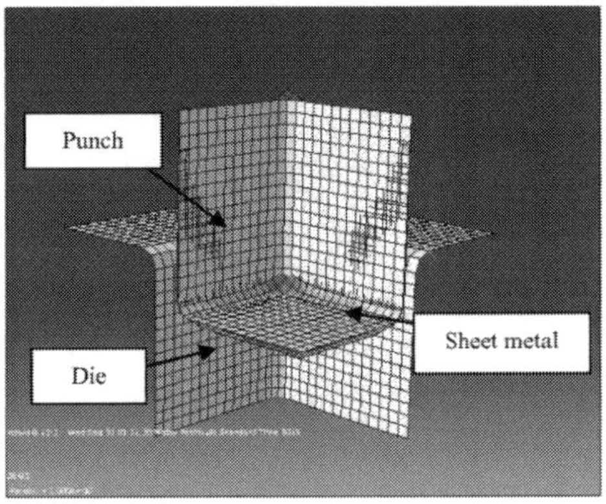

(b)

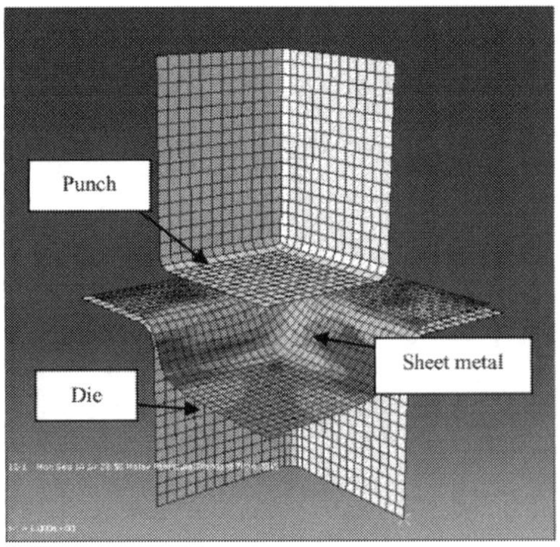

(c)

Figure 1.Finite element model
(a) Initial stage (b) Loading stage (c) Unloading stage

3 Results and discussions

The FE result is compared to investigate its reliability and deep drawing process was carried out experimentally. **Figure 2a** and **2b** show the tools and the hydraulic press machine that has been used.

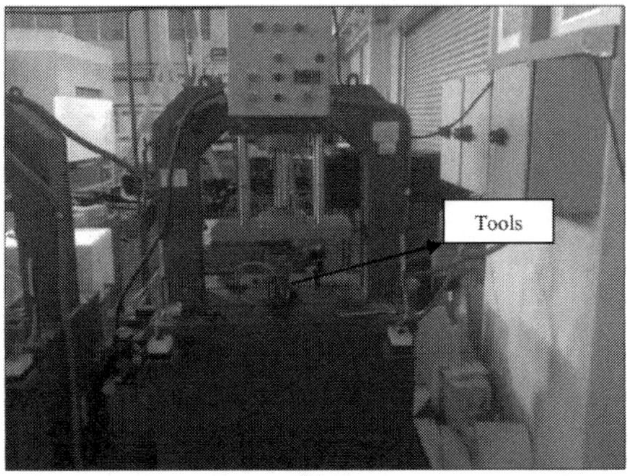

(a)

(b)

Figure 2: Deep drawing experiment set up (a) Hydraulic press machine (b) Tools used in the experiment

DOI: 10.1051/matecconf/20165403003

Figure 3 shows the experiment product from the process. The stainless steel sheet metal was successfully drawn for about 13mm depth into the die cavity.

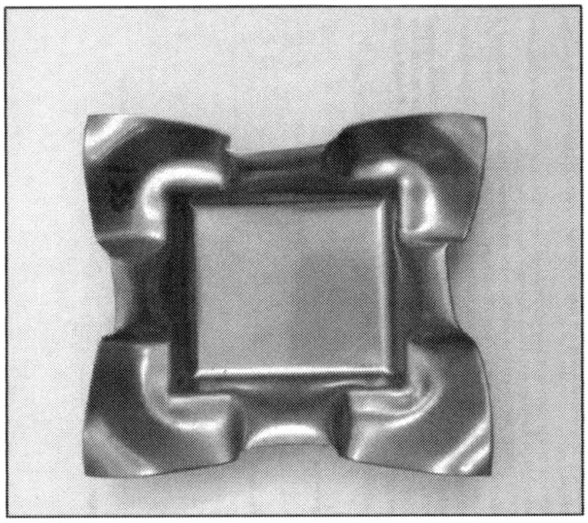

Figure 3. Experiment product

Figure 4a and **4b** show the comparison result for both simulation and experimental results. A qualitative good agreement between the experiment and the simulation has been shown throughout the figures. The wrinkle formation at the flange and cup wall looks similar for both results. The blank rises towards the side wall of punch with the formation of wrinkling along the flange. This happens due to zero holding force applied. When the holding force is less than needed, there is a big tendency for the wrinkles to occur[5]. The other factor that could contribute to wrinkling defect in deep drawing process is the drawing depth. The greater die cavity depth, the more blank will pulled down into the die cavity and the greater risk of wall wrinkling will occur. The wrinkling occur once the punch increasingly moves down into the die cavity and the most affected region is at the cup corner [4].Friction coefficient is another factor that influences wrinkling in deep drawing process. Surface quality of finish product, tool life and drawability of sheet metal are well dependent on presence of good friction coefficient between contact surfaces. The friction condition enhances the flow of material into the die cavity. An appreciable increase on the frictional force between the blank and the tools is starting when the blank passes the die radius. As the contact condition change, this is a tendency of increasing friction force and enabling a smooth flow of the blank into the die cavity, hence can prevent the wrinkling defect.

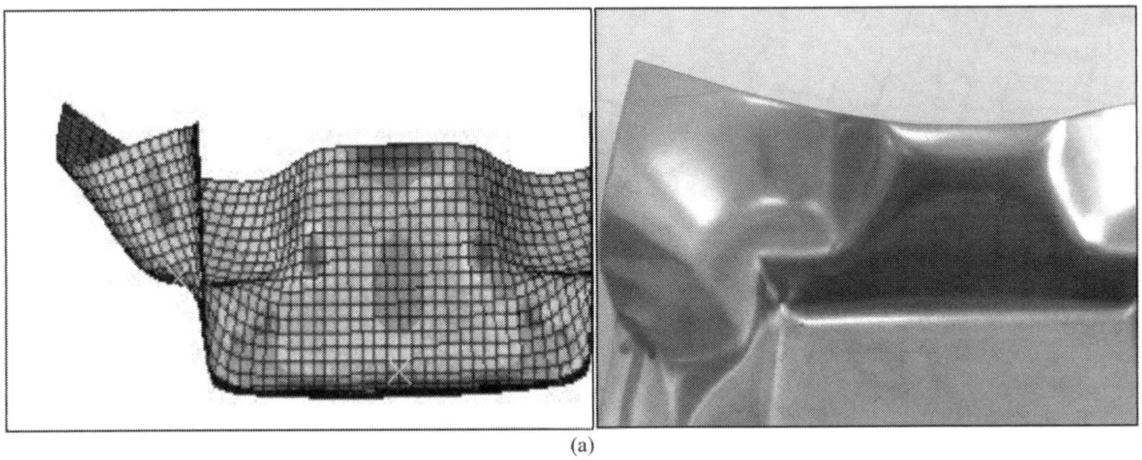

(a)

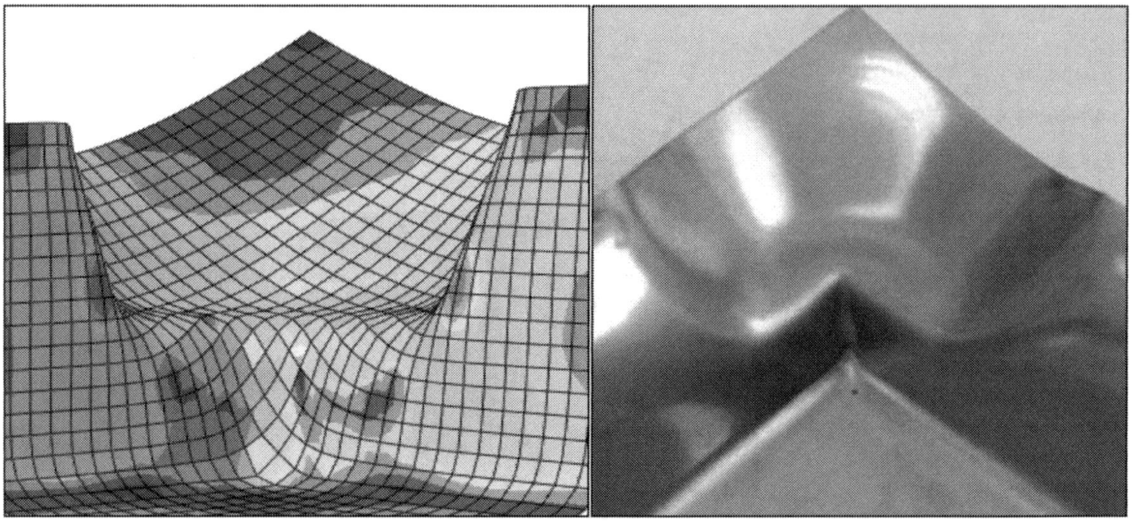

(b)

Figure4. Simulation and experimental results compared (a) half deformable part (b) one quarter deformable part

As a conclusion from this study, the finite element result was compared to the experimental result visually for any similarity. Further studies will be continued on testing the reliability of kinematic and mixed hardening model in predicting the wrinkling defect.

4 Further works

To improve the FE simulation in prediction of wrinkling in deep drawing process, there is a need to consider a constitutive model that takes into account the bending and unbending condition. For this purpose, kinematic and mixed hardening model had been acknowledged to provide the answer. Another study is already in final stage to test the reliability of kinematic and mixed hardening model in predicting the wrinkling behaviour in deep drawing process and the result will be publish in another publication.

References

1 R. Padmanabhan, M. C. Oliveira, J. L. Alves, and L. F. Menezes, "Numerical simulation and analysis on the deep drawing of LPG bottles," *Journal of Materials Processing Technology*, **200**, pp. 416-423, (2008).

2 M. Volk, B. Nardin, and B. Dolšak, "Application of Numerical Simulations in the Deep-Drawing Process and the Holding System with Segments' Inserts," *Strojniški vestnik – Journal of Mechanical Engineering*, **57**, pp. 697-703, (2011).

3 A. Wifi and A. Mosallam, "Some aspects of blank-holder force schemes in deep drawing process," *Journal of Achievements in Materials and Manufacturing Engineering*, **24**, pp. 315-323, (2007).

4 F. Ayari and E. Bayraktar, "Parametric Finite Element Analysis for a square cup deep drawing process," *Journal of Achievements in Materials and Manufacturing Engineering*, **48**, pp. 64-86, (2011).

5 S. S. Han and H. Huh, "Modified-membrane finite-element simulation of square cup drawing processes considering influence of geometric parameters," *Journal of Materials Processing Technology*, **48**, pp. 81-87, (1995).

 DOI: 10.1051/matecconf/20165403004

Robust Control of Magnetic Bearing in the Inertially Stabilized Platform for Aerial Remote Sensing

Qingyuan Guo, Gang Liu and Tong Wen

School of Instrument Science and Optoelectronics Engineering, Beihang University, Beijing 100191, China

Abstract. Comparing with the normal mechanical inertial stabilized platform (ISP), the novel ISP with magnetic bearing to suspend the yaw gimballing owns the ability of isolating the vibration and disturbance of external gimballing. However, on the one hand, complex structure of the magnetically suspended gimballing introduces the parametric uncertainties into modelling of magnetic suspension force, and the disturbance between the magnetic fields brings the un-modeled force into dynamics of magnetically suspended gimballing. On the other hand, the external disturbances such as wind drag affect the stability of magnetically suspended gimballing. Therefore, to weaken the negative influence on stability and improve the robustness of system, the H_∞ robust controllers for translation and tilting of gimballing are designed. The experimental results indicate that the robust controller of magnetically suspended gimballing owns the excellent robustness when the load of magnetically suspended gimballing changes.

1 Introduction

The ISP for remote sensing is promising to be applied in the observation and photography system fixed on the airborne platform and ship-based platform [1][2]. Usually, the loads (include POS and camera) are installed on the yaw gimballing, so stability and control precision of yaw gimballing are critical to the quality of image achieved by camera [3]. For the ordinary mechanical ISP with three axis frames, the motions of roll angle, pitch angle and inner yaw angle are separated by the mechanical levitation methods such mechanical bearing and gear[4][5]. However, for the mechanical levitation methods between gimballing, which is touched between the external gimballing and internal gimballing, the disturbance of external gimballing possibly transfer from the external gimballing to internal gimballing, so the control precision and stability of yaw gimballing are affected. Therefore, one kind of novel ISP with magnetic suspension technology was designed [6-8], the magnetic suspension method is untouched so that the disturbance can be effectively isolated. Furthermore, to provide a stable environment to the yaw gimballing, there were a series of relevant research to improve the robustness of magnetically suspended system, a gain scheduled H_∞ robust control scheme with free parameters was proposed to eliminate the unbalance vibration in a rotor system supported by magnetic bearings [9]. And, for an active radial homopolar magnetic bearing system, the robust control of rigid rotor vibration was designed analytically, and the initial responses/transient responses and robustness of the designed controller was confirmed [10]. In addition, aiming at the difficulty of obtaining the accurate model of magnetic bearing, the robust stabilization of a voltage-controlled three pole active magnetic bearing was considered [11]. However, those papers only focused on control of the magnetically suspended rotor with high-speed and light-weight, and the magnetically suspended rotor did not need to bear load. There is no publication about magnetically suspended gimballing with great rotary inertia, great bearing ability and wide load range until now. Therefore, in this paper, considering the parametric uncertainties, external disturbance and change of load, the robust control of magnetically suspended gimballing is designed, and the suspension performance under the situations of different loads is tested.

2 Model of the magnetically suspended gimballing

Figure 1. Structure of ISP with magnetic bearings. (a) photograph, (b) exploded view.

DOI: 10.1051/matecconf/20165403004

The configuration of ISP with magnetic bearings is shown as Fig.1, the whole ISP consists of three gimballing---roll gimballing, pitch gimballing and yaw gimballing. And, yaw gimballing, as the innermost gimballing, all load components are installed on it, its control precision and stability are critical to the quality of imaging. In this paper, magnetic bearings, instead of mechanic bearings and gears, are applied to levitate the yaw gimballing in order to absolutely isolate disturbance from outer gimballing. For the pitch gimballing, it is outsider the yaw gimballing, the roll gimballing is the outermost gimballing.

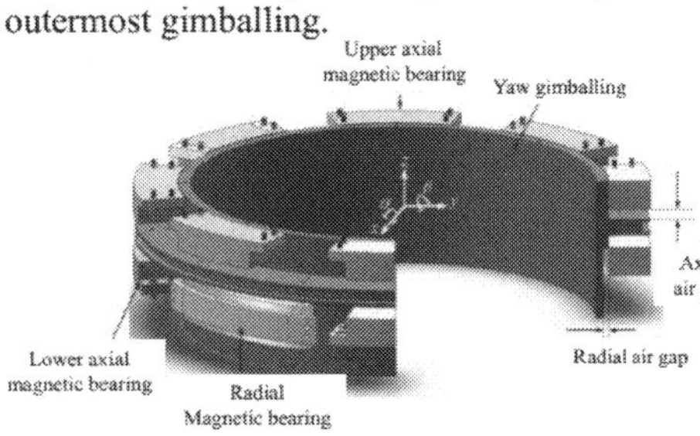

Figure 2. Structure of magnetically suspended gimballing.

As illustrated as Fig.2, the yaw gimballing is suspended by magnetic bearings in radial and axial directions, the structure of magnetic suspended gimballing is composed with radial magnetic bearing and axial magnetic bearing. Eight same axial magnetic bearings are located at the upper surface of yaw gimballing, four same axial magnetic bearings are fixed on lower surface, which are mounted on circumferential position as the four of upper axial magnetic bearings to control the axial translation. And, other four magnetic bearings control the tilting of gimballing around X and Y axis. In addition, radial translation of magnetically suspended gimballing is controlled by two pairs of magnetic bearings which are installed in radial direction. As a consequent, the suspension of yaw gimballing on five degree of freedom is realized.

2.1 *The translational model of magnetically suspended gimballing*

Virtually, the translational control of the magnetically suspended gimballing is the control of magnetic bearing in radial direction and axial direction, the position and velocity of inner yaw gimballing are controlled by adjusting the force generated by magnetic bearing.

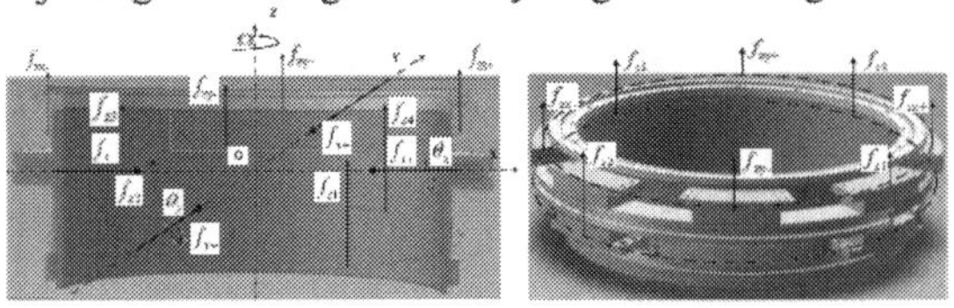

Figure 3. Analysis of force received by magnetically suspended gimballing.

The radial control of magnetically suspended gimballing is to realize the stable suspension, according to the analysis in Fig.3, the dynamic model of magnetically suspended gimballing in the radial and axial directions is

$$\begin{cases} m\ddot{x} = f_x \\ m\ddot{y} = f_y \\ m\ddot{z} = f_z \end{cases}$$
(1)

The translational model of magnetically suspended gimballing is expressed as the transfer function of

$$\begin{cases} G_x(s) = \dfrac{X(s)}{I_x(s)} = \dfrac{k_{ix}}{ms^2 - k_{dx}} \\ G_y(s) = \dfrac{Y(s)}{I_y(s)} = \dfrac{k_{iy}}{ms^2 - k_{dy}} \\ G_z(s) = \dfrac{Z(s)}{I_z(s)} = \dfrac{k_{iz}}{ms^2 - k_{dz}} \end{cases} \tag{2}$$

Furthermore, the state variable of model is selected as $X_p = (x_p \ \ y_p \ \ z_p \ \ \dot{x}_p \ \ \dot{y}_p \ \ \dot{z}_p)^T$, input variable is $U_p = (i_x \ \ i_y \ \ i_z)^T$, output variable is $Y_p = (x_p \ \ y_p \ \ z_p)^T$, and state space function of translational system is

$$\dot{X}_p = A_p X_p + B_p U_p \tag{3}$$

And the output function of translational system is

$$Y_p = C_p X_p + D_p U_p \tag{4}$$

where

$$A_p = \begin{bmatrix} 0 & 0 & 0 & 1 & 0 & 0 \\ 0 & 0 & 0 & 0 & 1 & 0 \\ 0 & 0 & 0 & 0 & 0 & 1 \\ \frac{k_{dx}}{m} & 0 & 0 & 0 & 0 & 0 \\ 0 & \frac{k_{dy}}{m} & 0 & 0 & 0 & 0 \\ 0 & 0 & \frac{k_{dz}}{m} & 0 & 0 & 0 \end{bmatrix}, \quad B_p = \begin{bmatrix} \frac{k_{ix}}{m} & 0 & 0 \\ 0 & \frac{k_{iy}}{m} & 0 \\ 0 & 0 & \frac{k_{iz}}{m} \end{bmatrix},$$

$$C_p = \begin{bmatrix} 1 & 0 & 0 & 0 & 0 & 0 \\ 0 & 1 & 0 & 0 & 0 & 0 \\ 0 & 0 & 1 & 0 & 0 & 0 \end{bmatrix}, \quad D_p = \begin{bmatrix} 0 & 0 & 0 \\ 0 & 0 & 0 \\ 0 & 0 & 0 \end{bmatrix}.$$

3 The robust control of magnetically suspended gimballing

3.1 Robust control

Captions should be typed in 9-point Times. They should be centred above the tables and flush left beneath the figures.

The robust control directly addresses the robustness of control system by designing robust controller, the standard H_∞ robust control diagram is showed as the Fig.4,

DOI: 10.1051/matecconf/20165403004

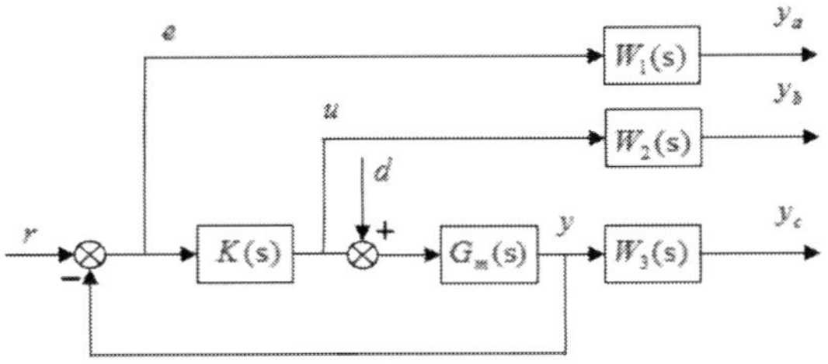

Figure 4. Robust control diagram.

where r is the reference input signal, e is the tracking error, u the control input, d measurement disturbance, y the output of system, $W_1(s)$ is the performance weighting function, $W_2(s)$ is the input weighting function and $W_3(s)$ is the output weighting function.

Given $G_m(\infty)=0$, the closed-loop transfer function from r to e, u and y is, respectively

$$S=\frac{E(s)}{R(s)}=\frac{1}{1+G_m(s)K(s)} \tag{5}$$

$$R=\frac{U(s)}{R(s)}=\frac{K(s)}{1+G_m(s)K(s)} \tag{6}$$

$$T=\frac{Y(s)}{R(s)}=\frac{G_m(s)K(s)}{1+G_m(s)K(s)} \tag{7}$$

where S is the sensitivity function, R is the penalty sensitivity function, and $S+R=I$.

During the procedure of designing suboptimal H_∞ robust controller, we try to find a stabilizing controller $K(s)$, which ensure the closed-loop system stable, and $\|P\|_\infty$, the closed-loop transfer function of enhanced object model, must satisfy

$$\min\|P\|_\infty=\min\left\|\begin{matrix}W_1(s)S\\ W_2(s)R\\ W_3(s)T\end{matrix}\right\|_\infty=\gamma_0\le\|P\|_\infty=\gamma \tag{8}$$

where γ is a minimal value of the optimization factor, and smaller γ is, the stronger the robustness.

3.2 Design of weighting matrix

A, The selection of $W_1(s)$

For the magnetic suspended gimballing, the frequency of disturbance usually locates at the low frequency domain, so the gain of S should be small to inhibit the disturbance. Consequently, the weighting value should be high, $W_1(s)$ owns the characteristics of low-pass filter [12].

B, The selection of $W_2(s)$

$W_2(s)$ is usually selected as constant to decrease orders of controller.

C, The selection of $W_3(s)$

$W_3(s)$ expresses the norm bound of multiplicative perturbation, it is selected as the function which own the characteristic of high-pass filter, and the rising slope should be high. Therefore, the gain of S would be decreased in the low frequency domain, and the gain of T would be inclined in the high frequency domain.

In our case, the nominal function of magnetically suspended gimballing in radial direction (X and Y axis) and axial direction (Z axis) is respectively,

$$G_{r0}(s)=G_x(s)=G_y(s)=\frac{300}{23s^2-480000} \tag{9}$$

$$G_{a0}(s)=G_z(s)=\frac{2000}{23s^2-3000000} \tag{10}$$

And, the nominal tilting function of magnetically suspended gimballing is

$$G_{t0}(s)=G_{zx}(s)=G_{zy}(s)=\frac{2000}{12s^2-3000000} \tag{11}$$

Based on the selection principle of weighting function, taking the radial translation of magnetically suspended gimballing as example, its weighting functions are selected as

$$\begin{cases}W_{r1}(s)=\dfrac{0.005s+1010000}{s+10}\\ W_{r2}(s)=\dfrac{0.0001s+5}{0.01s+100000}\\ W_{r3}(s)=\dfrac{s+500}{0.01s+1000}\end{cases} \tag{12}$$

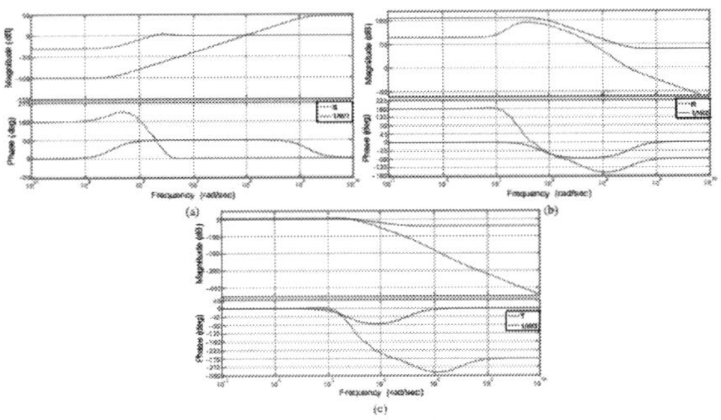

Figure 5. Characteristics of Weighing functions. (a) $W_{r1}(s)$, (b) $W_{r2}(s)$, (c) $W_{r3}(s)$.

And the amplitude-frequency characteristics and phase-frequency characteristics of weighting functions $W_{r1}(s)$, $W_{r2}(s)$ and $W_{r3}(s)$ are shown in Fig.5, and transfer function of the H_∞ robust controller is

$$K_r(s)=\frac{5.492\times10^6s^4+5.547\times10^{13}s^3+5.505\times10^{18}s^2+1.255\times10^{21}s+6.664\times10^{22}}{s^5+2.792\times10^5s^4+1.901\times10^{10}s^3+1.112\times10^{14}s^2+1.281\times10^{17}s+1.27\times10^{18}} \tag{13}$$

Similarly, for the axial translation of magnetically suspended gimballing, the weighting functions of axial translation are chosen as following,

$$\begin{cases} W_{a1}(s) = \dfrac{0.001s + 100200}{2s + 10} \\ W_{a2}(s) = \dfrac{0.0003s + 1.5}{0.01s + 100000} \\ W_{a3}(s) = \dfrac{2s + 700}{0.03s + 1200} \end{cases}$$
(14)

And the transfer function of the H_∞ robust controller is

$$K_a(s) = \frac{9.175\times10^{5}s^{4} + 9.212\times10^{12}s^{3} + 3.708\times10^{17}s^{2} + 1.541\times10^{20}s + 7.716\times10^{21}}{s^{5} + 2.114\times10^{5}s^{4} + 8.054\times10^{9}s^{3} + 4.449\times10^{13}s^{2} + 5.247\times10^{16}s + 2.613\times10^{17}}$$
(15)

In addition, for the tilting function of magnetically suspended gimballing, the weighting functions are chosen as

$$\begin{cases} W_{t1}(s) = \dfrac{0.001s + 1010200}{s + 10} \\ W_{t2}(s) = \dfrac{0.0001s + 1.2}{0.01s + 100000} \\ W_{t3}(s) = \dfrac{3s + 800}{0.001s + 10000} \end{cases}$$
(16)

The transfer function of the H_∞ robust controller is selected as following,

$$K_t(s) = \frac{3.273\times10^{6}s^{4} + 6.546\times10^{13}s^{3} + 3.274\times10^{20}s^{2} + 2.383\times10^{23}s + 3.73\times10^{25}}{s^{5} + 1.023\times10^{7}s^{4} + 2.274\times10^{12}s^{3} + 1.923\times10^{16}s^{2} + 5.467\times10^{19}s + 5.447\times10^{20}}$$
(17)

4 Experimental results

4.1 Experimental setup

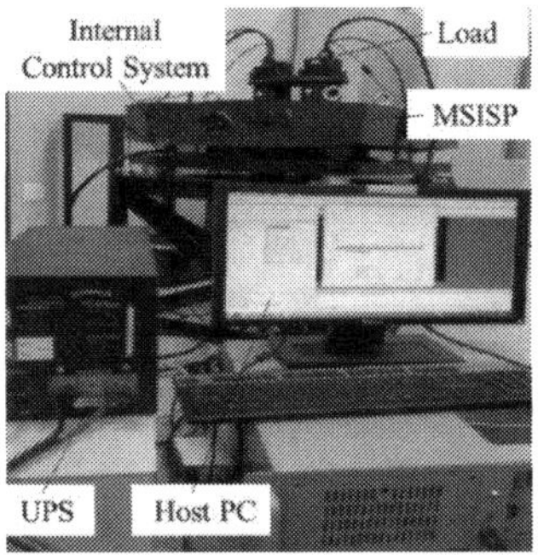

Figure 6. Experimental setup.

The experimental setup of magnetically suspended ISP is shown in Fig.6, the whole control circuit is embedded inside of magnetically suspended gimballing, and it consists of a digital signal processor TMS320F28335 with 12-bit A/D converter to control magnetic bearing, the digital PWM amplifier with 20KHz to drive the magnetic bearing, eddy current sensors to measure the displacement of rotor. In addition, the sampling frequency of whole control system is20KHz, the power supply provided by UPS is28V, other system parameters of magnetically suspended gimballing are listed in the Table.1.

Table 1. System parameters of magnetically suspend gimballing

Symbol	Value	Symbol	Value
k_{ix}	300 N/A	k_{dx}	480N/mm
k_{iy}	300 N/A	k_{dy}	480N/mm
k_{iz}	2000N/A	k_{dz}	3000N/mm
J_x	6.25N·m^2	J_y	6.25N·m^2
m	23kg	l	520mm

4.2 Suspension performance in translational direction

When the load of magnetically suspended gimballing are 15kg, 30kg and 45kg, the axial suspension performance of magnetically suspended gimballing is tested. As shown as Fig.7, the red dashed line illustrates the displacement of PID controller, the blue line shows the displacement of robust controller. Initially, there is no load installed on the magnetically suspended gimballing, the gimballing stably suspend at the axial equilibrium point.

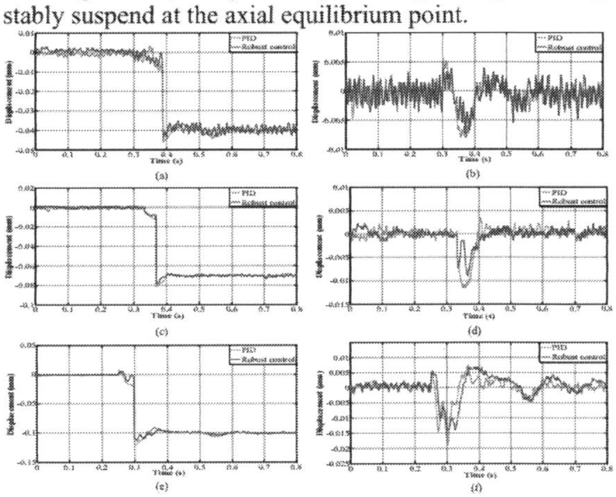

Figure 7. Translational suspension performance with loads, (a) axial suspension with 15kg, (b) radial suspension with15kg, (c) axial suspension with 30kg, (d) radial suspension with 30kg, (e) axial suspension with 45kg, (f) radial suspension with 45kg.

Then, the different loads are installed on the magnetically suspended gimballing, the comparison of radial drift between the robust control and PID is presented in the Table.2, the axial overshoot comparison is illustrated in Table.3. Thus, it is concluded that displacement overshoot of robust control is quite smaller than that of PID control when the loads are different.

Table 2. Maximum radial drift comparison between PID and robust control

	15kg	30kg	45kg
PID	0.0075mm	0.012mm	0.02mm
Robust Control	0.006mm	0.009mm	0.012mm

Table 3. Maximum axial overshoot comparison between PID and robust control

	15kg	30kg	45kg
PID	0.043mm	0.076mm	0.11mm
Robust Control	0.046mm	0.082mm	0.12mm

5 Conclusion

In this paper, we have analyzed the characteristics of magnetically suspended gimballing, the magnetically suspended gimballing owns greater bearing ability and wider bearing range, and the load of magnetically suspended gimballing will change according to the requirements. Consequently, those requirements introduce parametric uncertainties (unmodeled force, unmodeled dynamics), and the change of load cause a robustness problem to the magnetically suspended gimballing. Therefore, robust controllers of translation and tilting are designed to improve the robustness of magnetically suspended gimballing system. The experimental results demonstrate that robust controller has strong robustness when the load of the magnetically suspended gimballing changes.

Acknowledgement

This work is supported by the National Natural Science Foundation of China under Grant 61374211.

References

1. Hurak, Z. and M. Rezac. Image-based pointing and tracking for inertially stabilized airborne camera platform. IEEE Transactions on Control Systems Technology; **20(5)**: 1146-1159(2012).
2. Moon, J. and S.Y. Jung. Implementation of an image stabilization system for a small digital camera. IEEE Transactions on Consumer Electronics; **54(2)**: 206-212. (2008)
3. Mokbel, H.F., et al. Modeling and optimization of Electro-Optical dual axis Inertially Stabilized Platform. in Optoelectronics and Microelectronics (ICOM), 2012 International Conference on. IEEE: 372-377. (2012)
4. Hilkert, J.M. and B. Pautler. A reduced-order disturbance observer applied to inertially stabilized line-of-sight control. SPIE Defense, Security, and Sensing, International Society for Optics and Photonics, (2011).
5. Xiangyang Zhou, Hongyan Zhang, Ruixia Yu. Decoupling control for two-axis inertially stabilized platform based on an inverse system and internal model control. Mechatronics; **24(8)**: 1203-1213. (2014)
6. Zhuchong Lin; Kun Liu. Inertially stabilized line-of-sight control system using a magnetic bearing with vernier gimbaling capacity. Photonics Asia. International Society for Optics and Photonics: 92720Q-92720Q-11. (2014)
7. Jiancheng Fang, Chune Wang, and Tong Wen. Design and optimization of a radial hybrid magnetic bearing with separate poles for magnetically suspended inertially stabilized platform. IEEE Transactions on Magnetics; **50(5)**: 1-11. (2014)
8. Mu Q, Liu G, Lei X. A RBFNN-based adaptive disturbance compensation approach applied to magnetic suspension inertially stabilized platform. Mathematical Problems in Engineering, (2014).
9. Fumio Matsumura, Member, IEEE, Tom Namerikawa, Member, IEEE, Kazuhiko Hagiwara, and Masayuki Fujita, Member, IEEE. Application of gain scheduled H∞ robust controllers to a magnetic bearing. IEEE Transactions on Control System Technology; **4(5)**: 484-493(1996).
10. Zdzislaw Gosiewski, Arkadiusz Mystkowski. Robust control of active magnetic suspension:Analytical and experimental results. Mechanical Systems and Signal Processing. **22**:1297–1303. (2008)
11. Shyh-Leh Chen, Member, IEEE, and Cheng-Chi Weng. Robust Control of a Voltage-Controlled Three-Pole Active Magnetic Bearing System. IEEE/ASME Transactions on Mechatronics. **15 (3)**: 381-388. (2010)
12. Masayuki Fujita, Toru Namerikawa, Fumio Matsumura, and Kenko Uchida. μ-Synthesis of an Electromagnetic Suspension System. IEEE Transactions on Automatic Control **40(3)**: 530-536. (1995).

MATEC Web of Conferences **54**, 03005 (2016) DOI: 10.1051/matecconf/20165403005
MIMT 2016

Brainwave Controlled Electrical Wheelchair

Sim Kok Swee, Lim Zheng You ,Kho Teck Kiang

MMU University, Faculty of Engineering & Technology, 75450 Melaka, Malaysia

Abstract. Electrical wheelchair provides mobility to the people who lost their moving capability. To allow most of the people to be able to control the electrical wheelchair, the best solution is by directly using the brain instead of using joystick or buttons. The first objective of this project is to provide the basic mobility by constructing an electrical wheelchair. The second objective is to process the electroencephalogram (EEG) signals for controlling purpose. The EEG device which used in this project to capture the signals is the Emotiv EPOC headset. A computer is used to receive and process the EEG signals, then send the control signal to the microcontroller of the electrical wheelchair. All the hardware are connected wirelessly via Bluetooth.

1 Introduction

Nowadays in Malaysia, the number of disabled individuals are increasing every year. According to the statistics of Department of Social Welfare Malaysia, it states that the numbers of physical disabled individuals are increasing from the year of 2008 to 2011[1]. There are 445,006 people with disabilities registered In Malaysia as of 2012[2]. This also shows that the number of persons who lost their mobility are significant. To regain the mobility, the most common solution is by using a wheelchair. However today, most of the wheelchair requires user's hand to maneuver. Even for the electrical wheelchair, it still requires user's finger to control the joystick or button [3]. Therefore, some patients who completely lost their hands or those who suffer from Poliomyelitis disease will face the difficulties to control the wheelchair. In this way, they are unable to navigate themselves around. In order to overcome this problem, an alternative way that will enable most of the people to navigate the wheelchair is directly controlled by the brain. Therefore, this will bring an extremely high impact, especially for disabled individuals who cannot otherwise physically communicate [4].

In order to allow the brain to directly controlling the wheelchair, it requires the information of the brain activity. There are several ways to retrieve the signals of brain activity. They are Positron Emission Tomography [5], Single Photon Emission Tomography, Electroencephalography and Functional Magnetic Resonance Imaging. Among these methods, EEG is the best method to retrieve the signal of brain activity. This is because the EEG device is portable and non invasive [6].

Electroencephalogram (EEG) is a method which records electrical waveform from the brain by using electrodes appropriately placed on the head. It measures voltage fluctuations resulting from ionic current within the neurons of the brain [7]. This method records electrical motivations from the nerves of the brain. [8]. According to the human thoughts, different electrical waves are produced. Even for a muscle contraction, it will also generate a unique electrical signal [9]. These signals are further processed to determine the mental commands from the user's brain.

After signal processing, the mental commands such as forward, backward, left and right[10] can be used as controlling signals to control the movement of electrical wheelchair. This is a method which translate brain signals into control commands, also known as Brain-computer interfaces (BCIs) [11].

2 Problem statement

To provide basic mobility to the people who completely lost their hands and unable to move by themselves, brainwave controlled electrical wheelchair is the best solution. However, nowadays, a good EEG device is still relatively costly. Therefore, more researches have to be done for producing a suitable but affordable EEG device.

3 Design

The basic hardware design is the design of the electrical wheelchair mechanical structure. This project also involves designing electronic circuits for the electronic system of the electrical wheelchair. Software development is required for retrieving the signals from EEG device and further processing the signals to determine the mental commands. The software development also involves writing the program of the Arduino microcontroller on the electrical wheelchair so that electrical wheelchair performs the correct movement.

3.1 Hardware Design

Most of the wheelchairs available in the market have dimension around 36 inches (3 feet) long and 24 inches (2 feet) wide. In order to meet the standard size of the wheelchairs, the size of the base of prototype is 36 inches (3 feet) long and 24 inches (2 feet) wide. To achieve stability but considering the weight of the wheelchair, the material used to construct the wheelchair base and frame is steel hollow tube. The armrest, backrest and seat are covered by PVC leather for comfort purpose.

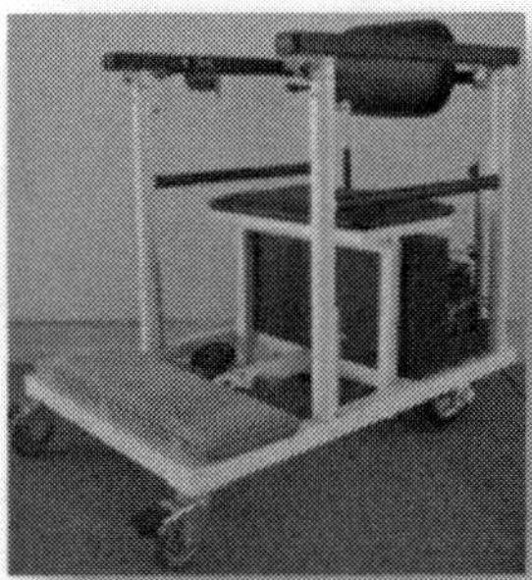

Figure 1. The Structure of Wheelchair.

The electrical wheelchair is designed with back wheel differential driven. Since the motors need to carry a person and also the whole structure, high torque motor are required. Therefore, scooter motors are chosen.

Figure 2. Scooter Motor with Gear

Arduino Uno microcontroller board is chosen as the main controller of the electrical wheelchair. This microcontroller is chosen as it has enough I/O pins for controlling the entire electrical wheelchair system and is relatively small size.

Figure 3. Arduino Microcontroller

Since scooter motor is used, a high current motor driver is required for driving the scooter motor. This is because the scooter motor will draw high current up to 10 ampere. A relays circuit board is also used for controlling the other motors such as power window motor on the electrical wheelchair.

Figure 4. High Current Motor Driver

Bluetooth module HC06 is used for wireless communication. This module connects to the Arduino microcontroller. This allows PC and smartphone to directly connect to it and receive the command signals.

Figure 5. HC-06 Bluetooth module.

Emotiv EPOC is the EEG device used to record and transmit the EEG signals. Emotiv EPOC is the official headset from Emotiv. It contains 14 channels, which means 14 sensor electrodes.

Figure 6. Emotiv EPOC headset.

3.2 Software Design

For schematic diagram design and PCB design, the Easy Applicable Graphical Layout Editor (EAGLE) is used. The openness of EAGLE design resources, such as its extensive and fully-open component libraries, ease the design process for all. [12]

Arduino Integrated Development Environment (IDE) [13] is used for compiling and uploads the source code to the Arduino Uno microcontroller. This software is open-source and is easy to upload code as it only requires USB cable, no extra circuit needed.

Microsoft Visual C# [14] is the IDE supports Visual C# with a full-featured code editor, compiler. It is a used to write the program for receiving the data sent from EEG device, processing the data and sending the command signal to the microcontroller via Bluetooth.

Android Studio is used to write an Android application which can be installed in smartphone to send the command signal to the microcontroller via Bluetooth without using the PC.

3.3 Methodology

The first step of the project is to select a best EEG device for reading the EEG signals. There are several EEG devices that can be found in the market. The comparison between the devices is done in table 1. From the comparison, it is found that the most suitable EEG device is the Emotiv EPOC headset as it contains the most channels among all these devices.

Device	Open EEG	Mindflex	Neurosky Mindset	Emotiv EPOC
Description	Building the EEG device and software from scratch	Levitating ball game from Mattel	Official headset from Neurosky	Official headset from Emotiv
Attention/ Meditation Value	No	Yes	Yes	Yes
Developer Tool	Yes and free	No	Yes, included with the purchase	Yes, included with the purchase
EEG channel	Depends on developer	1 channel	1 channel	14 channel
Costs	More than $200	$50	$199	$399

Table 1 Comparison between EEG devices which found in commercialized market

The following step is the development of the electrical wheelchair. This development process involves the construction of mechanical structure of the wheelchair and also electronics circuit design. Besides, this process also involves software programming design for the microcontroller board so that microcontroller board will send appropriate control signals to the motor driver.

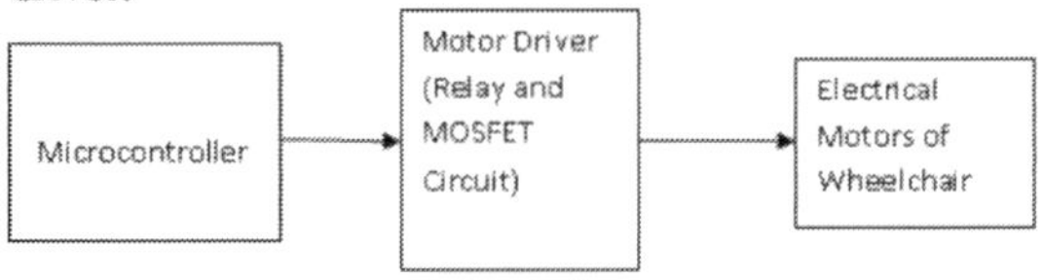

Figure 7. Block diagram of Electrical Wheelchair System

The entire electronic system includes few main devices. They are EEG device, computer and microcontroller of the electrical wheelchair. The interconnection of these devices is shown in the figure 8.

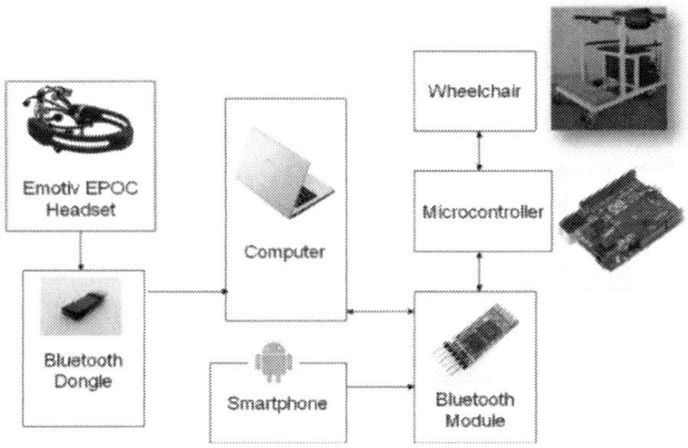

Figure 8. Block diagram of the complete brain controlled wheelchair system

The Emotiv EPOC headset is placed on the user's head for recording the EEG signals of the brain. The Bluetooth dongle is plugged in to the PC. It is used to connect with the Emotiv EPOC headset. In this way, the PC is able to receive the data sent from the Emotiv EPOC headset. The received data is further processed by using the application that wrote by using Microsoft Visual C#. The data is further processed to determine the mental command that is thinking by the user. According to the mental commands, output command signal is generated. Then, the command signal is sent to the microcontroller of the electrical wheelchair via Bluetooth. In this way, the electrical wheelchair is able to be directly controlled by the user's brain. Besides, an Android software can be also used to control the movement of the electrical wheelchair without require PC for sending command signal. However, the Android application is still not able to receive and process the EEG signals.

The program flowchart (shown in figure 9) shows that the electrical wheelchair is able to operate in two modes. One is by using the brainwaves to control and another mode is by using an Android application to control. The smartphone Android application is an extra feature that allows the user to control the electrical wheelchair like a common electrical wheelchair instead of the electrical wheelchair is only workable under the brainwaves control. Therefore, this program flow involves few software developments. First, a Windows

DOI: 10.1051/matecconf/20165403005

based application is developed by using Microsoft Visual C# for receiving and processing the incoming signals from EEG device. After processed the signals, the application generates control signal and send to the microcontroller of the electrical wheelchair via Bluetooth. Therefore, the program of the Arduino microcontroller have to be written properly so that it will send appropriate signals for controlling the wheelchair movement. This program is written and compiled by using Arduino IDE.

.

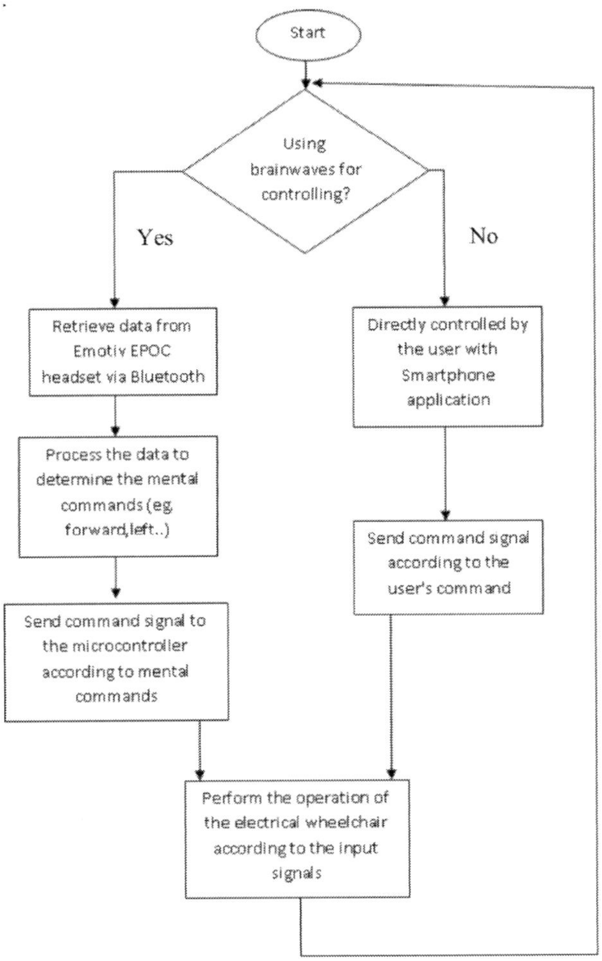

Figure 9. Main program flow of the brainwave controlled wheelchair system.

4 Results

The developed electrical wheelchair is able to be controlled directly by the brain. In addition, this electrical wheelchair is also able to work as normal electrical wheelchair by using the developed Android application. This project provides a new and effective solution for the people who lost their hands and mobility.

References

1. Public Service Department Malaysia, "What Is The Definition Of Senior Citizens In Malaysia?"[Online].Available: http://www.jpapencen.gov.my/english/senior_citizen.html. [Accessed: 01-Dec-2013].
2. Unicef, "Situation in Malaysia." [Online]. Available: http://www.unicef.org/malaysia/sowc_sowc_13-situation_in_malaysia.html. [Accessed: 29-Sep-2015].
3. L. Bi, X. A. Fan, Y. Liu "EEG-Based Brain-controlled Mobile Robots: A survey", *Human-Machine Systems*, **43**, 161-176, (2013).
4. A. Nijholt, D. Tan, G. Pfurtscheller, C. Brunner, J.del R. Mill´an, B. Allison, B. Graimann, F. Popescu, B. Blankertz, and K.-R. M¨uller, "Brain–computer interfacing for intelligent systems.", *IEEE Intelligence System*, **23**, 72–79. (2008).
5. WebMD. "Positron Emission Tomography (PET)." [Online]. Available: http://www.webmd.com/cancer/lymphoma/positron- emission- tomography. [Accessed: 9-Sep-2014].
6. T. F. Collura. "EEG Frequency Bands." *The Measurement, Interpretation, and Use of EEG Frequency Bands.* (1997).
7. E. Niedermeyer & B. F L. Silva. "Electroencephalography: Basic Principles, Clinical Applications and Related Fields." (2005).
8. K. H. Solanki, H. Pujara. "Brainwave Controlled Robot.", *International Research Journal of Engineering and Technology (IRJET)*, **2**, 609 - 612. (2015).
9. K. Yendrapalli , S. S. N. P. K. Tammana The "Brain Signal Detection for Controlling the Robot.", *International Journal of Scientific Engineering and Technology*, **3**, 1280-1283. (2014).
10. B. Rebsamen, C. Guan, H. Zhang, C. Wang, C. Teo, M. H. Ang, Jr., and E. Burdet, "A brain controlled wheelchair to navigate in familiar environments.", *IEEE Trans. Neural Syst. Rehabil*, **18**, 590–598. (2010).
11. K.. R. M¨uller and B. Blankertz, "Toward noninvasive brain–computer interfaces," *IEEE Signal Processing Magazine*, **23**, 125–128. (2006).
12. CSEagle, "What is EAGLE." [Online]. Available: http://www.cadsoftusa.com/eagle-pcb-design-software/about-eagle/ [Accessed: 29-Sep-2015].
13. Arduino LLC, "Arduino Software IDE." [Online]. Available: http://www.arduino.cc. [Accessed: 27-Aug-2015].
14. "Microsoft Visual C#," *Microsoft*, 1989. [Online]. Available: http://msdn.microsoft.com [Accessed: 27-Aug-2015].

MATEC Web of Conferences **54**, 03006 (2016) DOI: 10.1051/matecconf/20165403006
MIMT 2016

An Analysis for 2.4mm-2.4mm RF Connector Insertion Loss Measure From 45MHz Until 50GHz by Using Electronic Calibration Module and Mechanical Calibration kits in a Network Analyzer

Ming hui, Tan [1], Ahmad Yusairi Bani Hashim[2] and Mohd Rizal Salleh[3]

[1]*Radio Frequency Calibration Laborotary, National Instruments Malaysia Sdn Bhd, Malaysia*
[2,3]*Faculty of Manufacturing Engineering, Universiti Teknikal Malaysia Melaka, 76100 Durian Tunggal, Melaka, Malaysia*

Abstract. This paper is a study regarding the Radio Frequency (RF) connector characteristic (insertion loss) in applications at 50 Ω Impedance from 45MHz until 50GHz by using the Network Analyzer and two type of calibration kits that are Electronic Cal (E-Cal) Module and Mechanical Calibration (M-Cal) Kits. RF connector will always assume as idle (lossless) in a calibration system. It can exist in varying types in dimension and length (delay line). The purpose of this study is to determine the significant and impact of the RF connector loss against its frequency range. The objective is to gain more knowledge about the RF connector characteristic, network analyzer and the two type of calibrations kits setting and usage. One port and full two ports calibration techniques are used to determine the loss of RF connector. This study will analyze the differences between both techniques (one port and full two-port calibration) and calibration kits (E-Cal and M-Cal kits) in the RF connector insertion loss.

1 Introduction

Radio frequency (RF) connectors are found in a broad range of electronic equipment for extending the dynamic range of measuring equipment. Characterizing RF or microwave circuits and devices requires an accurate attenuation measurement. RF connector can be found in any dimension and frequency range. The problem is the RF connector it does not come to the reading of dB against the frequency range. The objective of this work is to create a new process to characterize RF connector using a network analyzer [1]. It is to determine the insertion loss and port match of the RF connector. This also covers the calibrations kit usage on the network analyzer. At the end of the project will implement an excel spreadsheet to be filled in all the parameters and information. The spreadsheet will guide how to determine the spec, system buy-off and calculating EN Ratio.

2 Background

RF connector is designed and manufactured a wide variety to meet the needs of nearly every application. The major specifications are the characteristic impedance, frequency range, and quality. Each connector has its unique parameters as well as its cautions and techniques for making reliable measurements [2]. Table 1.1 below reflects a variety of connectors typically used around in the field.

Table 1. RF Connector Usable Frequency Range

Connector type	Range in GHz
APC-7	Up to 18
Type-N	Up to 18
3.5mm	Up to 33
2.4mm	Up to 50

Precision Slotless Connector (PSC) is of higher quality than the Precision Variety (Slotted) and can make a better repeatable measurement [3]. A PSC is also defined as a metrology grade connector. It is expensive, reliable, repeatable and precision in dimension. The RF signals are much more stable in frequency and power than ocean waves. The results of the reflections are also more stable. In fact, they are stuck in a particular location on the RF path and are thus called Voltage Standing Wave Ratio (VSWR).

3 Method

3.1 Network Analyzer Set Up Process and Procedure

The N5230C is a Network Analyzer that can measure S-parameter up to 50GHz. It is suitable to replace the older version of Network Analyzer model 8510C. It is recommended that a user is familiar with the calibration techniques and terminologies to get the maximum understanding from the service manual or application note. The type of measurement calibration selected by the user depends on the device to be measured. For example one port or full two-port device. The calibration standards are available. The extent of accuracy enhancement desired is refered to the standards. Devices such as an adapter removal calibration for a different type of connectors use a combination of calibratons in measuring them. The accuracy and stability of a test equipment determine the accuracy of subsequent device measurements. The accuracy of the calibration standard in the network analyzer and the calibration method either one port or full two-port calibration was used in conjunction with the error correction model by using an automated E-Cal Kit [4]. It is important to carry out the calibration process according to the manufacturer recommended requirements. The network analyzer set up as below for S-parameter one port and full two-port calibration.

i. Preset the network analyzer
ii. Start frequency:45MHz
iii. Stop frequency:50GHz
iv. Averaging factor:1
v. IF bandwidth: 100 Hz
vi. Number of points: 201
vii. Data: S11

Then go to the network analyzer toolbar and select "Cal"

i. Do not preset the network analyzer anymore
ii. Select "Cal Wizard"
iii. "Smart Cal" (guided calibration) and one port / full 2 port calibration
iv. DUT port 1 is APC2.4 female
v. DUT port 2 is APC2.4 female (only for full two-port calibration)
vi. Select the cal kit, for this case, is "85056A 2.4mm calibration kits / E-Cal kit"
vii. Follow the pop-up message and make the calibration by connecting the M-Cal / E-Cal kit to the corresponding port.
viii. After finishing the calibration and save it to the directory name "cal 1"

After complete calibration, the network analyzer is ready to measure DUT. The new network analyzer calibration is more users friendly, and it is easy to perform the calibration. It only requires 20 minutes or less to complete the calibration as shown in Figure 1.

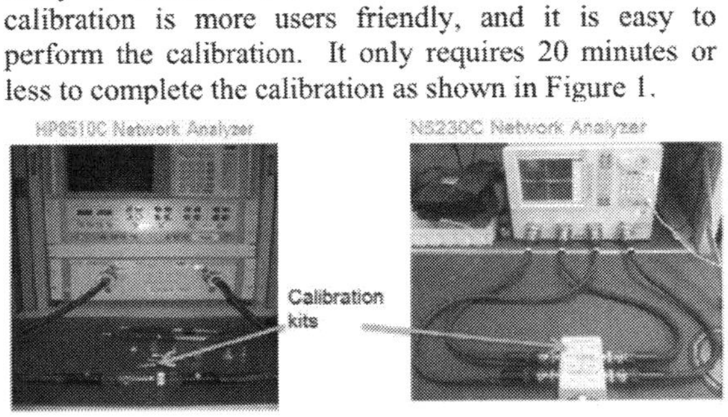

Figure 1. An Example of the VNA HP8510C and PNA N5230C with the connection.

3.2 S-parameter Single One Port (S1p) Calibration

After the setting in 3.1 completed, choose the right connector for port one network analyzer. This technique only consists of one port calibration only as shown in Figure 2 whereby the port 1 of the NA connected to the calibration kits. It takes 10 minutes to complete the calibration. It does not require any specific skills to do the S11 one port calibration by using the calibration kits.

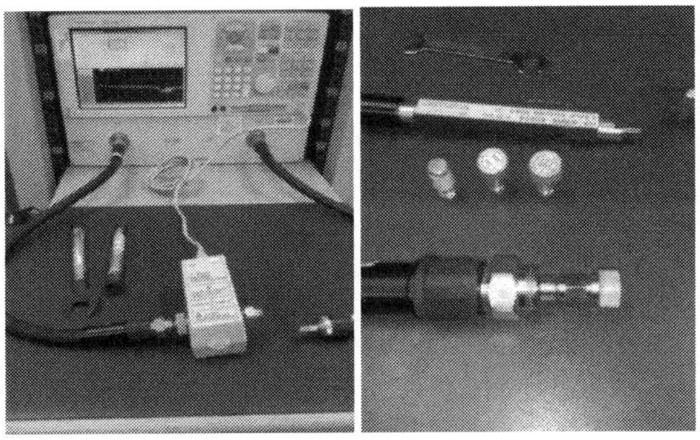

Figure 2. One Port Calibration using E-Cal (left) and M-Cal (right)

3.3 S-parameter Full Two Port (F2p) Calibration

This technique consists of both ports one and two at the network analyzer for calibration as shown in Figure 3. The time consumed for full two ports are 20 minutes to complete the calibration by using the E-Cal or M-Cal [5] module. It requires high skills to do the full two-port calibration.

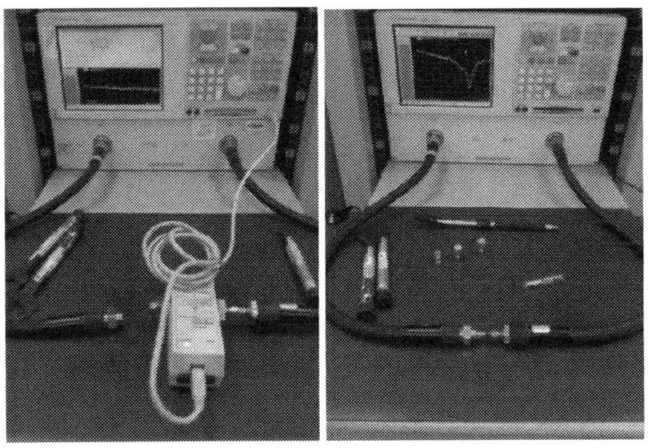

Figure 3. Full Two Port Calibration using E-Cal (left) and M-Cal (right)

3.4 Calculate Error Normalize Ratio (EN Ratio)

In the final results, the pass and fail condition will determine by the Error Normalize Formula (EN ratio). The EN ratio is performance statistic stands for Error Normalized. It is derived by dividing the difference between a DUT measured value and the reference assigned values by the square root of the sum of the square of the DUT's uncertainty and reference uncertainty that are pre-determine in the network analyzer calibration system. Equation 1 defines the EN ratio [6].

$$EN\ ratio = \left(\frac{ILC - REF}{\sqrt{U_{ILC}^2 - U_{REF}^2}} \right) \quad (1)$$

Where:
ILC is the Device under Test measure value
REF is the reference assigned value
U^2_{ILC} is the DUT measurement uncertainty squared
U^2_{REF} is the reference assigned measurement uncertainty squared.

4 Results And Discussion

4.1 S-parameter Single One Port (S1p) Calibration

In the overall results for Linear Magnitude, it can reflect in statistical theory. It is any statistical relationship between two sets of random variables or two sets of data. Correlation refers to any of a broad range of statistical relationships involving dependence. A high correlation applies to both data and they are are well related. A less relationship, however, will reflect the opposite direction, or it is not well related. Refer to the Figure 4, the Linear Magnitude for S1p are high correlation from 45MHz until 17GHz. After 17GHz, the results show that E-Cal S1p and M-Cal S1p have its trend that is less correlation. Beyond 17GHz, both results cannot determine which technique show the more accurate and precise measurement. More sets of measurement need to be taken to further study the results.

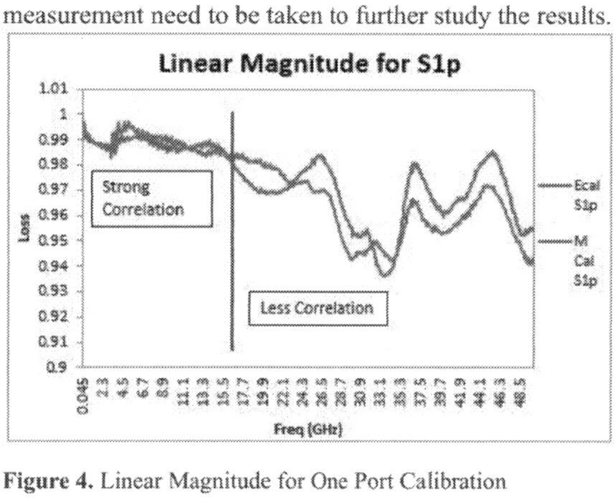

Figure 4. Linear Magnitude for One Port Calibration

Figure 5 show the loss in dB for the S1p technique. E-Cal S1p measured the RF connector having the maximum loss -0.564dB at 33GHz while M-Cal S1p measured the maximum loss -0.501dB at 29.7GHz, -0.539dB at 34.5GHz and -0.524dB at 49.3GHz. M-Cal S1p measure more peak loss compare to E-Cal S1p only found one peak loss in dB.

The maximum loss different for both techniques are 0.025dB. It is considered very close to each other. However both M-Cal S1p and E-Cal S1p measure noise and harmonic from 45MHz until 22GHz frequency range. Both sets of data also show that 22GHz until 50GHz having a clean and less harmonic measurement. The harmonics occur along the frequency is due to the averaging that is needed to be set higher instead of the averaging is set to 1 in this project.

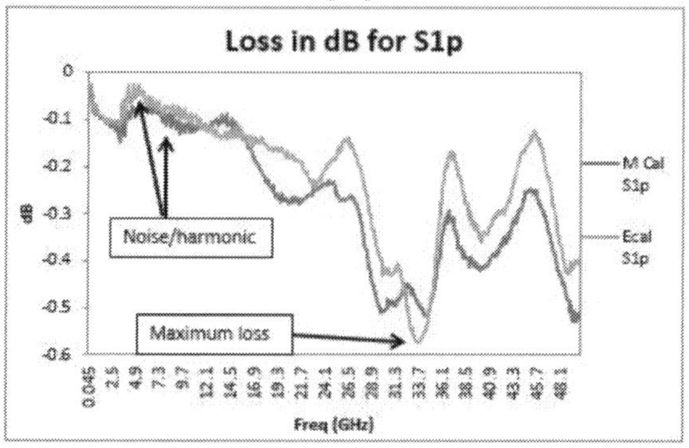

Figure 5. Loss in dB for One Port Calibration

4.2 EN ratio for Single One Port (S1p) Calibration

In this experiment, EN ratio was chosen to use to determine the pass and fail condition. EN ratio calculated more than 1 is a fail. If the EN ratio less than one is a pass. Table 2 show the RF connector results comparing the M-Cal S1p and E-Cal S1p. Both measurements was found passed for all the frequency range.

Table 2. EN Ratio for Single One Port Calibration

EN ratio S11 one port cal								
	M Cal S1p		E Cal S1p					
Freq GHz	Linear Mag	Uncertainty	Linear Mag	Uncertainty	Delta	RSS	EN Ratio	Status
0. 046	0.9969	0.0220	0.9968	0.0220	0.0001	0.0311	0.0028	PASS
1	0.9895	0.0220	0.9900	0.0220	0.0005	0.0311	0.0161	PASS
5	0.9911	0.0220	0.9950	0.0220	0.0039	0.0311	0.1254	PASS
10	0.9872	0.0220	0.9886	0.0220	0.0014	0.0311	0.0462	PASS
15	0.9878	0.0220	0.9867	0.0220	0.0011	0.0311	0.0345	PASS
20	0.9692	0.0480	0.9794	0.0480	0.0102	0.0679	0.1503	PASS
25	0.9727	0.0480	0.9790	0.0480	0.0063	0.0679	0.0926	PASS
30	0.9432	0.0860	0.9516	0.0860	0.0084	0.1216	0.0691	PASS
35	0.9463	0.0860	0.9472	0.0860	0.0009	0.1216	0.0072	PASS
40	0.9536	0.0860	0.9596	0.0860	0.0060	0.1216	0.0495	PASS
45	0.9720	0.0860	0.9847	0.0860	0.0127	0.1216	0.1045	PASS
50	0.9431	0.0860	0.9545	0.0860	0.0114	0.1216	0.0935	PASS

Refer to the Table 2 above, it can conclude that the RF connector was measured under very consistent and stable condition. The amount of change was too small. The maximum EN ratio was measured 0.1503 at 20GHz. It shows that at 20GHz, both data were found less correlation at this point.

4.3 S-parameter Full Two Port (F2p) Calibration

Adapter delay is specially only measured in a full two-port calibration system. It's measure the time delay between port one and port two of the network analyzer. The adaptor delay is significant because the network analyzer will automatically offset the delay time before measuring the DUT. Hence, the measurement makes more accurate and precision. Equation 2 defines the adapter delay time [7].

$$\text{Adapter Delay} = \pm \left(\frac{1}{4 X f} \right) \text{Second} \quad (2)$$

The adaptor delay only exists in the Programmable Network Analyzer series while the older generation of network analyzer requires another technique call Adaptor Removal method to measure the non-insertable device such as 2.4 mm female to 2.4mm female adapter. By using the equation 2 to manual calculate the 50GHz adaptor delay is approximate 0.005nsec as shown in Table 3. It is the best adapter delay value for a 2.4mm RF connector by using the formula to calculate.

Table 3. An Adaptor delay manually calculated at 50GHz

Frequency	50000000000	Hertz
Constant Value	1	-
Constant Value	4	-
Adaptor delay	5E-12	Seconds
Approx	0.005	nsec

Figure 6 show the E-Cal kit and Figure 7 show the M-Cal kit adapter delay measured in the network analyzer.

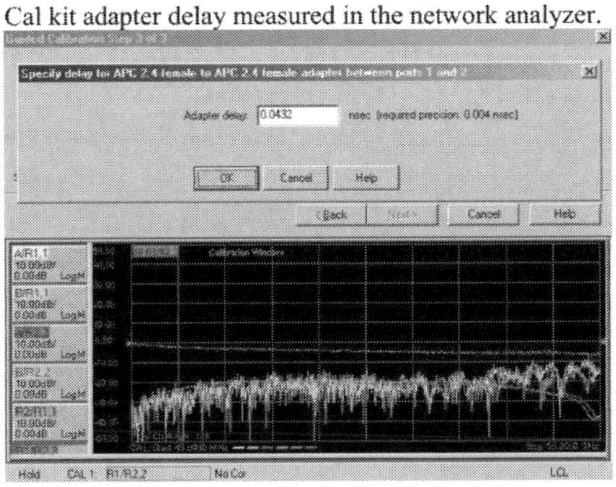

Figure 6. E-Cal Kit Adapter Delay in Network Analyzer

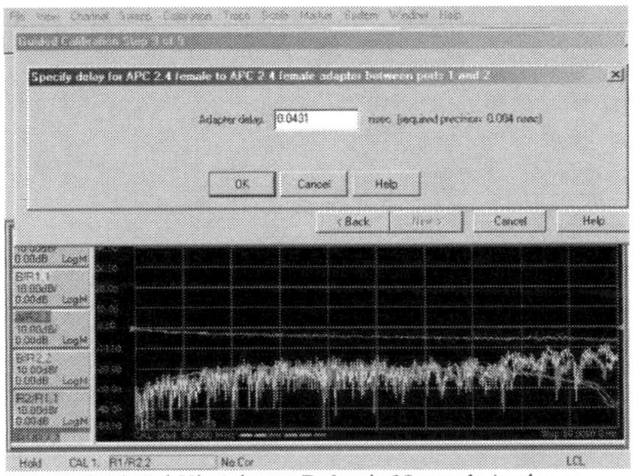

Figure 7. M-Cal Kit Adapter Delay in Network Analyzer

Refer to the Figure 8, the Linear Magnitude for F2p are high correlation from 45MHz until 50GHz. However at 28GHz and 32GHz, the E-Cal F2p has run off from M-Cal F2p trend before it comes to a spike at 36GHz. It shows E-Cal F2p at 28GHz measure less compensates, and 32GHz had been over compensated. It is because the E-Cal Module unable to identify the accurate reading for its open, short and load value at 28GHz and 32GHz. It does not perform well at that two frequency points.

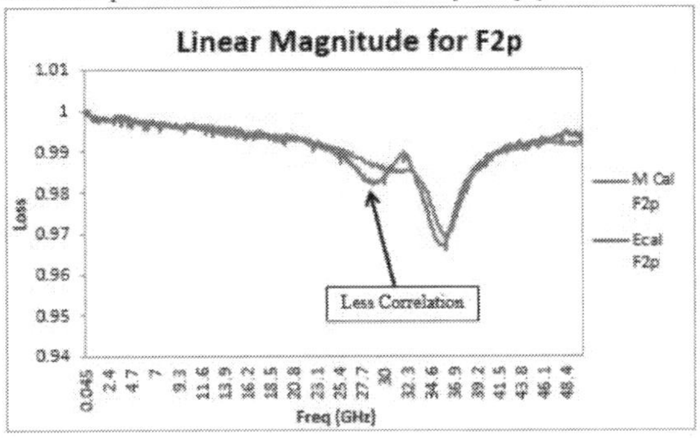

Figure 8. Linear Magnitude for Full Two Ports Calibration

Figure 9 show the loss in dB for the F2p technique. M-Cal F2p measured the RF connector having the maximum loss -0.274dB at 35.1GHz while E-Cal F2p measured the maximum loss -0.302dB at 36GHz. The maximum loss different for both techniques is 0.028dB. It is considered very close to each other. However, the M-Cal F2p measure more clean and linear than E-Cal F2p. E-Cal F2p shows noise and harmonic along the entire frequency range. Averaging causes the harmonics to occur along the frequency range. It needs to be set higher instead of value 1. Thus, M-Cal F2p produces a more accurate result than E-Cal F2p.

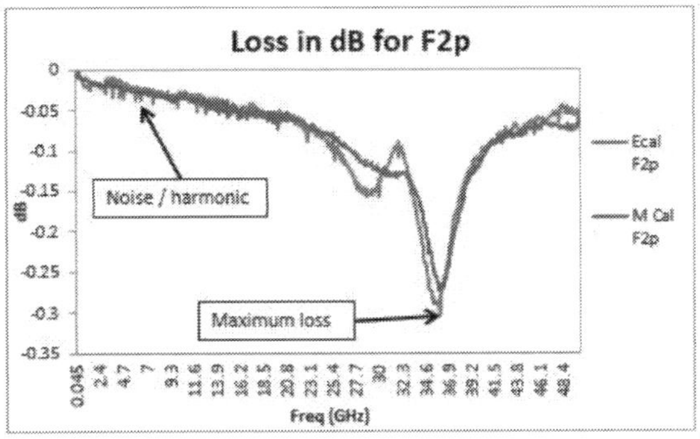

Figure 9. Loss in dB for Full Two Ports Calibration

4.4 EN Ratio for Full Two Port (F2p) Calibration

Refer to the Table 4, shows the RF connector results comparing the M Cal F2p and Ecal F2p. Both measurements were found passed for all the frequency range. Each of the measurements measured was very close. It can be concluded that the RF connector was measured under a very consistent and stable condition. The maximum EN ratio was measured 0.0501 at 15GHz. It shows that at 15GHz, both data M-Cal and E-Cal were found less correlation.

Table 4. EN Ratio for Full Two Port Calibration

EN ratio full two port cal								
	M Cal F2p		E Cal F2p					
Freq GHz	Linear Mag	Uncertainty	Linear Mag	Uncertainty	Delta	RSS	EN Ratio	Status
0. 045	0.9991	0.0220	0.9996	0.0220	0.0005	0.0311	0.0167	PASS
1	0.9986	0.0220	0.9978	0.0220	0.0008	0.0311	0.0263	PASS
5	0.9973	0.0220	0.9977	0.0220	0.0004	0.0311	0.0122	PASS
10	0.9957	0.0220	0.9947	0.0220	0.0010	0.0311	0.0309	PASS
15	0.9941	0.0220	0.9957	0.0220	0.0016	0.0311	0.0501	PASS
20	0.9935	0.0480	0.9916	0.0480	0.0019	0.0679	0.0280	PASS
25	0.9909	0.0480	0.9891	0.0480	0.0018	0.0679	0.0263	PASS
30	0.9858	0.0860	0.9821	0.0860	0.0037	0.1216	0.0302	PASS
35	0.9742	0.0860	0.9690	0.0860	0.0052	0.1216	0.0428	PASS
40	0.9866	0.0860	0.9862	0.0860	0.0004	0.1216	0.0030	PASS
45	0.9916	0.0860	0.9925	0.0860	0.0009	0.1216	0.0075	PASS
50	0.9923	0.0860	0.9922	0.0860	0.0001	0.1216	0.0006	PASS

5 Conclusion

Conclude that this project guides the operation of a network analyzer and calibration kits to perform the calibration job according to the standards and procedures. Network analyzer will not limit to measure the RF connector. An EN ratio had been calculated between the Mechanical Cal Kits and Electronic Cal Kits are well establishing to evaluate the technique of ports calibration. The S1p and F2p techniques also successful documented in this project. It is a unique process to calibrate an RF connector measured from 45MHz until 50GHz. The results also showed that the F2p very correlate along the wide bandwidth, but the S1p only correlate well in low frequency until 17GHz. Beyond the frequency of 17GHz, both techniques for S1p will start to lose the strong correlation.

Table 5. Overall Results for M-Cal and E-Cal Kits

Overall Results				
	Mechanical Cal Kit		Electronic Cal Kits	
	M Cal S1p	M Cal F2p	Ecal F2p	Ecal S1p
Time	5 minutes	30 minutes	20 minutes	10 minutes
Speed	Very fast	Slow	Medium	Very fast
Accuracy	Low	High	Medium	Low
Precision	Low	High	Medium	Low
Skills	Low	High	Low	Low
Investment	Expensive	Expensive	Expensive	Expensive

Table 5 is the overall conclusion of this study. It compares the technique of S1p and F2p against the time, speed, precision, skills and investment budget. Each technique has its advantages and disadvantages. If the method need apply in a mass production, M-Cal S1p is the most applicable. It is very fast but not accurate and not precise. It does not involve any high skills worker to perform the calibration job. Each RF connector only consumes 5 minutes per calibration compare to M-Cal F2p is the longest calibration time. It is 6 times longer than M-Cal S1p, but M-Cal F2p shows more accurate and precision measurement. M-Cal F2p technique suitable for high accuracy and particular workstation to measure the RF connector insertion loss and port match. It is also required high accurate, precision and skills workers to carry the calibration task. Electronic cal kit would be more users friendly. It does not require too much connection compare to Mechanical cal kit. It is all in one module to perform the entire calibration task. It is less connections and allowed the worker to multitasking while waiting for the Electronic cal kit module performing the calibration with the network analyzer. It is a great tool for low volume but wide variety production because no monitoring process is required along the calibration.

References

1. Keysight Technologies, Keysight Technologies Network Analyzer Selection Guide, USA: 5989-7603EN, August 19, 2015.
2. HUBER+SUHNER, RF Connector Guide - Understanding connector technology, Switzerland, 4th edition, 2007.
3. Keysight Technologies, Keysight 11904S 2.4 mm/2.92 mm Adapter Set, USA: 11904-90009, Edition 5, August 2014.
4. Keysight Technologies, Keysight Electronic Calibration Modules, USA: N4693-90001, Edition 2, July 29, 2015.
5. Keysight Technologies, Keysight Technologies 85056A 2.4 mm Precision Calibration Kit, USA: 85056-90020, October 2014.
6. Asia Pacific Laboratory Accreditation Cooperation, Calibration Interlaboratory Comparisons APLAC PT 001 APLAC PT 001, North Melbourne VIC 3051 ,Australia: NATA, 03/08.
7. Agilent Technologies No.16000-95024, Agilent E5070B/E5071B ENA Series RF Network Analyzers Adapter Characterization First Edition, Japan: Agilent Technologies Japan, Ltd., September 2002.

Novel Sliding Mode Control for Ball Screw Servo System

Rongrong Qian[1], Minzhou Luo[2], Jianghai Zhao[2] and Tao Li[2]

[1]*School of Information Science and Technology, University of Science and Technology of China, Hefei, Anhui, China*

[2]*Institute of Advanced Manufacturing Technology, Hefei Institute of Physical Science, Chinese Academy of Sciences, Changzhou, Jiangsu, China*

Abstract. This paper presents a novel sliding mode control to achieve a high positioning accuracy for ball screw servo system. A two-degree-of-freedom (2-DOF) model with time-varying parameter uncertainties and disturbance is firstly built to describe the first axial mode resonance for the ball screw servo system. A novel nonlinear sliding mode control is designed for the system with better transient response of fast response and non-overshoot. The stability of the nonlinear sliding-mode surface and controller are verified based on the Lyapunov function. The frequency response experiment result indicates that the system modelling is accurate enough and the simulation result shows that the tracking accuracy for ball screw servo system is improved with the proposed controller.
Keywords. ball screw servo system; sliding mode control; nonlinear sliding-mode surface; modeling

1 Introduction

Ball screw servo system is widely used in industrial high positioning facilities due to its high transmission stability and low sensitivity to external forces and inertia variations [1]. Since the high accuracy with high speed and high acceleration are the main demands and developing trends, it is particularly important to design the superior control law on the trajectory tracking for ball screw servo system.

As for the topic of tracking control system design, a large number of control strategies emerge to acquire the high performance tracking accuracy [2-5]. However, sliding mode control is employed frequently in the tracking control. For example, integral sliding mode controller [6] and adaptive sliding mode controller [7, 8] are presented to achieve high-precision positioning performance with repeatability and robustness. However, in order to improve the system convergence characteristics and transient response performance, Bijnan Bandyopadhyay et al. present a novel nonlinear sliding surface according to the composite nonlinear feedback principle [9]. With this nonlinear sliding-mode surface, the system can have a small damping ratio with fast response and along with the system getting close to the desired trajectory the damping ratio is getting bigger to avoid the overshoot. Apparently, it has the advantages of settling quickly without overshoot comparing to linear sliding mode controller.

In this paper, a control scheme is proposed to improve the sliding mode control based on the nonlinear sliding-mode surface for the ball screw servo system. The remainder of this paper is organized as follows. In section 2, the structure of the ball screw servo system is given. A two-degree-of-freedom (2-DOF) model describing the first axial mode resonance which has the most influence for the positioning accuracy in the ball screw servo system is built. In section 3, a novel sliding mode controller (NSMC) based on nonlinear sliding mode control is presented and derived for the 2-DOF model of ball screw servo system with bounded parameter uncertainties and external disturbances. The stability of the nonlinear sliding-mode surface and the NSMC are verified according to the Lyapunov function. In the end, the frequency response experiment is conducted to identify the 2-DOF model for ball screw servo system and the controller simulation of NSMC compared with traditional linear sliding mode control for the system is given.

2 Two-DOF system modeling

The structure of ball screw servo system analyzed in this paper is shown in Figure 1. It consists of ball screw driven by an AC motor through a coupling and assembled with the machine base by deep groove ball bearings. The worktable is mounted to the ball nut and constrained to the bilateral linear guides in order to move axially.

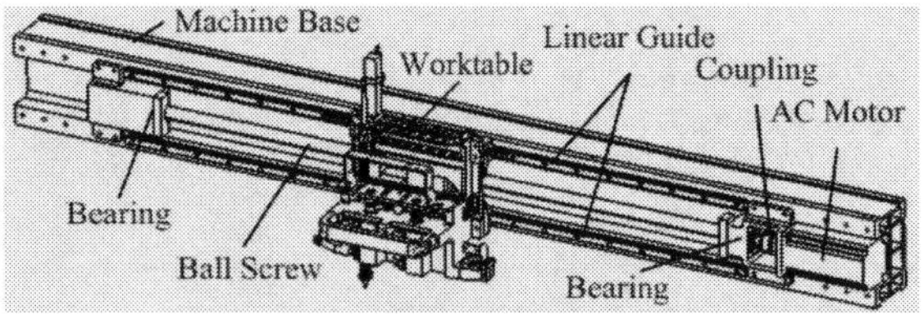

Figure 1. The structure of ball screw servo system

In the ball screw servo system, the first axial mode resonance has the most impact on the worktable linear positioning accuracy. The flexible characteristic of the translating components is the main reason for the first axial mode resonance. It includes the inherent flexibility of the ball screw and bearings and the elastic interaction characteristics between the worktable, nut and ball screw. A simplified model of two mass drive system shown in Figure 2 is used to simulate this flexible characteristic to capture the first axial dynamic for ball screw servo system and the 2-DOF mathematical model of the system is expressed as Eq. (1),

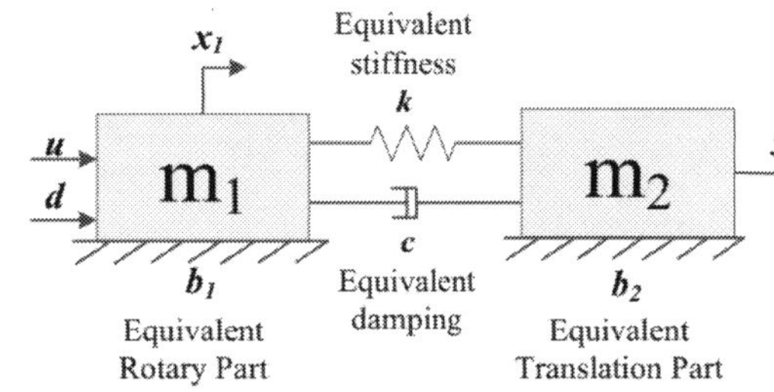

Figure 2. Simplified model for ball screw servo system

$$\begin{aligned} m_1\ddot{x}_1 &= -b_1\dot{x}_1 + k(x_2 - x_1) + c(\dot{x}_2 - \dot{x}_1) + u(t) + d(t) \\ m_2\ddot{x}_2 &= -b_2\dot{x}_2 + k(x_1 - x_2) + c(\dot{x}_1 - \dot{x}_2) \end{aligned} \quad (1)$$

Where m_1 is the rotary inertia equivalent to the motor shaft and m_2 is the equivalent inertia of worktable. b_1 is the viscous friction between the motor shaft and rotary bearing. b_2 is the viscous friction of the bilateral guides. k represents the total equivalent axial stiffness and c represents the damping in the preloaded nut. $u(t)$ is the control input voltage used to generate the motor torque to actuate the ball screw platform. $x_1(t)$ is the equivalent rotary displacement of the motor shaft and $x_2(t)$ is the equivalent rotary displacements of worktable multiplied by the screw pitch. $d(t)$ represents the bounded parameter uncertainties and equivalent external disturbances of the system. In this paper, $d(t)$ is regarded as the matched disturbance to the input.

The transfer function of 2-DOF model and the relation between $x_1(t)$ and $x_2(t)$ are derived from (1) as follows,

$$G_1(s) = \frac{x_1(s)}{u(s)} = \frac{m_2s^2 + (b_2 + c)s + k}{a_4s^4 + a_3s^3 + a_2s^2 + a_1s} \quad (2)$$

$$G_2(s) = \frac{x_2(s)}{u(s)} = \frac{cs + k}{a_4s^4 + a_3s^3 + a_2s^2 + a_1s} \quad (3)$$

$$\frac{x_2(s)}{x_1(s)} = \frac{cs + k}{m_2s^2 + (b_2 + c)s + k} \quad (4)$$

Where $a_4 = m_1m_2$, $a_3 = m_1(b_2+c)+m_2(b_1+c)$, $a_2 = k(m_1+m_2)+b_1b_2+b_1c+b_2c$, $a_1 = k(b_1+b_2)$.

The transfer functions (2)-(4) are used to identify the 2-DOF model parameters according to the frequency response measuring from the linear encoder which is to be discussed later.

The state space model transformed from Eq. (1) with the state vector $x(t) = [x_2(t), x_1(t), \dot{x}_2(t), \dot{x}_1(t)]^{\mathrm{T}}$ is inferred as follows,

$$\begin{bmatrix} \dot{x}_2(t) \\ \dot{x}_1(t) \\ \ddot{x}_2(t) \\ \ddot{x}_1(t) \end{bmatrix} = \begin{bmatrix} 0 & 0 & 1 & 0 \\ 0 & 0 & 0 & 1 \\ -\dfrac{k}{m_2} & \dfrac{k}{m_2} & -\dfrac{b_2 + c}{m_2} & \dfrac{c}{m_2} \\ \dfrac{k}{m_1} & -\dfrac{k}{m_1} & \dfrac{c}{m_1} & -\dfrac{b_1 + c}{m_1} \end{bmatrix} \begin{bmatrix} x_2(t) \\ x_1(t) \\ \dot{x}_2(t) \\ \dot{x}_1(t) \end{bmatrix} + \begin{bmatrix} 0 \\ 0 \\ 0 \\ \dfrac{1}{m_1} \end{bmatrix} u(t) + \begin{bmatrix} 0 \\ 0 \\ 0 \\ \dfrac{1}{m_1} \end{bmatrix} d(t) \quad (5)$$

The state vector is divided into two parts as $z_1(t) = [x_2(t), x_1(t), \dot{x}_2(t)]^{\mathrm{T}}$ and $z_2(t) = \dot{x}_1(t)$. The corresponding state space model is,

$$\begin{bmatrix} \dot{z}_1(t) \\ \dot{z}_2(t) \end{bmatrix} = \begin{bmatrix} A_{11} & A_{12} \\ A_{21} & A_{22} \end{bmatrix} \begin{bmatrix} z_1(t) \\ z_2(t) \end{bmatrix} + \begin{bmatrix} \mathbf{0} \\ B \end{bmatrix} u(t) + \begin{bmatrix} \mathbf{0} \\ B \end{bmatrix} d(t) \quad (6)$$

Where,

$$A_{11} = \begin{bmatrix} 0 & 0 & 1 \\ 0 & 0 & 0 \\ -k/m_2 & k/m_2 & -b_2 - c \end{bmatrix}, A_{12} = \begin{bmatrix} 0 \\ 1 \\ c/m_2 \end{bmatrix},$$

$$A_{21} = \begin{bmatrix} \dfrac{k}{m_1} & -\dfrac{k}{m_1} & \dfrac{c}{m_1} \end{bmatrix}, A_{22} = -\frac{b_1 + c}{m_1}, A = \begin{bmatrix} A_{11} & A_{12} \\ A_{21} & A_{22} \end{bmatrix},$$

$$B = \frac{1}{m_1}.$$

3 Novel sliding mode control

The novel sliding mode control (NSMC) based on nonlinear sliding-mode surface is presented here to suppress the first axial resonance of the ball screw servo system with bounded disturbance and parameter uncertainties. The design and stability verification of the control strategy are derived based on the Lyapunov function.

3.1 Nonlinear sliding-mode surface and stability analysis

The desired state trajectory for system (6) is defined as,

$$z_{ref} = \begin{bmatrix} z_{1ref} & z_{2ref} \end{bmatrix}^{\mathrm{T}} \quad (7)$$

$$z_{1ref}(t) = \begin{bmatrix} x_{2ref}(t) & x_{1ref}(t) & \dot{x}_{2ref}(t) \end{bmatrix}^{\mathrm{T}} \quad (8)$$

$$z_{2ref}(t) = \dot{x}_{1ref}(t) \quad (9)$$

In this paper, the continuous acceleration reference trajectory is used as x_{2ref}. x_{1ref} is acquired from equation (4) with the nominal parameters.

The system state errors are given as,

$$\begin{aligned} e_1(t) &= z_1(t) - z_{1ref}(t) \\ e_2(t) &= z_2(t) - z_{2ref}(t) \end{aligned} \tag{10}$$

The nonlinear sliding-mode surface $s(z,t)$ for the system is defined as,

$$s(z,t) = \sigma(y,r)z(t) = \begin{bmatrix} F - \psi(y,r)A_{12}^{\mathrm{T}}P & 1 \end{bmatrix} \begin{bmatrix} e_1(t) \\ e_2(t) \end{bmatrix} \tag{11}$$

Where $\sigma(y,r) = \begin{bmatrix} F - \psi(y,r)A_{12}^{\mathrm{T}}P & 1 \end{bmatrix}$ and r is the desired worktable equivalent rotary displacement. F is chosen to satisfy the condition that A_{11}-$A_{12}F$ has stable eigenvalues and the dominant poles have a very low damping ratio [9]. P is the positive definite symmetric matrix satisfied the Lyapunov equation below,

$$(A_{11} - A_{12}F)^{\mathrm{T}}P + P(A_{11} - A_{12}F) = -W \tag{12}$$

Where W is a known positive definite matrix.

$\psi(y,r)$ is a non-positive function and continuously differentiable with respect to y. The use of $\psi(y,r)$ is to increase the damping ratio of system by changing itself from 0 or a tiny value to a negative constant as the actual value is gradually moving closely to the expected value. In order to satisfy these requirements, $\psi(y,r)$ is chosen as follows,

$$\psi(y,r) = -\beta(e^{-\rho(y-r)^2}) \tag{13}$$

Where β is used to adjust the damping ratio and satisfies $\beta > 0$ and ρ should be big enough to make $\psi(y,r)$ to be a tiny value initially.

On the basis of the control principle of sliding mode control, the system states will be forced to the sliding-mode surface (11) by the control law in finite time and the following equation is satisfied,

$$s(z,t) = 0 \tag{14}$$

Combining equations (6), (10) and (11) gives that,

$$e_2(t) = -Fe_1(t) + \psi(y,r)A_{12}^{\mathrm{T}}Pe_1(t) \tag{15}$$

$$\dot{e}_1(t) = (A_{11} - A_{12}F + A_{12}\psi(y,r)A_{12}^{\mathrm{T}}P)e_1(t) \tag{16}$$

Therefore,

$$e_1(t) = 0 \Rightarrow e_2(t) = 0 \tag{17}$$

The following Lyapunov function is given to analyze the stability of the system states,

$$V_1(t) = e_1^{\mathrm{T}}(t)Pe_1(t). \tag{18}$$

Then,

$$\begin{aligned} \dot{V}_1(t) &= \dot{e}_1^{\mathrm{T}}(t)Pe_1(t) + e_1^{\mathrm{T}}(t)P\dot{e}_1(t) \\ &= e_1^{\mathrm{T}}(t)(A_{11} - A_{12}F)^{\mathrm{T}}Pe_1(t) + e_1^{\mathrm{T}}(t)P(A_{11} - A_{12}F)e_1(t) + \\ &\quad 2e_1^{\mathrm{T}}(t)PA_{12}\psi(y,r)A_{12}^{\mathrm{T}}Pe_1(t) < 0 \end{aligned}$$

From above, it can be seen that the error state $e_1(t)$ is asymptotic convergence as well as $e_2(t)$.

3.2 NSMC design and stability analysis

The NSMC is designed to drive the system (5) move to the nonlinear sliding-mode surface (11). The exponential approaching law (19) is combined here to design the controller.

$$\dot{s}(z,t) = -k_1 s(z,t) - k_2 \operatorname{sgn}(s(z,t)) \tag{19}$$

Where $k_1 > 0$ and $k_2 > 0$.

The derivative of sliding-mode surface Eq. (11) is as follows,

$$\begin{aligned} \dot{s}(z,t) &= \begin{bmatrix} -\frac{d}{dt}\psi A_{12}^{\mathrm{T}}P & 0 \end{bmatrix} \begin{bmatrix} e_1(t) \\ e_2(t) \end{bmatrix} + \begin{bmatrix} F - \psi A_{12}^{\mathrm{T}}P & 1 \end{bmatrix} \begin{bmatrix} \dot{e}_1(t) \\ \dot{e}_2(t) \end{bmatrix} \\ &= -\frac{d}{dt}\psi A_{12}^{\mathrm{T}}Pe_1(t) + (F - \psi A_{12}^{\mathrm{T}}P)\dot{e}_1(t) + \dot{e}_2(t) \\ &= -\frac{d}{dt}\psi A_{12}^{\mathrm{T}}Pe_1(t) + (F - \psi A_{12}^{\mathrm{T}}P)(\dot{z}_1(t) - \dot{z}_{1ref}(t)) + (\dot{z}_2(t) - \dot{z}_{2ref}(t)) \\ &= -\frac{d}{dt}\psi A_{12}^{\mathrm{T}}Pe_1(t) + \sigma(t)Az(t) - \sigma(t)\dot{z}_{ref} + Bu(t) + d(t) \end{aligned} \tag{20}$$

The controller could be designed as follows by combining with Eq.(19),

$$\begin{aligned} u(t) = -B^{-1}(&-\frac{d}{dt}\psi(y,r)A_{12}^{\mathrm{T}}Pe_1(t) - \sigma(t)\dot{z}_{ref}(t) \\ &+ \sigma(t)Az(t) + k_1 s + k_2(t) * \operatorname{sgn}(s)) \end{aligned} \tag{21}$$

Where k_2 should be chosen big enough to satisfy $k_2 \geq |d(t)|$.

In order to analyze the stability of proposed controller, the Lyapunov candidate function is considered as follows,

$$V_2(t) = \frac{1}{2}s^2(z,t) \tag{22}$$

The derivative of equation (22) is,

$$\begin{aligned} \dot{V}_2(t) &= s(z,t)\dot{s}(z,t) \\ &= s(z,t) \cdot (-k_1 s(z,t) - k_2 \operatorname{sgn}(s(z,t)) + d(t)) \\ &= -k_1 s(z,t)^2 - k_2 |s(z,t)| + d(t)s(z,t) \\ &< -k_1 s(z,t)^2 - k_2 |s(z,t)| + d(t)|s(z,t)| \\ &= -k_1 s(z,t)^2 + (d(t) - k_2)|s(z,t)| \\ &< 0 \end{aligned}$$

Therefore, the system state errors of Eq.(10) can be driven to the sliding-mode surface and converge to zero by the control law (21).

A saturation function shown in Eq.(23) with a certain threshold value is used here to instead of the sign function in equation (21) to weaken the chattering of the system,

$$\operatorname{sat}(u) = \begin{cases} -1, & u < -\ell \\ \frac{1}{\ell}, & |u| \leq \ell \\ 1, & u > \ell \end{cases}. \tag{23}$$

3.3 Advantage of system transient response by NSMC

In order to make the point that the transient response of system controlled by NSMC is improved comparing with the traditional linear sliding mode controls (LSMCs), a step response of permanent magnet synchronous motor (PMSM) simulation system is used here to compare the three controllers of the LSMC with under-damping (LSMC-UD) and LSMC with over-damping (LSMC-OD) and NSMC. The step response results are shown in Figure 3 and it can be concluded evidently that NSMC control algorithm can acquire the excellent transient response performance with fast response and non-overshoot which is conducive to ball screw servo system.

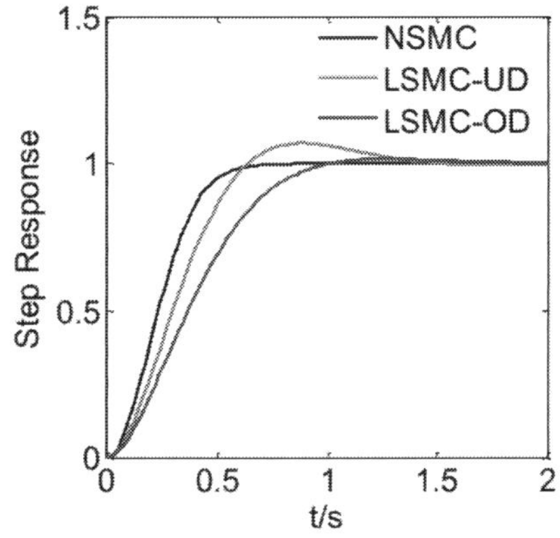

Figure 3. Step response for PMSM simulation system with three controllers

4 Simulation and experiment results

The ball screw servomechanism is shown in Figure 4. The worktable is supported by two THK SHS20C linear ball rolling guide-ways and driven by a THK precision ball screw of WGF2040-3G0+1170LC3 with 40mm pitch and 20mm diameter. 1.3 kW Mitsubishi servo motor is used with the driver MR-J3-70A to actuate the ball-screw through a MISUMI diaphragm coupling CPDT40-12-19. This setup is applied in the surface mounting machine. The high-resolution rotary and linear encoders are used in the detection process. The rotary encoder is in the motor which is setting to generate 40000 sinusoidal signals per revolution. The Renishaw optical incremental linear encoder RGH22Q50D00A is used to detect the translational displacement of the worktable. Both encoders are setting to have the position measurement resolution of 1 μm. The frequency response experiment is implemented at 20 kHz sampling frequency on dSPACE DS1104 system.

Figure 4. Ball screw drive servomechanism

4.1 Frequency response experiment modeling

In order to acquire the first axial mode resonance, the frequency response experiment is conducted by exciting the drive motor with a torque command which is a sinusoidal input voltage with amplitude 0.2 V at frequency ranges from 1Hz to 400Hz gradually increasing by 1 Hz increment. The 2-DOF model is obtained by the identification of Eq.(2)-(4) from the experiment result by the method of pole placement principle and least square method and shown as follows,

$$m_1 = 5.4741\times10^{-4}(\text{V/rad/s}^2),$$

$$m_2 = 10.0842\times10^{-4}(\text{V/rad/s}^2),$$

$$k = 353.2717(\text{V/rad}),\ c = 0.020(\text{V/rad/s}),$$

$$b_1 = 0.0013(\text{V/rad/s}),\ b_2 = 0.0013(\text{V/rad/s}). \quad (24)$$

Figure 5 shows the comparison of frequency response between the 2-DOF model and experiment measurement. It can be seen that the frequency responses of model are in good agreement with the first axial resonant frequency of the test bed.

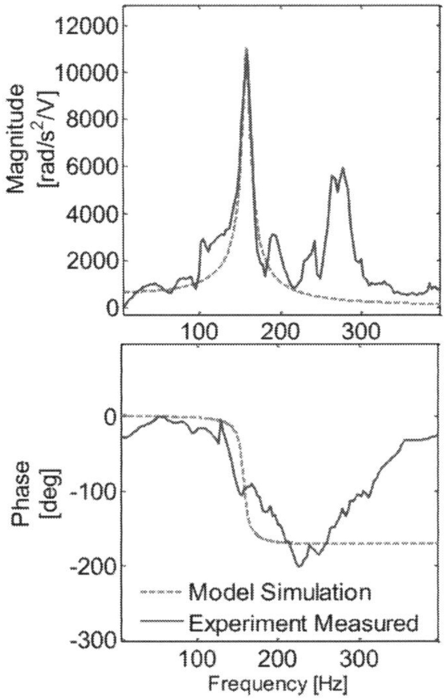

Figure 5. Frequency response comparison of model simulation and experiment measured

4.2 Controller simulation

In order to verify the effectiveness of the proposed NSMC in the ball screw servo system, two LSMC controllers with different damping ratios are used for comparison. The acceleration continuous signal shown in Figure 6 is used as reference trajectory here. The maximum acceleration of the reference is 10 m/s^2 and maximum velocity is 1.1 m/s. The equivalent motion of the worktable with the reference trajectory is a back and forth motion in the ball screw servo system.

The damping ratios of LSMC_1 and LSMC_2 are 0.5 and 0.99 respectively and these values are the initial and ultimate damping ratio in succession for NSMC controller.

The system parameter perturbations and disturbances in simulations are hypothesized as a whole as follows,

$$d(t) = 10\sin(100t) \quad (25)$$

Figure 7 shows the tracking errors and control signals for the three controllers for ball screw servo system. It can be seen that the tracking error is reduced considerably by NSMC algorithm compared with the LSMCs. Meanwhile, the control voltage of the NSMC is decreased in somehow and this is beneficial for the power reduction.

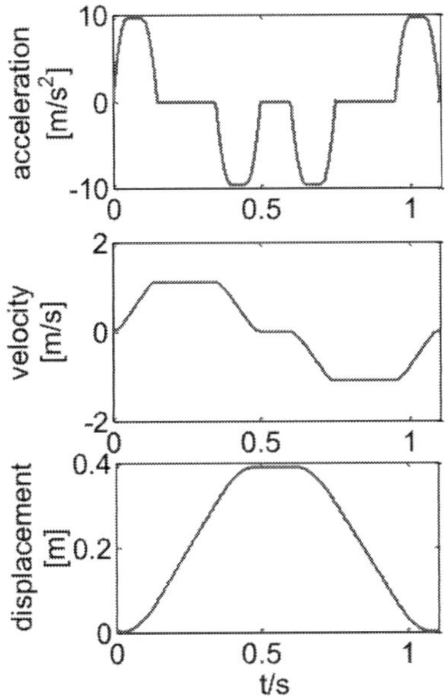

Figure 6. Reference Trajectory

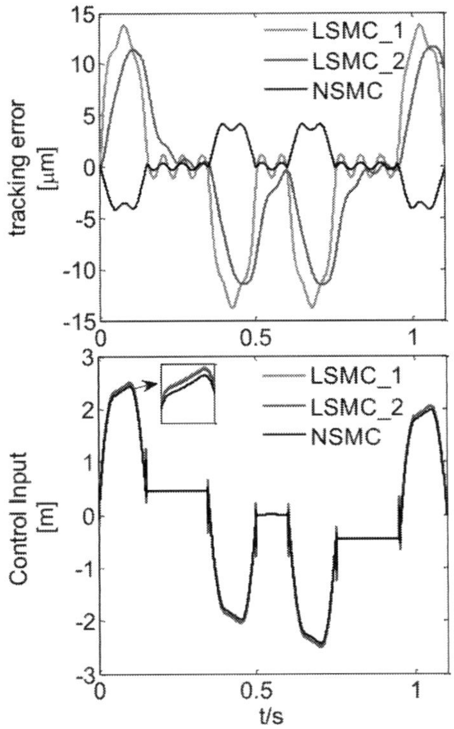

Figure 7. Three control strategies simulation comparison

5 Conclusion

The modeling and trajectory tracking control for a high speed ball screw servo system has been researched. A novel sliding mode controller based on nonlinear sliding-mode surface is designed to take full advantages of high transient response performance with fast response and non-overshoot on the ball screw servo system. The experiment and simulation results show that the system modelling is accurate enough and the tracking accuracy of the system is improved appreciably by the proposed control strategy. The designed control strategies could be applied in more plants to improve the transient response of the systems.

Acknowledgements

This paper is supported by the Project of National Nature Science Foundation of China (No. 51405469) and the Project of National Science-Technology Support Plan of China (No. 2015BAK06B02).

References

1. D. A. Vicente, R. L. Hecker, F. J. Villegas, & G. M. Flores, Modeling and vibration mode analysis of a ball screw drive. International Journal of Advanced Manufacture Technology, **58**(1): 257-265 (2012)
2. H. Zhu & H. Fujimoto, Mechanical deformation analysis and high-precision control for ball-screw-driven stages. IEEE/ASME Transactions on Mechatronics, **20**(2): 956-966 (2015)
3. H. Fujimoto & T. Takemura, High-precision control of ball-screw-driven stage based on repetitive control using-times learning filter. IEEE Transactions on Industrial Electronics, **61(7):** 3694-3703 (2014)
4. T. Yamamoto, M. Naito, G. Chen, et al. Robust control design for ball screw system via descriptor representation and loop shaping. in IEEE 2014 Proceedings of the 6th International Conference on Modeling Identification & Control (ICMIC), **2014:** 82-87 (2014)
5. M. Hanifzadegan and R. Nagamune, Tracking and structural vibration control of flexible ball-screw drives with dynamic variations. IEEE/ASME Transactions on Mechatronics, **20**(1): 133-142 (2015)
6. C. L. Chen, M. J. Jang, and K. C. Lin, Modeling and high-precision control of a ball-screw-driven stage. Precision Engineering, **28**(4), 483-495 (2004)
7. K. Erkorkmaz and A. Kamalzadeh, High bandwidth control of ball screw drives. CIRP Annals – Manufacturing Technology, **55**(1), 393-398 (2006)
8. L. Dong & W. C. Tang, Adaptive backstepping sliding mode control of flexible ball screw drives with time-varying parametric uncertainties and disturbances. ISA Transactions, **53**(1), 110-116 (2014)
9. B. Bandyopadhyay, D. Fulwani, and L. Fridman, Sliding mode control using novel sliding surface. Springer (2007).

The Influence of Couple Stress Components and Electrostatic Actuation on Free Vibration Characteristics of Thin Micro-Plates

Amir R. Askari, Masoud Tahani

Department of Mechanical Engineering, Faculty of Engineering, Ferdowsi University of Mashhad, Mashhad, Iran

Abstract. The present paper deals specifically with the frequently asked question of are there any differences between classical non-actuated mode-shapes of plate-type MEMS and the electrically deformed ones obtained by the non-classical modified couple stress theory. To this end, a Kirchhoff's thin plate model together with the modified couple stress theory is considered to extract the eigenvalue problem which governs on free vibrations of the system. Using the free vibration equation of the system, it is proved that neither the electrostatic actuation nor the couple stress components influence on linear mode-shapes of the system. To emphasis on the validity of the present proof, the extended Kantorovich method is employed to provide analytical solutions for an electrically actuated micro-plate with clamped boundary conditions. The solutions, whose accuracy is verified by available results in the literature, also indicate that the linear mode-shapes of an electrically actuated micro-plate are totally independent from the influences of size and electrical loading.

1 Introduction

Technology of micro-electro-mechanical systems (MEMS) has experienced many improvements in recent years. The building blocks of these systems are some sort of electrically actuated mechanical structures such as bars, beams or plates [1]. In general, an electrically actuated micro-plate is a conductive and elastic thin plate suspended over a stationary rigid electrode and deflects toward its substrate by applying the external voltage [2]. One of the most important phenomena associated with electrically actuated micro-plates is pull-in instability which occurs when the input voltage reaches a critical value called pull-in voltage. In this manner, the elastic restoring force of the micro-plate cannot resist against the Coulomb attraction and it suddenly collapses toward the substrate underneath it [3].

Due to the coupling between electrical and structural physics as well as the strong non-linearities arising in the field of electrically actuated micro-structures, modelling of these systems is so challenging [4]. To date, variety of mathematical procedures has been employed to uncover the static and dynamic behaviors of such systems [5-7]. Amongst all of these mathematical techniques, reduced order modelling does not suffer from long run-time and has a unique advantage in providing the ability of presenting analytical solutions for MEMS problems [3]. In general, creating reduced order models (ROMs) eliminates the spatial dependence in the governing equations and performs a transformation from the physical coordinates of the device to a set of generalized coordinates. In this way, the governing differential equations of equilibrium are reduced to a set of algebraic equations which can be solved more easily than the governing equations themselves. It is to be mentioned here that utilizing the Galerkin weighted residual method for eliminating the spatial dependence is so usual in the creation of ROMs [3]. In this way, it is proved that employing linear mode-shapes of the system accelerates the convergence of the procedure and significantly reduces the computational costs [2, 8, 9].

In recent years, it has experimentally been observed that the mechanical behavior of structures in small scales significantly depends on their size [10-13]. In spite of the fact that the classical continuum mechanics is the most common theory for investigating the mechanical behavior of structures, it is incapable of capturing their size-dependent behavior at micron and sub-micron scales. To remove this incapability of classical theory (CT), some higher-order theories of elasticity, which account for the effects of higher-order stress and strain components in the strain energy expression, are employed. Amongst all of the size-dependent higher-order theories, the modified couple stress theory (MCST), due to its simplicity and accuracy, has been frequently utilized to study the size-dependent behavior of MEMS devices [14-18].

As it was mentioned above, employing the linear mode-shapes of the system in the creation of ROMs significantly accelerates the convergence of the procedure. Hence, the main object of the present paper is to investigate the effects of size and electrostatic loading on the free vibration characteristics of electrically actuated thin micro-plates as one of the most important building blocks of MEMS devices. To this end, the size-dependent

Kirchhoff thin plate model [19] is employed to develop the equation which governs on the linear free vibration of the system. Using this governing equation, it is proved that the micro-plate linear mode-shapes are totally independent of size and electrostatic loading. Also, it is shown that the MCST to CT frequency and pull-in voltage ratios of the system are totally independent of plate aspect ratio as well as its boundary condition and only depends on the Poisson ratio of the micro-plate. Finally, to emphasis on the independency of micro-plate mode-shapes from the influences of size and electrostatic loading, the extended Kantorovich method (EKM) is also employed to extract the size-dependent free-vibration characteristics of fully clamped micro-plates as the most applicable type of plate-based MEMS.

2 Theoretical formulations

Figure 1 shows a fully clamped electrically actuated rectangular micro-plate whose length, width, thickness and density are denoted by b, a, h and ρ respectively. A space-fixed Cartesian coordinate system is also attached to the center of the mid-plane of the micro-plate. That is, $\hat{x}$, $\hat{y}$ and $\hat{z}$ are the coordinates along the width, length and thickness, respectively. According to the MCST-based size-dependent thin plate model presented by Tsiatas [19], the governing equation of motion takes the form [16]

$$\rho h\frac{\partial^2 \hat{w}}{\partial \hat{t}^2}+\left(\frac{Eh^3}{12\left(1-\nu^2\right)}+\frac{Ehl^2}{2\left(1+\nu\right)}\right)\nabla^4\hat{w}-F_{es}=0 \quad (1)$$

where E, ν, l and $\hat{w}$ are the Young modulus, Poisson ratio, material length scale parameter and transverse deflection of the micro-plate, $\hat{t}$ is time and ∇^4 operator refers to

$$\nabla^4=\left(\frac{\partial^4}{\partial \hat{x}^4}+2\frac{\partial^4}{\partial \hat{x}^2\partial \hat{y}^2}+\frac{\partial^4}{\partial \hat{y}^4}\right) \quad (2)$$

The electrostatic actuation by polarized DC voltage V_{DC} (i.e. F_{es}) can also be expressed as [8]

$$F_{es}=\frac{\varepsilon V_{DC}^2}{2\left(d-\hat{w}\right)^2} \quad (3)$$

in which ε is the dielectric constant of medium. In addition d is the initial gap between the non-actuated micro-plate and its substrate underneath it. Introducing the dimensionless variables $x=\hat{x}/a$, $y=\hat{y}/b$, $w=\hat{w}/d$ and $t=\hat{t}/T$ the non-dimensionalized form of the governing equation of motion takes the form

$$\ddot{w}+\left(1+6\alpha_2\right)\left(\frac{\partial^4 w}{\partial x^4}+2\alpha_1^2\frac{\partial^4 w}{\partial x^2\partial y^2}+\alpha_1^4\frac{\partial^4 w}{\partial y^4}\right)-\frac{\alpha_3}{(1-w)^2}=0 \quad (4)$$

where the dot super-script refers to partial derivative with respect to dimensionless time. Also, the time scale T, which is determined such that the coefficient of $\ddot{w}$ becomes unity, is defined as follows

$$T=\sqrt{\frac{12\rho a^4\left(1-\nu^2\right)}{Eh^2}} \quad (5)$$

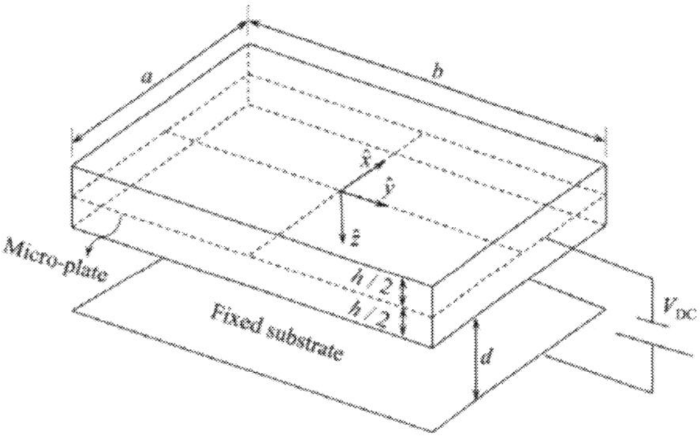

Figure 1. Schematic of an electrically actuated rectangular micro-plate.

The normalized parameters of the problem are also given by

$$\alpha_1=\frac{a}{b},\ \alpha_2=\frac{1-\nu}{\left(h/l\right)^2},\ \alpha_3=\frac{6\varepsilon a^4\left(1-\nu^2\right)V_{DC}^2}{Eh^3d^3} \quad (6)$$

Assuming simple harmonic motion for the linearized system, the eigenvalue problem associated with Eq. (4) which governs on the free vibrations of a micro-plate can be obtained as

$$\left(\frac{\partial^4\psi_{mn}}{\partial x^4}+2\alpha_1^2\frac{\partial^4\psi_{mn}}{\partial x^2\partial y^2}+\alpha_1^4\frac{\partial^4\psi_{mn}}{\partial y^4}\right)-\lambda_{mn}\psi_{mn}=0 \quad (7)$$

where λ_{mn} and ψ_{mn} represent the eigenvalue of the Eq. (7) and its associated eigenfunction, respectively. It is to be noted that the eigenvalue of Eq. (7) relates to the natural frequencies of the micro-plate using Eq. (8) as

$$\lambda_{mn}=\frac{\alpha_3+\omega_{mn}^2}{1+6\alpha_2} \quad (8)$$

As it is seen from Eq. (7), the size and electrostatic parameters (i.e. α_2 and α_3) are not appeared in this equation. Hence, it is obvious that variation of α_2 and α_3 does not change the eigenvalue and eigenfunction of this equation (i.e. λ_{mn} and ψ_{mn}). Furthermore, according to Eq. (7), it can simply be concluded that λ_{mn} and its associated eigenfunction (i.e. ψ_{mn}) will only be changed if the value of micro-plate aspect ratio (i.e. α_1) is varied. It is worth noting to mention here that the independency of thin micro-plates linear mode-shapes from electrostatic excitation and couple stress components has also been observed numerically by Tahani et al. [16] using the finite element method.

As it was mentioned above, λ_{mn} takes like values for micro-plates with same aspect ratio parameter. Hence, it is apparent that increasing the effect of size (i.e. the values of α_2) increases all natural frequencies of the system. Also, increasing the electrostatic attraction by making an increase in the values of input voltages decreases all of its natural frequencies.

In view of Eq. (8), the ratio of size-dependent natural frequencies obtained by the MCST to those determined through the CT becomes

$$\frac{\omega_{mn}^{\mathrm{MCST}}}{\omega_{mn}^{\mathrm{CT}}} = \sqrt{\frac{(1+6\alpha_2)\lambda_{mn} - \alpha_3}{\lambda_{mn} - \alpha_3}} \tag{9}$$

From Eq. (9), it is obvious that the MCST to CT natural frequency ratio in the absence of electrostatic field takes the form

$$\frac{\omega_{mn}^{\mathrm{MCST}}}{\omega_{mn}^{\mathrm{CT}}} = \sqrt{1+6\alpha_2} \tag{10}$$

which is independent of the micro-plate aspect ratio as well as its boundary condition and only depends on the Poisson ratio of the micro-plate. It is worth noting that this important conclusion has also been reported numerically by Tsiatas [19] and analytically by Askari and Tahani [20] through the assumed mode method.

Let us consider Eq. (8) again. From this equation, one can write

$$\omega_{mn} = \sqrt{(1+6\alpha_2)\lambda_{mn} - \alpha_3} \tag{11}$$

From Eq. (11), it is apparent that there exists a critical value for the electrostatic parameter in which the first natural frequency of the micro-plate is vanished. It is to be noted that this critical value of α_3 can represent a good approximation for the pull-in parameter of the system [8, 9, 16]. Therefore, in view of Eq. (11), one can obtain the MCST to CT pull-in parameter ratio as

$$\frac{\alpha_3^{\mathrm{MCST}}}{\alpha_3^{\mathrm{CT}}} = 1+6\alpha_2 \tag{12}$$

which is also independent of the micro-plate aspect ratio as well as its boundary condition and only depends on the Poisson ratio of the micro-plate. It is to be mentioned here that, according to Eq. (6), the MCST to CT pull-in voltage ratio also takes the form

$$\frac{V_{\mathrm{PI}}^{\mathrm{MCST}}}{V_{\mathrm{PI}}^{\mathrm{CT}}} = \sqrt{1+6\alpha_2} \tag{13}$$

In the next section, employing the EKM, an analytical solution for micro-plates with fully clamped boundary conditions as the most applicable case in MEMS structures is provided to emphasis on the independency of micro-plate linear mode-shapes from the influences of size and electrical loading.

3 Application of the EKM

According to the general procedure of the EKM [21], the micro-plate mode-shape can be considered as a multiplication of two separable functions as

$$\psi_{mn}(x,y) = f(x)g(y) \tag{14}$$

where $f(x)$ and $g(y)$ are un-known functions in terms of x and y to be determined. Substituting the separated form of the micro-plate mode-shape into Eq. (7) and constructing the variational form of this equation, we obtain

$$\int_{-1/2}^{1/2}\int_{-1/2}^{1/2}\left\{\left[\left(g\frac{d^4 f}{dx^4} + 2\alpha_1^2\frac{d^2 f}{dx^2}\frac{d^2 g}{dy^2} + \alpha_1^4 f\frac{d^4 g}{dy^4}\right) - \lambda_{mn}(fg)\right]\delta(fg)\right\}dxdy = 0 \tag{15}$$

The EKM is an iterative procedure in nature. Hence, it is assumed that g is a prescribed known function satisfied the fully clamped boundary conditions. In this case, the variation of ψ_{mn} would be just because of the variation of un-known function $f(x)$. Utilizing the fact that δf takes arbitrary values except on the boundaries and employing the fundamental lemma of variational calculus [22], we obtain

$$\frac{d^4 f}{dx^4} - I_1\frac{d^2 f}{dx^2} + (I_2 - \lambda_{mn})f = 0 \tag{16}$$

where

$$I_1 = 2\alpha_1^2\frac{\int_{-1/2}^{1/2}\left(\frac{dg}{dy}\right)^2 dy}{\int_{-1/2}^{1/2} g^2\, dy} \tag{17a}$$

$$I_2 = \alpha_1^4\frac{\int_{-1/2}^{1/2}\left(\frac{d^2 g}{dy^2}\right)^2 dy}{\int_{-1/2}^{1/2} g^2\, dy} \tag{17b}$$

For the next iteration step, utilizing the prescribed known function f obtained from Eq. (16), the ordinary differential equation (ODE) which governs on the unknown function g can be determined as

$$\frac{d^4 g}{dy^4} - I_1' \frac{d^2 g}{dy^2} + \left(I_2' - \frac{\lambda_{mn}}{\alpha_1^4} \right) g = 0 \tag{18}$$

where

$$I_1' = \frac{2}{\alpha_1^2} \frac{\int_{-1/2}^{1/2} \left(\frac{df}{dx} \right)^2 dx}{\int_{-1/2}^{1/2} f^2 \, dx} \tag{19a}$$

$$I_2' = \frac{1}{\alpha_1^4} \frac{\int_{-1/2}^{1/2} \left(\frac{d^2 f}{dx^2} \right)^2 dx}{\int_{-1/2}^{1/2} f^2 \, dx} \tag{19b}$$

As it is seen, employing the EKM, the partial differential equation (PDE) which governs on the free vibration of the micro-plate is separated to two simple ODEs with constant coefficients. By choosing the initial guess function $g(y) = (4y^2 - 1)^2$, the solution of Eq. (16) takes the form

$$f(x) = C_1 \sinh(\vartheta x) + C_2 \cosh(\vartheta x) + C_3 \sin(\upsilon x) + C_4 \cos(\upsilon x) \tag{20}$$

where

$$\vartheta = \sqrt{\frac{I_1}{2} + \sqrt{\lambda_{mn} + \left(\frac{I_1}{2} \right)^2 - I_2}} \tag{21a}$$

$$\upsilon = \sqrt{-\frac{I_1}{2} + \sqrt{\lambda_{mn} + \left(\frac{I_1}{2} \right)^2 - I_2}} \tag{21b}$$

Imposing the boundary conditions and following some straight forward mathematical operations, we obtain

$$C_2 \cosh\left(\frac{\vartheta}{2} \right) + C_4 \cos\left(\frac{\upsilon}{2} \right) = 0 \tag{22a}$$

$$\vartheta\, C_2 \sinh\left(\frac{\vartheta}{2} \right) - \upsilon\, C_4 \sin\left(\frac{\upsilon}{2} \right) = 0 \tag{22b}$$

$$C_1 \sinh\left(\frac{\vartheta}{2} \right) + C_3 \sin\left(\frac{\upsilon}{2} \right) = 0 \tag{23a}$$

$$\vartheta\, C_1 \cosh\left(\frac{\vartheta}{2} \right) + \upsilon\, C_3 \cos\left(\frac{\upsilon}{2} \right) = 0 \tag{23b}$$

where Eqs. (22) and (23) are associated with symmetric and anti-symmetric mode-shapes, respectively. It is to be mentioned here that the eigenvalue of Eq. (16) can also be determined by equating the determinant of the coefficients of the sets of Eqs. (22) and (23), respectively, for symmetric and anti-symmetric mode-shapes. By continuing the iterative procedure, the solution of Eq. (16), which is presented in Eq. (20), can be utilized as the prescribed known function in Eq. (18) to find the unknown function $g(y)$ as

$$g(y) = C_1' \sinh(\vartheta' y) + C_2' \cosh(\vartheta' y) + C_3' \sin(\upsilon' y) + C_4' \cos(\upsilon' y) \tag{24}$$

where

$$\vartheta' = \sqrt{\frac{I_1'}{2} + \sqrt{\frac{\lambda_{mn}}{\alpha_1^4} + \left(\frac{I_1'}{2} \right)^2 - I_2'}} \tag{25a}$$

$$\upsilon' = \sqrt{-\frac{I_1'}{2} + \sqrt{\frac{\lambda_{mn}}{\alpha_1^4} + \left(\frac{I_1'}{2} \right)^2 - I_2'}} \tag{25b}$$

It should be noted that, due to the similarity of Eqs. (16) and (18), their eigenvalue-eigenfunction equations will be the same (for more details see Ref. [20]). It has been shown that continuing the present procedure would leads to the convergence of the results which is so rapid and usually occurred after the second iteration step [20].

As it can be observed from Eqs. (16) – (25), the eigen value of Eq. (7) and the characteristics of the linear mode-shape of the micro-plate (i.e. υ, ϑ, C_i, υ', ϑ' and C_i' $(i = 1, 2, 3, 4)$) only depend on the values of I_1, I_2, I_1' and I_2' which are independent of the size and electrostatic parameters. In other words, as it is expected, the eigenvalue and eigenfunction of Eq. (7) depend only on the micro-plate aspect ratio.

4 Results and Discussion

To validate the accuracy of the present EKM procedure, a convergence study is carried out in Table 1 for the first two eigenvalues of a square micro-plate. As it is seen from this Table, the convergence of the EKM is so rapid and usually occurs after the second iteration. It is noteworthy that, since the square root of the micro-plate eigenvalues are on the order of its natural frequencies, we have chosen to present them in Table 1 instead of the eigenvalues themselves to provide a better physical sense.

Table 1. Convergence of the first two eigenvalues of a square micro-plate.

Number of Iteration (i)	$\sqrt{\lambda_{11}}\big\|_{i,i-1}$	$\sqrt{\lambda_{11}}\big\|_{i,i}$	$\sqrt{\lambda_{12}}\big\|_{i,i-1}$	$\sqrt{\lambda_{12}}\big\|_{i,i}$
1	35.9995	35.9990	35.9995	73.5333
2	35.9990	35.9990	73.4054	73.4054
3	35.9990	35.9990	73.4054	73.4054
4	35.9990	35.9990	73.4054	73.4054

Verification is also performed here for the converged values of the first four eigenvalues of micro-plates with three different aspect ratios in Table 2. The square roots of these eigenvalues are compared with those reported by Arenas [23] for large scale plates in the absence of the electrostatic field using the CT. It is to be noted that, according to Eq. (8), the square root of the micro-plate eigenvalues equals with its classical natural frequencies in the absence of electrostatic field. As it can be observed from Table 2, the present findings are in excellent agreement with those published by Arenas [23].

Table 2. Comparison of the first four eigenvalues for a micro-plate with three different aspect ratios.

α_1		$\sqrt{\lambda_{11}}$	$\sqrt{\lambda_{12}}$	$\sqrt{\lambda_{21}}$	$\sqrt{\lambda_{22}}$
1	Present	35.9990	73.4054	73.4054	108.2359
	Arenas [23]	36.1087	73.7372	73.7372	108.8499
0.5	Present	24.5812	31.8334	63.9853	71.0813
	Arenas [23]	24.6480	31.9618	64.1724	71.4252
0.1	Present	22.4324	22.6112	61.7503	61.9810
	Arenas [23]	22.4419	22.6335	61.7650	62.0188

As it is shown in two previous sections, the MCST to CT frequency and pull-in voltage ratios depend only on the Poisson ratio of the micro-plate. This fact is illustrated in Figure 2. Figure 2 depicts the MCST to CT frequency or pull-in voltage ratios versus the variation of the normalized length scale parameter (i.e. *h*/*l*) for three different Poisson's ratios. As it is seen from this figure, the effect of size on micro-plate natural frequencies and its pull-in stability threshold can usually be ignored for cases in which the ratio of micro-plate thickness to its material length scale parameter is larger than 10. In addition, increasing the values of Poisson's ratio of the micro-plate decreases the MCST to CT frequency and pull-in voltage ratios.

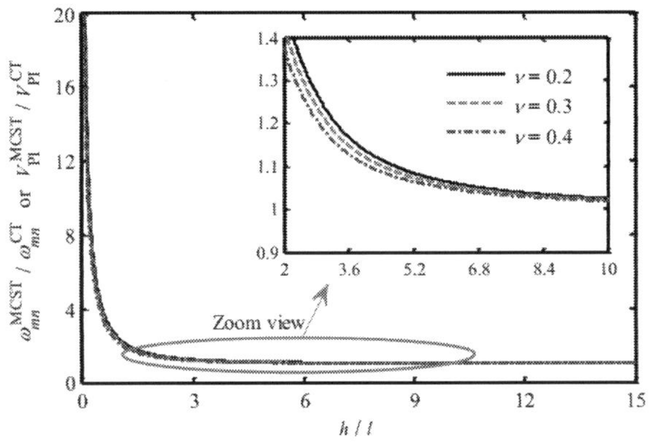

Figure 2. Influence of size on natural frequencies and pull-in instability threshold.

5 Concluding remarks

In the present paper, the influences of size and electrostatic actuation on free vibration characteristics of rectangular thin micro-plates were studied. It was proved that the linear mode-shapes of these structures are totally independent of size and electrostatic loading. Also, it was shown that the MCST to CT frequency and pull-in voltage ratios are totally independent of micro-plate aspect ratio as well as its boundary conditions and only depends on the Poisson ratio of the micro-plate. Furthermore, to emphasis on the independency of micro-plates linear mode-shapes from the effects of size and electrostatic loading, an analytical solution for micro-plates with fully clamped boundary conditions was extracted using the EKM. The accuracy of the EKM solution was also verified through the direct comparison between its findings and those published in the literature.

References

1. S. D. Senturia, *Microsystem Design*, Kluwer Academic Publishers, Dordrecht (2001)
2. R. C. Batra, M. Porfiri and D. Spinello, Smart Mater. Struct. **16**, 23-31 (2007)
3. A. H. Nayfeh, M. I. Younis and E. M. Abdel-Rahman, Nonlinear Dynam. **41**, (1), 211–236 (2005)
4. M. I. Younis, *MEMS Linear and Nonlinear Statics and Dynamics*, Springer, New York (2011)
5. C. Jin, Z. Jiang, and W. Xingtao, J. Micromech. Microeng. **14**, 1, 57 (2004)
6. H. Sadeghian, G. Rezazadeh, and P. M. Osterberg, J. Microelectromech. S. **16**, 6, 1334-1340 (2007)
7. J.-H. Kuang and C.-J. Chen, J. Micromech. Microeng. **14**, 4, 647 (2004)
8. R. C. Batra, M. Porfiri, and D. Spinello, Int. J. Solids Struct. **45**, 3558-3583 (2008)
9. R. C. Batra, M. Porfiri and D. Spinello, j. Sound Vib. **309**, 600-612 (2008)
10. N. A. Fleck, G. M. Muller, M. F. Ashby, and J. W. Hutchinson, Acta Metall. Mater. **42**, 475–487 (1994)
11. J. S. Stölken and A. G. Evans, Acta Mater. **46**, 5109-5115 (1998)
12. A. C. M. Chong and D. C. C. Lam, J. Mater. Res. **14**, 4103-4110 (1999)
13. A. W. McFarland and J. S. Colton, J. Micromech. Microeng. **15**, 1060-1067 (2005)
14. M. Rahaeifard, M. H. Kahrobaiyan, M. Asghari, and M. T. Ahmadian, Sensor Actuat. A-Phys. **171**, 370-374 (2011)
15. A. Mohsenzadeh, M. Tahani, and A. R. Askari, J. Nanoscience **2015**, 9 (2015)
16. M. Tahani, A. R. Askari, Y. Mohandes, and B. Hassani, Int. J. Mech. Sci. **94-95**, 185-198 (2015)
17. A. R. Askari and M. Tahani, Appl. Math. Model. **39 (2)**, 934-946 (2015)
18. K. F. Wang, T. Kitamura, and B. Wang, Int. J. Mech. Sci. **99**, 288-296 (2015)
19. G. C. Tsiatas, Int. J. Solids Struct. **46**, 2757–2764 (2009)
20. A. R. Askari and M. Tahani, J. Mech. Sci. Technol. **29**, 5, 2135-2145 (2015)
21. A. D. Kerr, Int. J. Solids Struct. **5**, 559-572 (1969)
22. J. N. Reddy, *Energy Principles and Variational Methods in Applied Mechanics*, John Wiley & Sons, New York (2002)
23. J. P. Arenas, J. Sound Vib. **266**, 912-918 (2003)

The Developing of the Boiler-Pipeline-Defects Magnetic Leakage Testing

Zhou Huang[1,2], Zhengwang Xu[1,2]

[1]*School of Electrical and Electronic Engineering, Hubei University of Technology, Wuhan, China*
[2]*Hubei Collaborative Innovation Center for High-efficiency Utilization of Solar Energy, Hubei University of technology, Wuhan, China*

Abstract. Aiming at the detecting of the boiler pipelines defects in power stations, the structure and the principle of the magnetic leakage testing system was introduced. The main contents are the producing and the detecting of the leaking magnetic field, the signal transmitting over a long distance, the acquiring and the processing of the data. The system has advantages of high speed and high precision in defects detecting. It can be used in different kinds of power stations.

1 Introduction

In power stations, there are a lot of boiler pipelines. They are working in the condition of high temperature and high pressure. That can make defects such as corrosion pits and/or flaws in the pipelines. The defects should be found and rehabilitated in time. Otherwise, the leaking or even exploding accident will make the safe producing of the power stations impossible [1].

Nowadays, there are different kinds of methods that can be used to detect the pipeline defects [2]. Such as penetration testing, magnetic leakage testing, magnetic particle testing, ultrasonic testing, eddy current testing, radiographic testing and so on. In 2014, Y. Liu [3] studied two current pipeline of electromagnetic nondestructive testing methods namely eddy current testing and magnetic flux leakage testing, and demonstrated the magnetic flux leakage detection technology compared to other detection technologies with clear advantages. H. Xia [4] analyzed the influence of the effect of magnetic flux leakage testing factors in order to improve the detection results. That suggests that magnetic flux leakage detection technology has become increasingly mature.

Based on the former works, the writers developed a pipeline defect detecting system based on the magnetic leakage testing method [5]. It has advantages of high speed and high precision at defects detecting.

2 The structure of the system

The detecting system consists of two main parts. One is a trolley which can climb along the pipelines, and another one is a controlling system which acquires and processes the signals. The two parts are connected by a cable. The structure is shown as Figure 1.

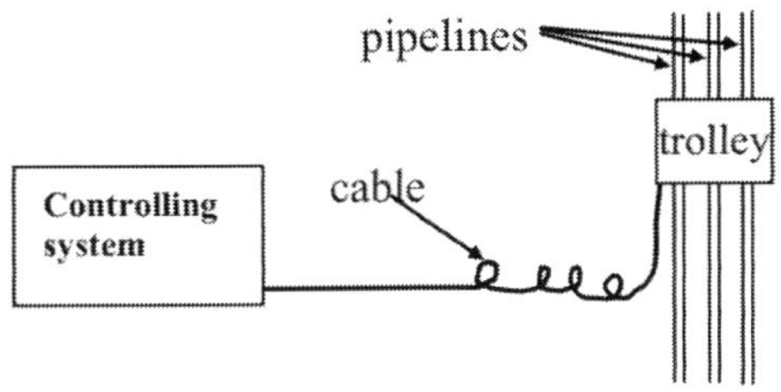

Figure 1. The structure of the system

There are permanent magnets mounted in the trolley. They enable the trolley attracting on the pipelines to climb vertically. The permanent magnets produce a magnetic field around the pipelines on the side to provide the condition for magnetic testing. Between the permanent magnets, hemicycle probes are mounted close to the pipelines. Eight hall elements are installed in each probe to detect the changing of leaking magnetic field. When the trolley climbs along the pipelines, the probes and the hall elements move accordingly.

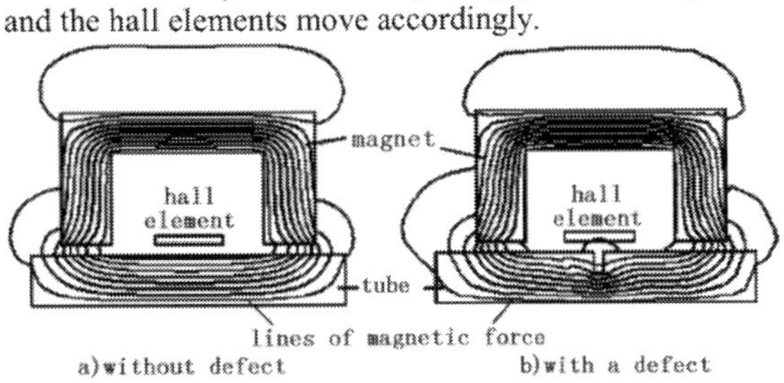

Figure 2. The illustration of magnetic leakage testing

The principle of magnetic leakage detecting is shown in Figure 2. Because the permanent magnets are mounted close to the pipelines, the lines of magnetic force are mostly closed between the magnets and the pipelines. If the pipelines don't have any defect, shown as Figure 2.a, the magnetic leakage is very little, so the signals picked up by the Hall elements are weak. Contrariwise, if there is a defect, shown as Figure 2.b, the magnetic resistance is large in the normal closed loop of the lines of magnetic force, so there is more magnetic leakage, the Hall elements can pick up stronger signals.

Then, the signals picked up by the Hall elements are amplified and processed. According to the processed signals, we can judge the health of the pipelines. In order to speed the detecting progress, three probes are installed in the trolley to detect 3 pipelines at one time.

3 The detail of the trolley

The structure of the trolley is shown in Figure 3. It is driven by a 24V DC motor. Its speed can be adjusted manually via the power supply voltage in the range of 1m/s ~ 3m/s. The trolley can move up and down with the power plant boiler pipe surface under the adsorption force of the magnetizing device along. Via special designing calculations and structural stress analyses, it has some capabilities of anti overturning and correction. It also has good sealing performance and the ability to overcome the effects caused by the floating ash and slagging. It can cross obstacles and turning smoothly. It will be a effective device in the power plant boiler pipelines testing automation and intelligent.

The specific structure is shown in Figure 3. It consists mainly of the DC motor 4, the synchronous transmission system 6, the roller pair (Part 8,9,10), the panel 2, the magnetization device 3 and the test probes 5, etc.

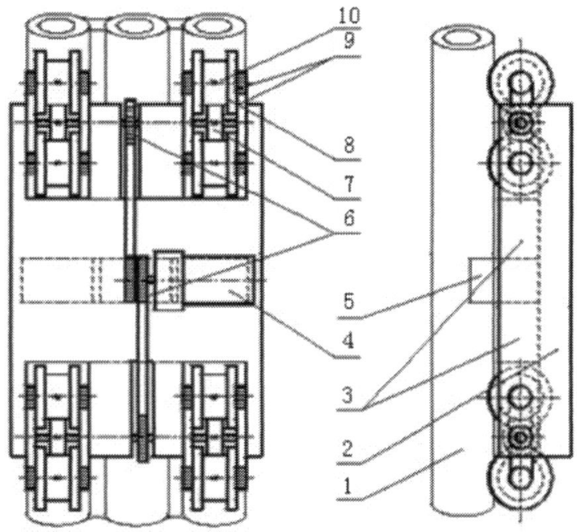

Figure 3. 1. Pipeline under testing; 2. Panel; 3. Magnetizer; 4. DC motor; 5. Probe; 6. Synchronous belt and synchronous wheel; 7. Drive wheel; 8. Roller; 9. Synchronizing wheel; 10. Gear

There are two Hall elements model SS495A1 mounted on each sensor in the test probe [6], as shown in Figure 4. They output two detecting signals respectively. The placement is shown in Figure 4. The detection range of any Hall element is from -680Gs to +680Gs. The amplifiers in the sensor amplify the signals come from the Hall elements and output voltage signals range from 0V to 5V. As the middle point, 0Gs is corresponding to 2.5V DC output signal.

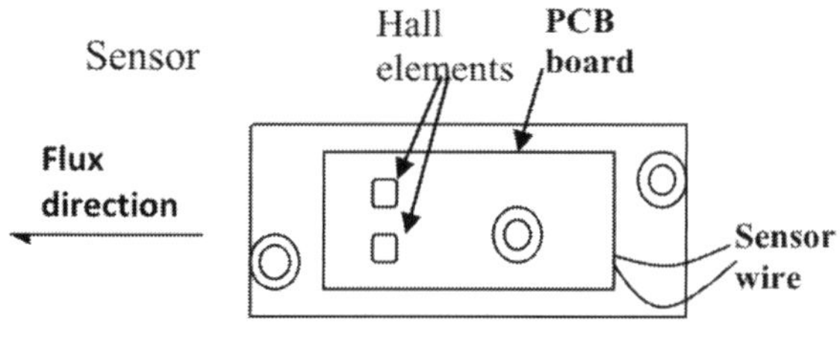

Figure 4. Sensor principle picture

The Hall sensor array consists of three semicircular magnetic flux leakage detection probes. There are four sensors mounted in each probe, shown as Figure 5. So there are 8 Hall elements in each probe. The probes form a 24-channel sensor array, which transforms magnetic flux leakage(MFL) signals into 24 channels of electrical signals. Every signal is processed by a channel independently.

The Hall sensor is used after calibration. That can make sure of the testing accuracy. But the angle of the Hall sensor may change in the course of use because of collisions, vibrations or other causes. That will decrease the accuracy of sensor. To avoid these risks, the sensors are encapsulated carefully. This treatment also ensures the strength of the wire at the connection with the circuit board. The sensor output signal is connected to the data acquisition module via shielded cable to avoid the interference outside. Sensor physical picture is shown in Figure 5.

Figure 5. Probe physical picture

4 The structure and the principle of the controlling system

The structure of the controlling system [7] is shown as Figure 6. It is mainly composed of a PC, the current-to-voltage circuit (shown as I/U), a data acquisition card and an adjustable power supply. All the parts are mounted in a metal box, shown in the dashed line box in Figure 6, except the PC. The current signal is converted into a voltage signal by the current-to-voltage circuit, specific works see below. The data acquisition card is used to collect the voltage signal converted by the current-to-voltage circuit. The received signal is transmitted through an USB cable to the PC for subsequent processings. Through the adjusting knob, the output voltage of the adjustable power supply can be controlled to change the DC motor's speed. Then the speed of the trolley is controlled accordingly. The SW is a change-over switch, which is used to change the polarity of the adjustable power supply output voltage. So it can control the motor reversing to control the trolley running up or down.

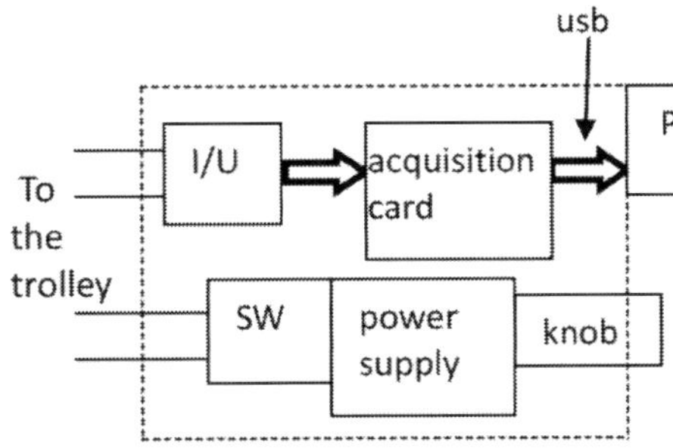

Figure 6. The controlling system

The PC runs a data processing software complied by LABVIEW. The processed data are stored and displayed. The stored data can be restored as needed for further processing or displaying. The locations of the defects can be found directly by observing the waveforms of the signals.

5 The signal transmitting over a long distance

Because the signals picked up by the Hall elements are very weak, and the pipelines are dozens of meters long or even longer, it is difficult to transmit the signals directly from the trolley to the controlling system on the ground. Some methods are needed to ensure the signal transmitting; otherwise the signal attenuation and the noise outside can destroy the detecting result.

In order to solve the signal transmitting problem, the writers developed a signal converting circuit. It consists of two parts. One part is installed in the trolley, which converts the original voltage signals picked up by the hall elements into current signals. Another part is installed in the controlling system, which converts the current signals back to voltage signals. The current signals are the intermediators which are transmitted from the trolley to the controlling system.

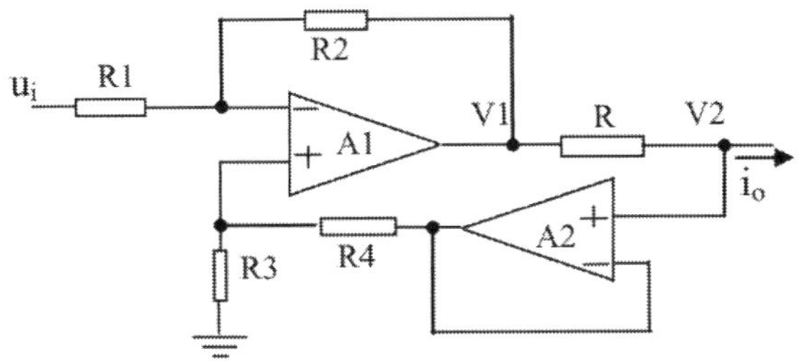

Figure 7. The principle of signal converting from voltage to current

In the first part, the principle of the signal converting circuit which converts the voltage signal into current signal is shown in Figure 7.

According to the working principle of operation amplifiers(OP) [8], the OP A2 forms a follower. And for the OP A1, we have

$$\frac{(u_i - V2\frac{R3}{R3+R4})}{R1} = \frac{(V2\frac{R3}{R3+R4} - V1)}{R2} \quad (1)$$

If $\frac{R2}{R1} = \frac{R4}{R3} = K$, the equation above will be

$$\frac{(u_i - V2\frac{1}{1+K})}{1} = \frac{(V2\frac{1}{1+K} - V1)}{K} \quad (2)$$

It can be simplified further as

$$V2 - V1 = Ku_i \quad (3)$$

So the output current of the circuit

$$i_o = \frac{V1 - V2}{R} = -\frac{K}{R}u_i \quad (4)$$

That is, the output current signal is proportional to the input voltage signal. The output current signals change accordingly when the voltage signals picked up by the hall elements change. So the signal converting from voltage to current is realized.

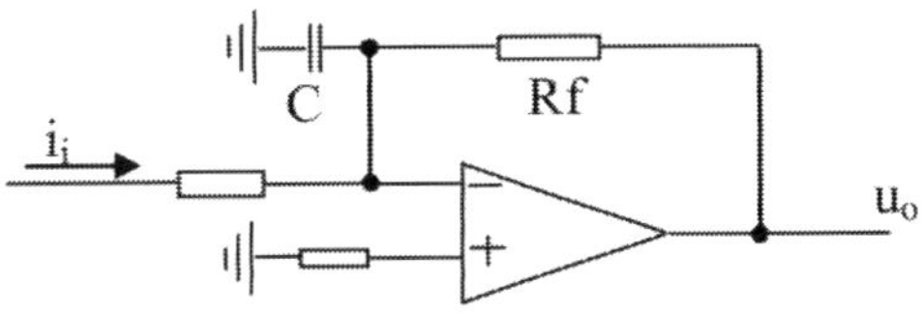

Figure 8. The principle of signal converting from current to voltage

In the controlling system, the converting is reverse from current to voltage. The principle is shown in Figure 8.

According to the working principle of OPs, we can see clearly

$$u_o = -i_i \times Rf \quad (5)$$

If the current signal at the output end of the circuit shown in Figure 7 is connected by a cable to the input end of the circuit shown in Figure 8, we have $i_i = i_o$. Together with the equations (4) and (5), we get

$$u_o = -(-\frac{K}{R}u_i) \times Rf = \frac{K \times Rf}{R}u_i \quad (6)$$

When $\frac{R}{Rf} = K$, we get $u_o = u_i$.

That is to say, the output signal of the converting circuit from current to voltage is equal to the input signal of the converting circuit from voltage to current. So, both the signal converting of two directions are achieved. Therefore, the aim is realized that to convert the original voltage signal to current signal for transmitting over a long distance.

There are two advantages to use current signals as the intermediator for signal transmitting over a long distance. First, the energy/power of normal outside noises is finite, they affect current signals little. Second, the current is the same in the same branch circuit, the resistance of the transmission line has no effect to the current signal when it is transmitted over a long distance. On the contrary, the voltage signal is sensitive to the outside noises, and the resistance of the transmission line attenuates the voltage signal in the manner of dividing voltage.

Furthermore, in order to enhance the anti-interference ability, a small capacitor C is installed in the converting circuit from current to voltage, as shown in Figure 8. The capacitor can filter noise further. So, the waveforms of the output voltage signals of the signal converting circuit are even better than the waveforms of the voltage signals observed directly on the outputs of the hall elements in the experiments. That proves sufficiently that the long distance signal transmitting circuit is successful.

Real things are worse than those in theory. In the project development process, a 12V DC power supply is used for the system at the begining. In the actual debugging, it is found that the circuits work well when a short wire is used to connect the controlling system and the trolley. But they don't work when a long wire is used. The final output voltage can not reach the theoretical value at the end of the current-to-voltage circuit.

After testing, we found that the presence of a large voltage drop on the singal tranmitting wire, that resulted in abnormal situations. In order to reduce the weight, the wire was chooded very thin. Its nominal diameter was 0.036mm, and its nominal DC resistance was 16790Ω /km at 20℃. In the experiments, the distance from the controlling system to the trolley was approximately 30 meters. So the round-trip resistance of the thin transmitting wire was

$$R = 0.03 * 2 * 16790 = 1007.4\ \Omega \quad (7)$$

The output current of the voltage-to-current circuit, io in Figure7, was around 12mA. Then the voltage drop on the thin wire was

$$U = 0.012 * 1007.4 = 12.0888\ V \quad (8)$$

Which was large than 12V, the power supply voltage. So the OP could not have sufficient voltage output signal. After analyzing the problem clearly, a 15V power supply was used to stand for the 12V one. Then the whole system can work well. Because the actual boiler pipes are long up to 60 meters rather than 30 meters in the experiments, the system power supply should be further increased for the real detecting. Except this problem, if the distance is very long and a high voltage power supply is needed, the withstand voltage of the chips should be considered carefully.

After ajusting, a lot of experiments were made. The 8 example waveforms acquired from an experimental pipeline are shown in Figure 9. The small protuberances are the signal changes affected by the defects. The minimum depth of the defect that can be distinguished is 1 mm when the diameter of the manmade corrosion pit is 8 mms.

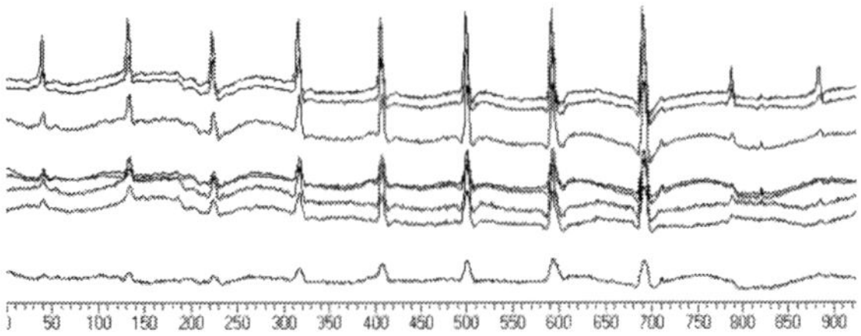

Figure 9. The example waveforms

In the trolley, there is an encoder which rotates together with a wheel of the trolley. The pulses sent out by the encoder are transmitted through the cable to trigger the data acquisition card. At each encoder pulse, the data acquisition card acquires data of the signals one time. So, the number of the points of the data is proportional to the moved distance of the trolley. We can compute the locations of the defects on the pipelines according to the number of the points in the signal waveform. The scale below the waveforms can help us to approach that. Therefore, the defect locations can be determined quickly for repairing.

6 Conclusions

Aiming at the detecting of the boiler pipelines defects in power stations, this paper introduced the structure and the principle of the magnetic leakage testing system. The main contents are the producing and the detecting of the leaking magnetic field, the signal transmitting over a long distance, the acquiring and the processing of the data, and the motion controlling method of the trolley. The system has advantages of high speed and high precision at defects detecting. It can be used in different kinds of power stations. And the system has characteristic of good adaptability, if the shape of the probe is modified properly to ensure the Hall elements are close to the detected materials, the system can be used for defect detecting of other kinds of ferromagnetic materials.

In the application process, in order to make better usage of magnetic flux leakage technology, we need to pay attention to some disturbing factors.

Because the magnetic flux leakage inspection needs some time to finish the magnetization, and different detection speed also affect the detection results, so in the detection process the speed should be maintained uniform and should not be too fast .

In the detection of the scanning process, although the design of the scanning plane is full coverage, due to save shape and form testing equipment and other reasons, it is not easy to achieve a comprehensive sweep surface. It is prone to missing defects. In addition, the sample ferromagnetic impurities, magnetization intensity, the flatness of detecting body and defect features will also affect the detection results.

In actual testing, it is neccery to choose a smaller detection probe, scan speed and to make the appropriate adjustments according to the actual testing situation.

Acknowledgment

This work was supported by the National Natural Science Foundation of China, under grant 50875077; was supported by the Natural Science Foundation of Hubei Province, under grant 2008CDA022; and was supported by Open Fund of the Key Laboratory of Nondestructive Test, Ministry of Education, Nanchang Hangkong University, under grant ZD200729009.

References

1. R. S. Moakhar, M. Mehdipour, M. Ghorbani, M. Mohebali, and B. Koohbor, Journal of Materials Engineering and Performance, **22**, 2691-2697 (2013)
2. X. Lei, Technology Supervision in Petroleum Industry, **7**, 28-30 (2013)
3. Y. Liu, S. Zhang, C. Yin, C. Qi, Mechatronics Engineering, **7**, 848-853 (2014)
4. H. Xia and C. Wang, Digital Technology and Applications, **10**, 181-182 (2014)
5. C. S. Angani, J. Kim, M. Le, and J. Lee, IEEE Transactions on Magnetics, **50**, 1-4 (2014)
6. J. Wang, L. Zhang, S. Zhou, and K. Song, NDT, **6**, 27-29 (2013)
7. X. Song and H. Zhang, Microcomputer Information, **13**, 101-102 (2009)
8. U. M. Perlis, *Analog Electronics* (Universiti Malaysia Perlis, 2011)

Development of an aerodynamic control system for the evaluation of the Open Core Engineering technology

Prof. Dr.-Ing. Elmar Engels, M.Eng. Alexander-Nicolai Köhler

University of Applied Sciences Fulda, Leipziger Str. 123, D-36037, Germany

Abstract. This paper describes the development of an aerodynamic control system used for the evaluation and demonstration of the award winning Open Core Engineering (OCE) technology. The aerodynamic control system was designed with physical properties that allow using it in rapid-control-prototyping (RCP) environments as well as with industrial motion-control components. Open Core Engineering is a quite modern technology invented by Bosch Rexroth AG that supports software development on several platforms for industrial automation applications. A 3D printed alternative design is explained as well.

1 Introduction and Objectives

Based on the fact that information technology merges more and more with production and process automation technologies the *Open Core Engineering* (OCE) technology was developed supporting the ideas of Industry 4.0 within the area of industrial drives and controls. Even though OCE was developed for industrial applications it can be beneficially used in mechatronics research and development projects as well. In order to evaluate the features and usability of the OCE an aerodynamic test bench was developed that fits very well to use cases of industrial and scientific applications.

The use cases could be human machine interface (HMI) development, IEC 61131-3 software development, RCP with MATLAB® / Simulink® or LabVIEW®, desktop software development like VBA for Microsoft Excel, Microsoft PowerPoint or Microsoft Access and remote control applications for mobile devices like iOS or Android based devices. From the perspective of mechatronic system developments these systems differ a lot in terms of programming paradigms, software development tools and languages as well as in performance classes. Performance class means that applications written in VBA for Excel or Java applications for smartphones cannot reach real-time properties as industrial motion-controllers - normally written in C, C++ or domain specific programming languages - can reach.

Thus the objective in this research project was to design a test bench that has a performance class not too restrictive for non-real-time applications and on the other hand being challenging enough to test the applicability of the OCE concept for higher performance industrial applications and to evaluate the usability. Fig. 1 shows a detail of the aerodynamic system in a horizontal orientation. The mechatronic requirements of its realization will be explain in the following sections.

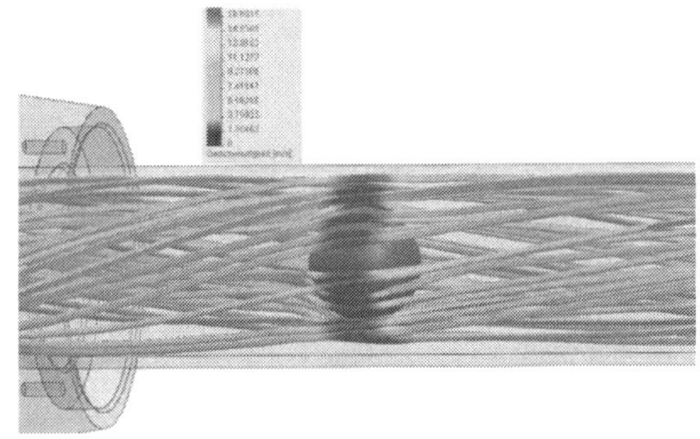

Figure 1: Aerodynamic control system Hoverball [1]

2 Requirements and Design Approach

There are a lot of mechatronic systems that could be considered to evaluate OCE with a test bench [2, 3]. However as mentioned above it had to cover several software domains and it was required to use a control application with a good visualization of the process for demonstration purposes. The general requirements for the test bench were to use it as demonstrator for applications from the domains of industrial automation, control theory, electric drives and obviously for software engineering.

To break these requirements a bit more down the objectives were to evaluate it in the area of industrial automation with project configuration and commissioning, software development with IEC 61131-3 programming languages, HMI design and use of industrial sensors and actors.

For the area of control theory it should be usable for typical tasks like modeling, analysis, simulation and identification of system parameters to mention only a few use cases. To perform these use cases the test bench should have some special characteristics like being instable or requiring control for stabilization, being non-linear and having the possibility to apply system perturbation, disturbances and to apply a fix set-point control as well as a servo control.

Evaluation of usability for industrial drives means that typical functionality of drives should be usable. This does not automatically mean that an industrial drive has to be used for the aerodynamic system. It could mean that typical functions like position control, velocity control or synchronization modes could be used. Sophisticated functions like cams would be desirable as well.

Software development is an integral part of all the mentioned areas but does not mean that software development theories are the focus. This means as explained above that different kind of programming paradigms like functional design, object oriented design and model driven design should be usable with the test bench but this is more determined by the programming languages themselves than by the test bench. However, the test bench must be programmable with these languages. So the test bench should have properties that support these software languages.

3 Motivation

Starting point of the development of this mechatronic system was the idea of creating a test bench based on aerodynamic systems presented in [4, 5]. The authors use micro-controller based electronics normally used for embedded system design. But they do not use components like industrial programmable-logic-controllers (PLC) or industrial drives that is a key target in this project. The principle of an aerodynamic levitated ball has several advantages. First of all the system includes moving components that visualize the process immediately. The system can be operated at extra-low-voltage and therefore in office environments. There are other test-benches like an inverted pendulum or a balanced ball that also visualize if control algorithms are applied properly. But a levitated ball is much slower in its reaction having at the same time a tendency to instability if the airflow is designed so.

Systems like magnetically levitated balls fall into the same class of application but they have the disadvantage that their movements are usually limited to a few millimeters meaning that the visible motion effect is pretty small compared to an aerodynamically levitated ball. Magnetically levitated balls do also have the disadvantage that the control cycle time has to be quite small within a range of typically 1ms to 0.1ms. Particularly slow update frequencies of big jitter during the control process jeopardize the usability with non-real-time systems. Therefore the decision was made for an aerodynamic levitated ball with a dedicated construction set-up. The test bench was called *Hoverball* and has several properties that make it very well suited for the evaluation of the OCE technology.

4 Realization of the Hoverball

In the following section the mechanical construction and its components will be explained without going into details of the aerodynamic system. Fig. 2 shows a transparent view of the main mechanical components. Basically the structure is composed of the aluminum body and the transparent acryl pipe. The acryl pipe guides the air flow generated in the body. Within that air flow a ball can be hovered if the air flow is controlled.

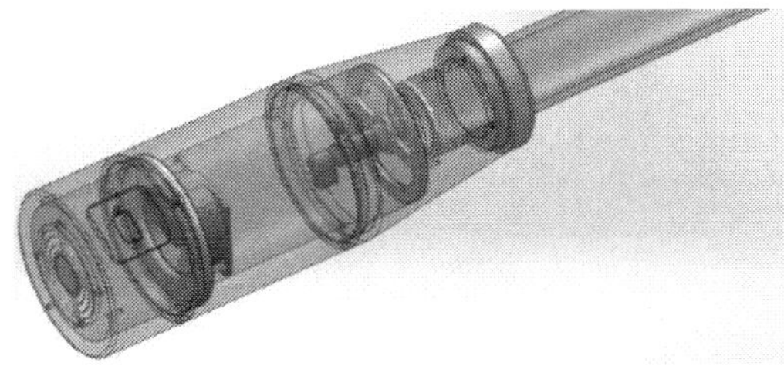

Figure 2: Mechanical concept of the Hoverball [1]

The body itself is split in the lower part, the center part and the top part. The lower part contains the electronic boards with power supply, signal conditioning and power electronics for a fan. Due to the position of the fan and the power electronics the fan has the side effect that it cools down the electronics.

The fan is hosted in the center part that has enough space to place additional components to regulate the air flow. Additional components could be a throttle flap to control the air flow or to make it even more turbulent or it could be a diffusor that would create a more laminar air flow. However the actual implementation was only used with and without diffusor. Obviously laminar air flow creates a more stable position of the ball and makes it slightly easier to find accurate controller parameters.

The upper part of the body contains the ultrasonic position sensor and a tightening mechanism for the acryl pipe. Normally it is difficult to place an ultrasonic positioning sensor in a small pipe as the radiation pattern of the sensor could detect edges of the construction or get interference of ultrasonic waves reflected at the cylindrical walls. The sensor used in this application has a radiation pattern that would exceed the diameter of the pipe by far but due to its position its beam enters the pipe without reflection and returns an acceptable signal quality for this application. The ultrasonic position sensor is programmable so the measurement range can be adjusted exactly to the length of the pipe and the sensor returns a 4..20 mA current signal proportional to the position of the ball.

As actuator to create the air flow a standard fan was used. It has the advantage that it keeps the noise level quite low and is pretty efficient versus a self-constructed motor with propeller. The fan speed can be modified due to a variation of the supply voltage. The operation voltage ranges from 6..24V which covers the complete air flow range needed to control the ball position.

The fan supply voltage is generated by a classic linear power operational amplifier circuit. The power operational amplifier has a high bandwidth and does not create any critical system dynamics.

The final important mechanical component is the ball itself that has to be levitated by the air flow. The ball is a standard but painted Ping-Pong ball. The painting of the ball has a mechatronics reason. The painting on the very regular ball causes desired irregularities. A standard Ping-Pong ball is precisely defined which cause arbitrary movements in the air flow. The painting modifies these characteristics. Particularly an inhomogeneous surface with a typical paint-drop at one point causes an asymmetric characteristic that can help modifying the stability of the ball. The effect obviously is very sensitive and hard to reproduce but the tendency is pretty much the same with painted balls. The effect that can be observed is that the center of gravity is properly defined in the direction of the paint-drop. When the ball is levitated the paint-drop is oriented downwards. Due to the air flow the ball has a tendency to rotate and to move lateral that is normally not intended. With the painted ball the rotation axis can be set to vertical. With additional air flow guidance within the metal body the ball can be influenced to move almost only vertically with almost no rotation.

5 System Design

As mentioned previously, the main objective was to evaluate the use of OCE with the test bench. The used system architecture is an industrial motion-controller hosting the motion-control firmware with motion-kernel and PLC-kernel connected to different execution environments. The motion-controller itself has a local bus interface that can be used to connect different kind of industrial peripherals as there are digital and analog inputs and outputs or communication devices. For the test set-up analog input and output channels were used to connect the Hoverball test bench. The analog output provides the set-point voltage for the fan speed and an analog input that is connected with the ultrasonic position sensor that measures the ball position within the acryl pipe. The motion-controller supports several additional fieldbus technologies like PROFIBUS DP, PROFINET, Ethernet/IP or sercos automation bus. During the evaluation the local bus interface was used only because of the additional time-delays that come with the different bus technologies. Communication dead-times destabilize the control system. Therefore it should be considered during modeling and control parameter settings but it was not in the center of interest during this project.

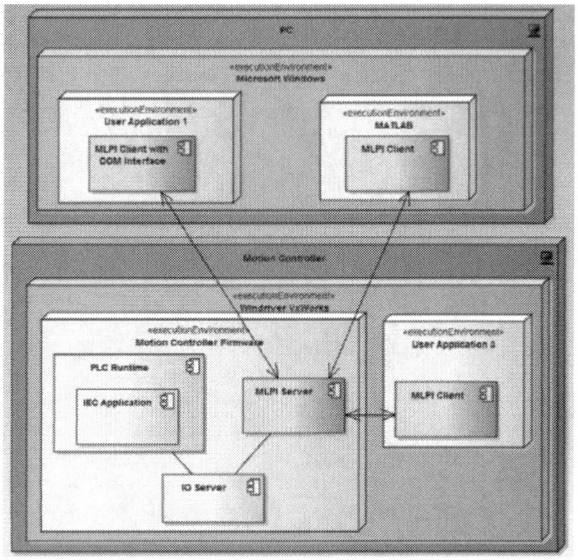

Figure 3: System architecture based on the Open Core Engineering

Fig. 3 shows a general overview of the system architecture in terms of an UML deployment diagram. The fundamental system architecture and some more details are described in [6, 7]. In the context of this project the key component is the execution environment *Firmware* at the lower left side. The classic approach of a motion-controller is the execution of a PLC program written in IEC 61131-3 languages as part of the software. This software is executed as real-time component. For the evaluation of the OCE with the Hoverball test bench explained above, the execution environment *Windows PC* was used but also substituted by or put in parallel to another execution environment *Android smartphone*. The *Ethernet* connection can be materialized by a local area network (LAN) or a wireless network (WLAN) and the execution environments can access the motion-controller in parallel.

First of all the test bench was put into operation with a standard IEC 61131-3 software program to determine a controller structure and to find reasonable control parameters as this is the reliable real-time environment. In parallel to this classic design approach a model of the Hoverball was created with MATLAB/Simulink. The system parameters were identified and converted to model parameters to simulate the ball behavior. With this model the parameters and structure of the control algorithm were optimized to get an acceptable system characteristic for the objectives of this project. Fig. 4 shows the measurement of the ball position as reaction to commanded sinusoidal set-points. The signal has a frequency of 80mHz and a magnitude of 50mm at an offset of 500mm. The ball position follows the commanded set-points with a phase-shift and an intended offset nicely. This means the process can be observed easily and modifications of the system of control parameters are very transparent. As control algorithm an adaptive PID-controller was used. As the controller was not in the center of interest for this paper it will not be explained here. However, it should be mentioned that the Hoverball does not have constant model parameters for all ball positions. The system can be separated in three areas. If the ball lies at the bottom of the acryl pipe or if it reaches the end of the pipe, the system changes significantly. This is because aerodynamics changes a lot at the air inlet and outlet. Within the area of 100mm to 700mm the system characteristics remain quite constant.

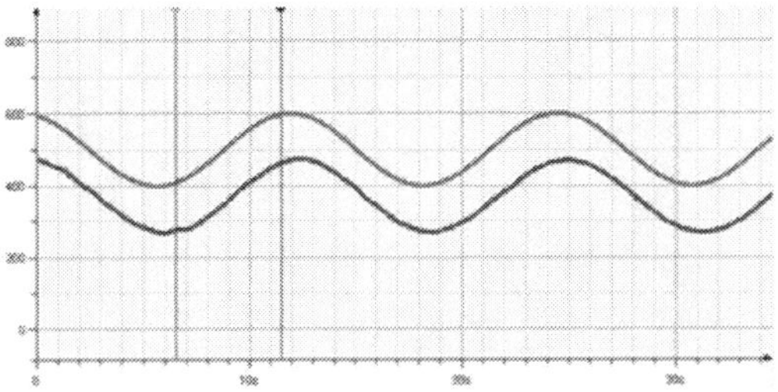

Figure 4: Reaction of the ball position (lower curve) to a commanded sinusoidal set-point (upper curve) in millimeters versus time on the horizontal axis [1]

In order to validate some of the functionality of OCE the control algorithm designed with Simulink was replicated with a MATLAB m-file and executed in a Windows PC execution environment (compare Fig. 3). A detailed description of the MATLAB programming interface for OCE was presented in [8]. The results that are not depicted here showed a similar result as shown in Fig. 4. However, as expected with a non-real-time (NRT) environment communicating via a LAN, the cycle time is not constant and the control algorithms work fine only if the system load on the NRT PC does not delay the execution of the control algorithm for more than 50ms. As a conclusion it is interesting that the control algorithm can stabilize the ball position if system load and communication traffic are respected. For demonstration and evaluation purposes this is acceptable for an industrial application it is not. However, the use of the NRT environment for hard-real-time applications is not the objective. For this purpose real-time environments with Open Core technology are available.

In order to evaluate the test bench with OCE running on a tablet PC or smartphone, an Android based smartphone was selected and a human machine interface for set-point inputs and actual value visualization was implemented in Java. As with the MATLAB control algorithms this implementation is a NRT environment with the additional challenge that communication uses a WLAN link that is shared with other communication nodes and therefore is non-deterministic and cannot be used for control algorithms needing almost stable cycle times of less than 50ms. This is why the set-point input and process visualization was implemented only. The Java app interfaces the real-time target running the IEC 61131-3 control algorithm and commands the set-points according to the user inputs. Fig. 5 shows the screen to move the ball position with finger swipes.

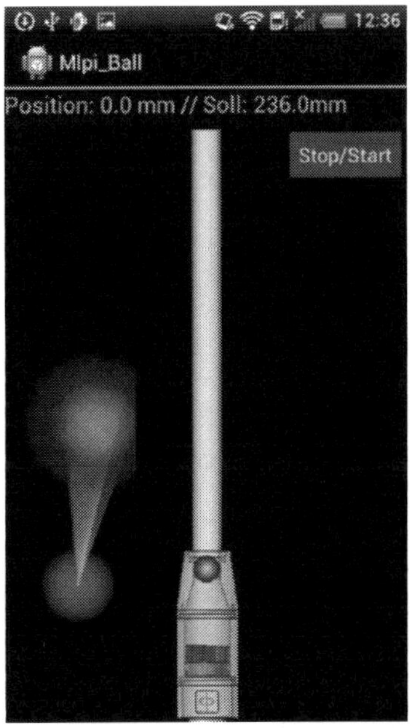

Figure 5: GUI to control the Hoverball with a smartphone [1]

6 Pre-study forImprovements

Following the positive results of the hard- and software tests the design and the construction of the Hoverball was reviewed regarding cost. The initial design was started with a target cost lower than 1000€ for the mechanical parts. The target was reached but it is pretty much for a prototype. Within the context of rapid-prototyping technologies a pre-study was conducted to evaluate the manufacturability with fused-layer-modeling (FLM) technology. Fused-layer-modeling is one of the low-cost 3D printing technologies. It basically describes an additive manufacturing technology based on 3-dimensional extrusion of thermoplastics like ABS, Nylon or PLA [10]. As the process is additive it allows production of workpieces that are slightly more complicated than classically produced workpieces. On the one hand side this gives more freedom for design aspects but on the other hand the technology comes with additional constraints, particularly for overhanging construction elements. Another constraint is robustness of the printed material.

Fig. 6 shows as a good example the Hoverball's lower part according to the original construction for production from aluminum and the redesign for manufacturing with FLM technology. The aluminum parts were design to be mounted on top of each other and fixed with screws which needs threading holes as shown in the figure. FLM components are made from layered thermoplastic which means that mounting screws parallel to a layer, which is the case here, is not very rigid. Instead of these threading holes so called screw nut boxes were designed. Before assembly screw nuts can be placed in these boxed and after assembly of parts these boxes fix the nuts and screws can be mounted from the exterior.

DOI: 10.1051/matecconf/20165403010

Another aspect is a overhanging element. To produce overhanging elements, FLM technology allows designing support material areas that create a support element bearing the overhanging element during production. After production this material can be removed. However, removing support elements usually causes visual marks that had to be avoided here due to aesthetic design requirements. Depending on production parameters overhanging elements can be produced if the angle between element and vertical axis is in the range of 60° or higher. In this design overhanging elements were reduced adding slopes with more than 60°.

The initial construction designed for aluminum material has a separate fan grid at the bottom side as it is easier and more rigid to mount a stainless steel grid than to mill the contour with curvatures in aluminum. This could be done in FLM as well but integrating the grid directly in the lower part is possible as well. It reduces the number of parts and assembly effort as well.

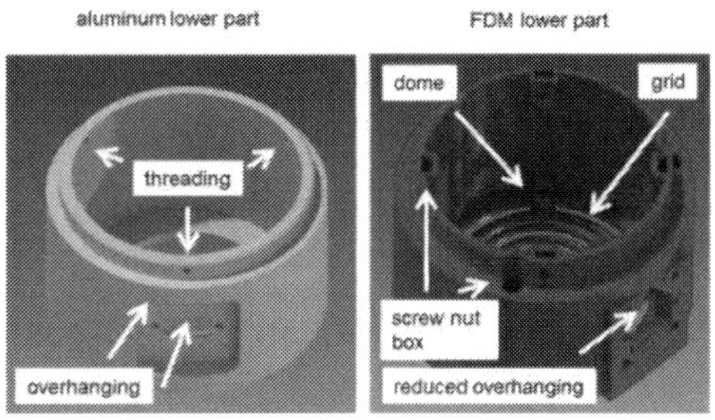

Figure 6: Lower part of the Hoverball in aluminum and lower part in FLM

This was targeted for the mounting of the PCBAs as well. Milling complex structured with small dimensions is hard to be produced with standard milling tools and machines. But FLM gives more flexibility and the workpiece can be more complicated. Therefore the mounting domes for the PCBAs were integrated similar to the fan grid at the bottom of the part. Integral keys were added as well to improve the rigidity of the domes.

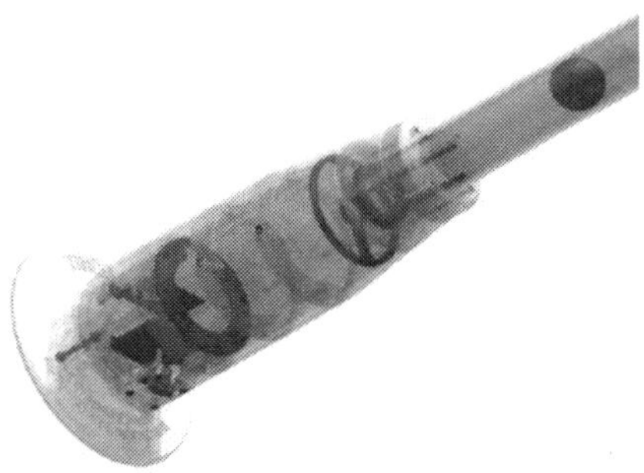

Figure 7: Transparent view of the FLM construction

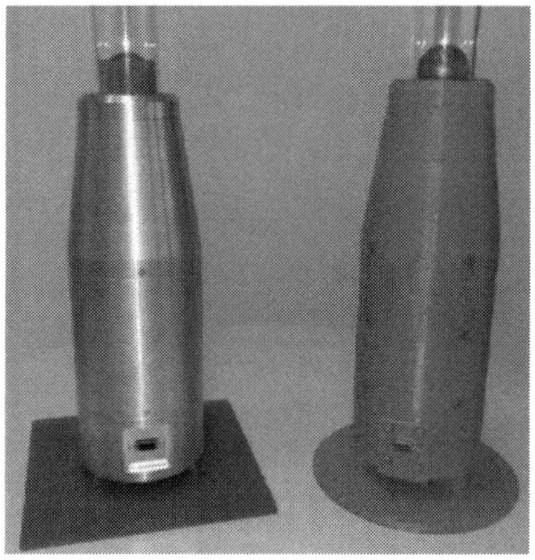

Figure 8: Final aluminum and FLM Hoverball

The lower part of the Hoverball is the most sophisticated FLM part. Other parts were constructed and produced similarly. Fig. 7 shows the complete construction of the FLM Hoverball in a transparent view. The aerodynamics of both constructions is similar but not exactly identical. On the one hand side this is mainly caused be the difference in the fan grids and on the other hand side by the slightly different surface structure of the milled aluminum compared to the rougher surface of FLM components which were not grinded after production.

The major objective designing the FLM Hoverball was the evaluation of the feasibility of producing it with FLM technology and secondly the reduction of production cost. As mentioned above, the aluminum mechanics cost several hundreds of Euros and it took about 4 weeks lead time to get parts. The FLM Hoverball was produced with about 2kg of PLA, which is - depending on material supplier - less than 100€ but the printing time of the components was about 80 hours.

7 Summary and Conclusion

The aerodynamic ball levitation test bench, the Hoverball, presented in this paper is the result of several research activities in the context of RCP in industrial automation partially described in [6, 7, 8, 9] extended by research results of a 3D printable version. It builds a control system that has system characteristics adapted to the performance of industrial motion-controllers not needing additional control electronics. With its tuned complexity and dynamics it is well designed for evaluation purpose but combines at the same time multi-physics like aerodynamics, ultrasonic measuring and a low voltage electric drive. The paper describes the construction and functionality of the Hoverball. It is used in three use cases evaluating the applicability of OCE for this test bench in non-real-time and real-time environments. The use cases cover the software development with IEC 61131-3 programming languages, MATLAB/Simulink modeling and a Java app for an Android smartphone.

Particularly the implementation of a stabilizing control algorithm running on a non-real-time execution environment like Microsoft Windows has shown that it is possible to use it but it is at the limit and requires dedicated conditions.

Finally the paper describes how the construction can be modified to turn it from a classical aluminum design into a 3D printable design. The design aspects for 3D printing are explained and a final version is shown as prove of the feasibility.

References

1. A. Kindler, Konstruktion, Inbetriebnahme und Implementierung eines Hoverball Labor-modells mit IndraWorks und Android. Bachelor thesis, UAS Aachen, unpublished (2013)
2. Quanser: Products and Solutions. http://www.quanser.com/Products/, accessed 10.03.2014
3. Festo: Lernsysteme. http://www.festo-didactic.com, accessed 10.03.2014
4. Vernier: PID Ping-Pong Ball Levitation. http://www.vernier.com/files/sample_labs/EPV-12-pid_ping_pong_ball.pdf, accessed 01.06.2013
5. M. Rawashdeh, B. Bowers, Levitating ball using a fuzzy control system http://markbowers.org/levitating-ball, accessed 01.06.2013
6. E. Engels, H. Schnabel, Rapid-Control-Prototyping of Industrial Drives for the sercos Automation Bus. Mechatronics (MECATRONICS), 2012 9th France-Japan & 7th Europe-Asia Congress on and Research and Education in Mechatronics (REM), 2012 13th Int'l Workshop on, pp.177-181, 21-23 Nov. 2012, IEEEXplore
7. E. Engels, S. Krauskopf, Innovation in Motion-Logic programming – a versatile interface, in: Proceedings to the 12th International Workshop on Research and Education in Mechatronics, REM 2011, Kocaeli, Turkey, 15.09.-16.09.2011, ISBN 978-975-8047-96-3
8. E. Engels, Industrieantriebe mit sercos Schnittstelle über MATLAB® steuern - Konzeption einer MATLAB® Toolbox für die Industrieautomation, AALE 2013, Stralsund, 28.02.-01.03.2013, in: Proceedings pp. 223-232, ISBN 978-3-8356-3364-3
9. E. Engels, Entwicklung einer aerodynamischen Regelstrecke für die multi-disziplinäre Lehre in der Industrieautomation, AALE 2014, Regensburg, 08.05.-09.05.2014, in: Proceedings pp. 31-39, ISBN 978-3-8356-7142-3
10. A. Gebhardt, Generative Fertigungsverfahren: Additive Manufacturing und 3D Drucken für Prototyping - Tooling – Produktion, Carl Hanser Verlag GmbH & Co. KG, ISBN 978-3446436510 (2013)

A Method Tuning Control System of Thermal Process in Startup Period

Do Cao Trung

Department of Automation and Control of Thermal Process, Hanoi University of Science and Technology (HUST), Vietnam

Abstract. The tuning control system of thermal process in startup period up today is very difficult problem due to difficulties are to model the process and to tune controller guaranteeing robustness of the system. In the paper the process is approximated second-order-plus-delay model defined by "cleft-over" algorithm [1]. And the parameters of controller defined based on required robustness of the system for startup case. Finally, the illustrative examples are presented.

1 Introduction

Let consider the typical control structure in thermal technology (Figure 1), where, O(s) and R(s) are transfer functions of process and controller.

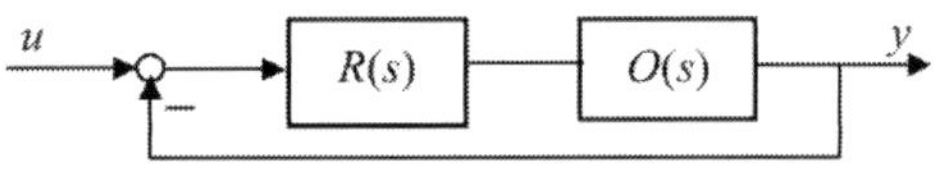

Figure 1. Typical control structure.

Currently, the tuning control system of the thermal process in start-up period is a difficult problem and unresolved up today.

Because of the existence, the startup and commissioning of thermal processes are often protracted, that makes implemented cost rises.

The obstacle is mostly the system identification in startup mode, i.e. in terms of change and lack of information. It is a difficult, complex problem that so far almost no research attention.

The article presents the problem of how to solve the tuning of control systems in start-up period consists of two phases: Phase 1 – system identification and Phase 2 – compute the parameters of controller.

2 System identification

2.1 Model definition

In thermal control engineering there are two main classes of control objects. The first includes dominant-lag processes with dead time and the other includes dominant-integral ones with dead time. The mentioned processes can be approximated by 2-nd-order lag plus delay model under the transfer function:

$$O_{Q2T}(s) = \frac{K}{(1+T_1 s)(1+T_2 s)}.e^{-\tau s} \tag{1}$$

Where: K – gain factor; T_1, T_2 – lag constants ($T_1 \neq T_2$); τ - dead time; s – complex variable.

The model (1) quite matches the dominant-lag processes [6]. In the other hand, for dominant-integral processes, the model (1) gives also very good approximation. Indeed, we may transform (1) to:

$$O_{Q2T}(s) = \frac{K/T_1}{(1/T_1 + s)(1+T_2 s)}.e^{-\tau s}$$

It is easy to see that if $1/T_1 \to 0$ when keep $K_0 = K/T_1 \to a < \infty$, then model becomes dominant-integral:

$$O_{Q2T}(s) = \frac{K_0}{s(1+T_2 s)}.e^{-\tau s} \tag{2}$$

The unit step response of (1) is

$$y(t) = h(t) = K\left[1 - \frac{T_1 e^{-\frac{t-\tau}{T_1}}}{T_1 - T_2} + \frac{T_2 e^{-\frac{t-\tau}{T_2}}}{T_1 - T_2}\right] \tag{3}$$

Where: t – time variable; $T_1 \neq T_2$.

In the time $t = t_i$, the response is

$$y(t_i, \mathbf{x}) = K\left[1 - \frac{T_1 e^{-\frac{t_i-\tau}{T_1}}}{T_1 - T_2} + \frac{T_2 e^{-\frac{t_i-\tau}{T_2}}}{T_1 - T_2}\right] \tag{4}$$

and the measuring process value is y_i. The parameters of vector $\mathbf{x} = \{K, \tau, T_1, T_2\}$ have to be calculated to warranty that the second-order lag plus delay (1) presents as exactly as possible the practical thermal process. This requirement will be satisfied if the below problem of optimization is solved.

$$F(\mathbf{x}) = \sum_{i=1}^{N}[y(t_i, \mathbf{x}) - y_i]^2 \to \min_{\mathbf{x}} \quad (5)$$

Here: N is the number of measuring points.

$F(\mathbf{x})$ is non-linear function of K, τ, T_1 and T_2 variables. Problem (5) will be solved by using "cleft-over" algorithm [1].

2.2 Optimizing model by cleft-over algorithm

We propose use cleft-over algorithm to solve the problem (5) summarized as below:

2.2.1 Specifying objective function [1, 2]

Problem (5) needs to be determined with constraints of parameters based on each practical process. The constraints are:

$$0 \le K;\ 0 \le T_1, T_2;\ 0 \le \tau \quad (6)$$

Problem (5)-(6) may be driven to an equivalent following unconstraint problem:

$$J(\mathbf{x}) = F(\mathbf{x}) + p^{*}\Psi(\mathbf{x}) \to \min_{\mathbf{x}} \quad (7)$$

Where $\Psi(\mathbf{x})$ is penalty function determined as:

$$\Psi(\mathbf{x}) = \left[|\tau| - \tau\right]^2 + \left[|T_1| - T_1\right]^2 + \left[|T_2| - T_2\right]^2 + \left[|K| - K\right]^2$$

$p = [10 \div 10^6]$ - the penalty coefficient.

2.2.2 Cleft-over step algorithm [1, 2]

The algorithm is built base on two concepts: moving direction of searching orbit and cleft-over step. The iterative equation of the algorithm is

$$\mathbf{x}^{k+1} = \mathbf{x}^k + \alpha_{k+1}\mathbf{s}^k$$

Where: $\mathbf{x}^k$, $\mathbf{x}^{k+1}$ - the begin and end of the k+1-step, α_{k+1} is the step length, $\mathbf{s}^k$ is moving direction.

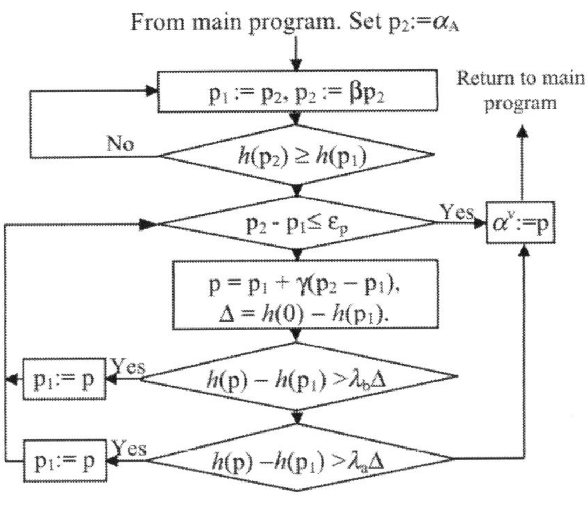

Figure 2. Flow chart of finding step length.

In the cleft-over-step algorithm the α_{k+1} defined based on "cleft-over principle" [2] that provides the begin and end points of each iteration lay on different sides of minimum point of objective function in $\mathbf{s}^k$ direction.

Length of each iterative step of the algorithm is defined by flow chart in Figure. 3.

For each iterative step, suppose:

$$h(\alpha) = J(\mathbf{x}^{k+1}) = J(\mathbf{x}^k + \alpha\mathbf{s}^k)$$

Each iterative step of the algorithm satisfies following condition:

$$\lambda_a[h(0) - h(\alpha^*)] < h(0) - h(\alpha^v)] < \lambda_b[h(0) - h(\alpha^*)] \quad (8)$$

Where: α^v – iterative cleft-over step; α^* - minimum point of $h(\alpha)$; λ_a, λ_b - coefficients: $0 \le \lambda_a \le \lambda_b \le 1$.

The algorithm of finding cleft-over step is shown as below:

Initial values: $\varepsilon_P > 0$, $\beta > 1$, $0 < \gamma < 1$, $\alpha_A > 0$.

- Step 1: Set $p_1 = 0$, $p_2 = \alpha_A$ and increase the values repeatedly by the rule: $p_1 := p_2$, $p_2 := \beta p_2$ until $h(p_2) \ge h(p_1)$, get the length (p_1, p_2) and go to step 2.
- Step 2: If the condition $p_2 - p_1 \le \varepsilon_P$ is satisfied, means not possible to find cleft-over step with exact level ε_P. Set $\alpha^v := p$ and end, return to main program. If not, go to the step 3.
- Step 3: Calculate $p := p_1 + \gamma(p_2 - p_1)$ and $\Delta = [h(0) - h(p_1)]$, go to step 4.
- Step 4: If $h(p) - h(p_1) > \lambda_b\Delta$, set $p_2 := p$; if $h(p) - h(p_1) < \lambda_a\Delta$, then set $p_1 := p$; repeat from step 2. The other case, $\lambda_b\Delta > h(p) - h(p_1) > \lambda_a\Delta$ and $\alpha^v = p$ is left-over step. Set $\alpha_A := \alpha^v$ and return to main program.

The direction $\mathbf{s}^k$ is determined to make the target function reduce after each iterative step, means $J(\mathbf{x}^{k+1}) < J(\mathbf{x}^k)$. The combination of cleft-over step adjusting rule and one move direction $\mathbf{s}^k$ will create one cleft-over step algorithm. The algorithm is shown in Figure. 3.

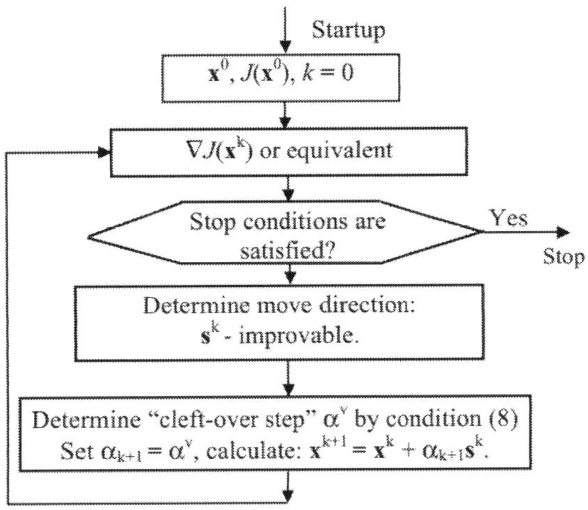

Figure 3. Cleft-over step algorithm.

The initial parameters are set as below:

$\gamma = 0{,}382$; $\lambda_a = 0$; $\lambda_b = 0{,}5$.

In the first step (k = 0), choose start point $\mathbf{x}^0$, the first step value $\alpha_A = 0{,}1$.

Normally, the first move vector is chosen: $\mathbf{s}^0 = -\nabla J(\mathbf{x}^0)$ (minus gradient value of target function in start point). The progress of optimization by using cleft-over step algorithm in step no. (k+1) consists of two parts as below:

- Calculate gradient $\nabla J(\mathbf{x}^k)$ and determine the direction s^k.
- Move by direction $\mathbf{s}^k$ until reach the cleft uptrend, means the condition (8) is satisfied. End the iterative step no. (k+1), get the new point: $\mathbf{x}^{k+1} = \mathbf{x}^k + \alpha_{k+1}\mathbf{s}^k$ with $\alpha_{k+1} = \alpha^v$ is cleft-over step.

The repeat is continuous until fixed conditions are satisfied. One of the most efficient moving directions is Affine [1].

3 Tuning controller

When the model of process is achieved, the calculation of controller parameters is able to fulfill by using some popular rules as Zigler-Nichol 1, SIMC and Robust-based.

3.1 Zigler-Nichol 1 method [3]

This method uses approximate model of process. With the unit step response is shown in Figure. 4, the chosen model is time delay integral:

$$O_{Z-N1}(s) = \frac{K}{s}.e^{-\tau s}$$

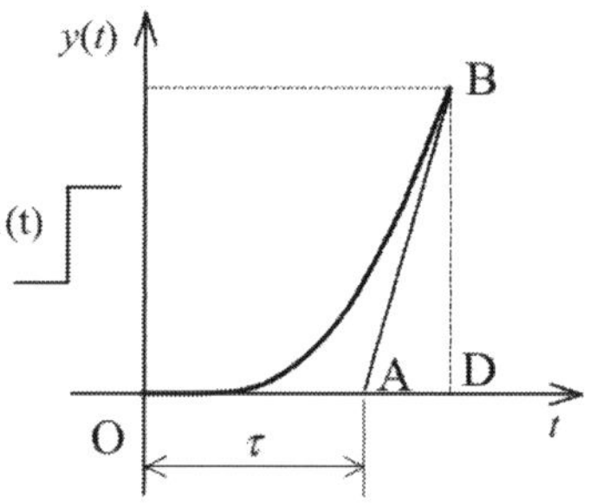

Figure 4. Process unit step response.

Here: K= BD/AD (Figure 4)

The PID controller equation is:

$$R(s) = k_P(1 + \frac{1}{T_I s} + T_D s).e^{-\tau s}$$

The parameters are set:

$$k_P = \frac{1{,}2}{K\tau}, T_I = 2\tau, T_D = \frac{\tau}{2}$$

3.2 SIMC method [4]

The original Internal Model Control based on model of process. The best improve of IMC is SIMC by Skogestad [4]. The PID controller is

$$R(s) = k_c(\frac{1+T_I s}{T_I s})(1 + T_D s)$$

Its parameters for model (1) are set as below:

$$k_c = \frac{1}{K}.\frac{T_1}{\tau_c + \tau}, T_I = \min\{T_1, 4(\tau_c + \tau)\}, T_D = T_2$$

Normally, the tuning parameter τ_c is chosen equal to time delay value τ of processes.

3.3 Robust-based method [5]

The method is based on model of process, with the general equation is

$$O(s) = \frac{B(s)}{A(s)}.e^{-\tau s}$$

The robust-based controller is defined:

$$R(s) = \frac{1}{\theta s}.\frac{A(s)}{B(s)}$$

The coefficient θ is lag time constant which is defined by the stable requirement of system. θ is chosen in the condition that the oscillation index of dominant pole of the control system varies in the range from 0,461 to 2.

4 Examples

4.1 Lag process

A superheated steam process of one thermal power plant has the transfer function:

$$O_{sh}(s) = \frac{1}{(1+54s)(1+22{,}8s)(1+22{,}8s)}.e^{-11{,}4s}$$

The unit step response of it is shown in Figure. 5.

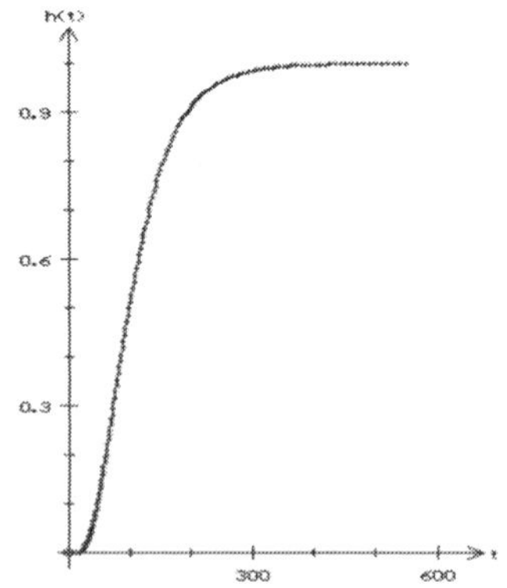

Figure 5. Unit step response of O_{sh}.

Collect the data from 0 to 30% of output at the time $t_{30\%}$.

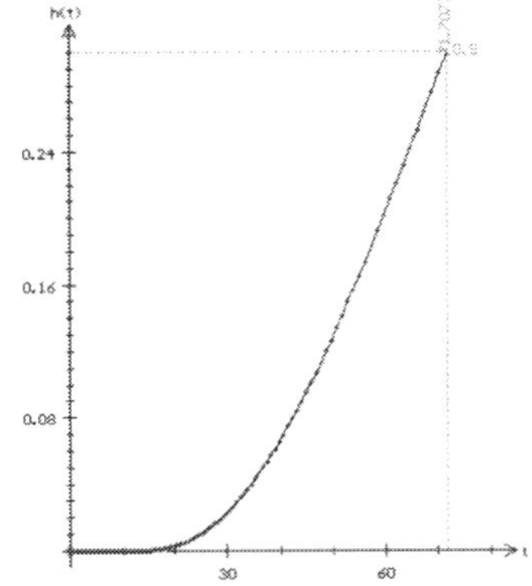

Figure 6. Initial data from characteristic of O_{sh}.

Identify the obtained data by model (1) with the limitation:

$$0 \le K;\ 0 \le T_1,\ T_2;\ 0 \le \tau \le t_{30\%} \le 74 \qquad (9)$$

Get 15 measuring points at the time $t_i = 0, 5, 10 \ldots 70$, values y_i is shown respectively in the Table 1.

Table 1. Fifteen measuring points.

t_i	0	5	10	15	20	…	65	70
y_i	0	0	0	0,001	0,003	…	0,243	0,286

Input the values in (5) will have function F(**x**), the penalty function Ψ(**x**) is determined from (8), coefficient p is set by 10^3. Finally, function J(**x**) in (7) is formed with variable **x** = {K, τ, T_1, T_2}.

Using cleft-over step algorithm to identify the system, after 37 iterative steps achieves the process model in startup period is

$$O'_{sh}(s) = \frac{31,813}{(1+2580s)(1+41,218s)}.e^{-16,565s}$$

By this model, we use the Zigler-Nichol 1 [3], SIMC [4] and robust-based [5] tuning methods to compute controllers.

After that input three obtained controllers in closed control loop of the original $O_{sh}(s)$. The closed-loop unit step responses of these different controllers are shown in Figure. 7.

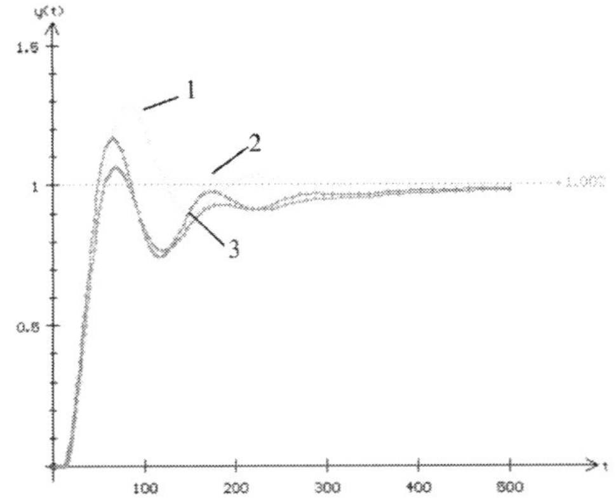

Figure 7. Closed-loop response with different controllers 1_Zigler-Nichol 1; 2_SIMC; 3_Robust-based.

The results express:

- All three controllers give good tuning quality.
- Three qualitative factors are used to compare are Decay, Overshoot and Steady time. The results of each method are expressed:
  + Decay: Robust-based is lower than SIMC, SIMC is lower than Zigler-Nichol 1.
  + Overshoot: Robust-based is lower than SIMC, SIMC is lower than Zigler-Nichol 1.
  + Steady time: Robust-based and SIMC are the same, both of them are a bit shorter than Zigler-Nichol 1.

4.2 Integral processes

A level drum process of one thermal power plant has the transfer function [7]:

$$O_{dr}(s) = \frac{0,055}{s(1+4,608s)^2}.e^{-34,671s}$$

The unit step response of it is shown in Figure. 8.

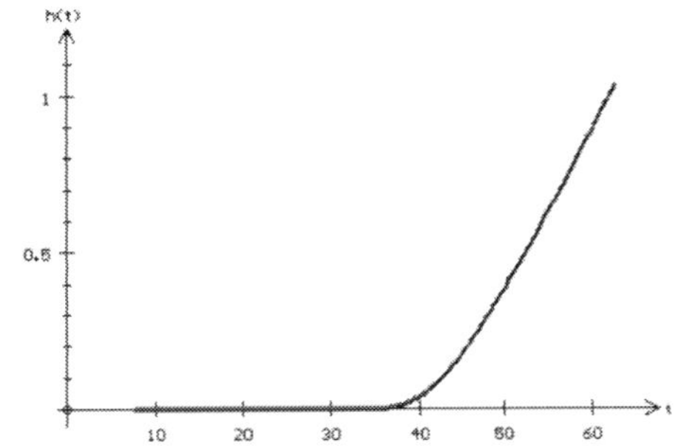

Figure 8. Unit step response of O_{dr}.

Collect the data from 0 to 30% of output at the time $t_{30\%}$.

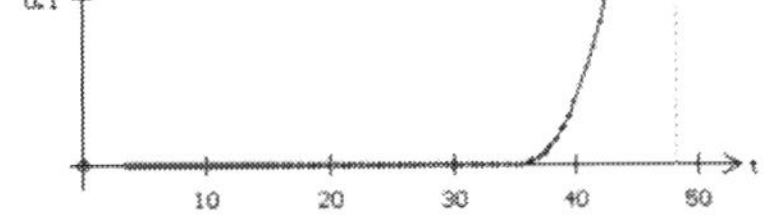

Figure 9. Initial data from characteristic of O_{dr}.

Identify the obtained data by model (1) with the limit conditions:

$$0 \le K;\ 0 \le T_1,\ T_2;\ 0 \le \tau \le t_{30\%} \le 49 \qquad (10)$$

Get 17 measuring points at the time $t_i = 0, 3, 6 \ldots 48$, values y_i is shown respectively in the Table 2.

Table 2. Seventeen measuring points.

t_i	0	3	...	36	39	42	45	48
y_i	0	0	...	0	0,056	0,163	0,228	0,293

Replace the values into (5) will have function F(**x**), the conditional function Ψ(**x**) is determined from (9), coefficient p is set by 10^3. Finally, function J(**x**) in (7) is formed by vector **x** = {K, τ, T_1, T_2}.

Using cleft-over step algorithm to identify the system, after 55 iterative steps achieves the process model in startup period is

$$O'_{dr}(s) = \frac{41,21}{(1+329,616s)(1+26,852s)}.e^{-35,763s}$$

By this model, we use the Zigler-Nichol 2 [3], SIMC [4] and robust-based [5] tuning rules to compute controllers.

After that, input three computed controllers into closed control loop of the root $O_{bh}(s)$. The closed-loop unit step responses of these different controllers are shown in Figure. 10.

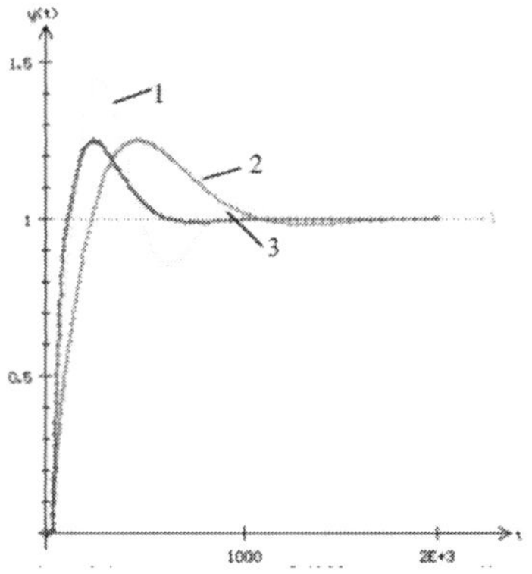

Figure 10. Closed-loop response of different controllers 1_Zigler-Nichol 1; 2_SIMC; 3_Robust-based.

The results express:

- All three controllers give good tuning quality.

- Three qualitative factors are used to compare are Decay, Overshoot and Steady time. The results of each method are expressed:

  + Overshoot: Robust-based is the same SIMC and much lower than Zigler-Nichol 1.

  + Decay: Robust-based is lower than SIMC, SIMC is much lower than Zigler-Nichol 1.

  + Steady time: Robust-based is much shorter than SIMC, SIMC is a bit shorter than Zigler-Nichol 1.

5 Discussion

- Thermal process identification algorithm is efficient.

- Second-order lag plus delay model is suitable for modeling thermal processes including dominant integral and lag classes.

- The characteristic of Second-order lag plus delay model is expressed via the relationship between T_1 and T_2. If they are so different, the model presents integral manner, otherwise it behaves as a lag model.

- The robust-based tuning rule gives general control quality better than SIMC and Zigler-Nichol 2 do in the two given examples.

- The private result of example is significant base in the choice of tuning rule for thermal process control systems.

- The identification of thermal process by Second-order lag plus delay model is very useful for tuning activity owing to the obtained controller in PID structure.

6 Recommendation

- The identification algorithm will give result faster and more exactly if the start values are sensibly chosen.

- The start vector should be set to satisfy: $\tau \approx T_2 \approx 10\%{*}T_1$ and $\tau + T_1 + T_2 \approx t_{30\%}$. The identification result will more exactly present practical process.

7 Conclusions

- The article proposes a tuning algorithm for control system of thermal process in startup period.

- The proposed identification method provides receive rather precise model of process.

- The method tuning controller for startup period should be robust-based method that is most flexible and able give desired robustness of the system.

- The proposed methodology of implementation is clear, comprehensible and general. It is able to effectively apply to tune thermal process control systems in startup period.

References

1. Nguyen Van Manh, MPEI, Moscow, *Diss. Doc. Sc.:* Optimizing *methods of uncertainty control* systems (1999)
2. Мань Н.В. Применение "Оврагоперешагового" метода нелинейной оптимизации для идентификации передаточной функции объектов управления //Теплоэнерге-тика. **№ 6**. С. 71-77, (1995)
3. Zigler, J.G.; N.B. Nichols, Trans. Am. Soc. Mech. Eng., *Optimum settings for automatic controllers*, **Vol. 64**, pp. 759-768 (1942)
4. Sigurd Skogestad, Chriss Grimholt, PID Control in the Third Millennium, Advances in Industrial Control, *The SIMC Method for Smooth PID Controller Tuning*, **Chapter 5**, pp. 147-175 (2012)
5. Nguyen Van Manh, The fifth Vietnamese national conference of Automation, Science report, *Robust synthesizing method of uncertainty control system*, **Vol. 5**, pp. 155-161 (2002)
6. Nguyen Van Manh, Thermal Energy Review, *A modeling method of thermal control processes by using second-order lag plus delay model*, **Vol. 125**, pp. 15-19 (2015)
7. Do Cao Trung, Hanoi University of Science and Technology, *MSc.: A method synthesizing optimal-robust PID for thermal processes* (2011)

MATEC Web of Conferences **54**, 04002 (2016) DOI: 10.1051/matecconf/20165404002
MIMT 2016

Embedded ECU Emulation Platform for Safe Adaptive Front-Light System Control Algorithm Development

Jiae Youn , Men Di Yin , Jeonghun Cho , and Daejin Park

School of Electronics Engineering, Kyungpook National University, 80 Daehak-ro, Bukgu, Daegu, Republic of Korea

Abstract. Safety-conscious design and implementation of automotive electronics is now being important in safe driving under severe electromagnetic disturbance. In this paper, we introduce our experience and development framework to develop a safe electronic control unit (ECU) automotive controller algorithm using hardware-in-the-loop (HILS) based on Matlab/Simulink models, which are connected to ECU hardware. We propose a design flow of Matlab/Simulink based on model-based algorithm implementation of ECU controller and chip level ECU function emulation methodology using dSPACE's MicroAutoBox™, which is an off the shelf embedded system for ECU software-hardware emulation. As a case study, we shows that fast design and implementation of the specific ECU hardware, which is integrated with custom-designed its controller algorithm for adaptive front-light system (AFLS) is easily evaluated by using the proposed development framework.

1 INTRODUCTION AND MOTIVATION

Nowadays, recent advance of design automation methodology and infrastructure for the rapid development of electronic devices has led to considerable growing number of electronic control units (ECUs) based on complex on-chip hardware-software cooperation, which are integrated in the automobiles as main controllers.

There are more than 100 ECUs in automotive, including more than 3 million software lines are embedded to control electronic systems. Considering the complexity of development and testing overhead to evaluate the complicated operations ECU, we adopted the model based development (MBD) [1] by which the hardware-software is partitioned into Matlab/Simulink-based soft building blocks and on-board emulation based hardware blocks.

This method can overcome the shortcomings such as difficult to modify the hardware blocks and to verify the designed functions by using traditional development approaches through manual programing in hardware level.

Even more important, MBD is utilized as a hardware-independent method, which can be done even if hardware does not exist. We adopted design methodology in the integration of soft-driven model verification, code generation and rapid prototyping validation. Our recent efforts show that the proposed methodology enables successful design and implementation the custom-designed AFLS controller ahead of schedule as shown in Figure 1. which shows the overall of ECU development simulation.

2 PROPOSED METHODOLOGY

The AFLS (Adaptive Front Light system) has been used as one of the driver assistance system for safety driving [2]. AFLS provides visibility during nighttime driving, by lightening hidden area such as blind spot in corner.

Original AFLS function controls the angle of lights and turns off the adaptive control mode to prevent mal-functions under receiving the disturbance such as external noise injection on ECU. To improve the side-effects in this exception case, ECU developers will try to modify hardware data-path and its software algorithm to control AFLS more precisely in emergency status. We focus to reduce these iterative design efforts to modify and improve the pre-designed building blocks including hardware and software.

As a case study, we show the development experience of ECU hardware modification and on-board or on-chip software emulation by using the proposed ECU emulation framework. The successfully designed ECU, which continuously control headlamp by LUT (Lookup Table) of the backup AFLS presented in previous work [3], could be easily evaluated by the proposed ECU emulation platform based on the cooperation of software-driven Matlab/Simulink models and hardware-level target board. The backup AFLS is monitoring AFLS status real-time. If main AFLS enters into an abnormal condition, the headlamp is automatically intervened by the backup AFLS.

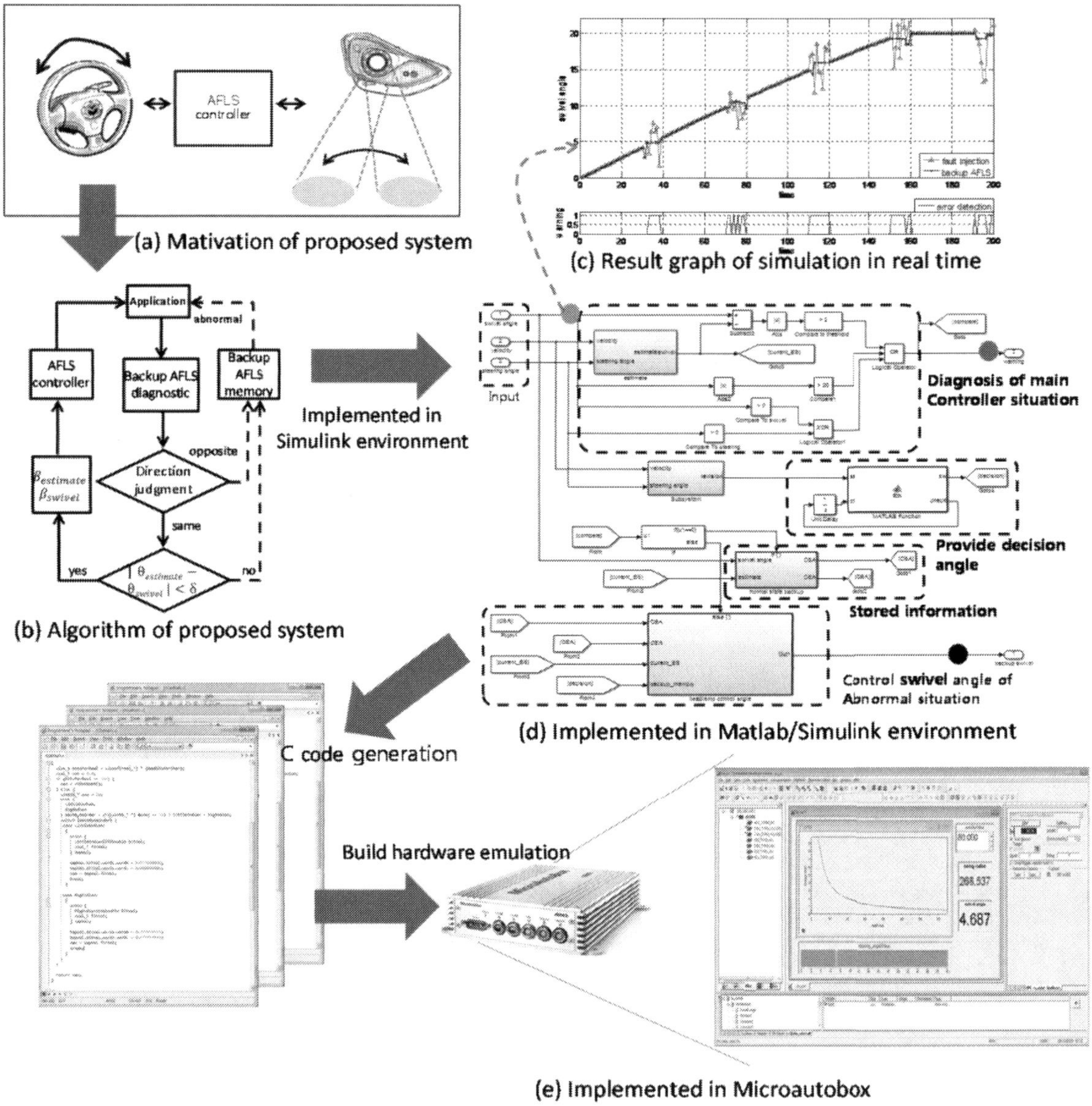

Figure 1. Overall of ECU development

Figure 1. show the overall of ECU development simulation. Figure 1. (a) show the motivation of proposed method. Figure 1. (b) shows propose backup AFLS algorithm. Then we implemented in Matlab/Simulink environment which is shown in model based design of Figure 1. (c). The requirement functions of AFLS can be easily modeled and evaluated by the methodology of model-in-the-loops (MIL) test.

After MIL test, we can easily hardware-in-the-loops (HIL) simulation and test which shows Figure 1. (e). In order to perform HIL test, we must generate target-specific firmware using Simulink configuration, which can be downloaded and executed in runtime on ECU chip. After the code generation is successfully build, we can HIL test using MicroAutoBox.

2.1 Model-In-the-Loop Simulation

The user defined AFLS control systems and its algorithm are developed using the proposed emulation method, which is based on Simulink-based model. An ECU designer in automotive company can easily modify the internals of AFLS systems by implementing user defined AFLS in soft-driven model level and compare result of simulation on the hard-driven target board level.

When AFLS is under fault situation, conventional AFLS control algorithm turns off the system modes. As a case study, we assume that an ECU designer tries to modify the functionality of the target-specific backup AFLS to monitor the status of the main AFLS controller and improve the safety of AFLs control operations. Newly designed AFLS in Figure 1. (d) follows the proposed model development in which the external environment such as steering angle, vehicle speed and swivel angle of main controller is virtually simulated required in order to diagnosis the main AFLS controller. This study assumes the maximum steering angle and swivel angle of -35° ~ 35° and -20° ~ 20 as a case study.

When main AFLS controller is abnormal situation, Backup AFLS can continuously control headlamp using LUT corresponding to steering angle range and information of normal situation in main controller. If main controller is normal situation, main controller status is saved in buffer. When main controller is abnormal situation, we use backup memory within steering angle range. Headlamp is statically controlled by LUT corresponding to steering angle range. So we use both LUT and backup information to control headlamp.

To validate the performance of user defined backup AFLS, abnormal error conditions is virtually injected into main AFLS controller, which is implemented by Matlab/Simulink model. The envrionmental disturbance such as noise can be easily described by using the components in Matlab/Simulink model and retargeted into the ECU emulation board.

Figure 1. (c) show the runtime waveform in the designed AFLS model in case which the wave signals

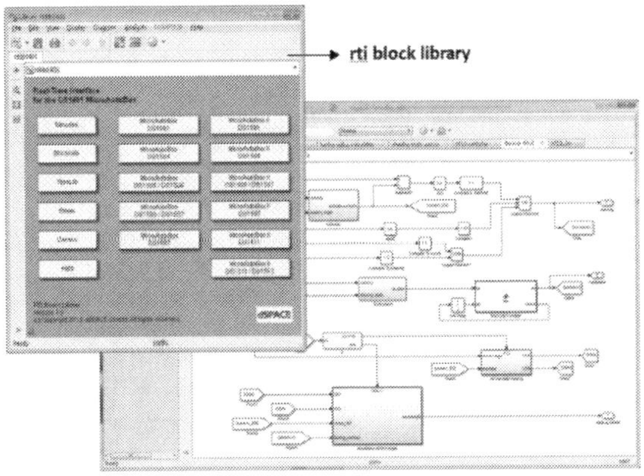

Figure 2. Using rti block labrary

shows swivel angle according to time. The green line shows the result of an abonormal AFLS controller. When error is detected by backup AFLS, the headlamp is controlled by backup memory. The black line shows the waveform of swivel angle controlled by backup memory. The second graph shows the main controller status. Result of simulation is evaluated by using scope block or plot compile in real-time.

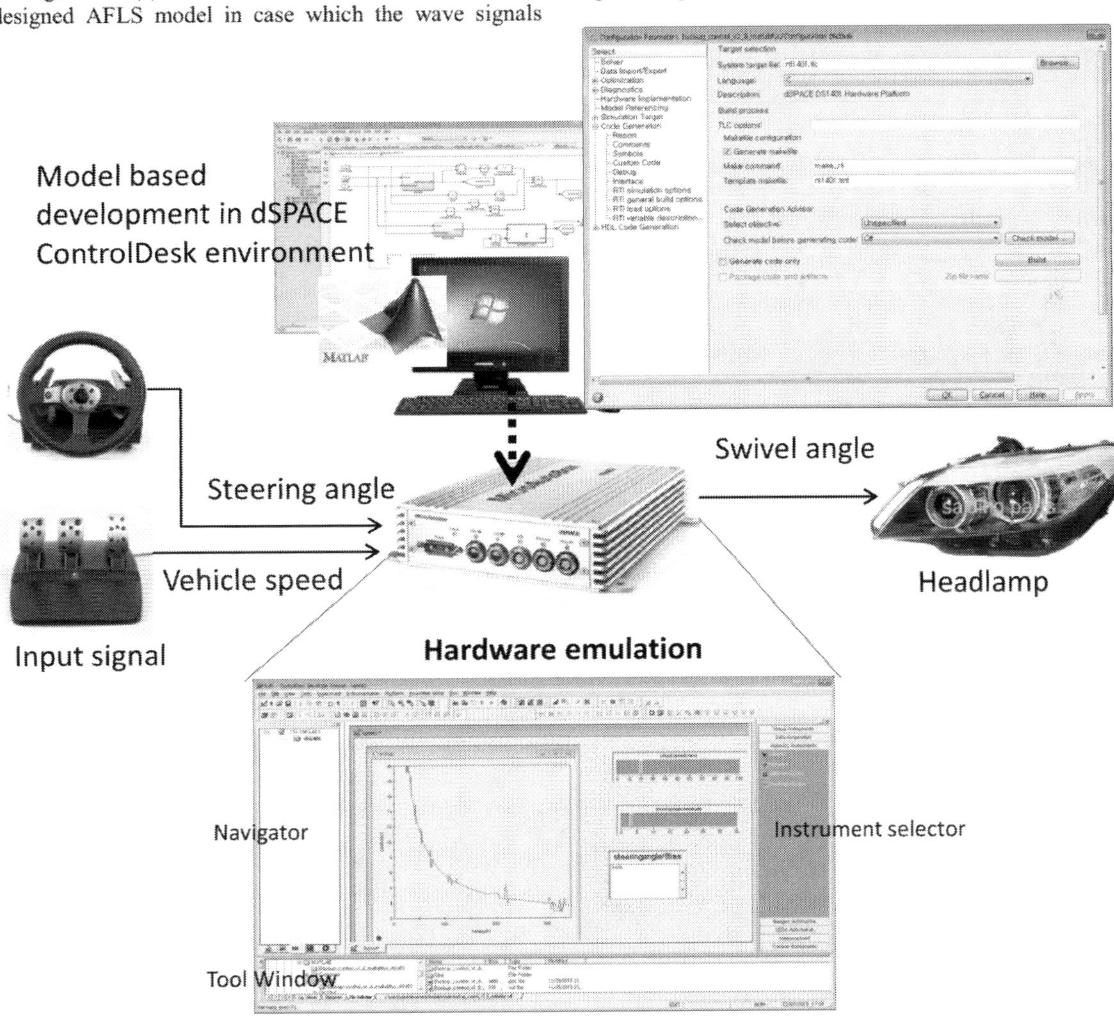

Figure 3. Hardware in the loop environment

2.2 Hardware-In-the-Loop Simulation

In our case study, the hardware-in-the-loop simulation (HIL) is implemented by using dSPACE MicroAutoBox^{TM} environment [4-5]. The controller and its algorithm in ECU chip has been designed and simulated using Matlab/Simulink block. As shown in Figure 2. In order to control input/output signal in MicroAutoBox, user must use rti block library.

Figure 3. shows HILS environment used in modeling and simulating custom-designed AFLS. Simulink-based AFLS controller can be build in MicroAutoBox [6]. As shown in head of Figure 3, Simulink-driven model implementation is successfully compiled into MicroAutoBox. After building simulink models, simulator-specific configuration files, such as *ppc* file and *sdf* file, are generated in ControlDesk and downloaded into the target hardware, which is configured with user-defined parameters.

ControlDesk software enables to control the MicroAutoBox from PC in real time. By using the ControlDesk, we could easily control parameters of user-designed model in simulink.

We can check that output of the model simulation according to input. In first step we create layout file using instrument selector. When layout is generated in ControlDesk, designer needs to located the instrumented variables in the variable browser and then probes by using the instruments block [7]. We use layout file to check result of algorithm testing, which is shown in bottom of Figure 3. We check the swivel angle using *plotterarray* block which shows swivel angle corresponding to the turning radius used in proposed algorithm. Then, we check the result of swivel angle corresponding to velocity which is changed using instrument block. So we can easily control parameters and test algorithm.

MicroAutoBox is used as a real-time system for prototyping user-designed hardware and its algorithms test such as chassis control, body control and advanced driver assistance system [8]. MicroAutoBox is offer to adapt the I/O and signal condition using ControlDesk. The previously verified design can be integrated in MicroAutoBox, and parameters various is easily changed in real-time. So user can easily hardware in the loop simulation. We can check output signal according to input using ControlDesk. Which has graphics user interface, and then we can check in real-signal using MicroAutoBox Zero Insertion Force (ZIF) connector.

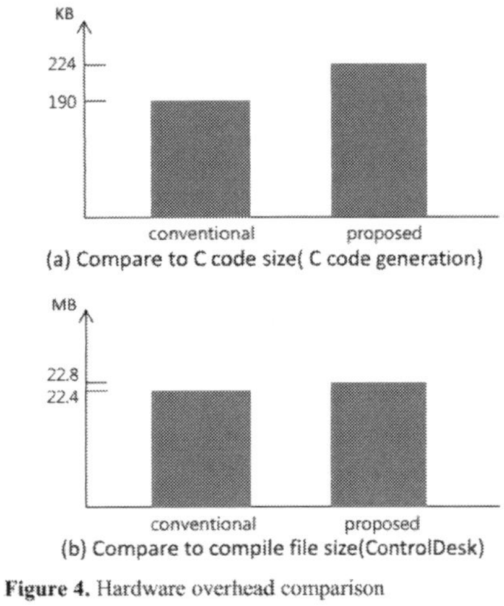

Figure 4. Hardware overhead comparison

2.3 Design case of Backup AFLS

AFLS is one of advanced driver assistance system (ADAS). AFLS improves visibility of driver during night-time driving. However, when conventional AFLS is abnormal situation, AFLS turns off the control mode and just performs fundamental lighting without movement of the lamp [9].

Backup AFLS monitor main AFLS status real-time and receives information such as velocity, steering angle, and current swivel angle according to main controller by BCM and main AFLS. Backup AFLS determines the error status saves the main AFLS information of normal status into backup memory, such as swivel angle corresponding to steering angle range.

If backup AFLS detect abnormal situation, backup AFLS read backup memory and LUT in order to continuously control the headlamp. If it is normal situation, saves the current swivel angle and information. If main AFLS enters abnormal state, backup AFLS notifies that the main controller is under abnormal situation. The backup AFLS monitor AFLS status in real-time and automatically intervened the main AFLS to control the system operation in emergency situation.

So we had to design a newly-designed backup AFLS based on fail-safety, which requires design efforts and implementation costs. We chose the HILS methodology to reduce the complexity and design time cost.

Our initial results in designing backup AFLS have been presented [3]. By injecting abnormal simulation through the Matlab/Simulink models, diagnosis functions of main AFLS fault situation have been successfully evaluated. In addition, the HILS evaluation, which is connected through the hardware-link between the PC0side Matlab/Simulink models and ECU-level emulation, results in more confidence for functions of the designed building blocks.

By using this methodology, Figure 4. show result of code size. Compile code size of proposed system is about 18 percent increase from conventional AFLS. However, hardware code size is just about 2 percent increase from conventional system.

3 CONCLUSION

This paper introduces our experience in terms of development framework to develop user-designed ECUs including hardware data-path and its software algorithm. We verified the implemented algorithm based on MIL test and HIL test by using Matlab/Simulink and ControlDesk toolbox in MicroAutoBox. We adopted Matlab/Simulink-based HILS methodology to verify and test in design step, which leads to the reduced time and cost in modeling, simulation and validating the user-defined AFLS hardware and software. As a case study, we evaluated using the proposed methodology for model design procedures to effectively compensate for the control error that may occur in AFLS ECU. We successfully could compare check result of our soft-driven implementation with real ECU hardware.

References

1. J. Ghidella and J. Friedman, "Model-based design streamlines development of body electronics systems," 2005
2. D. Kim, "Study of optimized tuning in full afls head lamps," in proceedings of the FISITA 2012 World Automotive Congress. Springer, 1719 (2013)
3. J. A. Youn, M. D. Yin, J. Cho, D. Park, *GCCE* (2015)
4. Wang, Lingchang, "Development of a Hardware-In-the-Loop Simulator for Battery Management Systems," The Ohio State University, 2014
5. Gomez, Martin, "Hardware-In-the-loop simulation," Embedded Systems Programming **14**, 38 (2001)
6. M. H. Horter, C. Stiller, *Intelligent Vehicle Symposium*, 299 (2009)
7. N. Quijano, K. Passino, *A tutorial introduction to control systems development and implementation with dSPACE* (2002)
8. "http://www.dspace.com/en/pub/home/products/hw/micautob.cfm.
9. R. Ma, *Automotive adaptive front-lighting system reference design, Texas Instruments* (2013)

Improving Efficiency and Reliability of Protection System of Tarbela Power Station

Muddasar Nawaz[1],Dr.FarhanZulfiqar[2], Dr. Mohammad Bilal[2]and Akif Zia Khan[2].

[1]*Student of Centre of Energy system NUST University, Islamabad, Pakistan.*
[2]*Lecturer in Centre of Energy System NUST University, Islamabad, Pakistan.*

Abstract. Protective relaying system plays a vital role in ensuring the proper and smooth functioning of the Power plant. As the generator is the most expensive part of the plant that's why the provision of a healthy protection system is must. Protection system guarantees the reliability and safe operation of the plant and is called the brain of the power system. Tarbela is the largest power plant of Pakistan and considered as a back bone of power system of Pakistan. In this paper, the generator protective relay of Tarbela power plant is tested and their results are presented. All the protective relays of Tarbela Power Plant are of electromechanical types which were manufactured in 1970-80. Setting, calculation and test result of these relays were then analyzed and recommendation is made for improving the reliability of the protection system of the said plant. Features of new numerical relays and comparison with the installed relays are also presented and finally the description of upgradation/retrofitting of installed electromechanical relay are discussed for smooth transition to new technology.

1 Introduction

Pakistan is facing major energy crisis now a days. In these circumstances availability of existing installed power plants is very important. Tarbela power plant has an installed of 3478MW which will be enhanced to 6300MW in near future.

Protective relays provide a vital role in the power system as it ensures the safety of the electrical equipment [1]. Main function of relay is to detect the faulty part the equipment selectively and initiate command for the quick isolation of that apparatus. The main objective of the protection system is to only isolate the faulty equipment and thus maintain the continuity of power from the other equipment. That's why protection system is called the brain of the power system.

2 Background

In Tarbela power plant all the protective relays are of electromechanical types which were manufactured in 1970-80's. All the generator protective relays were tested for checking the response and recommended for the possible solution for improving the reliability of the power plant.

This system is divided into three main components.

- •Instrument Transformer
- •Protective relays
- •Circuit Breaker

Instruments transformer first step down the voltage and current level to predetermined value for the protection relays. From these voltage and current protection calculates or senses the fault and initiate the trip by ordering to circuit breaker which breaks the current path of the faulty circuit [2].

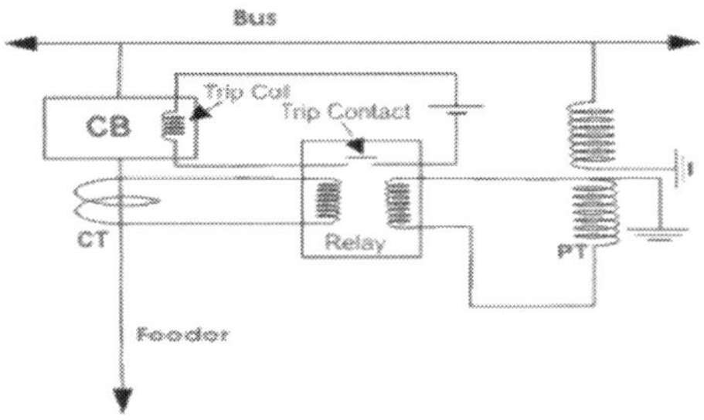

Figure1. Basic Diagram of Protection System

3 Calculations of secondary values

Principal Data of Generating Unit:
Power=175MW (10 Units) and 432MW (4 Units)
Rated Voltage: 13.8KV
Power factor: 0.85 of Unit 01-04
First the primary current of tarbela Unit is calculated as:

$$\text{Power} = \sqrt{3}\, V\, I \cos\emptyset \quad (1)$$

$$175 \times 10^6 = \sqrt{3} \times 13.8 \times 10^3 \times 0.85$$

$$I_p = 8614\ A \quad (2)$$

CT ratio: 12000/1

The rated secondary current will be=$I_s = {8614}/{12000} = 0.7178\ A$(3)

PT ratio: 14400/120 V

The rated voltage at secondary side is=

$$V_s = \frac{V_p}{Pt\ ratio} = \frac{13800}{120} = 115\ V (4)$$

4. Testing of generator protective relay

The testing of protective relays is very essential as protection works only in abnormal/fault condition. If there is any internal fault in the electromechanical relay, this defect cannot be detected until the occurrence of the system fault. So periodic maintenance testing should be carried out at least once a year.

The purpose of this test is to check whether the protection system from the relay input terminal onwards working properly or not. This can be done by applying the voltage or current from the test set to the relay terminal and check the relay contacts for alarm/trip signal. This test is known as secondary injection testing.

4.1 Generatorovervoltage relay

This protection is provided with the generator for the excessive voltage generated due to some problem in the generator exciter (AVR) or due to over speeding of the generator after the load rejection. Hydroelectric generator may go to 150% or more of rated speed when the full load is rejected and thus voltage reach to up to a very high level. This overvoltage can damage the insulation of the generator winding if sustained.

For Overvoltage protection, Protective relay is only fed with generator voltages through generator potential transformer. Relay constantly monitor the generator voltage and initiate trip in case of overvoltage. For overvoltage protection the general setting of voltage is between 105%-120%. In our case the setting is 112.5% which is 15.525KV.

The primary rated voltage is: 13.8KV and secondary side:$V_s = 115\ V$

Test Result:

Sr.No.	Voltage Injected	Time(sec)
01	135V (Pickup)	15.5 sec
02	202.5V	8.00 sec
03	270V	5.00 sec
04	337.8	2.40 sec

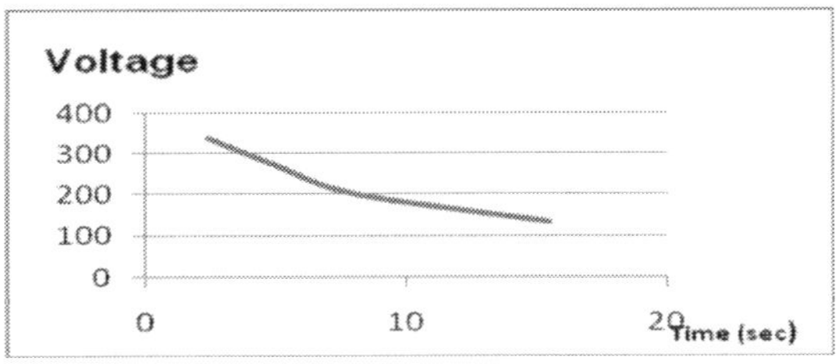

Figure2. Graph of Pickup Voltage Vs Time

The result of this relay reveals that the overvoltage relay is operating almost linear with the time rather than time inverse characteristics. It will be better to use the time inverse response relay as if the voltage increases rapidly; it should trip the generator instantaneously.

4.2 Generatordifferential protection

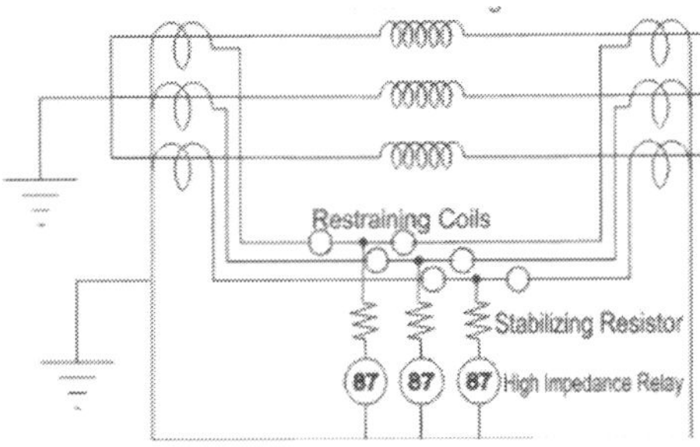

Figure3. Differential Relay CT connections

Differential is the main protection of the generator stator winding internal fault. The input of this protection relay consists of two sets of CT's. One CT set is installed on Generator output side and other on the neutral side. In normal condition the current on the both side of the generator is equal. If there is any mismatch in the current due to internal fault of the winding, the relay detects the difference between the current and then emits the trip command.

The recommended setting for this relay is 10% difference of rated current.

Test Results

Phase.	Current Injected	Current injected	Time(msec)
R	103mA	200mA	432
Y	102mA	200mA	405
B	102mA	200mA	523
R	298mA	200mA	521

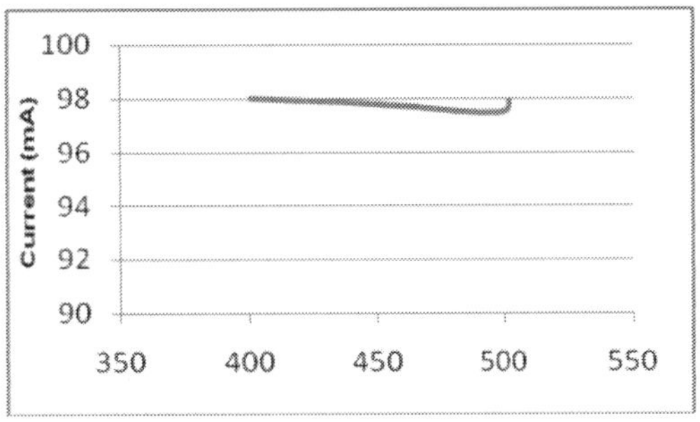

Figure 4. Pickup Current vs Time (milli second)

In the differential relay testing, 200mA was injected from the both current sources then the magnitude of the current of one side is slowly decreased and the relay operated when the difference was about 98mA. This electromechanical has a drawback. If there is a fault in the Red phase only, it will initiate the trip command, but there is no flag installed in the relay which tell the operator about the problem of the exact phase. So testing of the generator and protection circuit of all the three phases instead of one will be very time consuming.

4.3 Negative phase sequence

This relays detect the unbalancing load of the generator. The three phase power system is symmetrical in normal condition. When a fault occur on the power system, it will become unsymmetrical thus cause unbalance flow of current will cause the excessive temperature of the generator and undesired condition. This relay detects this undesired condition through the input current from the CT and initiate the trip.

The setting for the unbalance load is 7% for alarm and 20% for tripping.

Test Result:

R-phase	Y- phase	Blue-phase	Time(sec)
1.94mA	0	0	4.06 sec
0	1.90mA	0	4.5 sec
0	0	1.99mA	4.00 sec
3.99mA	2.0mA	2.0mA	3.99 sec

4.4 Generator overcurrent relay

Generator overcurrent protection protects the generator from overloading and short circuit and other fault which can increase the load on the generator. For this purpose one CT output is connected to a protective relay which is continuously monitoring the current of the generator. If the current level goes up due to any abnormal condition this relay operate and initiate the trip.

The setting of this relay is around 120-150% of rated current. In our case the relay setting is 140% approx.

Test Results

Sr.No.	Current Injected	Time(sec)
01	1A (Pickup)	3.044 sec
02	1.5A	2.350 sec
03	2.0A	1.340 sec
04	2.5A	0.792 sec

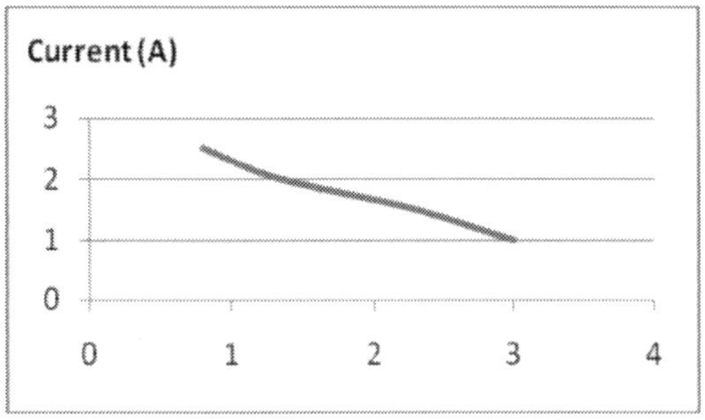

Figure 5. Pickup Current Vs Time(sec)

One overcurrent relay installed in the Unit 04 was tested earlier was found faulty. The disk of the electromechanical relay was moving but don't have the enough force for contact making. The contact resistance was in kilo-ohm. The same defective relay was replaced with this healthy one.

4.5 Generatorvoltage controlled overcurrent/ generator impedancerelay

This relay works on the same principal as of distance relay. If a short circuit occurs, the current will increases and voltage suddenly decrease. This relay checks impedance of the system by monitoring both the current and voltage of the system. This relay works as backup protection of generator. If the ratio of voltage and current drops to a certain value this relays operates and initiate the trip.

Setting of this relay can be calculated as:

$$\text{Generator impedance} = \frac{V}{I} \times \frac{1}{\sqrt{3}} \frac{115}{.7179 \times \sqrt{3}} \quad (5)$$

So Generator Impedance= 92.5Ω

In order to avoid operation of the relay due to the fault on the HV side of the transformer, the setting of this is kept very low. i.e. 12Ω per phase.

Test Results

Phase	Voltage Injected (V)	Current Injected (I)	Time(sec)
R	40	2.11	5.01
Y	40	2.130	5.05
B	40	2.150	5.09
R	60	3.110	2.05

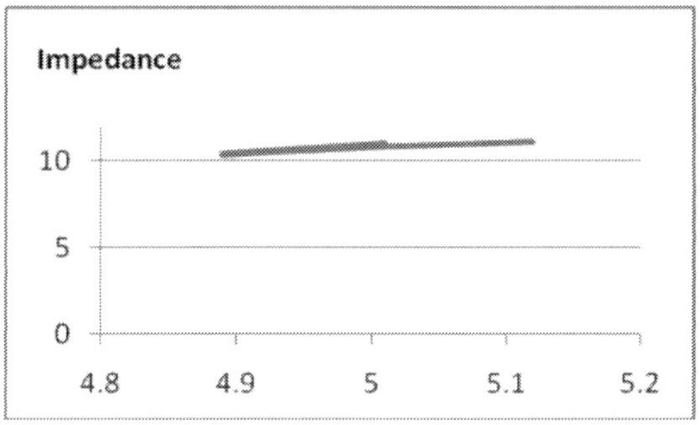

Figure 6. Impedance vs Time(s)

4.6 Transformer differential relay

The working principle of generator transformer differential relay is same as of generator differential relay. This relay checks the current of generator neutral and transformer output current. The transformer output current will not match exactly as the generator current. So a matching CT is used with the protective relay for matching the current of protective relays on both sides. If a fault occurs in the winding of generator or transformer, the matching of the current is disturbed and this relay order for the trip of the generator and transformer.

Test Results

Phase.	Current Injected	Current injected	Time(msec)
R	150mA	301mA	235
Y	150mA	301mA	205
B	150mA	300mA	245
R	250mA	400mA	205

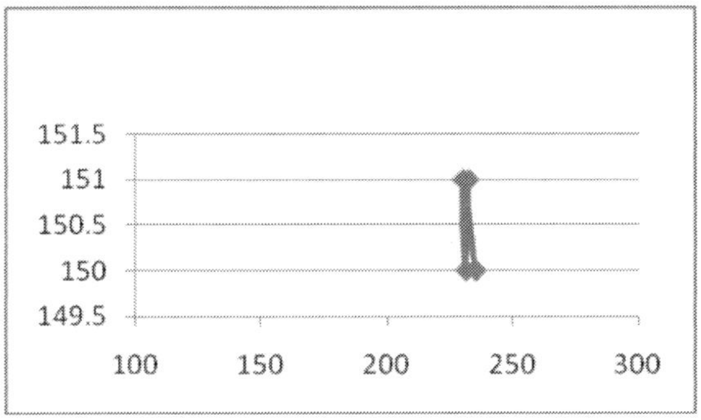

Figure 7. Pickup Current Vs Time

4.7 Generator antimotoring relay

This protection is provided with the generator for protecting the generator to run as motor. If the prime mover i.e.the water entering into turbine of the generator fails due to the governor failure or shutdown and the wicket gates of the turbine closes which causes the blockage of water flow to the turbine. As the generator is connected to the power system, for achieving the same frequency of the system the generator will run as a motor and move with the same speed. To avoid this undesired condition Antimotoring relay has been installed. This relay continuously measure the phase angle of the voltage and current, in case of this fault, the direction of flow of current is reverse which is detected by the relay and order the trip command.

The recommended setting for the reverse power relay of hydro electric power plants is 5% of the rated power.

Test Result

Voltage injected: $100\angle 0^{0}$
Current injected: 0.4A
Relay operated (Phase Angle): $90^{0} - 270^{0}$

4.8 Generator pt balance relay

This relay checks the generator voltages from the two PT's installed on the same generator. The correct status of the generator voltage is necessary because PT's are used for metering, protection and control of generator voltage. This relay continuously monitors the voltage of the generator and detect if any difference in the voltages of the PT's due to blown fuse or other fault, thus preventing improper function of other equipment which is associated with the PT voltage.

Recommended setting of this relay is 10% difference in PT's Voltage.

Test Result

Phase.	Voltage Injected (V)	Voltage injected (V)	Time(sec)
R	100	90	5.044
Y	100	90	5.034
B	100	91	6.008
R	100	89	5.031

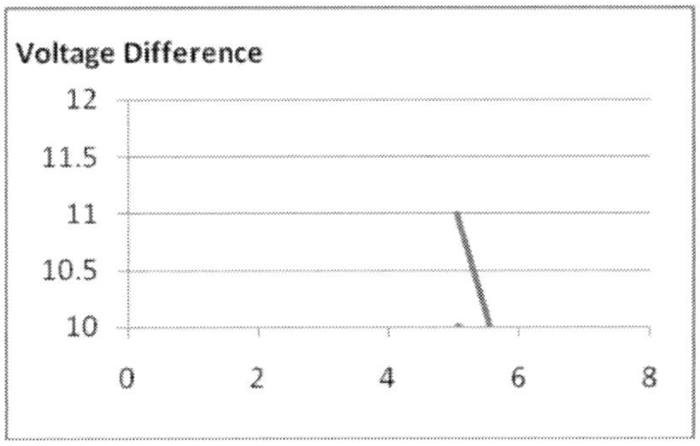

Figure 8. Voltage Difference Vs Time

5. Upgradationofelectromechanical relays

For upgradation of protection system, the main equipment to replace is the relays with the new state of the art relays. Other equipment like CT's,PT's, wiring and other allied equipment will be the same if functioning properly.

For upgradation process the most important work is the relay setting and interfacing which includes the old wiring of CT, PT, alarm and breaker trip wires connected to the old relay [4].Before the startup of this activity the wiring of the relays should be carefully checked with drawings because in many cases wiring circuit has been modified and not updated in the drawings. Before the dismantling, properly ferrules/tag the wires again for easy handling of work at later stages.

New Power plants i.e. Ghazi Barotha having a capacity of 1450MW is also using new state of the art protection system which comprises of numerical relays. Numerical relay are also for protecting the equipment used in transmission and distribution power system. These are multifunction relays having function of protection, metering, control. These relay can have feature of recording the faults and alarms for troubleshooting the fault easily and effectively. New numerical relays have the feature of multipurpose protection. i.e. one relay can monitor all the generator protection. But still it is recommendable to use two numerical relays for the generator protection of onegenerating unit. And both these relays must have a redundant processor and power supply which will increase the protection level to a very high level.

Electromechanical relays can be replaced with the numerical relays after setting, programming and testing of the parameters of new relay. The new relays require less space and wiring as compared to the older one as only one relay can perform all the protection functions.

6. Conclusion

The tested relay of Tarbela power station didn't show linear and stable response. The protection system installed in the power plant is no more reliable and efficient. Because the relay installed are of electromechanical type which were manufactured in the 70's. The burden of these electromechanical relays is very high thus causing extra burden on the current and potential transformer.

These relays now need periodic testing and calibration because with the passage of the time the reliability of these relays has been drastically decreased. And too many tests on these relays also effects the performance of these relays. These relays are composed of mechanical switches, springs, linkages which are prone to be weaken. In several occasion these relay generated some internal fault in it and cause a false tripping and thus cause a generation loss to the government.

Moreover these relays provide only a single function. i.e. one relay for one protection while new state of the art one numerical relay can provide all the functions of the generator protection[5]. New numerical relay response is very fast and precise as compared to the mechanical one. And it is more economical to use one relay instead of many[6]. So it is highly recommended to replace these electromechanical relays with the new numerical relays. As numerical relays contains many new features like self supervision and recording of disturbance in the power system.

References

1. Sykes, J.; Madani, V.; Burger, J.; Adamiak, M.; Premerlani, W., "Reliabilty of protection systems (what are the real concerns)," in Protective Relay Engineers, 2010 63rd Annual Conference for , vol., no., pp.1-16, March 29 2010-April 1 2010.
2. Tian; Qi-Wei Hu; Yong-Sheng Bai; Yan Tian, "Notice of Retraction Reliability analyses of digital protection relay system," in Quality, Reliability, Risk, Maintenance, and Safety Engineering (QR2MSE), 2013 International Conference on , vol., no., pp.1308-1311, 15-18 July 2013.
3. Wright, S.A.; Kangala, A.; Sahni, M.; Lee, W.J.; Kenarangui, R.; Threlkeld, H.F.; Martinez, M.A., "Benefits of upgrading protection schemes for hydroelectric power plants," in Industrial and Commercial Power Systems Technical Conference, 2005 IEEE , vol., no., pp.218-225, 8-12 May 2005.
4. Sperl, J.; Carper, B.; Coe, W., "Upgrading your protective relays — when theory meets reality," in Protective Relay Engineers, 2010 63rd Annual Conference for , vol., no., pp.1-8, March 29 2010-April 1 2010.
5. Bogdanov, D., "Redundant design of relay protection system for power plant unit with synchronous generator," in Electrical Apparatus and Technologies (SIELA), 2014 18th International Symposium on , vol., no., pp.1-4, 29-31 May 2014.
6. Novak, J.J.; Kirby, R.D., "Better, faster, and more economical integrated protective relaying and control using digital bits and logic," in Petroleum and Chemical Industry Technical Conference, 2008. PCIC 2008. 55th IEEE , vol., no., pp.1-12, 22-24 Sept. 2008

MATEC Web of Conferences **54**, 04004 (2016) DOI: 10.1051/matecconf/20165404004
MIMT 2016

Generalized Equations for Potential Water Savings from Rainwater Tanks in Adelaide under Different Climates

Upendra Paudel and Monzur A. Imteaz

Department of Civil & Construction Engineering, Swinburne University of Technology, Melbourne, VIC 3122, Australia.

Abstract. With ever-increasing water demand, authorities around the world are considering different options including water recycling and rainwater harvesting to minimize potable water demand. Among all the alternative options, rainwater harvesting system is the most feasible to minimize potable water demand, This paper presents development of generalised equations for the quantifications of potential water savings under different climatic conditions for an Australian city, Adelaide. An earlier developed daily water balance model, eTank, which can calculate potential water savings in three climatic conditions (dry, average and wet) was used for this purpose. Several relationship graphs of water savings were produced through model calculations for different input parameters, i.e. roof area, tank volume and rainwater demand for each climatic condition. From the produced graphs, three (one for each climatic condition) generalised equations were developed, where water savings were presented as a function of roof area, demand and tank volume. Results from the developed equations were compared with model calculated results under different conditions and it is found that results from the generalised equations are very close to the model calculated results. Such equations are expected to be very helpful for general end users and will encourage them for implementing rainwater harvesting with prior knowledge.

1 Introduction

Water shortage is a major problem in many parts of the world. Regions which depend on groundwater, due to over-extraction ground water table dropping down significantly. Moreover, due to the impacts of climate change some parts of the world are experiencing less amount of rainfall than average rainfall. As such it is imperative for the scientist, water supply engineers, policy makers and government authorities to think on the reliability and sustainability of the town water supply which is mostly rely on ground/surface water. To reduce the over extraction of ground water and cope with climate change, different stakeholders are emphasizing to use stormwater harvesting systems as one of the alternative options. Among all the stormwater harvesting systems, rainwater harvesting system is getting much more attention because it is easy to harvest and needs lesser or minimal treatment for laundry, toilet flushing and outdoor uses.

There have been several studies around the globe quantifying potential rainwater collection and potable water savings. Vaes and Berlamont [1] developed a model to determine the effectiveness of rainwater tanks and stormwater runoff using long term historical rainfall data. Coombes and Kuczera [2] found that for an individual building with a 150 m2 roof area and 1-5 kL tank in Sydney can yield 10-58% mains water savings (depending on the number of people using the building). According to Coombes and Kuczera [2], depending on roof area and number of occupants, rainwater tank use can result in mains water annual savings of 18-55kL for 1kL sized tanks and 25-144kL for 10kL sized tanks. In Sweden, Villarreal and Dixon [3] investigated water savings potential of stormwater harvesting systems from roof areas. Villarreal and Dixon [3] noted that a mains water saving of 30% can be achieved using a 40m3 sized tank (toilet and washing machine use only). Ghisi et al. [4-5] investigated the water savings potential from rainwater harvesting systems in Brazil and found that average potential for potable water savings to be 12-79% per year for the cities analysed.

Khastagir and Jayasuriya [6] using historical daily rainfall data analysed reliability of rainwater tanks and presented contours of optimum tank sizes for surrounding areas of Melbourne for a supply reliability of 90%. Muthukumarran et al. [7] found that use of rainwater inside a home in regional Victoria (Australia) can save up to 40% of potable water use. Farreny et al. [8] examined the quantity and quality of rainwater harvesting in Spain and found that sloping smooth roofs may harvest up to about 50% more rainwater than flat rough roofs. Mun and Han [9] developed a design and evaluation method for a rainwater harvesting system on the basis of water balance equation and found that a design based on sensitivity analysis and proper management of a rainwater harvesting system should be emphasized to improve the operation efficiency. Some researchers (Aladenola and

DOI: 10.1051/matecconf/20165404004

Adeboye) [10] used monthly rainfall data for the analysis of rainwater savings potentials and rainwater tank design. However, Imteaz et al. [11] through a case study in south-west Nigeria using a daily water balance model, has shown that analysis using monthly rainfall data greatly overestimates the required tank size.

Berwanger and Ghisi [12] conducted feasibility analysis for a city in Brazil and commented that rainwater tank will be feasible for only selective cases depending on water demand and roof area. Jung et al. [13] conducted economic feasibility of rainwater tanks for seven major cities of South Korea considering continuous supply of rainwater demand and concluded that to be able to achieve continuous supply the required rainwater tank size is not economically feasible. Due to this fact often a smaller tank is used which requires augmented supply from townwater supply or other sources.

Most of the studies used continuous simulations of historical daily data for a long period (depending on data availability) and eventually making an average of cumulative historical savings (or other model variables). Through such analysis of averaged variables/parameters, rainwater tank users may not get an adequate insight of the expected realistic situation(s) in regards to variability of outcomes as a particular year might have an unusual rainfall pattern compared to usual pattern of occurrences (i.e. sporadic bursts and/or longer dry periods). With the impacts of climate change, such ranges of realistic outcomes are expected to be widening further. To overcome this issue, Imteaz et al. [14] developed a daily water balance model (eTank) for the analysis of rainwater tank outcomes under three different climatic conditions (i.e. dry, average and wet). Through the statistical analysis of historical rainfall data, dry year is defined as annual rainfall value closer to 10th percentile value whereas 50th value is average year and annual rainfall value closer to 90th percentile value is defined as a wet year. Imteaz et al. [15] introduced a factor, *Rainwater Accumulation Potential (RAP), which is* a ratio of roof area and water demand; and using eTank presented relationships of *RAP* with rainwater reliability under different climatic conditions for Melbourne city. Incorporating another factor to earlier proposed RAP, this paper presents a modified form of *Rainwater Accumulation Potential (RAP),* which is the ratio of water demand to roof area multiplied by tank volume and shows relationships of modified RAP with annual savings under different climatic conditions for Adelaide. Eventually, established relationships are presented in the form of generalized equations having only 'roof area' and RAP as independent variables.

2 Methodology

The daily water balance model proposed by Imteaz et al. [16] was the basis of eTank development. The model considers daily rainfall, contributing catchment (roof) area, losses (due to leakage, spillage and evaporation), storage (tank) volume and water demand for calculating rainwater tank outcomes; annual rainwater savings, annual overflow, annual townwater use and reliability. In the model, the primary input value is the daily rainfall amount for three differeent years (dry, average and wet years). The daily runoff volume is calculated from daily rainfall amount by multiplying the rainfall amount with the contributing roof area after deducting the losses. For this study, from the produced runoff, a 10% deduction was applied to account for several losses (first flush, leakage, spilling and evaporation). Generated runoff is diverted to the connected available storage tank. Available storage capacity is compared with the accumulated daily runoff. If the accumulated runoff is bigger than available storage volume, excess water (overflow) is deducted from the accumulated runoff. Amount of water use(s) is deducted from the daily accumulated/stored runoff amount, if sufficient amount of water is available in the storage. In a situation, when sufficient amount of water is not available in the storage, model assumes that the remaining water demand is met from the town water supply. The detailed mathematical procedures, formulations and logical sequences are outlined in Imteaz et al. [17].

The model was used to produce series of curves showing expected annual water savings for different input conditions (roof area, demand, tank size and climate condition). From the generated curves, a set of equations for different roof areas and a particular climate condition were derived using best-fit technique. Best-fit equations for a particular condition was selected with the aim of achieving further relationships among the equations' co-efficients and exponents. Eventually, 3 generalised equations were derived each for a particular climate condition. These derived equations are presented in the 'Results' section. As a particular dry/average/wet year may have unusual rainfall patterns (over the year), this study used 5 years' rainfall data for each of the dry, average and wet conditions.

3 Data

For the current study, a raingauge station in Adelaide Airport was selected. Daily rainfall data for the station was collected from the Bureau of Meteorology website (http://reg.bom.gov.au/climate/data/). For the station daily rainfall data of 57 years (1956 to 2012) were available. Through statistical analysis of total annual rainfall data, three separate years (1965, 1988 and 1963) were selected as dry year, average year and wet year. For the selection of 5 years' data, for each of the conditions four additional years were selected in a way that out of these four years, two years are having annual rainfalls immediately higher and the other two years are having annual rainfalls immediately lower than the rainfall amount of above selected years. Selected years and corresponding annual rainfall amounts are shown in Table 1.

4 Results

Daily water balance model was simulated with the daily rainfall data for the above-mentioned years to evaluate the annual water savings for different tank sizes (2500L,

5000L, 7500L and 10,000L) with roof area ranges from 100~300 m^2 and demand from 200 to 500 L/day. Then graphs of annual water savings with *RAP* were drawn for different roof areas and climatic conditions (dry, average and wet).

4.1 Generalized equations

Figure 1 shows annual water savings as a function of RAP for average year having roof area of 200 m^2.

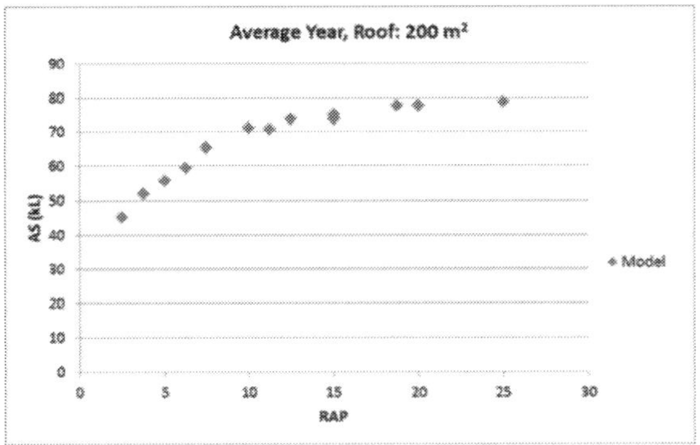

Figure 1. Model calculated annual water savings for different RAP values in an average year having 200 m^2 roof areas.

Curve on the Figure 1 can be expressed as a single equation using best-fit technique. Using best-fit technique, five equations each for a particular roof area for the selected average year were derived. The equations are as follow:

Roof Area 100 m^2: $y = 35.245 \times (RAP)^{0.0324}$ (1)

Roof area 150 m^2 : $y = 38.444 \times (RAP)^{0.1407}$ (2)

Roof Area 200 m^2: $y = 37.824 \times (RAP)^{0.2486}$ (3)

Roof area 250 m^2: $y = 39.956 \times (RAP)^{0.3}$ (4)

Roof area 300 m^2 : $y = 42.857 \times (RAP)^{0.3262}$ (5)

Where 'RAP' is the Rainwater Accumulation Potential calculated as:

$$RAP = (D \times V)/RA$$

Where, 'D' is the rainwater demand in Litre (L), 'V' is the tank volume in cubic metre (m^3) and 'RA' is the roof area in Square metre (m^2). Coefficients (35.245, 38.444, 37.824, 39.956 and 42.857) and exponents (0.0324, 0.1407, 0.2486, 0.3 and 0.3262) of the above set of equations can be correlated with the associated roof areas. Using best-fit technique, above-mentioned coefficients can be expressed as Equation 6 and above-mentioned exponents can be expressed as Equation 7:

$$5.97 \times \ln(RA) + 7.66 \quad (6)$$

$$0.28 \times \ln(RA) - 1.25 \quad (7)$$

Table 1. Selected rainfalls and corresponding years

Climate	Year	Annual Rainfall (mm)
Dry	2008	292
	2002	323
	1965	326.2
	1994	326.6
	1961	328.5
Average	1995	420.4
	2004	426.3
	1988	441.6
	2011	444.2
	1998	450.6
Wet	1960	538.7
	1968	570.7
	1963	573.3
	1983	576.2
	1974	586.3

Applying above-mentioned procedure, five equations can be combined into a single equation (Equation 8) having independent variable of 'roof area' and *RAP*. So, annual water savings equation for average year:

$$WS = (5.97 \times \ln(RA) + 7.66) \times (RAP)^{(0.28 \times \ln(RA) - 1.25)} \quad (8)$$

Where, WS is the annual water savings in Kilolitres (kL).

Similarly, following the same procedure for the other climatic conditions, two more equations (Equation 9 for dry year and Equation 10 for wet year) for annual water savings were developed:

For dry year:

$$WS = (0.09 \times \ln(RA) - 9.23) \times \ln(RAP) + (6.01 \times \ln(RA) + 0.84) \quad (9)$$

For wet year,

$$WS = (24 \times \ln(RA) - 105.25) \times \ln(RAP) + (-7.15 \times \ln(RA) + 64.89) \quad (10)$$

4.2 Comparison of results and discussion

Figure 2 shows the comparisons of model calculated annual water savings values with the developed equation results for dry, average and wet years having roof area of 150 m^2. Similarly, figure 3 shows the comparisons of model calculated annual savings results with derived equation produced results having roof area of 250 m^2 for dry, average and wet years. The comparisons show the equation developed annual savings results are very close to the results produced by the model (eTank). Only for the roof area of 250 m^2 having *RAP* values of 12.0 and 7.5 for wet year show small deviations from the model produced results. These deviations are not significant as anyway in the wet years there will be enough water to fulfil non-potable water demands. These developed equations are valid for roof areas ranging from 100 to 300

m^2, tank sizes 2500L to 10,000L and daily water demands 200L to 500L per day. For other scenario, these equations may not be applicable, although these are the usual scenario practiced/implemented in Australia.

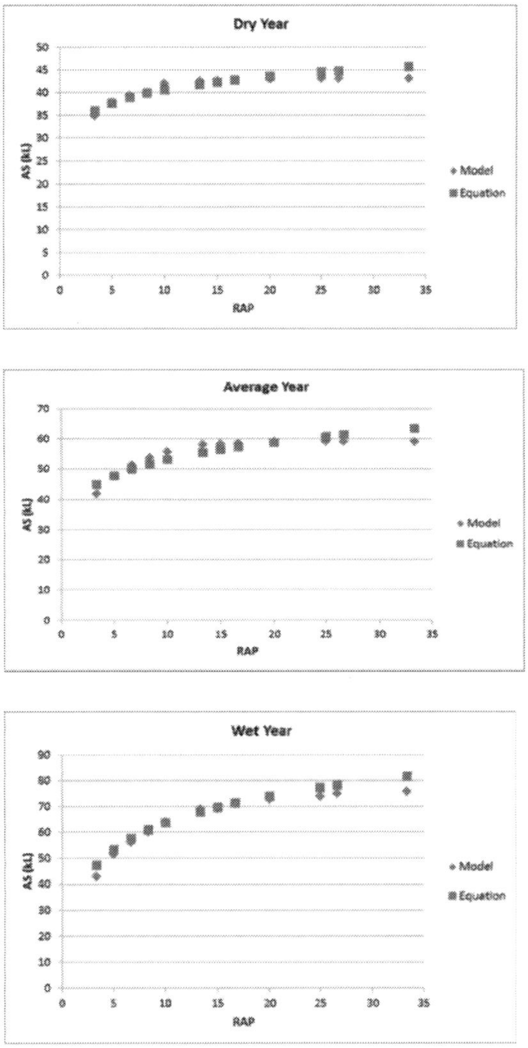

Figure 2. Comparison of model calculated annual water savings with derived equation results in dry, average and wet year for roof areas 150 m^2.

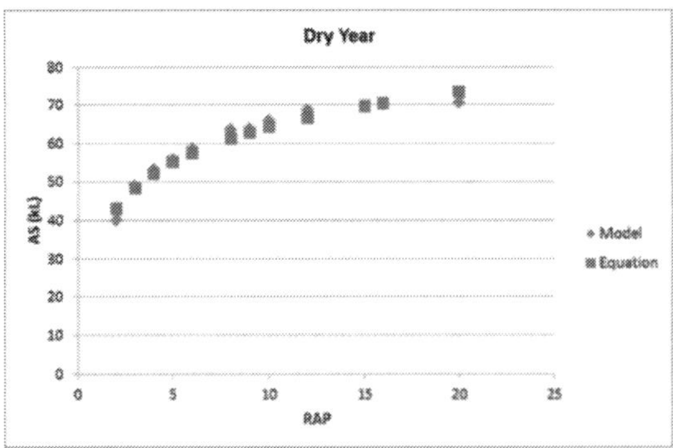

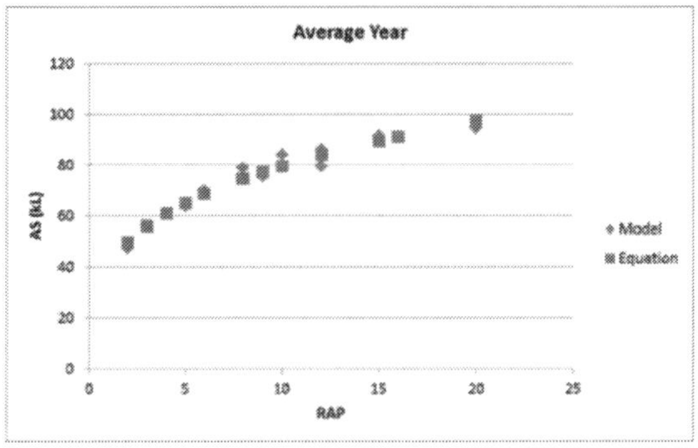

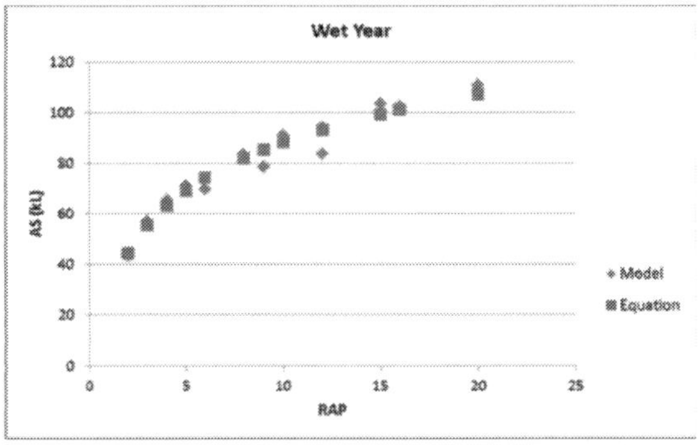

Figure 3. Comparison of model calculated annual water savings with derived equation results in dry, average and wet year for roof areas 250 m^2.

5 Conclusions

Most of the countries around the world are under severe water stress. Impacts of climate change caused to worsen the situation. Water authorities in those countries struggling to maintain ever-increasing potable water demand and considering different alternatives to minimize potable water demand. For the countries having moderate to high rainfall amounts, among all the alternative options rainwater harvesting is most feasible. However, proper quantification of potential water savings are often misleading. Many studies considered monthly rainfall data for the quantification of water savings; however analysis with monthly rainfall data is unable to produce accurate quantifications, as with the monthly data it is not possible to consider overflow loss, which is likely to happen after consecutive decent rainy days. It is recommended that at least daily (subject to availability) rainfall data must be used for proper quantification of potential water savings. Nonetheless, some researchers used sub-daily (even minute) time-step data. Some other researchers used complex tools/algorithms for such analysis. General end users do not get much benefit through such complex analysis. Moreover, almost all of those complex analysis provide a single answer in regards to potential water savings, which is not realistic. In reality, a range of water savings are expected depending on climatic conditions. eTank is a great tool, which provides expected rainwater savings under three climatic conditions (dry, average and wet). Still many end users will struggle to select and upload daily rainfall data from the website provided by the authority. Developed generalised equations are very helpful for such users. General users can easily calculate water savings with the help of a calculator or spreadsheet using these equations. Even these equations can be easily converted to 'mobile app', which will promote uses of rainwater tanks through prior knowledge on potential water savings. It is to be noted that the developed generalized equations are only valid for a particular location (in this instance Adelaide). For other locations, applying the similar procedure different sets of equations can be developed. Generalizing three different climatic equations into single equation can be target for future research.

References

1. G. Vaes, J. Berlamont, Urban Water J, **3**, 303-307 (2001)
2. P. Coombes, G. Kuczera, *28th International Hydrology and Water Resources Symposium*, Wollongong, NSW, The Institution of Engineers, Australia, **2**, 235-242 (2003)
3. E.L. Villarreal, A. Dixon, Build Environ, **40**, 1174-1184 (2005)
4. E. Ghisi, D.L. Bressan, M. Martini, Build Environ, **42**,1654-1666 (2007)
5. E. Ghisi, D.F. Tavares, V.L. Rocha, Resour Conserv and Recy, 54, 79-85 (2009)
6. A. Khastagir, N. Jayasuriya, J Hydrol, **381**, 181-188 (2010)
7. S. Muthukumaran, K. Baskaran, N. Sexton, Resour Conserv and Recy, **55**, 945-952 (2011)
8. R. Farreny, T. Morales-Pinzon, A. Guisasola, C. Taya, J. Rieradevall, X. Gabarrell, Water Res, **45**, 3245-3254 (2011)
9. J.S. Mun, M.Y. Han, J Environ Manage, **93**, 147-153 (2012)
10. O.O. Aladenola, O.B. Adeboye, Water Resour Manag, **24**, 2129–2137 (2010)
11. M.A. Imteaz, O. Adeboye, S. Rayburg, A. Shanableh, Resour Conserv and Recy, **62**, 51-55 (2012)
12. H. Berwanger, E. Ghisi, IJSHD, **2**(3), 104-114. (2014)
13. K. Jung, T. Lee, B.G. Choi, S. Hong, Water Resour Manag, **29**(3), (2014)
14. M.A. Imteaz, R. Karki, A. Shamseldin, and C. Matos, IJCAET (article in press).
15. M.A. Imteaz, A. Ahsan, A. Shanableh, Resour Conserv and Recy, **77**, 37-43 (2013)
16. M.A. Imteaz, A. Rauf, A. Aziz, *International Congress on Modelling and Simulation MODSIM*, Perth. December. 3300-3306 (2011)
17. M.A. Imteaz, A. Shanableh, A. Rahman, A. Ahsan, Resour Conserv and Recy, **55**, 1022-1029 (2011)

Unlock Screen Application Design Using Face Expression on Android Smartphone

Rhio Sutoyo, Jeklin Harefa, Alexander, Andry Chowanda

Bina Nusantara University, Jakarta, Indonesia

Abstract. Nowadays, smartphone has been shifted as a primary needs for majority of people. This trend is making people do activities on their smartphone such as exchange information, playing games, paying bills, discussing ideas, etc. In almost every smartphone, security features (i.e. password, pin, face unlock) are already built in which are intended to protect the data inside the smartphone. Nevertheless, those security features are not perfect and have flaws either technically or by human errors. This research aim to provide an application that uses people' facial expressions to improve the security of smartphone, namely Android. We implement face detection and expression recognition in this application as an improvement from the existing unlock screen, which is face unlock. The results that are achieved from this research is an unlock screen application on Android devices that utilizes people' facial expression changes to open their smartphone. This application can eliminates the technical errors from existing face unlock screen application which can be deceived by photos.

Keywords: unlock screen, face expression recognition, biometric.

1 Introduction

The growing of globalization era makes smartphone as a part of today's life for most of the people. Not only adults, but young and elder people also generally have smartphone. One of the growing smartphone in the world is Android. Based on the pie chart in the Fig. 1 below, it can be seen that Android users was increasing significantly, starting in 2009 that only about 4% from market share, to 67% in 2012. In 2015, it is predicted that Android users are still on the first rank of the market share.

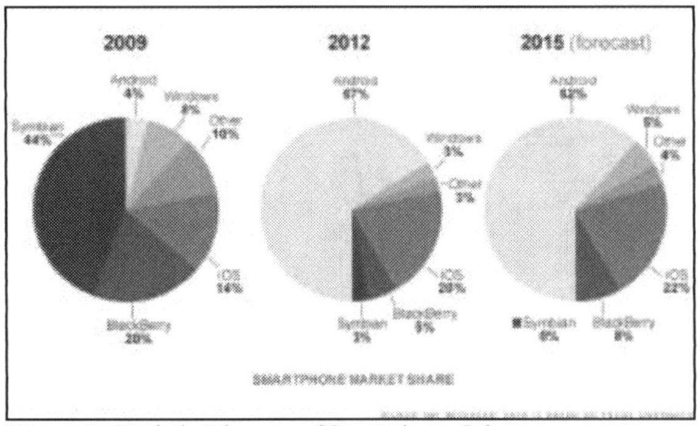

Figure 1. Statistic Diagram of Smartphone Sales

Based on that statistic, there is no wonder if many mobile vendors would prefer Android smartphone. Furthermore, the features in Android are growing as the increasing of Android demand and production. One of the feature is the security, which was originally just "a pin", now it has become more varied and sophisticated. For instance, pattern unlock, face unlock, and face and voice unlock. One of the most secure security system is a system that using biometric authentication [1]. Biometric is the study of the measurable biological characteristic. Biometric has a uniqueness of each individual. Android has implemented biometric on its face recognition lock. Face recognition has become an important topic in many application, such as security system, credit card verification, and identifying criminals. For example, face recognition is able to compare the face from the face database so it can assist in identifying criminals [2]. However, there is a weakness on that system which is if the users show photos of their face, the security system will detect that photos as the real user's face and the smartphone will be automatically opened. In conclusion, this flaw opens up strangers using users' photos to open their smartphone. Considering the weaknesses in the face recognition system, we believe we need a new kind of security system which can ensure that the faces shown in the system are not photos, but the real face of smartphone owners. In this work, we are building applications to eliminates those flaws by creating an unlock screen application with computer vision technology which uses people' face expression.

2 System Overview

Face recognition is a technology of artificial intelligence which allows computer to recognize people' face. It starts by capturing face photos through the camera. Then, the images will be processed so that face can be identified accurately by computer [3]. To the best of our knowledge, there are no face recognition tools that are perfectly able to distinguish between real faces and photos. As we have mentioned before, the goal of this study is to create a mechanism to overcome the inability of computer to recognize the real faces and photos by using facial expression changes. After successfully recognizes the face and the facial expression changes, the locked smartphone will be opened. To facilitate understanding of the process, a system overview will be presented in the following flow chart. There are two major processes, which are *Face Recognition Training* and *Unlock Screen*.

2.1 Face Recognition Training

The first major process is face recognition training. The purpose of this step is to acquire the data of users' face photos, process them, and stored them which will later be used for the unlock screen process. The overall system can be seen in the Fig. 2 below:

Figure 2. Face Recognition Training System

a. Face Detection

At the first step, face detection is done by using the front camera in the smartphone. The users must show their faces in the front of the camera and then the application will detect the face. In this process, we use the Viola Jones algorithm. For the classifier, we use data stored in XML files to decide how to classify each image location.

Figure 3. Users is Locating Their Faces in the Circle

In order to improve the accuracy, users are asked to locate their face in the shown circle on their smartphone (Fig. 3). Once the face is detected, the camera will take a few pictures of the face. Furthermore, the images will be cropped on the face area and the results will be resized to a size of 200 x 200.

b. Image Processing

Before we process the captured photos into the training step, they need to be treated first to decrease the training error rates.

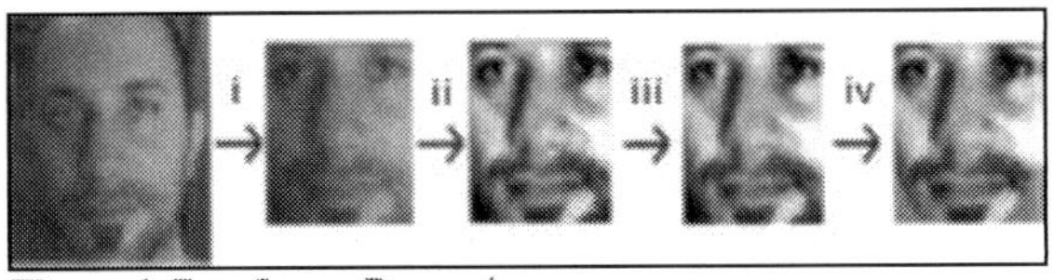

Figure 4. Face Image Processing

Therefore, the face images will be processed in the following process [4]:

i. Geometrical Transformation and cropping.
   This process would include scaling, rotating, and translating the face image so that the eyes are aligned, followed by the removal, chin, ears and background from the face image.
ii. Separate histogram equalization for left and right sides
   This process standardizes the brightness and contrast on both left and right sides of the face independently.
iii. Smoothing
   This process reduces the image noise using a bilateral filter.
iv. Elliptical mask
   The elliptical mask removes some remaining hair and background from the face image.

The example result of this process can be seen in the Fig. 4 where each photo is different and better than the previous photo.

c. Face Recognition Training

The processed photos from the image processing step will be used for data training in the current step. We conducted several test on various algorithms for the face recognition training such as: Eigenface, Fisherface, and LBPH (Local Binary Pattern Histogram). The result of this training process is stored in the form of an XML files which will be used later for identifying faces. For this step, we implemented the LBPH (Local Binary Pattern Histogram) algorithm because it has the best value of accuracy and FPS based on our test. The comparison of those algorithms will be discussed in the research result and discussion section.

2.2 Unlock Screen

Unlock screen is the second major process in the application. The purpose of this step is to compare the captured photos from the smartphone with the data training which is resulted from the face recognition training. If the photos are matched, the process will continue to the face expression detection. In that step, the users' face will be locked and they have to change their expression from natural expression to smile expression. The smartphone will be unlocked when those criteria are completed. The overall system of unlock screen can be seen in the Fig. 5 below:

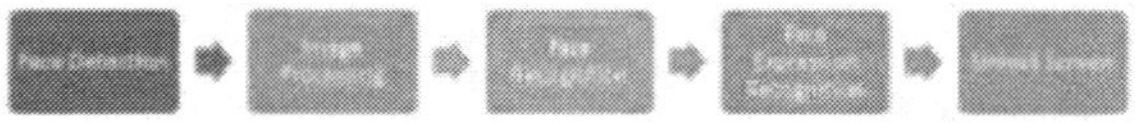

Figure 5. Application Framework when Opening the Screen

a. Face Detection and Image Processing

Both of these steps are similar with the processes in the face recognition training. These steps are done again in the unlock screen step to detect the users' face and process the captured photos before the system continue to the next step.

b. Face Recognition

The function of this step is to determine whether the captured and processed photos from the previous step is matched with the photos in the data training or not. The prediction will be determined by the number value of accuracy from the algorithm that we have implemented, which are LBPH (Local Binary Pattern Histogram).

c. Face Expression Recognition

Facial expression recognition consists of three major steps, including: pre-processing, feature extraction, and classification. The pre-processing step relates to face localization, tracking, and registration to remove the variability due to changes in head pose and illumination. The feature extraction aims to convert pixel data into a higher-level representation of shape, motion, color, and texture, which minimizes within class variations while also maximizing between class variations. The classification step aims to detect the action unit in the face [5]. In this work, we perform the face expression recognition after the face recognition step is succeed. The process is done by analyzing the facial point of users' face. When the expression recognition process is performed and the users change their expression (i.e. from natural to smile), the unlock screen function will be called. Furthermore, in this step the users' face will be locked after the face recognition step. Thus, if the face detection and face recognition step break (i.e. the users move their faces from the circle shown in the Fig. 3), the process will be back to face detection and face recognition again. We implement this method in order to avoid cheating (e.g. swapping photos of the users from natural expression to smile expression). In this step, we also have compared several algorithms such as Viola Jones, ASM and SVM, and Principal Component Analysis (PCA) in order to find the best algorithm for our application. Based on our tests, we choose the PCA algorithm. Our test results will be discussed in the research result and discussion section.

d. Unlock Screen

This function will be called when all of the constraints from the steps above are fulfilled. When this function is called, the system will unlock the phone screen and the users are able to use their phone normally.

3 Research Result and Discussion

We performed our tests with Samsung I9300 Galaxy S III that have Android OS, version 4.1 (Jelly Bean) installed. Moreover, it has Quad-core 1.4 GHz Cortex-A9 processor and have 1GB of RAM. The front camera specification is 1.9 MP, 720p@30fps. The resolution is 720 x 1280 pixels, 4.8 inches (~306 ppi pixel density) and it contains 5 MB of storage.

3.1 Face Detection

In this research, we apply Viola-Jones algorithm to detect the face in camera. There are five cascade classifiers that are tested, and one of them that most optimal both in speed and accuracy will be chosen. The results of each cascade is as follow:

Table 1. Comparison of Classifier Based on FPS and Accuracy

Classifier Name	FPS Range	Accuracy
Haarcascade_frontalface_alt_tree	~2-3 fps	96.296%
Haarcascade_frontalface_alt	~2-4 fps	96.296%
Haarcascade_frontalface_alt2	~3-4 fps	100%
Haarcascade_frontalface_default	~3-4 fps	100%
Lbpcascade_frontalface	~4-5 fps	100%

** Higher frame per second (FPS) means faster speed*

Based on the table above it can be seen that the best classifier to perform face detection is *lbpcascade_frontalface*, which in this classifier, obtained the highest FPS with good accuracy rate when performing face detection.

3. 2 Face Recognition

There are three algorithms that we have tested, namely Eigenface, Fisherface, and LBPH (Local Binary Pattern Histogram). This process begin with an image training of the users' face. The output of this training is XML files that are used to identify the owner of the face. Once the training is done, the face image that want to be recognize will be tested and will gives prediction value as the output of the test. The prediction value will be used for determine the threshold value that serves as determinants whether the face recognized or not.

Table 2. Eigenface, Fisherface, and LBPH Algorithm Test Results

Comparison Subject	*Eigenface*	*Fisherface*	*LBPH*
Value prediction when testing with the same face	4633.81	318.59	29.32
Smallest value prediction when testing with different faces	2004.2	61.42	71.88
Biggest value prediction when testing with different faces	8360.78	2805.77	367.5
FPS Range	0.67	1.23	6.58

** Lower prediction value means better face recognition*
** Higher frame per second (FPS) means faster speed*

Based on the Table 2, it can be seen that the best algorithm to recognize face is LBPH algorithm. It obtained the highest FPS and accurate prediction value so that it can perform a better face recognition.

3.3 Face Expression Detection

At facial expression recognition, we compare Viola Jones with ASM (Active Shape Model) and SVM (Support Vector Machine), and also with PCA (Principal Component Analysis). The test conducted by calculating the accuracy and the FPS speed of the camera for each algorithm when the face expression detection is performed. We detail the result in the table below:

Table 3. Face Recognition Technique Test Results

Comparison Subject	Viola Jones	ASM and SVM	PCA
Accuracy	44 %	56.67 %	86.67%
FPS	1.74 – 4.36 fps	4.67 – 6.31 fps	4.13 – 5.57 fps

* *Higher frame per second (FPS) means faster speed*

Based on the tests, the PCA is the best technique for recognizing facial expression. It is because of PCA has the higher speed and the best accuracy among the others to perform facial expression recognition.

4 Related work

4.1 Enhanced Local Texture Feature Sets for Face Recognition under Difficult Lighting Conditions

Tan and Triggs [6] tried to perform face recognition when light conditions are not controlled. They overcome this by combining powerful lighting normalization, local texture-based representation of the face, adjusting the distance transformation, kernel-based feature extraction, and the unification of many features. In their research, Tan and Triggs perform image processing which can eliminate the effects of light as much as possible but still be able to keep the details of the image that will be needed in the process of face recognition. Tan and Triggs also introduce a Local Ternary Patterns (LTP), a development of LBP which can distinguish the pixels with better and quite sensitive to noise in the flat areas. They showed that by changing the ratio of local spatial histogram transformations that have similar distance metrics, it will improve the accuracy of face recognition using LBP or LTP. Furthermore, they also improve their methods by using Kernel Principal Component Analysis (PCA) when extracting features and combines Gabor wavelets with LBP. A combination of Gabor wavelets with LBP produces better accuracy than the Gabor wavelets or LBP alone. The results of their research was tested on a dataset FRGC-204 (Face Recognition Grand Challenge version 2, experiments to-4) and produces an accuracy rate of 88.1% and 0.1% received error rate (False Accept Rate).

4.2 Fully Automatic Recognition of the Temporal Phases of Facial Actions

Valstar and Pantic [7] conducted a study in the facial expression recognition based on Action Unit. Action Unit is the smallest facial movements that can be seen visually. In their research, a video will be processed to find the position of the face in the image. Then after getting the position of the face in the picture, the system will finds 20 facial fiducial points automatically by using Gabor feature based classifier. Those facial points will be monitored to detect the expression changing. Facial movement data points will be classified by Support Vector Machine (SVM) and stored as an Action Unit. Valstar and Pantic has developed three versions to detect the movement of Action Unit, which are version 2005, 2006 and 2007. Valstar and Pantic successfully detect 22 Action Unit and detect automatically in 2007. Furthermore, for each Action Unit detected on the face, there will be a model of transient activation derived from several stages (neutral, onset, apex, offset). Action Unit at some stage is classified by the SVM to determine facial expressions. The results of research conducted by Valstar then tested and resulted in the introduction of Action Unit percentage of 95.3% on the image display facial expressions and by 72% in the image display spontaneous facial expressions.

4.3 A Dynamic Appearance Descriptor Approach to Facial Actions Temporal Modelling

In this study [5,8], there is a new approach to the analysis of the temporal dynamics explicitly produced by facial movements using a descriptor that can describe the dynamic changes in the face which is called Local Phase of Three Orthogonal Planes Quantisation (LPQ-TOP). Until now, most of the facial expression recognition system using only a static descriptor, where if there is a change in the movement of the face and other relevant information is ignored. This journal discusses these limitations by expanding LPQ static descriptor into Three Orthogonal Planes (TOP) which is inspired by a similar expansion in the Local Binary Patterns, combined with Gabor filter on the image histogram, LGBP-TOP [9]. These experiments have shown that the LPQ-TOP is very suitable for the detection problem Action Unit (AU) temporary. Moreover, it also can achieve detection of activation of Action Unit. However, when tested on a dataset GEMEP-FERA, Jiang et al concluded that the head does not pose an obstacle varies need to be resolved.

5 Conclusions and Future Work

In this paper, we have presented a novel approach to unlock phone with facial expression recognition method. Our approach utilizes the best algorithms for face detection, face recognition, and face expression detection based on our tests. These tests are important for the performance of our application because the application cannot detect and recognize the face correctly if the accuracy is bad and the FPS speed is slow. In the face detection tests, we notice that the Haar classifier has good accuracy but is too heavy to be implemented on Android smartphone. On the other hands, LBP classifier has a pretty good FPS even though the accuracy is not as good as Haar classifier. Moreover, LBPH algorithm can be used to solve the problem of face recognition on Android smartphone. It is because the accuracy and FPS of LBPH is good enough to run on the smartphone. We also have implemented PCA algorithm for the face expression recognition by taking a picture of users' mouth and matched it with the existing facial expression. For the future work, we hope other general expressions such as surprised, angry, sullen, etc. can also be covered by the system. The reason for this is because not all people are able to perform good smiles which can easily captured by the system.

References

1. F. Ahmad, A. Najam, & Z. Ahmed, , Image-based Face detection and recognition: "State of the art", IJCSI International Journal of Computer Science Issues, **9**, issue 6, no. 1, 169-172, (2012).
2. G. Dashore, & V.C. Raj, An efficient method for face recognition using principal component analysis. International Journal of Advanced Technnology & Engineering Research, **II,** 2, 23-28, (2012).
3. P. V Saudagare & D. S. Chaudhari, "Facial expression recognition using neural network–An overview," International Journal of Soft Computing and Engineering (IJSCE), **2**, 1, 224-227, (2012).
4. D. L. Baggio, et al., Mastering OpenCV with Practical Computer Vision Projects. Packt Publishing, (2012).
5. B. Jiang, M. F. Valstar, B. Martinez, and M. Pantic, "A dynamic appearance descriptor approach to facial actions temporal modelling," IEEE Transactions of Systems, Man and Cybernetics-Part B, **44**, 2, 161–174, (2011 and 2014).
6. X. Tan and B. Triggs, "Enhanced local texture feature sets for face recognition under difficult lighting conditions," IEEE Transactions on Image Processing, **19**, 1635-1650, (2010).
7. M. F. Valstar and M. Pantic, "Fully automatic recognition of the temporal phases of facial actions," IEEE Transactions on Systems, Man, and Cybernetics, Part B: Cybernetics, **42**, 1, pp. 28-43, (2012).
8. B. Jiang, M. F. Valstar, & M. Pantic, "Action Unit detection using sparse appearance descriptors in space-time video volumes," IEEE Int'l. Conf. Face and Gesture Recognition (FG'11), 314-321, March (2011).
9. T. R. Almaev & M. F. Valstar, "Local gabor binary patterns from three orthogonal planes for automatic facial expression recognition," Proc. Affective Computing and Intelligent Interaction, 356-361, (2013).

MATEC Web of Conferences 54, 05002 (2016) DOI: 10.1051/matecconf/20165405002
MIMT 2016

3D Catalogue based on Augmented Reality in Android Operating System

Natalia Chandra, Fredy Purnomo, Michael Yoseph Ricky, Christianto Leonard, Arvin Verdian, Hendrick

School of Computer Science, Bina Nusantara University, Jakarta, 11480, Indonesia

Abstract. Along with the continued development of Smartphone and the era of globalization, human needs are increasing for technology. They continue to look for ways that can ease their lives and it is up to the explanation of the visualization that closer to the actual visualization and support fast, precise and clear retrieval of interactive information. In order to realize the visualization, augmented reality technology is being used. Augmented reality is a technology that uses camera technology to recognize real world, images, objects, and environments and superimposes virtual information and data onto reality in real time. It combines a virtual object into two or three dimensions in a real three-dimensional environment. By developing an application with augmented reality, it can help user to look more detail into a product and give advantages to company to save the brochure production cost. The development of this application is using client-side and server-side, where the client-side will be run on android smartphone and server-side using a php based web service. The goal of this paper is to develop a 3D catalogue that helps to depict a product in interactive ways with zoom, rotate, choose color, and animation features.

Keywords: Augmented Reality, Android, Catalogue, Marker, Smartphone, 3D Model.

1 Introduction

Many companies and organization nowadays promote their products conventionally such as advertising in the media, posters or brochures, catalog and those cost a lot of money. Augmented reality is a type of computer vision that uses camera technology to recognize real world images, objects, and environments and superimposes virtual information and data onto reality in real time. It turns the smart devices into a lens for adding video, audio, 2-D content, 3-D content and location-based information [1]. Carmigniani describe about the technology that used to do augmented reality, which are computer vision that divided into 2, featured based and model based, AR devices like display, input devices, tracking, and computer, AR interfaces like tangible AR interfaces, collaborative AR interfaces, hybrid AR interfaces, and multimodal AR interfaces, and AR Systems [2]. Schmalstleg, Langlotz, dan Billinghurst describe about augmented reality 2.0 concept, which are the combination of augmented reality 2.0 and web 2.0, where the augmented reality itself is developed in mobile [6].

In the last few years, many developers tried to apply augmented reality technology to give interactive interaction for user in terms of information search, especially in the smartphone technology. Based on International Data Corporation (IDC) shown in fig. 1, android continue to dominate the global smartphone market, with over 255 million units shipped and 85% of the market share in the second quarter of 2014 [3].

By looking the development of both technologies, many companies use those to reduce production cost and increase the product sale. Using augmented reality, allow customer to look the product picture more real so it can represent the product intact. Combining augmented reality with the internet can connect customer and companies anytime and anywhere with Smartphone.

The next section describes about reporting tool requirements. Section III describes the comparison of reporting tools. The detail of architecture is described in section IV. The last section is evaluation and summary.

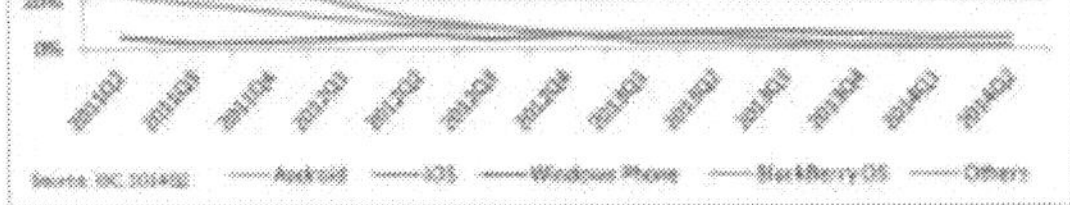

Fig.1: WorldWide Smartphone Market Share

2 Comparison between similar augmented reality application

In this section, we will compare some recent augmented reality application that developed before, which are: Brother AR and Go Sweat Go Ion.

2.1 Brother AR

This is an augmented reality application that published by Brother International Singapore Pte Ltd as an interactive way to promote their product.

The application was claimed as the first 3D mobile augmented reality (AR) application to experience printers and Multi-Function Centre (MFC) at a different level from Brother. This is a free application that can be downloaded from play store. In the application, there is a virtual setup feature that allow user to simulate setting up the printer at home or office and give the idea how the product will fit into a space [4]. Fig. 2 shows the user interface of Brother AR application.

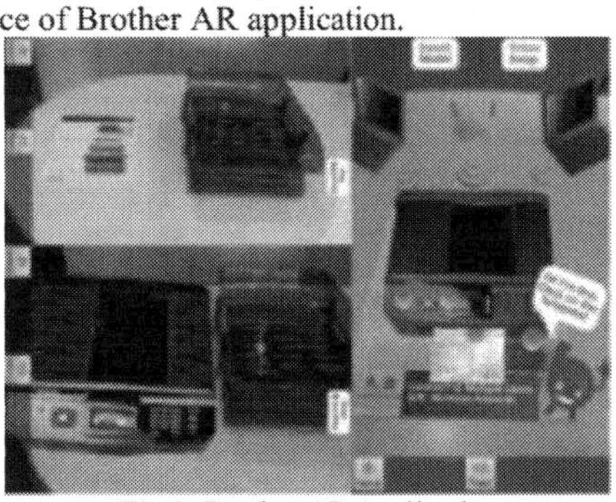

Fig.2: Brother AR Application

2.2 Go Sweat Go Ion.

This is an application that published by Amerta Indah Otsuka Company as a sole proprietor for pocari sweat drinks in Indonesia. This application is made for the purpose of giving education to society of how important exercise and consume their drink.

Besides that, there are a lot contents inside that require the user to keep moving, such as Capturion game that allow user to catch the ions using hand [5]. Fig.3 shows the user interface of Go Sweat Go Ion application.

Fig. 3: Go Sweat Go Ion Application

3 Augmented reality 3D catalogue detail and architecture

The application will be developed on client-side and server-side where the client-side will be run on Smartphone and the server-side will be run on a web service that connect the admin and database server. All the features are developed to synchronize the data between the client-side to the database so that the application can be run offline and the feature will run online to download the data from database server.

In this section, there are five features that will be explained which are views the 3D model, choose the colour, zoom, rotate, and run the 3D model.

3.1 View 3D Model

In this feature, user can view the 3D model by scan the marker through their smartphone camera. Fig. 4 shows the detail architecture on how this feature works:

a. User opens the application and point the smartphone camera to the 2D marker.
b. The 2D marker will be search by the sensor ID in database.
c. If the 2D marker detected by the server, the system will get the 3D object and texture. The user can view the 3D model through their smartphone.

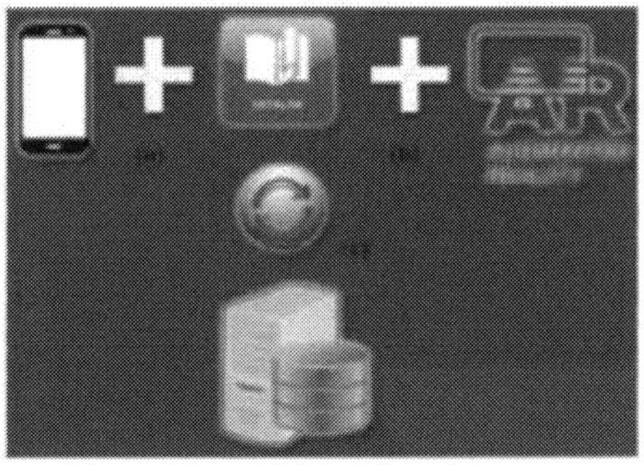

Fig. 4: View 3D Model Architecture

Fig. 5 shows the 3D model in smartphone view after user scan the 2D marker. There is guidance for user on how to use the features.

Fig. 5: View 3D Model in Smartphone View

3.2 Choose The Colour of 3D Model

In this feature, users are able to look available different colours of a product. The architecture of this feature is start when user able to look the 3D model on the screen like shown in fig.5, the system will get the 3D object and all the textures based on the product. The detail architecture shows in table 1, the alt-step explain about when the product only have one colour, then system only show the available colour to user.

Table 1: Change Colour Detail Architecture

Actor Action		*System Response*	
Step-1	User press Prev and Next button.	**Step-2**	Get the 3D texture based on the product ID.
		Step-3	Put the texture to 3D model and show the 3D model to user.
Alt step-2		If there is no texture on that product, then the 3D model will not change colour.	

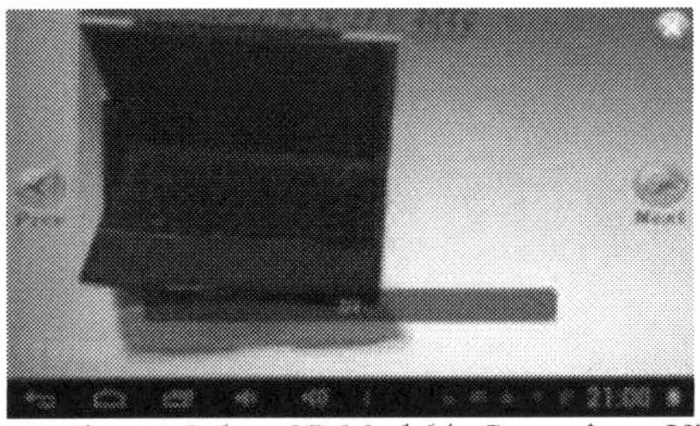

Fig. 6: Choose Colour 3D Model in Smartphone View

Fig. 6 shows the user interface in smartphone view. There are two buttons which are prev and next, the users are able to choose the colour of a product by click the prev and next button in the smartphone screen.

3.3 Zoom the 3D Model

In this feature, users are able to look the product in different sizes by pinch action in smartphone screen. The detail architecture shows in table 2.

Table 2: Zoom 3D Model Detail Architecture

Actor Action		*System Response*	
Step-1	Users do the pinch in or pinch out to the screen.	**Step-2**	Do the zoom in or zoom out and show it to the 3D model.

Fig. 7: Zoom the 3D Model in Smartphone View (Pinch Out)

Fig. 8: Zoom the 3D Model in Smartphone View (Pinch in)

Fig. 7 shows the user interface how the 3D object is zooming out by pinch out and Fig. 8 shows the user interface how the 3D object is zooming in by pinch in, in the smartphone view.

3.4 Rotate the 3D Model

In this feature, users are able to rotate the 3D model with swipe action to smartphone screen. The rotation process will focus on the X and Y axis. The process will be:

a. If user swipe up the 3D model, then the system will add the X-axis (x++) like shows in fig. 9.

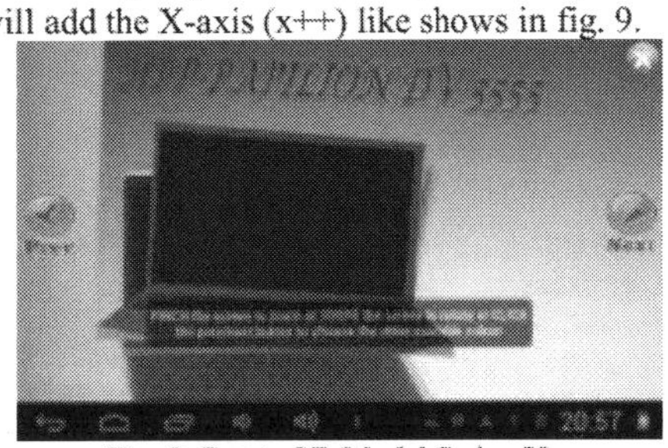

Fig. 9: Rotate 3D Model Swipe Up

b. If user swipe down the 3D model, the system will minus the X-axis (x--) like shows in fig. 10.

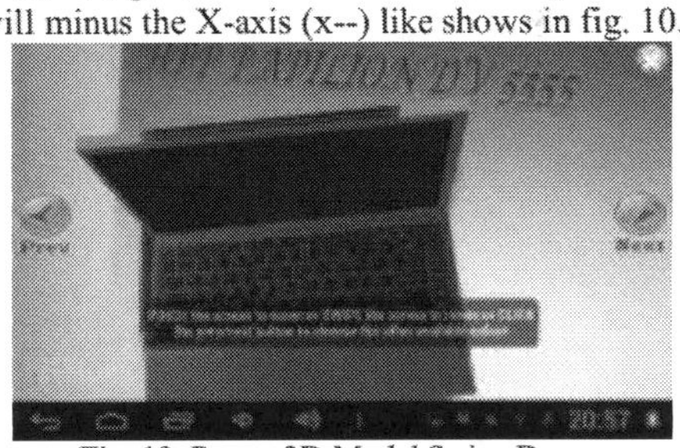

Fig. 10: Rotate 3D Model Swipe Down

c. If user swipe left the 3D model, the system will add the Y-axis (y++) like shows in fig. 11.

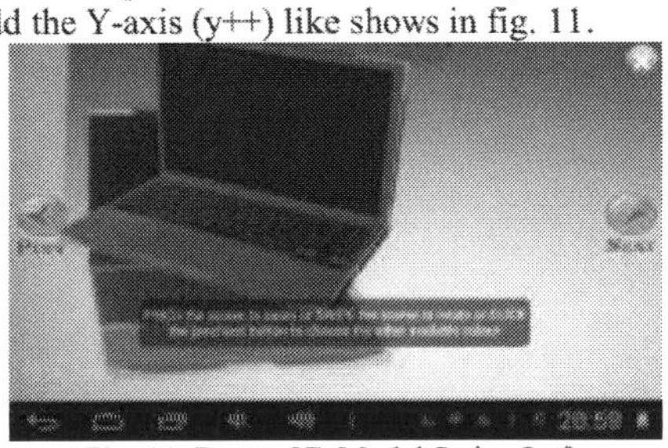

Fig. 11: Rotate 3D Model Swipe Left

d. If user swipe right the 3D model, the system will minus the Y-axis (y--) like shows in fig. 12.

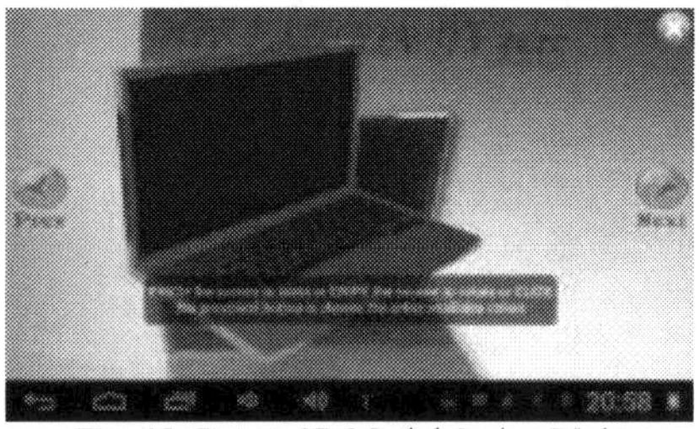
Fig. 12: Rotate 3D Model Swipe Right

3.5 Run the 3D Model Animation

In this feature, users are able to view the animation of a 3D model by click on to it. The system will search through the product ID, if there is available animation on the product, the system will return it to the user. The detail architecture shows in Table 3, how the user action and the system response. The alt-step is explained about when there is no animation data, the system will show nothing to the smartphone screen.

Table 3: Run 3D Model Animation Detail Architecture

Actor Action		***System Response***	
Step-1	User click or touch on 3D model.	**Step-2**	Run the animation that save on 3D model.
Alt step-2	If there is no animation saved in 3D model then there is no animation will be run.		

4 Conclusions

Based on research of two applications, Brother AR and Go Sweat Go Ion features. Both of applications have similar features which are view 3D model, animation and markerless. 3D Catalogue application give more features and make user able to look the product in different perspective by using zoom, rotate, and color chooser features. Brother AR and Go Sweat Go Ion application focused on one product, but using 3D catalogue can give the company to promote more of their products. There are several future work that will be done in the future by authors to make 3D catalogue application more complete such as develop it in another platform, connect to company social media and notification to user every time company has new product.

References

1. Augmented Reality, http://www.metaio.com/technology/, Metaio GmbH 2014.
2. J. armigniani, B. Furht, M. Anisetti, P. Ceravolo, E. Damiani, and M. Ivkovic, *Augmented Reality Technology, Systems and Application*. Springer Science+Business Media, **51**, 341-377, (2010).
3. IDC: International Data Corporation, Smartphone OS Market Share, Q2 2014, http://www.idc.com/
4. Brother AR: http://play.google.com, Brother International Singapore, 2012.
5. Go Sweat Go Icon: http://play.google.com, PT. Amerta Indah Otsuka, 2013.
6. D. Schmalstleg, T. Langlotz, and M. Billinghurst, *Augmented Reality 2.0*. Springer-Verlag/Wien, **10**, 13-37, (2011).

MATEC Web of Conferences **54**, 05003 (2016) DOI: 10.1051/matecconf/20165405003
MIMT 2016

Weight Value and Map Complexity in Theta*

Phuc T.H. Le and Ki-dong Lee

Robot lab, Department of Computer Engineering, Yeungnam University, Gyeongsan-si, South Korea

Abstract. A performance of an experiment to analyze the relationship between weight value and map complexity. That means there is a trial to find which weight value performs the best on maps with different complexity. And then in this article there is an application of hierarchical and dynamic weight method into Theta* path-finding algorithm. The dynamic weight value is picked from the first experiment. For different abstract map, a weight value is changed base on the complexity of the abstract map. This article contains a re-test of new implementation of Theta* and compare to the Mr. Nash's Theta* to collect the positive outcome. And by applying the dynamic weight method with different weight based on map complexity, the dynamic weight theta* perform better in time costing

1. Introduction

Path-finding algorithm is a process of finding a road between two given points – start and destination ones - in a random environment. In this study field, people used this technique in many other life fields such as: robotic, game design and network. In the past, path-finding algorithm had to solve two issues: find a way between two points in a map and the optimal shortest path. In the current day, people create one more problem: calculate an accurate path with in the short time. There are several path-finding algorithms have been introduced recently [1]. For example, Deep-First search, Breadth-First search, Best-First search, Dijkstra, A* and the most recent path-finding algorithm – Theta* [1, 2, 3, 4, 5, 6, 7, 8, 9, 10, 11, 12, 13, 14 and 15]. While some of the algorithms search for the optimal path, others find a path within the shortest time. There are many implementing methods to reduce the time consuming such as using different types for database structures [6, 7, 8, 9, 10, 11 and 12]. In this paper, there is an application of dynamic weight to boost the speed of Theta* [3, 4 and 14]. But the found weight value will related to the map complexity of the environment. In order to complete our experiment, the experiment process firstly run different weight values on different maps with size 10x10 with specified maps' complexity. The weight value which gives a best outcome will be the value that is applied for typical map abstracts. Before us, there is no research on the relationship between weight value and map complexity.

2. Related Works

Dijkstra (DA) was invented by Edsger Dijkstra. By giving a start (source) point, DA can search a shortest path in the map [1]. DS is a type of path-finding algorithm that applies the comparison condition to find the path. However, the high cost of computation time is a weighted issue of Dijkstra.

A* is a search algorithm which is expanded from Dijkstra by applying a heuristic value [3]. The found path seems to be made by the non-human object [4]. The formula of A* is the formula (1)

$$F = G + H \tag{1}$$

where G is the distance from the start point to the current point and H is the estimated distance from the current point to the goal point.

Theta* - post-smoothing process of the A* algorithm - was created by Alex Nash in 2007 [4]. The problem of Theta* is that computation time is more than A* but the path is not always guaranteed to be the optimal one [14].

Weighted Theta* is a variant of Theta* algorithm [13]. In this algorithm, a new parameter called weight will be added to Theta*'s F formula (2)

$$F = G + W * H \tag{2}$$

, where G and H is the same as A* in formula (1), and W is a weight value ($0<W<+\infty$).

HPA* stands for hierarchical path-finding A* algorithm [15, 16]. This method reduces the complexity of the map by dividing the map into the smaller abstract maps. Botea et.al proved that HPA* saves up to 10 times faster than A* but the increasing in path-length is within 1% [11]. But the path from HPA* is similar to A* which looks unnatural.

In 2015, there is a research article which applied the dynamic weight and HPA* method into Theta* to reduce the time consuming [15].

However on these previous researches, there is no article defining a specific weight value on each type of map. Therefore, a performance of an analysis on the Theta* determines the relationship between weight value and map complexity. From the collected results, a conclusion of a new version of Theta* which can find the path in faster way comparing to the original Theta* is drawn.

3. Specific Weight - Dynamic Weight Theta*

The main idea of this article is to specify which weight value is good for a specific type of map. And when the abstract map matches to that type of map, then corresponding weight value is applied to that map. Moreover, a run on the dynamic weight theta* on different sizes map is made. The big map is divided into smaller abstract maps with size 10 points x10 points. Then for each abstract map, a scan is made for the map's environment to detect its complexity and the use of statistic data to determine the applied weight value for that map area. By doing this, the value is determined the best for a specific type of map. Then a performance is made on the dynamic weight Theta* in 12 cases.

In this paper, an application is created that user can pick a map containing obstacle points, a start point and goal point. The app is written in Java and tested on Intel Core 2 with 2GB Ram. In the app, a class in Java called Node, which has some properties: x, y represents the longitude and latitude position of the point in the grid map, color to present the status of the map such as: obstacle, goal, start, and path. In order to make these points are easier to see, there is an assignment of a certain size for a point (30 pixels). All of these points will be combined into a square 2D map size n. As in Fig. 1, the blue point has an arrow with the word "goal/ destination point" is a destination point. The green point has an arrow with the word "start point" is a start point. These points with white color are the free points that can be passed through. And the rest points in the map are these obstacles that can't be passed. This map is a point map type. And when the path is found by the application it is displayed a java window frame. On the result window, the found path is connected with series lines. All these points are belonged to the path found by the Theta* algorithm. At each case, two pieces of information ae collected: the computation time of the process from the beginning of the process to the end of the process in milliseconds, and the path-length of the solution path, which is count by the sum of each path point's Euclidean distance to their parents.

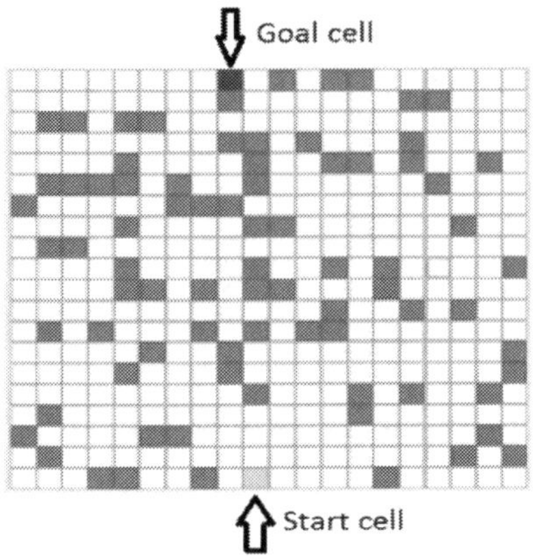

Figure 1. Example of Map

3.1 Weight on Specific Map Type

The implemented dynamic weight Theta* is called as "DW Theta*." And a base for comparison is the original Theta*, "Author Theta*" from Nash et.al. [3, 4 and 14] On general, 30 different maps are used to evaluate the weight value. The main purpose of these activities is to figure out which what weighted value makes the algorithm perform the fastest in speed in which type of map complexity. In this article, the time unit is millisecond. At the beginning of our experiment, the "DW Theta*" runs on 6 cases. In all the six cases the map sizes are always 10 points x 10 points, but the percentages of obstacle are increased from 0% to 50%. For each case, a test the algorithm is made on 5 different maps.

In the next experiment, there is a pick of many different types of map with same size (100 points: 10x10). The tested types of map includes: maps with 0% obstacles, maps with 10% obstacles, maps with 20% obstacles, maps with 30% obstacles, maps with 40% obstacles and maps with 50% obstacles. These test weight values are: 1, 2, 3, 5, 10 and 100. On each map the weighted Theta* run with the same weight value from the beginning of the search to the end. The repeat this same test with the same map but with the different weight values is made. Then a collection of the data is made to determine which weight value is the best for this type of map. The test result can be found from these figures. The best weight value in case 1, 2 and 3 is 5. The best one in case 4 is 3. The best one in case 5 is 2 and best one in case 6 is 1. That means if the complexity of the abstract map is less than 30%, we use the weight value equals to 5. If the complexity of the abstract map is more than or equals to 30% but less than 40%, we use the weight value equals to 3. If the complexity of the abstract map is more than or equal to 40% but less than 50%, we use the weight value equals to 2. And if the complexity is greater than or equals to 50% the weight value is equal 1.

3.2 Dynamic weighted Theta*

In this experiment, square maps with size n x n, which is divided by 10 and has 0 in reminder, are used to perform the dynamic weight theta* and divide the input map into smaller abstract map size (100 points: 10x10). For example, if users input the map size 20x20(400 points), they have 4 abstract maps size 10x10(100 points each). The weight value of the abstract map is decided based on the complexity of that map. The complexity of the map is calculated by counting the percentage of number of obstacles on the total points of the abstract map. The two bellowing algorithms are tested with different cases: case 1 - map 20x20 with 30% obstacles, case 2 - map 30x30 with 30% obstacles, case 3 - map 40x40 with 30% obstacles, case 4 - map 50x30 with 30% obstacles, case 5 - map 20x20 with 40% obstacles, case 6 - map 30x30 with 40% obstacles, case 7 - map 40x40 with 40% obstacles, case 8 - map 50x30 with 40% obstacles, case 9 - map 20x20 with 50% obstacles, case 10 - map 30x30 with 50% obstacles, case 11 - map 40x40 with 50% obstacles and case 12 - map 50x30 with 50% obstacles.

After determining which value is the best for each type of map's complexity, a test on "DW Theta*" is made once again to compare with the "author Theta*" on different 12 cases as which are listed on the previous paragraph. And from all 12 cases, the positive data is collected to draw some conclusions.

Table 1. Time consuming in the first 6 cases

	Case 1	Case 2	Case 3	Case 4	Case 5	Case 6
Author Theta*	6.66	10.73	5.69	18.18	1.15	2.82
DW Theta*	1.81	2.63	0.82	2.62	0.80	2.71

Table 2. Time consuming in the last 6 cases

	Case 7	Case 8	Case 9	Case 10	Case 11	Case 12
Author Theta*	4.6	8.96	0.35	1.36	1.93	1.97
DW Theta*	3.88	5.36	0.28	0.75	1.32	0.85

Table 3. Path-length in the first 6 cases

	Case 1	Case 2	Case 3	Case 4	Case 5	Case 6
Author Theta*	105.81	332.55	264.96	771.28	121.29	265.93
DW Theta*	183.9	365.84	424.39	1201.98	142.17	318.17

Table 4. Path-length in the last 6 cases

	Case 7	Case 8	Case 9	Case 10	Case 11	Case 12
Author Theta*	595.95	525.61	131.14	373.36	316.94	811.82
DW Theta*	950.54	1640.24	197.16	547.53	405.16	1034.81

4. Experiment Results & Discussion

At these figures 2 to 7, the x-axis represents for the time-consuming, which is calculated in seconds; and the y-axis is used to express the weight values. From these figures 2 to 7, a conclusion is made to determine that the weight value which equals 5 makes the theta* perform faster than other weight value with maps having a percentage of obstacles less than or equal 20%. And for the maps having a percentage of obstacles more than 20% but less than or equal 40%, the best weight value is 2. And finally, for the maps having a percentage of obstacles more than 40%, the best weight value is 1. There is no consideration for these maps with percentage of obstacles which are higher than 60% because these cases are hard to find a solution path. If there is an addition of a weight value which is greater than one into the calculation of F value the runtime will perform faster but the path is not optimal because of the increasing in its path-length. And if there is an addition of weight value which is smaller than one into the calculation of F value the runtime will perform slower but the path is closer-to-optimal because of the increasing in accuracy.

This experiments looks like the part of Nash et. al's research [4 and 14]. But instead of using the stable and unchanged weight value, there is a use of the dynamic weight values which depend on the map's complexity. There is an exchange between the path-length and time cost. In this paper, the dynamic weight is applied on different abstract maps but generally all these weight values are greater than 1. When the applied "DW Theta*" algorithm is compared to the original Theta* - "author Theta*." The "DW Theta*" algorithm performs faster than the "author Theta*" in all 12 cases, as it is on table 1 and table 2. On the other hand, these solutions have their path-length be longer than the "author Theta*" as table 3, Table 4. Nash et.al mentions that Theta* - "author Theta*" performs faster than weighted Theta*. From the experiment, our Theta* - a dynamic weight Theta* performs faster the original Theta*. To explain for this fact, the complexity of A*, which is $O(b^d)$, where b is the average number of successors per state and d is the length of the solution, is analyzed. The number of successors in weighted A* is reduced because when if user applies the weight, they limit many unsuitable cells. This fact cause the b in weight A* is less than the A*. This method might cause a drawback that it has an ability to eliminate the cell or point that should be included on the solution path. Therefore the d value might increase a lot and make the time.

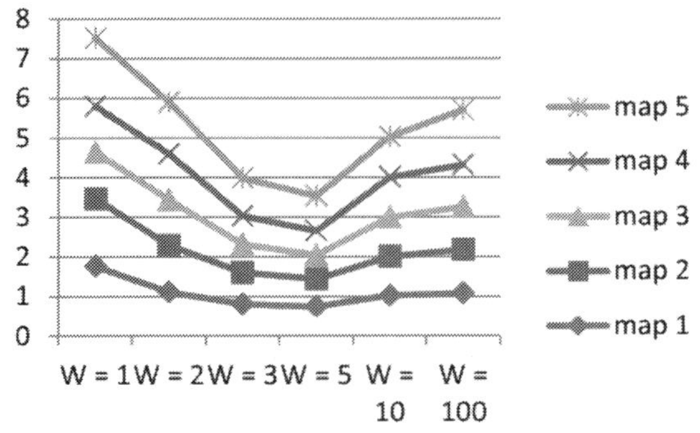

Figure 2. Time costing in case 1 with 5 different maps

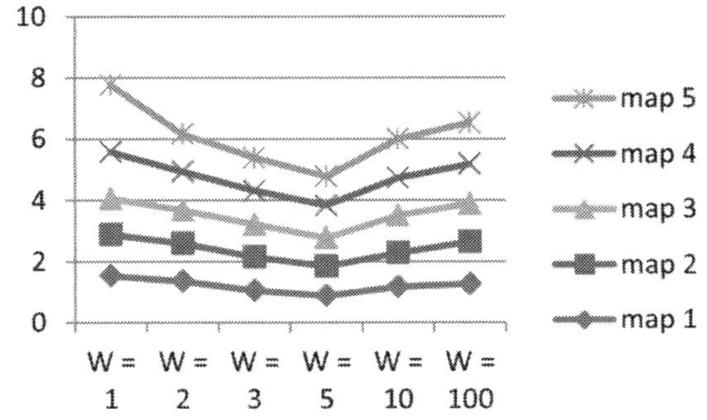

Figure 3. Time costing in case 2 with 5 different maps

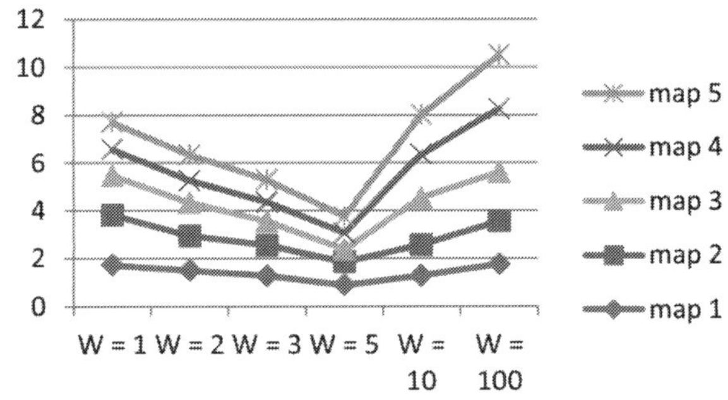

Figure 4. Time costing in case 3 with 5 different maps

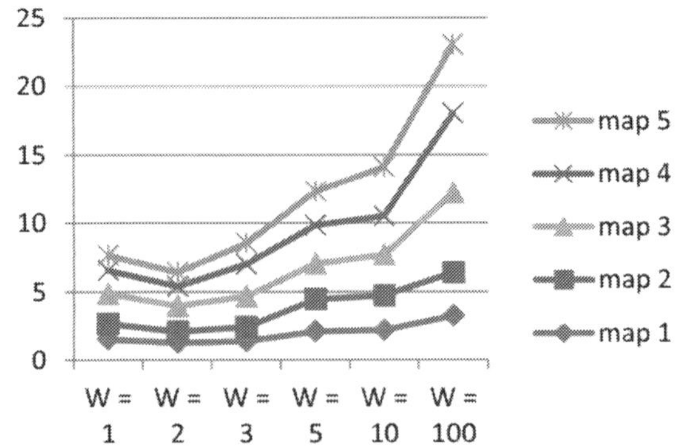

Figure 5. Time costing in case 4 with 5 different maps

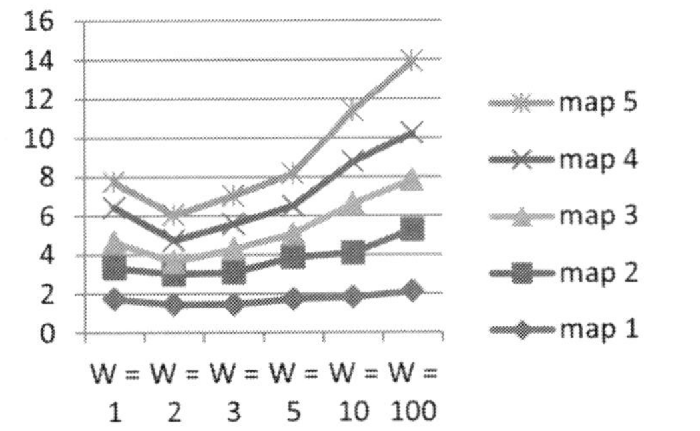

Figure 6. Time co sting in case 5 with 5 different maps

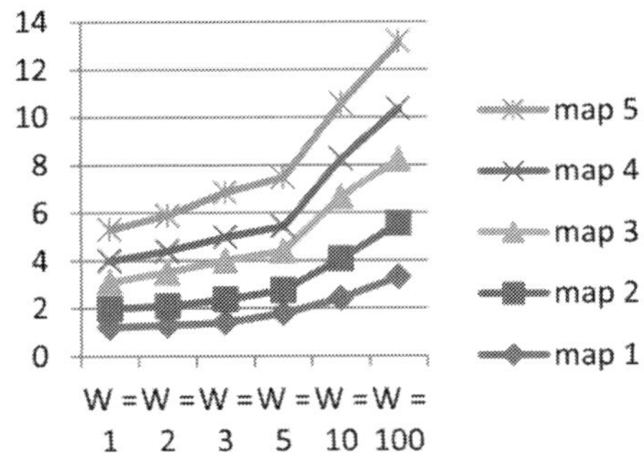

Figure 7. Time costing in case 6 with 5 different maps

5. Conclusion and Future Works

In future work, the application of dynamic weight from 0 to infinite on different map areas based on the percentage of blocks based on the idea of Mr. Wilt in his article [13] is considered. Moreover, the idea of combining dynamic weight theory with the hierarchical method can be applied to these variants of Theta*, such as AP- Theta* or lazy Theta*. As the result of these experiments, weight value makes the Theta* performs well on specific map's complexity is determined. In additional, in this article, an introduction of a way to increase the performance by applying weighted method and hierarchical method from A* to Theta* using the dynamic weight value based on the map complexity is stated. The applied weight values are from 1 to infinite. The new method is tested in many cases to collect data to draw some strong conclusion that the implementation makes the Theta* path-finding process run faster. Applying weight method to heuristic searching algorithms is making a trade-off between path-length and time-consuming. So when dynamic weight method is implemented to Theta*, the same result is showed. The explanation of the trade-off is displayed in the complexity of its – $O(b^d)$, which depends on both the number of expansion neighbor and the length of solution path.

References

1. T.H. Cormen, C.E. Leiserson, and R.L. Rivest, , *Introduction to Algorithm*, McGraw-Hill Companies, Inc., ch.3, (2009)
2. S.E. Putri, and N. Normalina, "Implementation and Analysis of Depth-First Search (DFS) Algorithm for Finding the Longest Path." In International Seminar on Operational Research (InteriOR), (2011)
3. A. Nash, S. Koenig, and A. Felner, "Theta*: Any-Angle Path Planning on Grids," *In Proceeding of the AAAI Conference on Artificial Intelligent*, pp. 1177-1183, (2007)
4. K. Daniel, , A. Nash, S. Koenig and A. Felner, "Theta*: Any-Angle Path Planning on Grids," Journal of Artificial Intelligence Research 39, pp. 533-579, (2010)
5. T. Li, L. Qi, and D. Ruan, "An Efficient Algorithm for the Single-Source Shortest Path Problem in Graph Theory," *Proc. of 3rd International Conference on Intelligent System and Knowledge Engineering*, **volume 1**, pp. 152-157, (2008)
6. C. Xi, F. Qi, and L. Wei, "A New Shortest Path Algorithm based on Heuristic Strategy," Intelligent Control and Automation, 2006. WCICA 2006. The Sixth World Congress on, **volume 1**, pp. 2531-2536, (2006)
7. K. Magzhan, and H.M. Jani, "A Review and Evaluation of Shortest Path Algorithms," International journal of scientific & technology research, **volume 2**, issue 6, (2013)
8. J. Freeman, "Parallel Algorithm for Depth-First Search," Technical Reports (CIS), University of Pennsylvania, pp. 428, (1991)
9. X. Cui, and H. Shi, "A*-based Path finding in Modern Computer Games," International Journal of Computer Science and Network Security, **volume 11**, No. 1, pp. 125-130, (2011)
10. A. Botea, and M. Muller, "Near Optimal Hierarchical Path-Finding," Journal of Game Development, **volume 1**, pp. 7-28, (2004).
11. R.M. Jansen, and M. Buro, "HPA* Enhancements," *AAAI Conference on* Artificial Intelligence and Interactive Digital Entertainment, pp. 84-87, (2007)
12. A.E. Hansen, and R. Zhou, "Anytime Heuristic Search," Journal of Artificial Intelligent Research 28, pp. 267-297, (2007)
13. C. Wilt, and W. Ruml, "When does Weighted A* Fail?" Proceeding *of the Symposium on Combinatorial Search (SoCS-12)*, pp. 137-144, (2012)
14. A. Nash, "Any-Angle Path Planning," Doctor of Philosophy Thesis (Computer Science) at the Faculty of the USC Graduate School University of Southern California. , (2012)
15. Phuc, Le T.H., "Applying Dynamic Weight on Theta Star Path-finding Algorithm in 2D Grid Map," *Proceeding of International Conference on Intelligent Computing, Electronics Systems and Information Technology (ICESIT-15)*, (2015)

Credit scoring with a feature selection approach based deep learning

Van-Sang Ha [1], Ha-Nam Nguyen [2]

[1] *Department of Economic Information System, Academy of Finance, Hanoi, Viet Nam*
[2] *Department of Information Technology, University of Engineering and Technology, Hanoi, Viet Nam*

Abstract. In financial risk, credit risk management is one of the most important issues in financial decision-making. Reliable credit scoring models are crucial for financial agencies to evaluate credit applications and have been widely studied in the field of machine learning and statistics. Deep learning is a powerful classification tool which is currently an active research area and successfully solves classification problems in many domains. Deep Learning provides training stability, generalization, and scalability with big data. Deep Learning is quickly becoming the algorithm of choice for the highest predictive accuracy. Feature selection is a process of selecting a subset of relevant features, which can decrease the dimensionality, reduce the running time, and improve the accuracy of classifiers. In this study, we constructed a credit scoring model based on deep learning and feature selection to evaluate the applicant's credit score from the applicant's input features. Two public datasets, Australia and German credit ones, have been used to test our method. The experimental results of the real world data showed that the proposed method results in a higher prediction rate than a baseline method for some certain datasets and also shows comparable and sometimes better performance than the feature selection methods widely used in credit scoring.

1 Introduction

The main purpose of credit risk analysis is to classify customers into two sets, good and bad ones [1]. Over the last decades, there have been lots of classification models and algorithms applied to analyse credit risk, for example decision tree [2], nearest neighbour K-NN, support vector machine (SVM) and neural network [3-7]. One important goal in credit risk prediction is to build the best classification model for a specific dataset.

Financial data in general and credit data in particular usually contain irrelevant and redundant features. The redundancy and the deficiency in data can reduce the classification accuracy and lead to incorrect decision [8-9]. In that case, a feature selection strategy is deeply needed in order to filter the redundant features. Indeed, feature selection is a process of selecting a subset of relevant features. The subset is sufficient to describe the problem with high precision. Feature selection thus allows decreasing the dimensionality of the problem and shortening the running time.

Credit scoring and internal customer rating is a process of accessing the ability to perform financial obligations of a customer against a bank such as paying interest or an original loan on due date, or other credit conditions for evaluating and identifying risks in the credit activities of the bank. The degree of credit risk changes over individual customers and is identified through the evaluation process. It is based on existing financial and non-financial data of customers at the time of credit scoring and customer rating.

Credit scoring is a technique using statistical analysis data and activities to evaluate the credit risk against customers. Credit scoring is shown in a figure determined by the bank based on the statistical analysis of credit experts, credit teams or credit bureaus. In Vietnam, some commercial banks start to perform credit scoring against customers but it is not widely applied during the testing phase and still needs to improve gradually. For completeness, all information presented in this paper comes from credit scoring experience in Australia, Germany and other countries.

Many methods have been investigated in the last decade to pursue even small improvement in credit scoring accuracy. Artificial Neural Networks (ANNs) [10-13] and Support Vector Machine (SVM) [14-19] are two commonly soft computing methods used in credit scoring modelling. Recently, other methods like evolutionary algorithms [20], stochastic optimization technique and support vector machine [21] have shown promising results in terms of prediction accuracy.

In this study, a new method for feature selection based on various criteria are proposed and integrated with a deep learning classifier in credit scoring tasks.

The rest of the paper is organized as follows: Section 2 presents the background of credit scoring, deep learning and feature selection. Section 3 is the most important section that describes the details of the proposed model. Experimental results are discussed in Section 4 while concluding remarks and future works are presented in Section 5.

2 Materials

2.1 Feature selection

Feature selection is the most basic step in data pre-processing as it reduces the dimensionality of the data. Feature selection can be a part of the criticism which needs to focus on only related features, such as the PCA method or an algorithm modeling. However, the feature selection is usually a separate step in the whole process of data mining.

There are two different categories of feature selection methods, i.e. filter approach and wrapper approach. The filter approach considers the feature selection process as a precursor stage of learning algorithms. The filter model uses evaluation functions to evaluate the classification performances of subsets of features. There are many evaluation functions such as feature importance, Gini, information gain, the ratio of information gain, etc. A disadvantage of this approach is that there is no relationship between the feature selection process and the performance of learning algorithms.

The wrapper approach uses a machine-learning algorithm to measure the good-ness of the set of selected features. The measurement relies on the performance of the learning algorithm such as its accuracy, recall and precision values. The wrapper model uses a learning accuracy for evaluation. In the methods using the wrapper model, all samples should be divided into two sets, i.e. training set and testing set. The algorithm runs on the training set, and then applies the learning result on the testing set to measure the prediction accuracy. The disadvantage of this approach is highly computational cost. Some researchers proposed methods that can speed up the evaluating process to decrease this cost. Common wrapper strategies are Sequential Forward Selection (SFS) and Sequential Backward Elimination (SBE). The optimal feature set is found by searching on the feature space. In this space, each state represents a feature subset, and the size of the searching space for n features is $O(2n)$, so it is impractical to search the whole space exhaustively, unless n is small.

2.2 Deep Learning

Deep learning (deep machine learning, or deep structured learning) attempt to model high-level abstractions in data by using multiple processing layers with complex structures or otherwise, composed of multiple non-linear transformations. There are several theoretical frameworks for Deep Learning, but this research focuses primarily on the feed-forward architecture used by H2O. The basic unit in the model is the neuron, a biologically inspired model of the human neuron. In humans, the varying strengths of the neurons' output signals travel along the synaptic junctions and are then aggregated as input for a connected neuron's activation. In the model, the weighted combination $\alpha = \sum_{i=1}^{n} w_i x_i + b$ of input signals is aggregated, and then an output signal $f(\alpha)$ transmitted by the connected neuron. The function f represents the nonlinear activation function used throughout the network and the bias b represents the neuron's activation thresh-old. Multi-layer, feed-forward neural networks consist of many layers of inter-connected neuron units, starting with an input layer to match the feature space, followed by multiple layers of nonlinearity, and ending with a linear regression or classification layer to match the output space. Multi-layer neural networks can be used to accomplish Deep Learning tasks. Deep Learning architectures are models of hierarchical feature extraction, typically involving multiple levels of nonlinearity. Deep Learning models are able to learn useful representations of raw data and have exhibited high performance on complex data such as images, speech, and text. The procedure to minimize the loss function $L(W,B \mid j)$ is a parallelized version of stochastic gradient descent (SGD). Standard SGD can be summarized as follows, with the gradient $\nabla L(W,B \mid j)$ computed via back propagation.

Parallel distributed and multi-threaded training with SGD in H2O Deep Learning

---

Initialize global model parameters W,B
Distribute training data T across nodes
Iterate until convergence criterion reached
For nodes n with training subset T_n, do in parallel:
Obtain copy of the global model parameters W_n ,B_n Select active subset $T_{na} \subset T_n$ (user-given number of samples per iteration)
Partition T_{na} into T_{nac} by cores n_c
For cores n_c on node n, do in parallel:
Get training example $i \in T_{nac}$
Update all weights $w_{jk} \in w_n$, biases $b_{jk} \in B_n$
$w_{jk} := w_{jk} - \alpha(\partial L(W,B \mid j))/\partial w_{jk}$
$b_{jk} := b_{jk} - \alpha(\partial L(W,B \mid j))/\partial b_{jk}$
Set $W,B := Avg_n\, W_n$; $Avg_n\, B_n$
Optionally score the model on train/validation scoring sets

3 The Proposed Method

Our method uses Deep Learning to estimate the performance consisting of the cross validation accuracy and the importance of each feature in the training data set. In a multi-node system this parallelization scheme works on top of H2O's distributed setup, where the training data is distributed across the cluster. Each node operates in parallel on its local data. After that, we determine best feature set by choosing the best of Average score + Median Score and the lowest SD. To deal with over-fitting problem, we apply n-fold cross validation technique to minimize the generalization error.

Step 1: Train data by Random Forest via 20 trails, calculate and sort median of variables important
Step 2: Add each feature with best variables important and train data by Deep Learning with the cross validation
Step 3: Calculate score for each feature F_i^{score} where i=1..n (n is the number of features in current loop).
Step 4: Select best feature using selection rules
Step 5: Back to step 1 until reach the desired criteria

In step 2, we use deep learning with n-fold cross validation to train the classifier. In the jth cross validation, we will obtain a set of (F_j, A_j^{learn}, $A_j^{validation}$) that are the feature importance, the learning accuracy and the validation accuracy respectively.

We will use those values to compute the score criterion in step 3.

In step 3 we use the results from step 1 and step 2 to build the score criterion which will be used in step 4. The score of feature i^{th} is calculated by:

$$F_i^{Score} = \sum F_{ij} \times (A_j^{learn} + A_j^{validation}) \quad (1)$$

The main of our algorithm is presented in step 4. In this step, we will select best of features using rules: the best of Average + Median Score and the lowest standard deviation (SD).

Rule 1: select features with the best of median score
Rule 2: select features with the best of average score
Rule 3: select features with the lowest SD

These rules guarantee us to get the highest accuracy and the lowest Standard deviation. This proposed method tends to find the smallest optimal set of features in order to reduce the number of output features as much as possible. Then, machine-learning algorithms are used to calculate the relevance of the feature. Based on the calculated value of conformity level, we find the subset of features having less number of features while maintaining the objective of the problem.

4 Experiment and results

Our proposed algorithm was coded using R language (http://www.r-project.org), using H2O Deep Learning package. This package is optimized for doing "in memory" processing of distributed, parallel machine learning algorithms on clusters. A "cluster" is a software construct that can be can be fired up on your lap-top, on a server, or across the multiple nodes of a cluster of real machines, including computers that form a Hadoop cluster. We tested the proposed algorithm with several datasets including two public datasets, German and Australian credit approval, to validate our approach.
In this paper, we used Random Forest with the original dataset as the base-line method. The proposed method and the base-line method were executed on the same training and testing datasets to compare their efficiency. Those implementations were repeatedly done 20 times to test the consistency of obtained results.

4.1. German credit approval dataset

The German credit approval dataset consists of 1000 loan applications, with 700 accepted and 300 rejected. Each applicant is described by 20 attributes. Our final results were averaged over these 20 independent trials (Fig. 1). In our experiments, we use the default value for the hidden parameter and the number of epoch parameter was tried with value of 10. The averages of classification results are depicted in Fig. 1.

The best subset contains 19 features and its accuracy is 74.68 %.

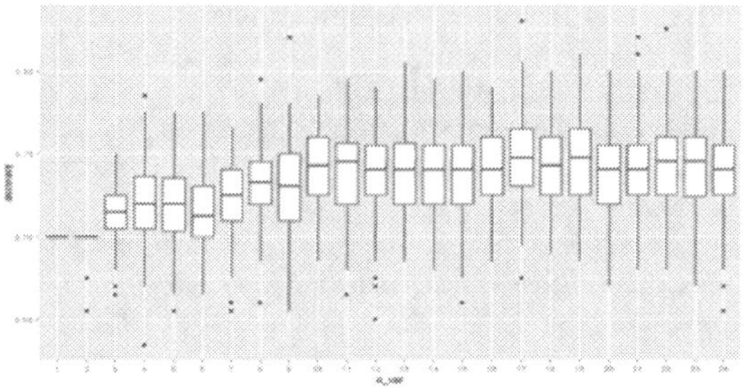

Figure 1. Accuracy in case of German dataset.

Table 1 shows the performances of different classifiers over the German credit datasets. Baseline is the classifier without feature selection. Classifiers used in [22] include: Linear SVM, CART, k-NN, Naïve Bayes, MLP. Filter methods include: t-test, Linear Discriminant analysis (LDA), Logistic regression (LR). The wrapper methods include: Genetic algorithms (GA) and Particle swarm optimization (PSO).

Table 1. Performances of different classifiers over the german credit dataset

Classifier	Filter methods			Wrapper methods		Baseline
	t-test	LDA	LR	GA	PSO	
Linear SVM	76.74	75.72	75.10	76.54	73.76	77.18
CART	74.28	73.52	73.66	75.72	74.16	74.30
k-NN	71.82	71.86	72.62	72.24	71.60	70.86
Naïve Bayes	72.40	70.88	71.44	71.56	74.16	70.52
MLP	73.28	73.44	73.42	74.03	72.54	71.76
RandomForests						73.40
Our method	**74.68**					

Comparing the performances of various methods in Table 1, we saw that the ac-curacy of deep learning on the subset of newly selected features obviously in-creases, and the number of features has been reduced by 21%. The average accuracy is 73.4% on the original data. After applying the feature selection, the aver-age accuracy increases to 74.68%.

Moreover, relying on a parallel processing strategy, time to run 20 trails with 5-fold cross validate taken by our method is only 5286 seconds (~88 minutes) while other methods must run several hours. This result highlights the efficiency in terms of running time of our method when filtering the redundant features.

4.2 Australian credit approval dataset

The Australian credit dataset is composed of 690 applicants, with 383 credit worthy and 307 default examples. Each instance contains eight numerical features, six categorical features, and one discriminant feature, with sensitive information being transferred to symbolic data for confidentiality reasons. The averages of classification results are depicted in Fig. 2.

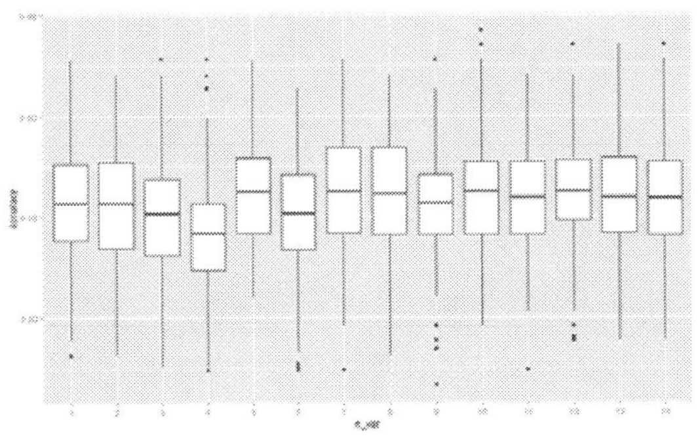

Figure 2. Accuracy in case of German dataset

Table 2 shows that the accuracy of Deep learning on a subset of 7 selected features obviously increases. The average accuracy is 85.82% on the original data. After applying the feature selection, the average accuracy increases to 86.24%. Relying on parallel processing, time to run 20 trails with 5-fold cross validate taken by our method is only 2769 seconds (~46 minutes).

Table 2. Performances of different classifiers over the Australian credit dataset

Classifier	**Filter methods**			**Wrapper methods**		**Baseline**
	t-test	LDA	LR	GA	PSO	
Linear SVM	85.52	85.52	85.52	85.52	85.52	85.52
CART	85.25	85.46	85.11	84.85	84.82	85.20
k-NN	86.06	85.31	84.81	84.69	84.64	84.58
Naïve Bayes	68.52	67.09	66.74	86.09	85.86	68.55
MLP	85.60	86.00	85.89	85.57	85.49	84.15
Random forests						85.82
Our method	**86.24**					

5 Conclusion

In this paper, we focused on studying feature selection and Deep Learning meth-od. Features selection involves in determining the highest classifier accuracy of a subset or seeking the acceptable accuracy of the smallest subset of features. We have introduced a new feature selection approach based on feature scoring. The accuracy of classifier using the selected features is better than other methods. Fewer features allow a credit department to concentrate on collecting relevant and essential variables. The parallel processing procedure leads to a significant decrement in runtime. As a result, the workload of credit evaluation personnel can be reduced, as they do not have to take into account a large number of features during the evaluation procedure, which will be somewhat less computation-ally intensive. The experimental results show that our method is effective in credit risk analysis. It makes the evaluation more quickly and increases the accuracy of the classification.

References

1. E. I. Altman and A. Saunders, "Credit risk measurement: developments over the last 20 years," *Journal of Banking and Finance*, vol. 21, no. 11-12, pp. 1721–1742, (1997)
2. Z. Davoodabadi and A. Moeini, "Building C ustomers' Credit Scoring Models with Combination of Feature Selection and Decision Tree Algorithms," vol. **4**, no. 2, pp. 97–103, (2015)
3. A. Khashman, "A neural network model for credit risk evaluation," *International Journal of Neural Systems*, vol. **19**, no. 4, pp.285–294, (2009)
4. T. Bellotti and J. Crook, "Support vector machines for credit scoring and discovery of significant features," *Expert Systems with Applications*, vol. **36**, no. 2, pp. 3302–3308, (2009)
5. F. Wen and X. Yang, "Skewness of return distribution and coefficient of risk premium," *Journal of Systems Science and Complexity*, vol. **22**, no. 3, pp. 360–371, (2009)
6. X. Zhou, W. Jiang, Y. Shi, and Y. Tian, "Credit risk evaluation with kernel-based affine subspace nearest points learning method," *Expert Systems with Applications*, vol. **38**, no. 4, pp.4272–4279, (2011)
7. G. Kim, C. Wu, S. Lim, and J. Kim, "Modified matrix splitting method for the support vector machine and its application to the credit classification of companies in Korea," *Expert Systems with Applications*, vol. 39, no. 10, pp. 8824–8834, (2012)
8. Liu, and Motoda, "Feature selection for Knowledge Discovery and Data mining," *Kluwer Academic Publishers*, 1998.
9. Guyon, and Elisseeff, "An Introduction to Variable and Feature Selection," *Journal of Machine Learning Research*, pp 1157-1182, 2003.
10. Oreski, S., Oreski, D., & Oreski, G., "Hybrid system with genetic algorithm and artificial neural networks and its application to retail credit risk assessment". *Expert Systems with Applications*, 39(16), 12605–12617, 2012
11. Saberi, M., Mirtalaie, M. S., Hussain, F. K., Azadeh, A., Hussain, O. K., & Ashjari, B. , "A granular computing-based approach to credit scoring modeling". *Neurocomputing*, 122, 100–115, (2013)
12. Lee, S., & Choi, W. S., "A multi-industry bankruptcy prediction model using back-propagation neural network and multivariate discriminant analysis". *Expert Systems with Applications*, 40(8), 2941–2946, (2013)
13. Ghatge, A. R., & Halkarnikar., "Ensemble Neural Network Strategy for Predicting Credit Default Evaluation", 2(7), 223–225, (2013)
14. Chaudhuri, A., & De, K., "Fuzzy Support Vector Machine for bankruptcy prediction. Applied Soft Computing Journal", 11(2), 2472–2486, (2011)
15. Ghodselahi, A., "A Hybrid Support Vector Machine Ensemble Model for Credit Scoring," *International Journal of Computer Applications*, 17(5), 1–5, (2011)
16. Huang L., Chen C., and Wang J., "Credit Scoring with a Data Mining Approach Based on Support Vector Machines," *Computer Journal of Expert*

Systems with Applications, vol. 33, no. 4, pp. 847-856, (2007)

17. G. Eason, B. Li T., Shiue W., and Huang H., "The Evaluation of Consumer Loans Using Support Vector Machines," *Computer Journal of Expert Systems with Applications*, vol. 30, no. 4, pp. 772-782, (2006)
18. Martens D., Baesens B., Gestel T., and Vanthienen J., "Comprehensible Credit Scoring Models Using Rule Extraction from Support Vector Machines," *European Computer Journal of Operational Research*, vol. 183, no. 3, pp. 1466-1476, (2007)
19. Wang Y., Wang S., and Lai K., "A New Fuzzy Support Vector Machine to Evaluate Credit Risk," *Computer Journal of IEEE Transactions on Fuzzy Systems*, vol. 13, no. 6, pp. 25-29, (2005)
20. S. Oreski and G. Oreski, "Genetic algorithm-based heuristic for feature selection in credit risk assessment," *Expert Syst*em. Appl., vol. 41, no. 4, pp. 2052–2064, (2014)
21. Y. Ling, Q. Y. Cao, and H. Zhang, "Application of the PSO-SVM model for credit scoring," Proc. - 2011 7th Int. *Conf. Comput. Intell. Secur*. CIS 2011, pp. 47–51, (2011)
22. Deron Liang, Chih-Fong Tsai, Hsin-Ting Wua, "The effect of feature selection on financial distress prediction," *Knowledge-Based Systems* 73 289–297 (2015)

An Intelligent Matching System for the Products of Small Business/Manufactures with the Celebrities

Junho Jeong[1], Yunsik Son[2], Seokhoon Ko[1] and Seman Oh[1]

[1]*Dept. of Computer Engineering, Dongguk University, Seoul, Korea*
[2]*Department of Brain and Cognitive Engineering, Korea University, Seoul, Korea*

Abstract. With the development of ICT, e-business have brought consumers opportunities to choose a wider range of products to purchase. However, these opportunities made it difficult for them to make a choice. Under this circumstance, product recommendation systems have arisen, helping consumers with purchase decision. In this paper, an intelligent matching system is proposed to connect small business/manufactures with celebrities. Previous matching/recommendation systems using narrowing approach based on limited information or using widening approach based on users' information. But these approaches have many restricts. The proposed matching system produces suitable celebrity candidate set to advertise target product of small business/manufactures based on relationship graph which can compare the similarity of each celebrity.

1 Introduction

Recently, there are various recommendation systems for matching products between celebrities to increasing promotion effects exist. The goal of these systems is increasing sales volume by the selecting suitable celebrities as an advertisement model. But, the existing systems did not work effectively, because its matching methodologies have simple approaches and using few features to matching the target product with celebrities.

This paper proposes the new method to improve promotion effect through advertising by suitable matched celebrities for the given product with using star relationship graph. The proposed method collets the information of the celebrities automatically on the web, and generating the star relationship graph as the relationship/similarity of each celebrities using KNN distance with hierarchical random graph model.

2 Related studies

2.1. Recommendation system

With the development of ICT, e-business have brought consumers opportunities to choose a wider range of products to purchase. However, these opportunities made it difficult for them to make a choice. Under this circumstance, product recommendation systems have arisen, helping consumers with purchase decision. In recent years, rarely can we find online stores without this system which deals with various kinds of products.

As an information filtering system, the recommendation system predicts the preference or estimation of the users for the recommended item. The recommendation system is largely fall into two categories: context based filtering [1, 2] and collaboration filtering [3].

Table 1. Categories of recommendation system

Category	Contents
Context based Filtering	- using the information of user or product's contents - using the techniques of the natural language processing or information searching - depend on product's description (metadata) - narrow recommendation range - needed small size information
Collaboration Filtering	- using the estimation of users - using the similarity of the products - using many users' estimation information - wide recommendation range - needed big size information - divide by user based method and item based method

Table 1 shows characteristics of the two kind of the recommendation system.

These recommendation systems used on various fields as product sales (Amazon, Netflix), news, video

 DOI: 10.1051/matecconf/20165405005

recommendation (Google, Youtube, Yahoo, Naver), social recommendation (Facebook).

2.2. Hierarchical random graph

A hierarchical random graph [4] is a random graph whose definition encodes some information about the expected community structure of the graph. The definition is given in terms of a dendrogram which gives information about the probability that any two nodes in the graph are connected. Each node has probabilities associated with other nodes, such that the probability for a node higher up the graph is strictly less than a node higher down the graph. In hierarchical random graph, any two vertices are joined by an edge with a probability corresponding to their lowest common ancestor in the dendrogram.

Because of the inequalities satisfied by the probabilities, this will naturally generate a graph with community structure (i.e. clusters of mutually connected vertices with a few connections to vertices in other clusters) and it called 'hierarchical' because, by the nature of the dendrogram, also it can describe the communities of communities, communities of communities of communities etc. Figure 1 shows example of data representation using the hierarchical random graph.

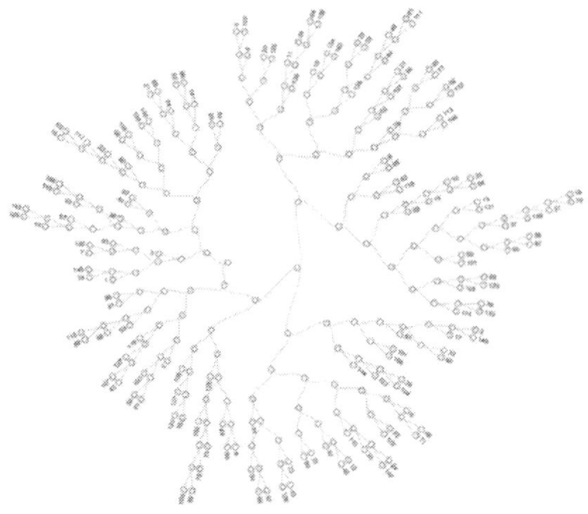

Figure 1. Visualization example of the comunity information represented by hierarchical rendom graph

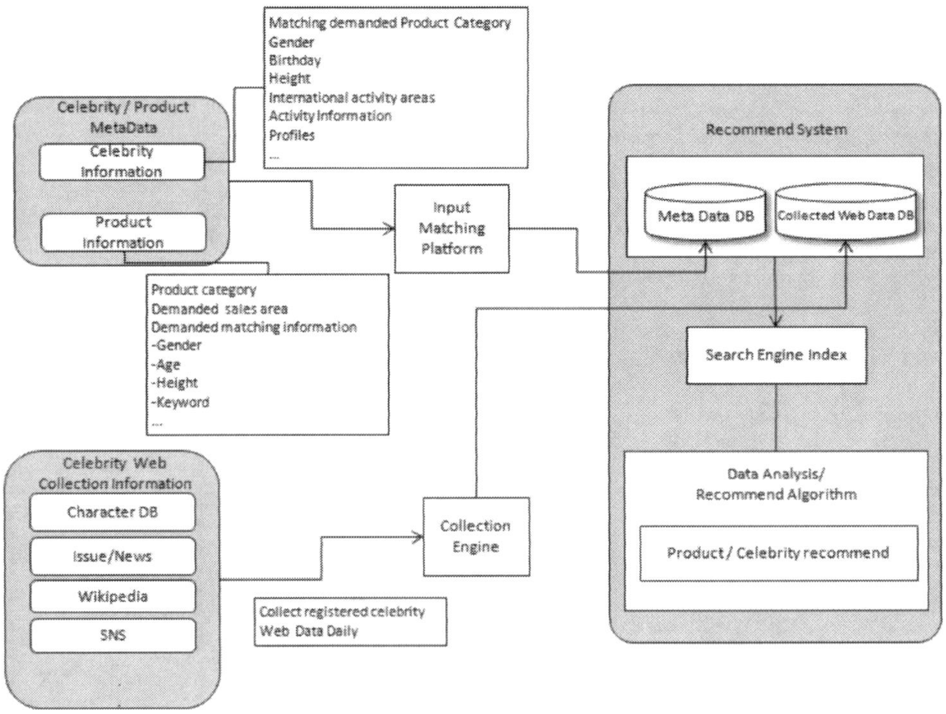

Figure 2. System structure of the intelligent matching system

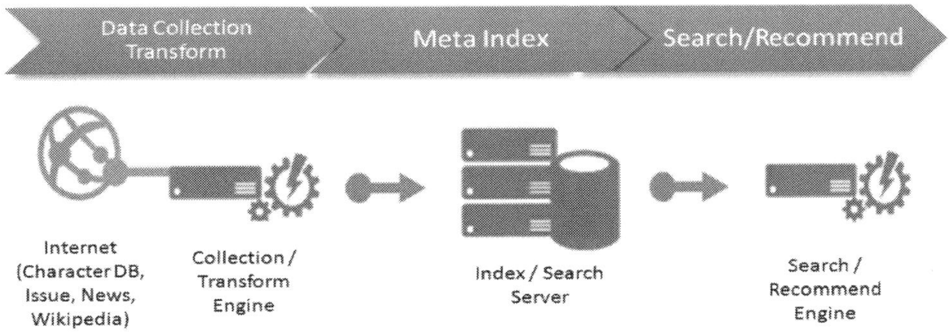

Figure 3. Data collection process for constructing star infromation

3 Proposed business matching system

Recommendation system is part of information filtering system. It is prediction and evaluation system about recommended item to consumer. Recommendation system depends on consumer, consumer's comments, natural-language processing using product's name, information retrieval and explanation of product. Therefore, the recommendation system can implement two methods. First is content-based filtering method. It recommends with small information but it has narrow recommendation range. Second is collaborative filtering. It uses consumer's evaluation to calculate similarity measure but it needs a large of information. Proposed system is a hybrid form [5] that takes advantage of the two implementations.

3.1. Matching system model

The proposed recommendation algorithm based on weight extracts effective matching result using small information. It is compatible recommendation system in the initial operating system. Figure 2 is a block diagram of a system like that proposed. Each system was constructed in modular function to handle the feedback type because of flexible data and matching result.

The system builds and index collected information to the database. The information is collected in two ways. One way is to collect from internet such as Character DB about celebrities, Issue/News, Wikipedia, and Social Network Services by Collection Engine. And other ways are to obtain Celebrities/Product meta data that made by manager of celebrity or product provider. The system recommends suitable celebrities for products by analyzing the index information to match the relationship between the product and the celebrities.

3.2. Data collection method

Data collection process gathering news information published on the web in order to collect information on the operating system constitutes the initial organization of the data as shown in Figure 3 shown collection steps.

An intelligent recommend system is required, such as Figure 2 is a meta data search engine as data collection/indexing/search due to the context based filtering. Celebrity metadata that is collected through the Web includes unstructured text data in addition to the formal data. Therefore, indexing engine is to generate a metadata index information by morphological analysis.

Typically, the default metadata of celebrity is managed directly by the himself or agency manager. However, there are practical limits to ensure immediate and effectiveness of data entry. Since Web collection engine to collect and classify the data in the various channels such as Character DB about celebrities, Issue/News, Wikipedia, and Social Network Services. The information is the metadata of the matching system provides the system operator and celebrities agency. As a result, by using this information it is able to be configured to increase the accuracy and probability of matching of the data for the matching system.

3.3. Star relationship graph based matching model

Proposed graph based matching model is using 2 kind of graph; first is PG (Product Graph) and next is SRG (Star Relationship Graph).

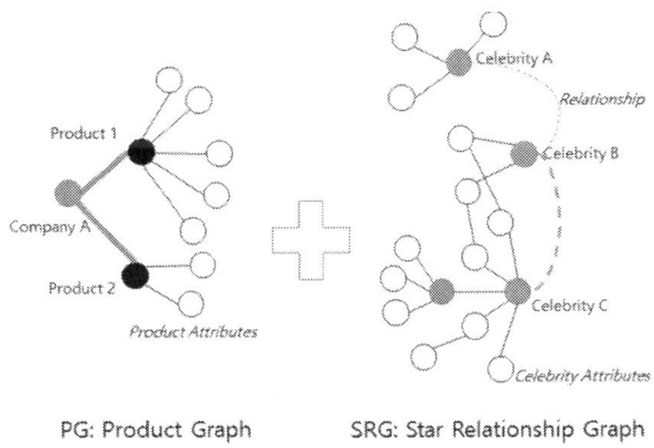

Figure 4. Graph models for macthing system; Product graph & star relationship graph

PG is constructed by attributes of the company's target products and SRG is combined by collected attributes by data collection process. Figure 4 shows the overall graph model. SRG based matching model as shown in Figure 5 was constructed celebrity as a model for improving the performance of a weight based algorithm, and others associated with the hierarchical graph.

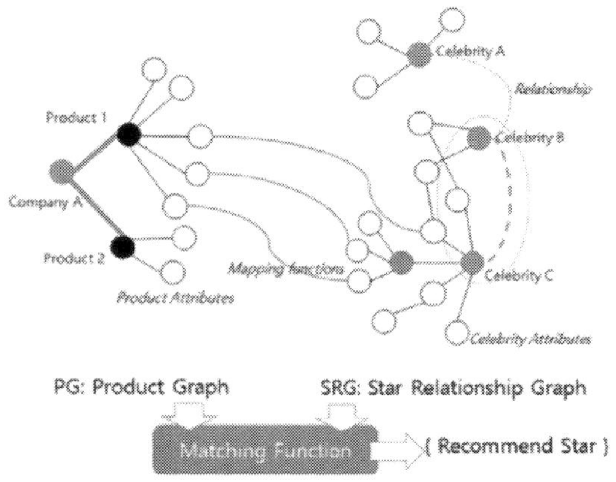

Figure 5. SRG besed matching model

It suggests a recommendation of celebrity group. It has two type of graph. PG (Product Graph) is made up classify product, prices, features, sales target and sales area node. SRG is made up celebrity of age, gender, active period, active state and the existing advertising items node. It generates matching result with graph mapping.

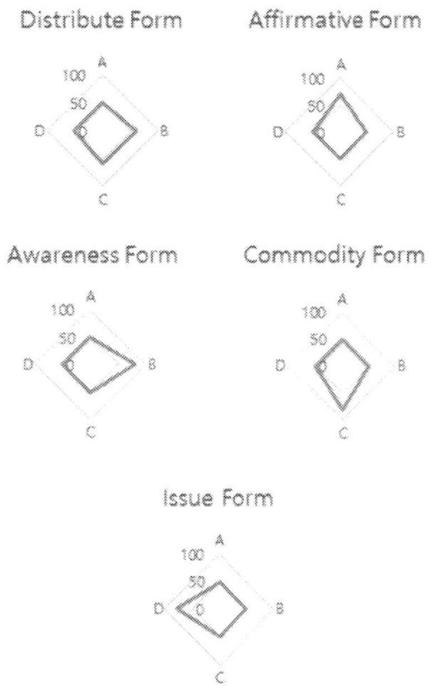

Figure 6. Specialized example forms using proposed metric

As Fig. 6, each celebrity is estimated by 4 metrics; Degree of Affirmative, Degree of Awareness, Degree of Commodity and Degree of Issue. Firstly, degree of affirmative is calculated by exposure rate of broadcasting, internet and number of SNS follower with posting. And next, degree of awareness is estimated by number of personal store and number of awards during one year. Degree of commodity is calculated using number of advertisement during one year, advertisement for competitors during one year, and estimated cost/profit. Finally, degree of issue is estimated by number of news article during one year and activated programs; TV, radio, films, and etc.

SRG generated using hierarchical random graph with KNN distance based on estimated celebrity information. Thus, each node of SRG can be find a similar node set and estimate the nearest with similar nodes.

Table 2 shows the matching process using the SRG and PG.

Table 2. Matching process using SRG

1. Configure the celebrity set SRG graph using the information related to celebrities
2. SRG represents celebrity relationships depending on the property with each of the celebrities
3. Represent the information of the product in PG: Product Graph
4. Selecting the highest matching rate celebrity of the SRG and PG according to the pre-defined matching function,
5. Extract the relevance celebrities linked to selected celebrities from SRG
6. Send expected to get higher matching rate celebrity with product

4 Conclusion and further research

In this paper, the matching system based on relationship graph model was proposed. Proposed method designed to improve promotion effect through advertising by recommend suitable matched celebrities for target product using the relationship graph which represents the properties of celebrities and their similarity.

In the future, a process to evaluate the implemented matching system is necessary. And study to be elaborated the matching system how to reflect the preference and matching rate score is needed.

Acknowledgements

This work was supported by the Technology Innovation Program (10048990, Development of business matching platform and collaboration product with celebrity and Korean stars for the domestic and overseas business activation of small business and manufactures) funded by the Ministry of Trade, industry & Energy (MI, Korea)

References

1. A. Ansari, S. Essegaier, R. Kohli. Internet Recommendations Systems, J. Mar. R. **37**, 363 (2000)
2. G. Adomavicius, T. Alexander. oward the next generation of recommender systems: A survey of the state-of-the-art and possible extensions, IEEE T. Know. Data Eng. **17**, 734 (2005)
3. M. Balabanović, Y. Shoham. Fab: content-based, collaborative recommendation, Comm. ACM, **40**, 66 (1997)
4. A. Clauset, C. Moore, M. E. Newman. Hierarchical structure and the prediction of missing links in networks, Nature **453,** 98 (2008)
5. S. Debnath, N. Ganguly, P. Mitra. Feature weighting in content based recommendation system using social network analysis, 17th int. Conf. WWW, 1041 (2008)
6. R. Burke. Hybrid Recommender Systems: Survey and Experiments, User Mod. User-Adapt. Int. **12**, 331 (2002)
7. Y. W. Chen. A collaborative filtering recommendation algorithm based on contents' genome, IET Int. Conf. Infor. Sci. Cont. Eng. 1 (2012)

MATEC Web of Conferences **54**, 05006 (2016) DOI: 10.1051/matecconf/20165405006
MIMT 2016

Pathloss Measurements and Prediction using Statistical Models

Zia Nadir[1], Mohammad Bait-Suwailam[2], Mohammad Idrees[3]

[1,2] *ECE, PC 123, Box 33, SQU, Muscat, Oman*
[3] *DOMAS, PC 123, Box 33, SQU, Muscat, Oman*

Abstract. There is a major expansion in the market of wireless communications services in urban areas. New regulatory environments as well as rivalry in the communications industry necessitate that these systems be arranged quickly and at low cost. Computer-based radio propagation prediction tools are strong candidates for this goal. This paper provides an extension of pathloss analysis in urban environments in Oman. The paper addresses the applicability of Okumura-Hata model in an area in Oman in GSM frequency band. Trigonometric Regression Model is used for data prediction at unknown nodes. These can be used to make desired forecasts. The results are acceptable, can be used primarily for OMAN, and can be extended for other regions.

1. Introduction

Mobile communications is currently at its fastest growth-period in history; due to enabling technologies, which permit wider deployment. Historically, growth in the mobile communications field has now become slow, and has been linked to technological advancements. The need for high quality and high capacity networks, estimating coverage accurately has become extremely important. The fundamental task is to predict the coverage of the proposed system. Propagation models are useful for predicting signal attenuation or path loss which may be used as a controlling factor for system performance or coverage so as to achieve perfect reception [1]. It has been found that the mechanisms behind electromagnetic wave propagation are diverse and characterized by certain phenomena such as reflection, refraction and diffraction of waves. These phenomena induces signal scattering, fading and shadowing along the signal path and their effects can best be described (in a large scale) by the path loss exponent which defines the rate of change of attenuation that the signals suffers as it propagates from the transmitter to the receiver. Growth in the mobile communications field has now become slow, and has been linked to technological advancements. Therefore, for more accurate design coverage of modern cellular networks, signal strength measurements must be taken into consideration in order to provide an efficient and reliable coverage area.

This article addresses the evaluations between the statistical and the experimental analysis at GSM frequency of 900 MHz using Okumara model which is most widely used. The key idea of modern cellular systems is that it is possible to serve the unlimited number of subscribers, distributed over an unlimited area, using only a limited number of channels, by efficient channel reuse.

2. Empirical Propagation Models

2.1 Okumura's Measurements

Okumura carried out extensive drive test measurements with range of clutter type, frequency, transmitter height, and transmitter power. It states that, the signal strength decreases at much greater rate with distance than that predicted by free space loss [2-4].

2.2 Hata's Propagation Model

Hata model was based on Okumura's field test results and predicted various equations for Pathloss with different types of clutter. It is well suited model for the Ultra High Frequency (UHF) band [5]. The limitations on Hata Model due to range of test results from carrier frequency 150 MHz to 1500 MHz, the distance from the base station ranges from 1 km to 20 km, the height of base station antenna (h_b) ranges from 30 m to 200 m and the height of mobile antenna (h_m) ranges from 1 m to 10 m. It was also observed that the signal strength is a function of distance and antenna height, as we can see in this work the highest antenna has less propagation path loss and as the distance increases the path loss also increases [6]. Hata created a number of representative Pathloss mathematical models, as illustrated in following Eq. 1. Okumura takes urban areas as a reference and applies correction factors as following. For urban areas:

$$L_{dB} = A + B Log_{10} R - E_{1,2,3} \tag{1}$$

Where

$A = 69.55 + 26.16 Log_{10} f_c - 13.82 Log_{10} h_b$

$B = 44.9 - 6.55 Log_{10} h_b$

$E_1 = 3.2\left(Log_{10}\left(11.7554 h_m\right)\right)^2 - 4.97 \quad for\ cities;\ f_c \geq 300MHz$

$E_2 = 8.29\left(Log_{10}\left(1.544 h_m\right)\right)^2 - 1.1 \quad for\ cities;\ f_c \leq 300MHz,$

$E_3 = \left(1.1 Log_{10} f_c - 0.7\right) h_m - \left(1.56 Log_{10} f_c - 0.8\right)$

for relatively smaller cities.

h_m ;MS antenna height [m]; d_m ;distance between the mobile and the building [km]; h_0 ; typical height of a building [m]; h_b ;BS height [m]; r; great circle distance between BS and mobile [m]; f_c=f x 10^{-6} carrier frequency [MHz]; λ ; free space wavelength [m].

The practical Pathloss can be calculated using the Eq. (2):

$$L_p(dB) = P_t - P_r \qquad (2)$$

Where P_t is the transmitted power which is equal to 47 dB and P_r is the received power. 'A' and 'E_3' are as mentioned above for small and medium cities [7-9]. The generation of such predictions is based on the assumption that the power of a signal decreases monotonically with the increase of the distance traveled by the signal [10].

3. Results and Discussions

Data Collection

An intensive drive test has been conducted along all pre-identified paths. GSM sample every 450 ms. This sampling rate is adequate to determine representative signal levels. The drive test process was conducted using TEst Mobile System, (TEMS) Investigation Data Collection V13.0 tool. The measurement reports of radio signal strength and many other RF parameters are collected by a special purpose mobile handset connected to the PC in which the TEMS program is installed. Furthermore, the positioning information is collected via GPS antenna terminal which have direct communication with GPS satellite

Data Processing

The measured data for each path has been recorded in terms of log files. These log files have been exported to a post processing tool called ACTIX for further processing and extraction of the required data. ACTIX Analyzer is a tool used for drive test post-processing, supporting network optimization and network troubleshooting. It is capable of troubleshooting RF problems automatically for GSM, Wi-MAX, WCDMA and LTE technologies. The data files collected from TEMS Investigation is post-processed from this tool. Also, it has been used to convert the recorded log files into graphical format for map viewing via Google Earth program (not shown). Moreover, it has been utilized to export coverage log files into Excel worksheets format for further manipulation. After that, the required data has been manipulated as the main aim of the conversion was to obtain the path loss and the link distance. After determining the Pathloss of the practical measurements for each distance, the study was carried in order to make a comparison between the experimental and theoretical data. Results (Fig. 1) show that the measured Pathloss is less than the predicted Pathloss by a major difference. However, there are several reasons that may have caused those significant differences. First of all, in Japan there are few areas virtually satisfying the conditions; and if any, they are narrow. Because of that reason Okumura selected the value for urban area as standard for open areas [8]. Moreover, the geographical situation of Japan is different from that in Oman due to geographical differences. As such, Root Mean Square Error (RMSE) was calculated between measured Pathloss value and those predicted by Hata model using the Eq. (3) [11-12]:

$$RMSE = \sqrt{\left(\sum \frac{\left(P_m - P_r\right)}{\left(N-1\right)}\right)} \qquad (3)$$

(2)

where;

P_m: Measured Pathloss (dB); P_r: Predicted Pathloss (dB); N: Number of Measured Data Points

The RMSE was found 113.459 dB but the acceptable range is up to 6 dB. Therefore, the RMSE is adjusted with the Hata equation for urban area. The modified result of Hata equation is shown in Fig. 2 and the RMSE in this case is less than 6dB, which is acceptable.

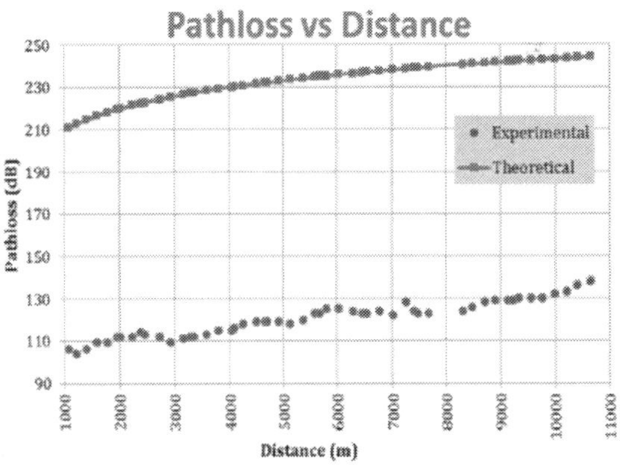

Fig. 1: Theoretical and Experimental Pathloss

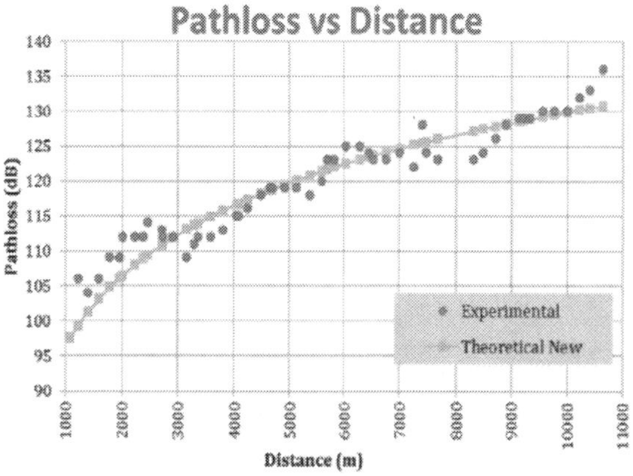

Fig. 2: Pathloss on experimental data set B.

Based on that practical data, the propagation Pathloss and the distance has been re-verified for another cell [7]. Theoretical simulation and the obtained experimental data is compared and analyzed further using a Trigonometric Regression model as in Eq. 4.

$$L_d = \beta_o + \beta_1 d + \beta_2 \mathrm{Sin}\left(\frac{\pi L}{\sqrt{d}}\right) + \beta_3 \mathrm{Cos}\left(\frac{\pi L}{\sqrt{d}}\right) + \varepsilon_d \quad (4)$$

The following regression model with trigonometric terms as in Eq. 7 was specified for estimation and prediction of Path Loses because the area under study has a pattern of obstacles such as high-rise buildings and towers, which could interrupt the signals in an irregular cyclical manner. The model has the form:

Where

d is the distance in meters from an antenna; L_d is the Path loss at distance d; L is the overall expected Path loss; ε_d is the random noise at distance d. The ε_d is assumed to be independently and identically distributed. The $\beta_0, \beta_1 \ldots \beta_5$ are the parameters of the model. These parameters were estimated by Ordinary Least Squares (OLS) method. The assumption of independence of ε_d was checked by Durbin Watson (DW) test as given below.

$$DW = \frac{\sum (E_d - E_b)^2}{(E_b)^2}$$

Where E_d is the estimate of the random noise at distance d. and E_b is the estimate of the random noise distance b which is distance just before d. The estimates of all these parameters along with their standard errors (SE), t-values and p-values as well as R are presented in Table. 1. This table shows that most of the distance significantly affects the Path loss but the Durbin Watson (DW) test value should be close to 2 to satisfy the assumption of independence of noise at successive distances. The DW value of 1.36 is significantly below the significant points at 1% and 5% levels which means there is at least first order autocorrelation in Path loss at successive spatial points.

Table: 1 Estimates of the parameters

Parameters	Parameters Estimates	SE	T	P
β_0	103.846	4.251	24.43	0.000
β_1	0.046497	0.008669	5.36	0.000
β_2	1.0072	0.8852	1.14	0.260
β_3	1.6997	0.7908	2.15	0.036

DW=1.36 and R2 = 54.6

Table: 2 Estimates of the parameters with an additional term

Parameters	Parameters Estimates	SE	T	P
β_0	71.87	14.48	4.96	0.000
β_1	0.02941	0.01106	2.66	0.010
β_2	0.3195	0.1346	2.37	0.022
β_3	0.6000	0.8754	0.69	0.496
β_4	0.9330	0.8567	1.09	0.281

DW=1.93 and R^2 = 56.2

Therefore, it is decided to include an autoregressive term Lb in the model to accommodate the serial correlation. The model Eq. 4, thus takes the form as presented in Eq. 5 as below with usual OLS assumptions. The DW value of 1.93 in this table is very close to 2 hence the model assumption of noise term is satisfied.

$$L_d = \beta_o + \beta_1 d + \beta_2 L_b + \beta_3 \mathrm{Sin}\left(\frac{\pi L}{\sqrt{d}}\right) + \beta_4 \mathrm{Cos}\left(\frac{\pi L}{\sqrt{d}}\right) + \varepsilon_d \quad (5)$$

The estimates of the parameters presented in table 2 were used to predict the Path loss at each given distance. Fig. 3 clearly reveals the adequacy of the proposed model.

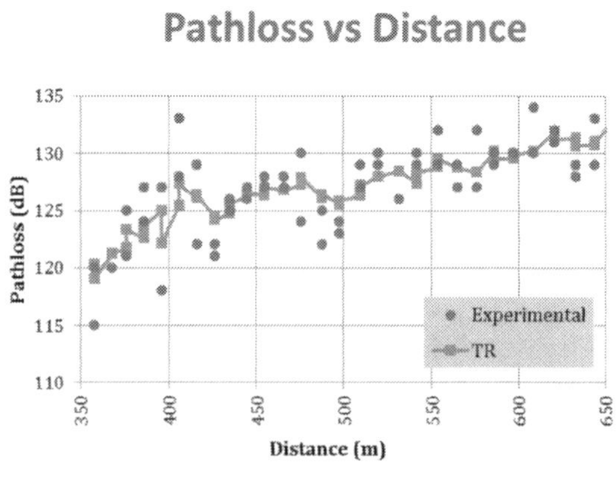

Fig. 3: Pathloss versus distance for Experimental, and Trigonometric regression modeling

Overall, by calculating the RMSE for the second cell and for most of the data sets it was found to be 3.2058 dB, which is an acceptable figure. However, few data points were a bit far from the interpolated values which are attributed to the nature of the cell with high rise buildings. Although there are many predictions methods that are based on deterministic processes through the availability of improved databases, the Okumura-Hata model is still mostly used [13-14]. That is because of the ITU-R recommendation for its simplicity and its proven reliability.

We also know that the obstacles in the path significantly influence the radio signal propagation [14]. Above all, wireless communication system avoids obstacles such as crossing objects owned by others. There are also many problems in a realization of wireless communication system in some applications. However, the current study of neural network and other methods might explain these that have influence in our data acquisition in business district area with high-rise buildings and curved roads.

4. Conclusion

This work focused on predicting the root mean signal strength in different areas. As most propagation models aim to predict the median Pathloss, existing prediction models differ in their applicability over different terrain and environmental conditions. The effects of terrain situation predicted at 900 MHz were analyzed. Experimental results of radio signals propagation for an urban area in Oman were compared with those predicted based on Okumura-Hata model. The contribution is the prediction by trigonometric regression and validation of experimental data. If environmental information is included in the model, better prediction results might be achieved. Trigonometric regression gave us also the missing experimental points showing a good agreement within acceptable limits.

Acknowledgment

The authors would like to thank SQU for the support to carry out this study and also Mr Mahmood, Mr Asaad and Mr Wahab for their valuable feedback.

References

1. Z. Nadir, M. Bait-Suwailam, M. Shafiq, "RF Coverage Analysis and Validation of Cellular Mobile Data using Neural Network", the International Journal of Neural Networks and Advanced Applications, (IJNN-NAUN), ISSN: 2313-0563, Vol. 1, pp. 30-36, 2014.
2. M. Kumari, T. Yadav, P. Yadav, "Comparative study of pathloss models in different environments," *International Journal of Engineering Science and Technology (IJEST)*, vol. 3, No. 4, pp. 2945-2949, Apr 2011.
3. M. Hata, "Empirical formula for propagation loss in land mobile radio services," *IEEE Transactions on Vehicular Technology*, vol. 3 VT 29, pp. 317-326, August 1980.
4. R. D. Wilson and R. A. Scholtz, "Comparison of CDMA and modulation schemes for UWB radio in a multipath environment," *Proceedings of IEEE Global Telecommunications Conference*, vol. 2, Dec 2003, pp. 754-758.
5. R. N., S. S. K., "A study on pathloss analysis for GSM mobile networks for urban, rural and suburban regions of Karnataka state," *International Journal of Distributed and Parallel Systems (IJDPS)* vol.4, No.1, DOI: 10.5121/ijdps. 2013. 4105 pp.53-66 January 2013.
6. D. A. Shalangwa, S. K. Singh, "Measurement and modeling of pathloss for GSM900 in sub urban environment over Irregular terrain," *International Journal of Computer Science and Network Security (IJCSNS)*, vol.10, No.8, pp. 268-274, August 2010.
7. Z. Nadir, N. Elfadhil, F. Touati, "Pathloss determination using Okumura-Hata model and spline interpolation for missing data for Oman," *World Congress on Engineering, IAENG-WCE-2008*, Imperial College, London, United Kingdom, 2-4 July 2008, pp. 422-425.
8. Z. Nadir, M. I. Ahmed, "Characterization of pathloss using Okumura-Hata model and missing data prediction for Oman," *IAENG Transactions on Engineering Technology*, vol. (5): special Edition of the *International multi-conference of Engineers and Computer Scientists 2009,* AIP conference proceeding, volume 1285, 2010, pp. 509-518.
9. Z. Nadir, "Empirical pathloss characterization for Oman," *IEEE Computing, Communications & Applications Conference 2012, (IEEE ComComAP 2012)*, HongKong University of Science and Technology, Hong Kong, China, 11-13 January 2012, pp.133-137.
10. F. Mourad, H. Snoussi, M. Kieffer, C. Richard, "Robust bounded-error tracking in wireless sensor networks," *16th IFAC Symposium on System Identification Brussels, SYSID'12* Brussels Belgium, 11-13 July 2012.
11. D. A. Shalangwa, & G. Jerome, "Pathloss propagation model for Gombi town Adamawa State Nigeria," *International Journal of Computer Science and Network Security (IJCSNS)*, vol.10, No.6, pp. 186-190 June 2010.
12. Z. Nadir, M. Ahmad, "RF Coverage and Pathloss Forecast Using Neural Network," *Advances in Intelligent Systems and Computing, (Book Series by Springer)*, ISSN: 2194-5357 (Print) 2194-5365 (Online), vol. 240, pp. 375-384, 2014.
13. Y. Okumura et al., "Field strength and its variability in VHF and UHF land-mobile radio service," *Review of the Electrical Communications Laboratory*, vol. 16, no. 9-10, September-October 1968.
14. Z. Nadir, M. Suwailam, "Pathloss Analysis at 900MHz for outdoor environment," *Proceedings of the 2014 International Conference on Communications, Signal Processing and Computers, (EUROPMENT 2014)*, Interlaken, Switzerland, 22-24 February 2014, pp.182-186.

Forecasting KOSPI using Elman network

Hongchul Ahn, Hotak Hong [a], Jongho Nang and Saejoon Kim

Department of Computer Science and Engineering, Sogang University, Seoul 121-742, Korea

Abstract. Due to the non-stationary nature of stock market index, making a prediction on its course is a truly challenging task. Research has been actively conducted to predict stock market indices by means of machine learning in recent years. In our research, we made a prediction of KOSPI for one week based on Elman Network. Based on the predictive result, we ran a simulation from which we obtained 3.16% yield over a period of one year. In this paper, we describe how we exploited Elman network to make predictions on stock markets, then we propose a method for using the predictive values for investment.

1. Introduction

Stock market indices are influenced by a great range of indicators hence making predictions on them is an extremely difficult task. As machine learning has been applied to many fields lately, many researchers are exploring to apply machine learning to predict stock prices. With so many indicators that affect the stock price, the choice of input parameters is critical to the prediction result [1, 2]. There are two major types of stock price analysis: fundamental and technical. While a firm's intrinsic values are studied in a fundamental analysis, past data such as stock price history and stock trade volume is reviewed to predict future stock price. There has been a widely accepted consensus among investors that investors cannot beat the stock market with technical analysis if the capital market operates efficiently. However, it has been demonstrated that information obtained with technical analysis is as powerful as data obtained with macroeconomic indicator analysis [3]. Technical analysis was also observed to be very effective in an uncertain market circumstances [4]. In experiments for this paper, we exploited technical analysis indices as input parameters. Features were selected with stepwise regression to prevent entering features repeatedly. Elman network, which is a type of RNN, was chosen as it shows solid performance on time series analysis and prediction was made on weekly KOSPI level, on which we base our investment decision.

The organization of this paper is as follows. In Section 2, related work is discussed. Section 3 explains our approach to input data selection and data processing. Section 4 demonstrates our experiment design including experiment length and machine learning method followed by experimental results. Section 5 concludes the paper with suggestions for future work.

a

This work was supported by the ICT R&D program of MSIP/IITP. [Open ICT Convergence Support Program (R0215-15-1001) and ICT/SW Creative Research Program (R2215-15-1003)]

2. Related Work

Machine learning has been actively embraced by more and more academic fields in recent years. One notable area that has adopted machine learning is stock market prediction. In the fields of business administration and economics, fundamental analysis and technical analysis have been produced [5, 6]. In statistics, numerous time series approaches [7]. Despite the existing many attempts to predict the stock market, controversy has persisted if stock market prediction would be valid or effective in the actual stock market. With the introduction of various machine learning methods, stock market prediction now shows satisfying performances.[8, 9] In this paper, we used Elman network, a type of RNN, for stock market prediction.

3. DATA

Many different types of information can be used to predict the direction of the stock market. There are two types of stock price analysis methods: fundamental analysis and technical analysis. The purpose of our experiment was predicting the stock market index of one week in the future, hence we excluded fundamental data and only used technical analysis data for input parameters, which we obtained from Daewoo Securities HTS. In order to prevent repeated input of input parameters, stepwise regression was used to make indicator selection. In our experiment, each input parameter is introduced and discussed on manipulation method. We used a statistics program R for experiment.

3.1 Stock Price Index

The following items are included when describing a stock: the opening and closing prices, the highest and lowest prices in a specific period of time. The price of a stock of a given period generally refers to the final price of the stock. Most technical indices used in this paper are calculated based on stock price and trading volume, hence a solid understanding of these indices is essential in successful configuration of input parameters.

3.2 Technical Indices for Experiment

The data frame we used consists of 14 features and yields one output variable. The features have been selected among meaningful technical indices of distinctive nature [10]. Input parameters were categorized as trend, oscillator, and momentum types, and representative technical indicators of each type were selected as our input parameters. Among the technical indices, moving average (2, 5, 20, 60) was chosen among trend-types, moving average convergence/divergence (MACD) and relative strength index (RSI) from oscillator types, and MO(3) and MO(9). We also added opening price, highest and lowest price, trading volume as well as sigma(20) in estimating stock prices. The opening and closing prices, highest and lowest prices at time point T are the most basic pieces of information in calculating the closing price of T+1. Trade volume and sigma were also added to our input parameter list as these add depth to estimation. From trend types, we chose moving average as an input parameter. MACD and RSI were chosen from oscillator type indices and MO from momentum types. As shown, we selected input parameters from trend, oscillator, and momentum types and weekly closing price at t+1 was estimated as output. Stepwise regression was adopted for final selection of indicators so as to suppress repeated impact in the experiment.

3.2.1 MA

The moving average during the period of N days is calculated as follows:

$$MA(t) = \frac{1}{N}\sum_{t=i}^{N-1} price(t-i) \qquad (1)$$

We used MA of 2, 5, 20, and 60 days

3.2.2 Momentum

Momentum is one of the most basic technical indices, calculated by comparing the current stock price to the stock price at a specific time in the past. Momentum of 0 indicates there was no change in the stock price over the course of a specific period of time while a positive value means the stock price has increased and a negative momentum means otherwise. Momentum is calculated as follows:

Table 1. Number of states for the indicators.

Label	Index
x1	open price (t)
x2	Highest price (t)
x3	Lowest price (t)
x4	Weekly close price (t)
x5	MA(2)
x6	MA(5)
x7	MA(20)
x8	MA(60)
x9	volume
x10	Momentum(3)
x11	Momentum(9)
x12	RSI(14)
x13	MACD(12,26)
x14	Sigma(20)
Y	Weekly close price (t+1)

$$MO(t) = price(t) - price(t-n) \qquad (2)$$

The term t is the closing price at day t. Comparing the current stock price and n days back shows whether the stock is going up or down in price. In our experiment, we used 3 day and 9 day momentum indices

3.2.3 RSI

Relative strength index (RSI) shows the magnitude of price increase of a stock. The value of RSI lies between 0 and 100. RSI values below 25 is interpreted as depressed increase while the stock is thought to be overheated if above 75. RSI is calculated as follows:

$$RSI(t) = \frac{A}{A+B} \times 100\ (0 \leq RSI \leq 100) \qquad (3)$$

The magnitude of each increase in the stock price over N days is added up and divided by the magnitude of each increase and decrease of stock price over N days to calculate RSI. In our calculation of RSI, 14-day period was used.

3.2.4 MACD

Moving average convergence/divergence (MACD) was devised by Gerald Appel when he observed short-term and long-term moving averages repeat convergence and divergence and address the MA only reflects the past and does not directly predict the future. MACD is calculated as follows:

$$MACD(t) = SMA(t) - LMA(t) \qquad (4)$$

where SMA stands for short-term moving average and LMA is long-term moving average. In our experiment, SMA(12) and LMA(26) were used.

4. Experiment

4.1 Data Processing

Our input parameters consist of a variety of technical indicators with different characteristics, and the indicators lie in different scales, hence each parameter needs to be normalized to be used as a feature. This is necessary because neural networks perform effectively when the values of input data are concentrated around zero. In our approach, all of the input parameters used as feature were normalized in [0, 1] as follows[11]:

$$\frac{X_t - Min(X)}{Max(X) - Min(X)} \quad (5)$$

The drawback of normalization as Equation (5) is that the minimum value becomes 0 hence data loss could incur. Hence we chose to add 0.00001 to all of the input parameter values to prevent this. The range of each feature after all is normalized within [0.00001, 1.00001].

4.2 Stepwise regression

Stepwise regression is a method for selecting variables that meaningfully affect the result in regression analysis. In selecting variables, a few methods are available such as forward selection, backward elimination, and stepwise selection. In our experiment, backward elimination was used, which starts with all of the candidate variables and deletes variables that least affect the model one by one. With backward elimination, we selected 7 out 14 parameters. Stepwise regression was performed in our experiment with the 7 input parameters.

Table 2. The 7 features selected with stepwise regression

Label	Index
x2	Highest price (t)
x3	Lowest price (t)
x6	MA(5)
x7	MA(20)
x9	volume
x12	RSI(14)
x13	MACD(12,26)
Y	Weekly close price (t+1)

4.3 Experimental period

In our experiment, our method was applied to the weekly data collected between January 2, 2000 and October 26, 2015. The 722 weekly data points between January 2, 2000 and October 28, 2013 were used for training; data for validation comes from between November 4, 2013 and October 27, 2014. Test was performed for the period between November 3, 2014 and October 26, 2015. In our approach, the initial model was formed with the training data and validated to identify parameters that improve performance to construct the final model.

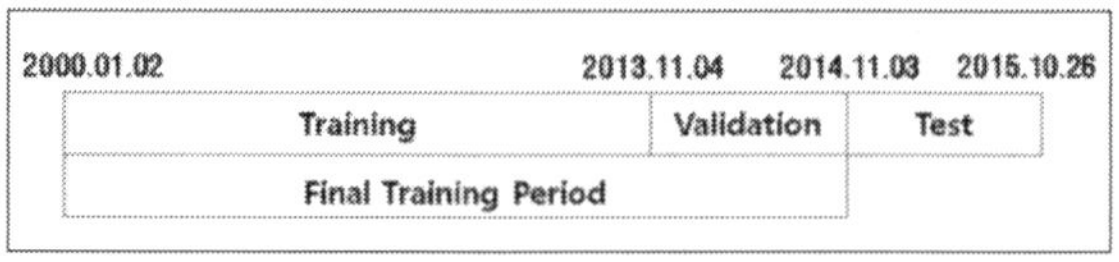

Figure 1. Data range for training, validation, and test.

4.4 Elman network

Elman network is known for satisfying performance for time-series data[12, 13], but it is a deep learning algorithm requiring a large volume of data. We resorted to Elman network, which is a simple recurrent network, for prediction since the amount of available data is not sufficiently large. There is a single hidden layer in a typical multi-layer perceptrons network, and a variety of data from the input parameters are extracted in the hidden layer before relayed to the output layer. A hidden layer consists of multiple nodes, and each node extracts different types of data from the input variables. An Elman network is similar to the multi-layer perceptrons model in many aspects such as the existence of a hidden layer consisting of multiple nodes as well as relay of input data from the input layer to the hidden layer and then to the output layer. Unlike the multi-layer perceptrons model however, in an Elman network, output variables are calculated at the same time when data is extracted in the hidden layer and fed as the input data for the next iteration. Figure 2 shows how a model is constructed in an Elman network by accumulating data over time. This characteristic was one major reason we opted for Elman network. To compare performance we also tested other machine learning algorithms such as SVM and multi-layer perceptrons with identical input data.

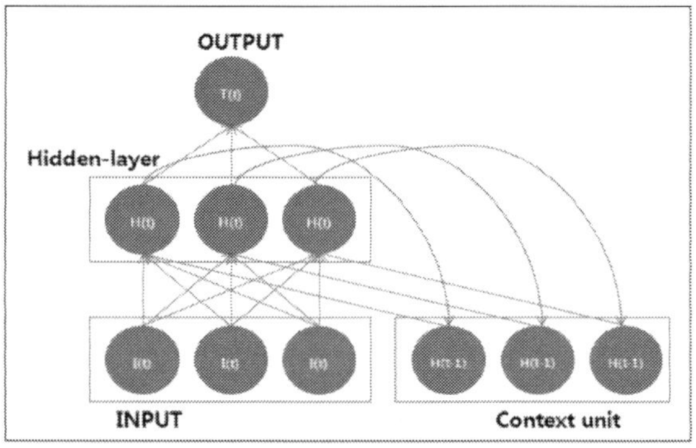

Figure 2. Elman network.

4.5 Forecast performance

Three different performance measuring methods were used in evaluating the performance of predicted values by each model. RMSE (root-mean-square error) (6) and MAE (mean absolute error) (7) evaluates the discrepancy between the estimated values by a model and the real data. Smaller values of RMSE and MAE indicate that the estimations are more accurate. MAPE (mean absolute percentage error) (8) expresses the difference between estimation and actual data in a ratio, and lower ratio indicates that the estimation is closer to the actual data.

$$RMSE = \sqrt{\frac{1}{N}\sum_{t=1}^{N}(X_t - F_t)^2} \quad (6)$$

$$MAE = \frac{1}{N}\sum_{t=1}^{N}|X_t - F_t| \quad (7)$$

$$MAPE = \frac{1}{N}\sum_{t=1}^{N}\frac{|X_t - F_t|}{X_t} \quad (8)$$

4.6 Experimental Results

Testing with each machine learning method showed that Elman network produces the best results for all RMSE, MAE, and MAPE compared to linear SVM or multilayer perceptron network, hence we chose Elman network in our method.

4.7 Profit simulation

With estimation data from the Elman network model, we simulated trading on the data.

Table 3. Experimental Performance.

Performance method	SVM(linear)	Neural Network (MLP)	Elman Network
RMSE	69.43	37.37	34.75
MAE	54.44	28.94	27.40
MAPE (%)	2.70%	1.45%	1.36%

Trading rules were defined based on the work[14] by Cheng et al.:

Rule 1 : Sell rule

$$IF\ |forecast(t) - actual(t)| < 0 \quad (9)$$

Rule 2 : Buy rule

$$IF\ |forecast(t) - actual(t)| > 0 \quad (10)$$

Our trading simulation was performed on an imaginary exchange-trade fund (ETF) that estimates stock prices with 100% accuracy and zero tracking error. For each week, a sell signal triggers buying an Inverse ETF and a buy signal triggers buying an Index ETF. If a sell signal is received for a stock currently owned, then all of the stock in question is all sold.

Definition of profit (11)

$$Profit = \sum_{t_s=1}^{p}\left(\frac{actual(t_s+1) - actual(t_s)}{actual(t_s)}\right) + \sum_{t_b=1}^{q}\left(\frac{actual(t_b) - actual(t_b+1)}{actual(t_b)}\right)$$

When we simulated stock trading based on these rules after making weekly predictions on KOSPI for one year starting in November 2014, we could obtain a profit of 3.16%. Our experiment demonstrates that investing in the stock market based on Elman network could yield profit.

5. Conclusion

Making weekly predictions on the stock market index based on machine learning was most accurate when the Elman network was used as evaluated with RMSE. Our trading simulation over one year of period starting in November 2014 produced profit at 3.16%. This is a meaningful result but it could have been affected by many other variables not considered in our experiment, and the result does not qualify as solid because the financial market is volatile. In the future work, we will work on making estimated values more accurate and applying the estimated results to stock trading.

References

1. M. C. Chan, C. C. Wong, and C. C. Lam, Financial Time Series Forecasting by Neural Network Using Conjugate Gradient Learning Algorithm and Multiple Linear Regression Weight Initialization, Computing in Economics and Finance, (2000)
2. R. J. Frank, N. davey, and S. P. Hunt, Time Series Prediction and Neural Networks, Jounal of Intelligent and Robotics System, Vol.**31** Issue1-3, pp.91-103, (2001)
3. Zhou, G., and Y. Zhu. An Equilibrium Model of Moving-Average Predictability and Time-Series Momentum. Working Paper, Tsinghua University and Washington University in St. Louis (2013)
4. Zhu, Y., and G. Zhou. Technical Analysis: An Asset Allocation Perspective on the Use of Moving Averages. Journal of Financial Economics, **92**, 519-522. (2009)
5. R. D. Edwards and J. Magee, Technical Analysis of Stock Trends, John Magee, Inc., (1974)
6. S. Cottle, R. F. Murray, F. E. Block, and D. L. Dodd, Graham and Dodd's *Security Analysis*(5th edition), McGraw-Hill. (1988)
7. S. M. Kendall and K. Ord, *Time Series*, Oxford, New York. (1997)
8. Vatsal H. Shah, Machine LEarning Techniques for Stock Prediction, Technical Report, New York University, (2007)
9. Li Deng. A Tutorial survey of architectures, algorithms, and applications for deep learning, APSIPA Transactions on Signal and Iformation Processing. Vol.3. (2014)

10. T. Y. Mieko., T. Seiji, Adaptive Use of technical indicators for the prediction of intra-day stock prices. (2007)
11. Adebiyi, A. A. and Ayo, C. K. and Adebiyi, M. and Otokiti, So, O. An Improved Stock Price Prediction using Hybrid Market Indicators, African Journal of Computing & ICT, Vol.5. No.5, pp.124-135. (2012)
12. H. Shi, X. Liu, Application on stock price prediction of Elman neural networks based on principal component analysis method, IEEE Transactions on Pattern Analysis and Machine Intelligence, Wavelet Active Media Technology and Information Processing (ICCWAMTIP), Vol.10, 411-414. (2014)
13. J Wang, J Wang, W. Fang, H. Niu, Financial Time Series Prediction Using Elman Recurrent Random Neural Networks, Hindawi Publishing Corporation Computational Intelligence and Neuroscience, Article ID 613073. (2015)
14. C. H. Cheng, L. Y. Wei, Y. S. Chen, Fusion ANFIS models based on multi-stock volatility causality for TAIEX forecasting, Neurocomputing, 3462-3468. (2009)

MATEC Web of Conferences **54**, 06001 (2016) DOI: 10.1051/matecconf/20165406001
MIMT 2016

Influence of the Shape of Explosive Charge on the Blast Wave Propagation

Yo-Han Yoo[1], Youngsik Choi[2], Joonwon Lee[2] and Kyung-Jae Yun[3]

[1]*Agency for Defense Development & University of Science and Technology, PO BOX 35-1, Yuseong-gu, Daejeon 305-600, Republic of Korea*
[2]*School of Mechanical Engineering, Chung-Ang University, Heukseok-ro 84, Dongjak-gu, Seoul 156-756, Republic of Korea*
[3]*Agency for Defense Development, PO BOX 35-1, Yuseong-gu, Daejeon 305-600, Republic of Korea*

Abstract. For several decades, a blast wave due to explosive detonation has been investigated extensively. Base on a significant amount of experimental data, the blast wave propagation has been predicted at specific conditions. However, only spherical shape of explosive has been considered in most studies. Recently, it was noted that the shape of explosive influences the blast wave propagation significantly. In this study, a finite element analysis was carried out to investigate the effect of the shape of explosive on the blast wave propagation. Two different shapes of explosive were compared in blast wave propagation; spherical and cylindrical shape. It was found that the spherical and cylindrical shape blast show different characteristics in the blast wave propagation. The spherical blast showed the isotropic wave propagation as verified by experimental data. However, the cylindrical blast showed more concentrated and faster propagation in the axial direction.

1 Introduction

In defense industry, the effects of explosive blast have been an important issue to address. A considerable amount of research was performed on the blast wave propagation due to explosive blast. There are many parameters which affect the blast wave propagation due to explosive blast. The medium where the explosive is detonated is one of the most important parameters [1].

From early stage of investigating the explosive blast, only spherical shape of explosive was investigated in terms of the pressure, impulse, arrival time and other parameters [2]. Although these empirical data enabled predicting the blast phenomena, the shape of explosive influences the blast wave propagation considerably [3,4].

Due to the lack of experimental data on various explosive shapes, a finite element analysis has been used to predict the blast wave propagation. With the aid of the computer, it is possible to calculate complex equations fast and efficiently. There are three methods to discrete meshes: Lagrangian, Eulerian and Arbitrary Eulereian Lagrangian(ALE). Each method has its own advantages in calculating the behaviour of solid structures or gas flows [5].

In this study, two different shapes of explosive were simulated by using a finite element analysis to investigate the influence of the shape of explosive charge on the blast wave propagation.

2 Free Air Blast

From the empirical data, researchers have found the fundamental parameters which explain the blast wave propagation due to the detonation of explosives such as TNT. Fig. 1 show the blast pressure pulse with relevant parameters [6].

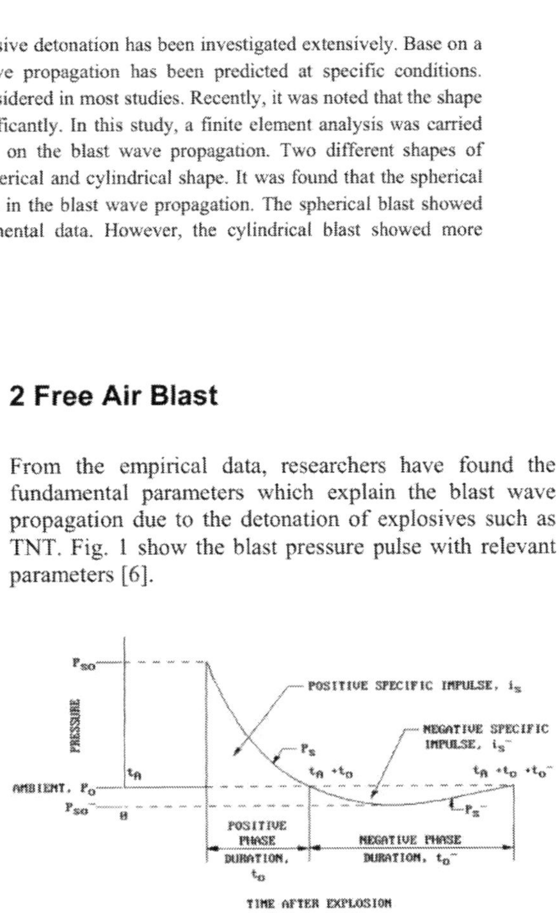

Figure 1. Blast pressure pulse [6].

From the ambient pressure, P_o, pressure increases rapidly to the peak pressure, P_{so} as the explosion pressure arrives. After the pressure reaches the peak value, pressure drops to ambient value. The time from the peak pressure to the ambient pressure is called a positive duration time. Subsequently, pressure drops to the negative value and restores to the ambient pressure. The time between these two points is called a negative duration time. The critical damage is mostly from a positive phase. However, a negative phase is also important in specific material [7]. The TNT shape influences the blast wave propagation. Hence, two different shapes of TNT are analysed to investigate the effect of TNT shape on the blast wave propagation. Fig. 2 shows two different shapes of TNT.

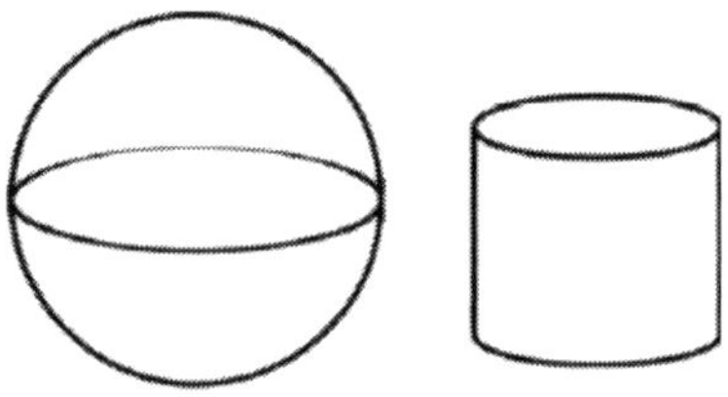

a) Sphere b) Cylinder

Figure 2. Two different shape of TNT

3 Simulation Setup

Fig. 3 shows the two different shapes of explosive model inside the medium area. 500mm x 500mm x 500mm of medium area (air) were generated by using 20mm mesh size. Both spherical and cylindrical explosives have a 1 kg of TNT equivalent mass. For TNT, Jones-Wilkins-Lee (JWL) equation of state (EOS) is used. See Eq. 1 where A, B, R_1, R_2 and ω are parameters for the JWL equation. The input parameters for TNT are shown in Table 1. All units are in mm, kg and ms.

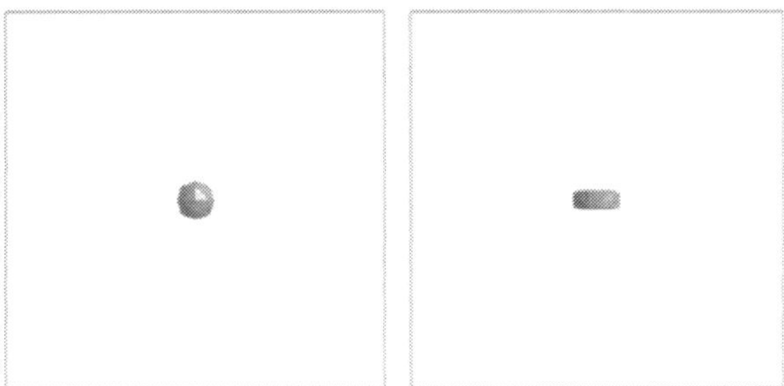

a) Sphere b) Cylinder

Figure 3. Shapes of explosive

The EOS for air is a linear polynomial. See Eq. 2 where C_0 to C_6 are coefficients for air. In case of ideal gas, C_4 and C_5 are 0.4, while the other coefficients are 0 [5]. The input parameters for air are shown in Table 2. All units are in mm, kg and ms.

4 Results

Fig. 4 shows the simulation results. Fig. 4a shows the initial stage of two different explosive model; spherical and cylindrical shapes in a quarter size. Each explosive contains 1 kg of TNT equivalent mass.

After detonation, the difference in pressure propagation is shown in Fig. 4b. The spherical blast shows the isotropic pressure propagation from the detonation point. However, the cylindrical blast shows a different trend. After detonation, the pressure propagates faster in the radial direction than the axial direction.

As the pressure propagates, this difference becomes more significant as shown in Fig. 4c. At 0.16 ms, the pressure propagation of the spherical explosive still shows the isotropic propagation, while that of the cylindrical explosive shows a significant difference between the radial and axial directions.

Fig. 5 shows the incident pressure distribution at a distance of 400mm from the detonation point. For the spherical blast, the pressure distribution is identical in the x and y directions. However, for the cylindrical blast, the pressure distribution shows a difference between the two directions. The pressure in the axial direction shows a larger peak pressure than the pressure in the radial direction. The peak pressure in the axial direction is 2.12 MPa, while the peak pressure in the radial direction is 1.62 MPa, which is same as the peak pressure in the x and y directions for the spherical blast.

$$P = A\left(1-\frac{\omega}{R_1V}\right)e^{-R_1V} + B\left(1-\frac{\omega}{R_2V}\right)e^{-R_2V} + \frac{\omega E}{V} \quad (1)$$

$$P = C_0 + C_1\mu + C_2\mu^2 + C_3\mu^3 + (C_4 + C_5\mu + C_6\mu^2)E \quad (2)$$

Table 1 Input parameters for TNT

Density	Detonation velocity	Chapman Jouget pressure				
1.658×10^{-6}	6930.0	21.0				
A	B	R_1	R_2	ω	E_0	V_0
373.76	3.7347	4.15	0.9	0.35	7.0	1.0

Table 2 Input parameters for air

Density	E_0	V_0				
1.2×10^{-9}	2.5×10^{-4}	1.0				
C_0	C_1	C_2	C_3	C_4	C_5	C_6
0.0	0.0	0.0	0.0	0.4	0.4	0.0

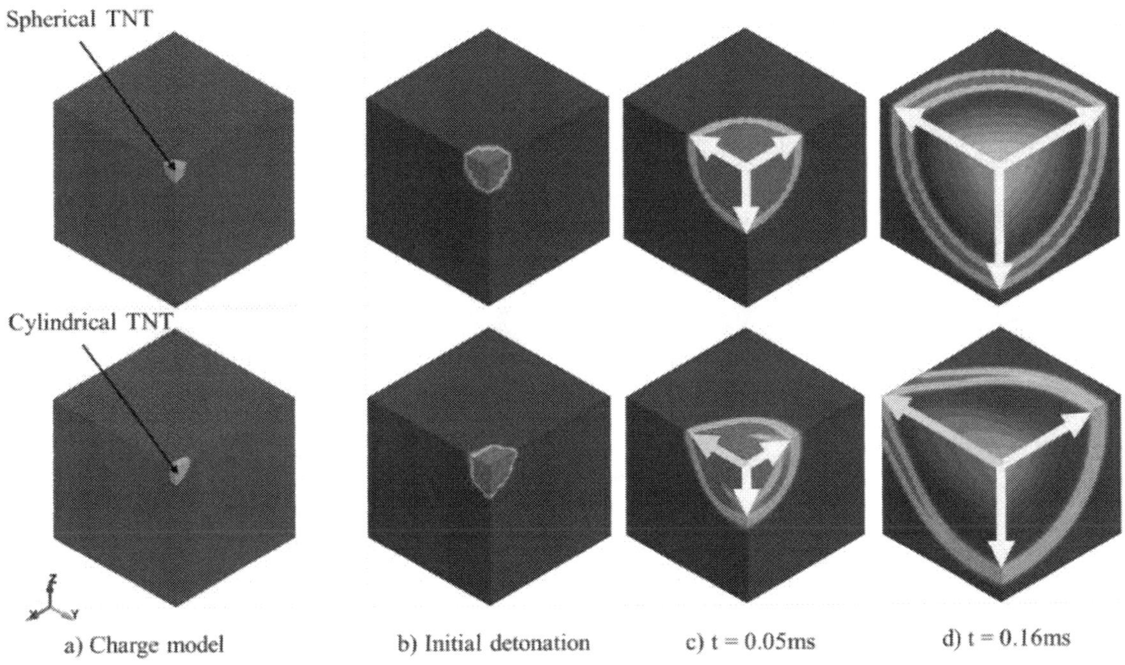

Figure 4. A quarter model of two explosive shapes and pressure propagations

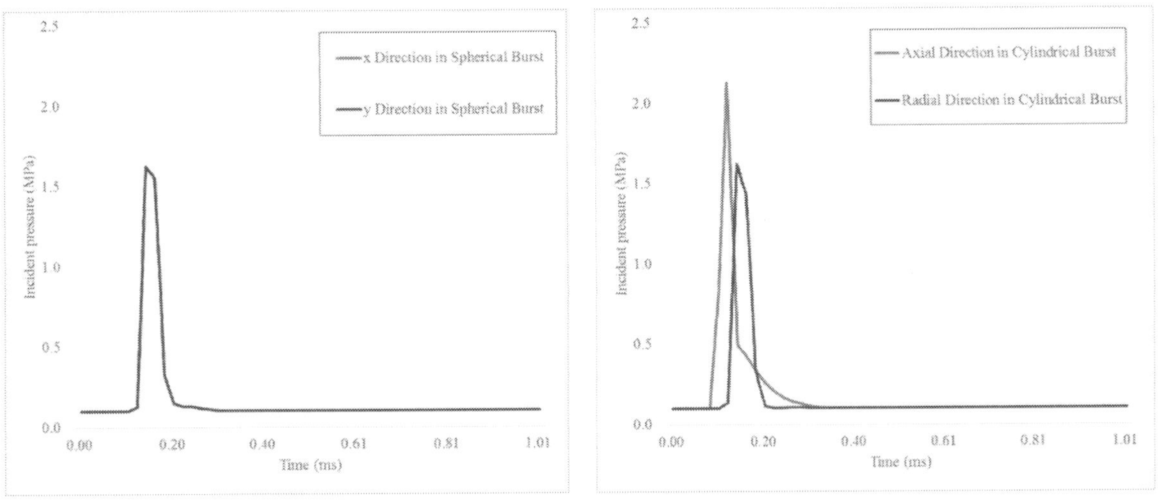

Figure 5. Incident pressure distribution at a distance of 400mm from detonation point

5 Summary

Two different shapes of explosive were simulated with an equal explosive weight. The results showed the difference in the blast wave propagation between the spherical and cylindrical shapes of explosive. The spherical blast showed the identical wave propagation in all directions. However, the cylindrical blast showed a different wave propagation depending on the direction. The pressure propagated faster and more concentrated in the axial direction than the radial direction.

For the spherical blast, the peak pressure values in the x and y directions were the same. For the cylindrical blast, the peak pressure value in the axial direction was larger than that in the radial direction. Consequently, it is conceivable that the influence of cylindrical blast is more significant in the axial direction than the radial direction, while the influence of the spherical explosive is isotropic.

Acknowledgements

The authors gratefully acknowledge the financial support provided by Agency for Defense Development (Grant Nos. ADD-13-70-04-05 and ADD-UE145116GD).

References

1. M. Held, Similarities of shock wave damage in air and in water, Propellants, Explosives, Pyrotechnics 15 (1990) 149-156.
2. G. B. Whitham, The propagation of spherical blast, Proceedings of the Royal Society of London: Series A. Mathematical and Physical Sciences 203 (1950) 571-581.
3. M. N. Plooster, Blast effects from cylindrical explosive charges: Experimental measurements, Denver Research Inst Co (1982)
4. C. Knock, N. Davies, T. Reeves. Predicting blast waves from the axial direction of a cylindrical charge, Propellants, Explosives, Pyrotechnics 40 (2014)
5. X. Quan, N. K. Birnbaum, M. S. Cowler, B. I. Gerber, R. A. Clegg and C. J. Hayhurst, Numerical simulation of structural deformation under shock and impact loads using a coupled multi-solver approach, Proceedings of 5th Asia-Pacific Conference on Shock and Impact Loads on Structures, (2003)
6. Department of the Army, Fundamentals of Protective Design for Conventional Weapons, Technical Manual TM5-855-1, November (1986)
7. T. Krauthammer, A. Altenberg, Negative phase blast effects on glass panels, International Journal of Impact Engineering 24.1 (2000) 1-17.
8. LS-DYNA® KEYWORD USER'S MANUAL, Livermore software technology corporation. (2012)

MATEC Web of Conferences 54, 06002 (2016) DOI: 10.1051/matecconf/20165406002
MIMT 2016

Impact Evaluation of a Decommissioned Landfill on Surrounding Groundwater Quality-a case study, Wang-Tien Landfill Site- southwest Taiwan

Hsien-Jui Yu, Chao-Shi Chen, Chia-Huei Tu, Ying Chia Ni, and Cheng-Wei Wu

Department of Resource Engineering, National Cheng Kung University, No.1, University Road, Tainan City 701, Taiwan

Abstract. Landfill leachate composes the potential risk of terrible environmental contamination due to the release of diverse pollutants leaking from the containment system. Hence, a study of contaminant transport is of interest for its suitable management. The aim of this paper was to evaluate groundwater pollution risks from contaminants near the decommissioned Wang-Tien landfill site using a numerical approach. Stable concentrations of various physical-chemical parameters were determined in surface and groundwater samples collected from the study area. Principal component analysis was applied and identified chloride concentrations as the main indicators of groundwater contamination caused by landfill leachate. In this study, Groundwater Modelling System (GMS) was used to predict the subsurface migration of contamination within 1, and 10 years. The numerical results of chloride have shown that the concentration in the adjacent creek of landfill was 782 mg/l. However, the Monitoring Standard Value of Environmental Protection Administration (EPA, in Taiwan) for Taiwan defined the maximum value of chloride in groundwater for 625 mg/l. The result shows that chloride migrated widely to other places, such as the creek, and has had an adverse effect on the environment.

1 Introduction

Landfills were generally not capped after closure to avoid the formation of leachate due to the infiltration of rainwater, nor were precautions taken to prohibit the spreading of leachate to underlying aquifers. Therefore, landfills need to be managed and controlled to avoid negative impacts on the environment after closure. Chloride is recognized as a pollutant that has lasting adverse on an ecosystem. High levels of chloride discharged into freshwater bodies may be harmful to fish and other aquatic organisms which are not adapted to live in saline environments. At the present time, GMS is widely used to model and simulate saturated groundwater flow. Many studies use the GMS software package and its features to predict and simulate conditions and conductivity of groundwater and the contamination of groundwater [1-6].

In this paper, a numerical modelling package, GMS (version 7.1.), was used to predict the fate and transport of chlorides in a decommissioned landfill located in Yong-Kang, Tainan City, Taiwan. Generally, as a first step in any predictive analysis, it is necessary to generate the observed conditions in model calibration. The process of calibrating flow and transport models typically involves several steps. The modelling of groundwater flow and contaminant transport at the Wang-Tien landfill site was begun before the field investigations had been completed; this allowed the model requirements to drive portions of the numerical model.

2 Materials and Methods

The procedure for applying a groundwater model includes the following steps: define study objectives, develop a conceptual model, construct a groundwater flow model, calibrate the groundwater flow model and perform sensitivity analysis, set a contaminant transport model, calibrate the model, make predictive simulations, document the modeling process, and perform a post audit. A modelling process is shown in Figure 1.

2.1 Description of study site

The study site, the Wang-Tien landfill, is located in the Yong-Kang Municipality of the Tainan region. The location of the study area lies between the following coordinates: 120° 16'1" E, 120° 16'22" E longitude and 23° 2' 29" N, 23° 2'36" N latitude, and it occupies a total area of 0.75 km^2. Figure 2 shows the location of the study area in the southwest of Taiwan. The elevation of the study area ranges from 25 to 28 m above mean sea level, with a mean land surface slope of 0.1 ‰ from southeast to northwest. The length of the landfill site measures 175 m from south to north, and 225 m from east to west, and is currently surrounded by an industrial zone.

Four of hydrogeological bores logs can be seen in Figure 3, showing the major geological layers in the vicinity of the closed landfill. The site is situated within the Hsu-Hsian creek watershed, which drains into the Yan-Shuei creek. From a regional water balance using historical meteorological data, the regional groundwater recharge rate was estimated to be within a range 0.14 to 0.75 cm/min. The landfill has been accepting municipal solid waste from the Tainan city since 1992. It was closed in 2002. The accumulative amount of solid waste reached approximately 773,970 m^3. The study area is located in a tropical climate zone with an annual average rainfall of 1828.4 mm and an annual average evaporation of 1476 mm.

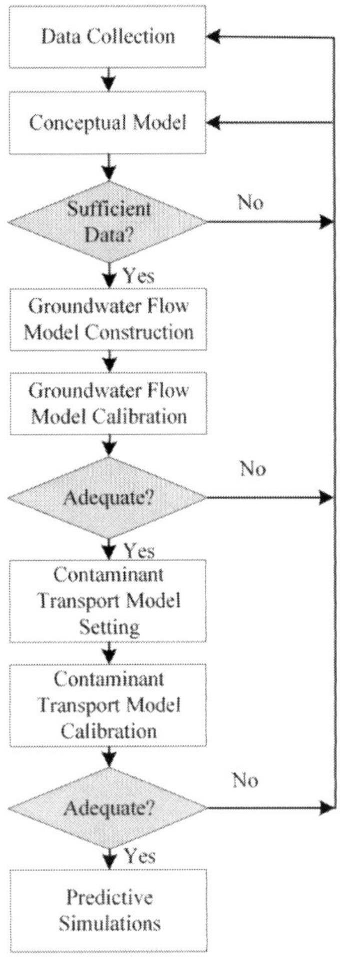

Figure 1. Flowchart of the modelling process

2.2 Groundwater Data

Certain parameters assigned in the numerical software for the simulation, including hydraulic conductivity, porosity, thickness and type of soil, were collected between 2013 and 2014 in the study area. Monitoring wells that were purged before sampling and groundwater that was collected by pumping, only after constant values of electrical conductivity and redox potential have been established.

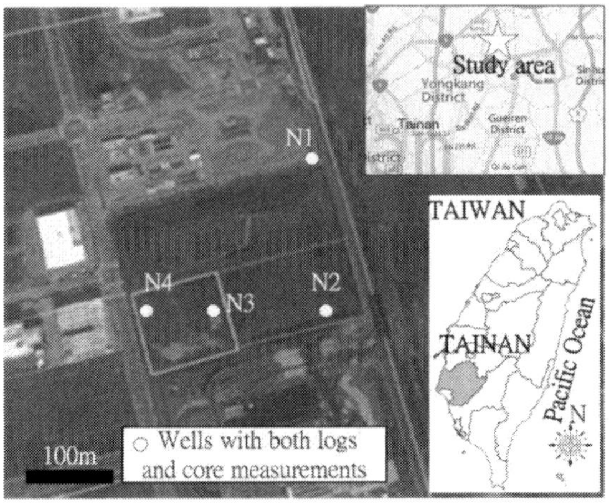

Figure 2. Location of study area.

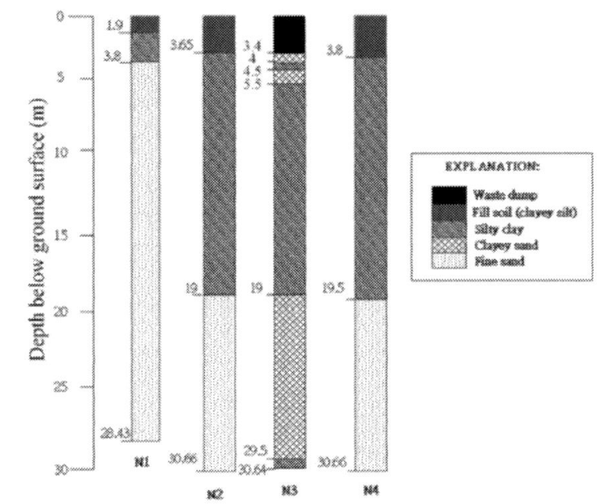

Figure 3. The major geological layers.

The measured physicochemical values of water samples from the study area can be seen in Table 1. The hydraulic conductivities for the sand layer were obtained from slug testing [7] in boreholes and the measured values of the aquifer from the study site. The hydraulic conductivity tensor in the numerical model was assumed to be isotropic.

2.3 Groundwater Flow Simulation Model

MODFLOW solves for the distribution of hydraulic head within the model domain, the velocity components of flow are calculated [8]. The key partial differential equation, as shown in Equation 1, used to predict flow is:

$$\frac{\partial}{\partial x}\left(K_{xx}\frac{\partial h}{\partial x}\right)+\frac{\partial}{\partial y}\left(K_{yy}\frac{\partial h}{\partial y}\right)-W=S_S\frac{\partial h}{\partial t} \qquad (1)$$

Where K_{xx}, and K_{yy} are the values of hydraulic conductivity along the x, and y coordinate axes; h is the potentiometric head; W is a volumetric flux per unit volume representing sources and sinks of water; S_S is the specific storage of the material; and t is time

Table 1. Physico-chemical parameters of water samples around the landfill in 2013.

Indicator of Water Quality	Unit	Level of Pollutants in Wells			
		N1	N2	N3	N4
Temperature	℃	25.5	25.1	26.4	25.2
pH		7.7	7.8	7.8	7.6
Electrical Conductivity	μS/cm	3990	2880	3110	5500
Total Dissolved Matters	mg/l	2100	1650	1600	2850
Total Hardness	mg CaCO/l	197	174	114	276
Ammonium Nitrogen	mg/l	4.31	5.43	2.32	0.15
Nitrite Nitrogen	mg/l	ND	ND	ND	ND
Nitrate Nitrogen	mg/l	0.14	0.10	0.01	ND
total Organic Carbon	mgC/l	3.1	13.0	1.9	2.6
Chlorides	mg/l	899	456	669	1440
Sulphates	mg/l	28.9	49.2	6.64	12.6
Arsenic	mg/l	0.0548	0.0699	0.0376	0.0336
Total Chromium	mg/l	ND	ND	ND	ND
Copper	mg/l	ND	<0.05	ND	ND
Manganese	mg/l	0.08	0.04	0.05	0.08
Ferrum Iron	mg/l	<0.05	0.14	0.06	0.06
Lead	mg/l	ND	ND	ND	<0.10
Zinc	mg/l	<0.01	0.13	<0.01	<0.01
Nickel	mg/l	ND	ND	ND	ND
Cadmium	mg/l	ND	ND	ND	ND
Mercury	mg/l	ND	ND	ND	ND
Average Hydraulic Conductivities	m/min	7.548×10^{-3}	4.304×10^{-3}	1.855×10^{-3}	1.441×10^{-3}

2.4 Contaminant transport model

To simulate the leaching of chlorides into groundwater, MT3DS was used. MT3DS is a modular three-dimensional transport model for the simulation of advection, dispersion, and the chemical reactions of dissolved constituents in groundwater systems. It uses cell-by-cell data, which is calculated and outputted by MODFLOW to establish the results [9]. The advection dispersion equation is written as follows:

$$\frac{\partial C}{\partial t}=\frac{\partial}{\partial X_i}\left[D_{ij}\frac{\partial C}{\partial X_j}\right]-\frac{\partial}{\partial X_i}[v_i C]+\frac{q_S}{\theta}C_S+\sum_{K=1}^{N}R_K \quad (2)$$

Where C is dissolved concentration; t is time; D_{ij} is dispersion coefficient; v_i is velocity component in ij direction.

2.5 Model boundaries

The boundaries assigned in numerical software contain: (i) Constant Head Boundary (CHB), (ii) evapotranspiration boundary, (iii) recharge boundary, and (iv) creek boundary. The assigned values recharge and evapotranspiration to the entire cells of mesh are 1828.4 mm/yr and 1195.8 mm/yr, respectively. Figure 4(a) shows the creek boundary, which is assigned in numerical mesh. The A-A' cross section was chosen to construct the contaminant vertical transport model (Figure 4(b)). The cross section was vertically discretized 10 layers in total with thickness of 2-3m. the bottom boundary of the model is restricted by depth of available investigation boreholes in this site and is assumed no flow. The flux flowing in or out in the lateral boundaries of the sand aquifer were estimated according the geometry of the boundary and the regional hydraulic gradient.

2.6 Model calibration

Model results were used during the calibration phase in order to verify between the simulated results and those observed in the groundwater monitoring wells. The water level measurements taken at the landfill site give the best overall contour of water levels throughout the aquifer.

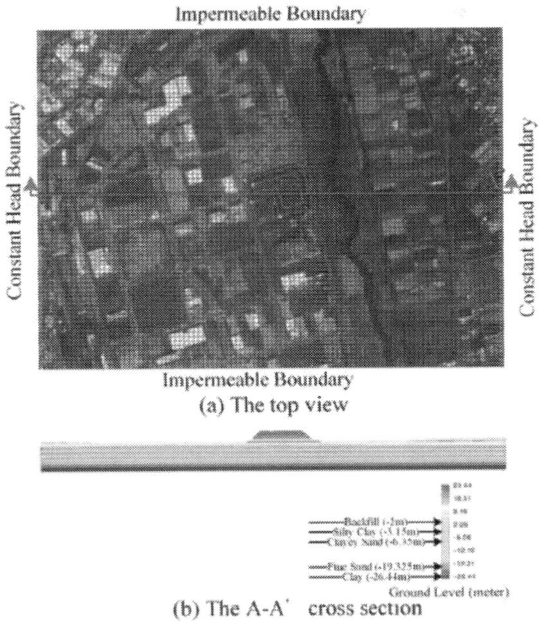

Figure 4. Location of creek boundary in study area

 DOI: 10.1051/matecconf/20165406002

The initial calibration study used assumed that there was a leachate mound within the landfill. This was based on groundwater level data from N1 to N4 obtained in 2013, as shown in Figure 5(a). Subsequently improved calibrations were applied and assumed that there was no leachate mound. They further assumed that the elevated water level at N1 to N4 was the result of perching mainly caused by the recirculation of leachate. A plot of the correlation between calculated and observed hydraulic heads indicates that the flow model calibration is representative of the physical conditions at the landfill site, as shown in Figure 5. To achieve the hydraulic head distribution in the landfill site, the finite difference mesh was set to a cell size of 20×20 m. The results of the calibration are shown in Figure 6, and it can be seen that, generally, good agreement between calculated and observed groundwater levels was achieved.

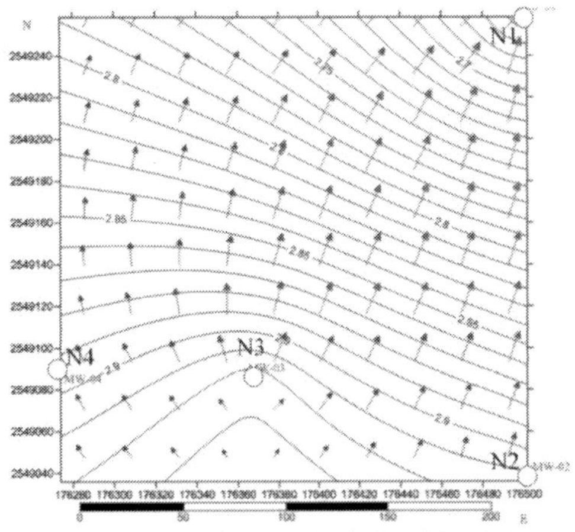

(a) Average groundwater hydraulic head distribution and flow vector in 2013

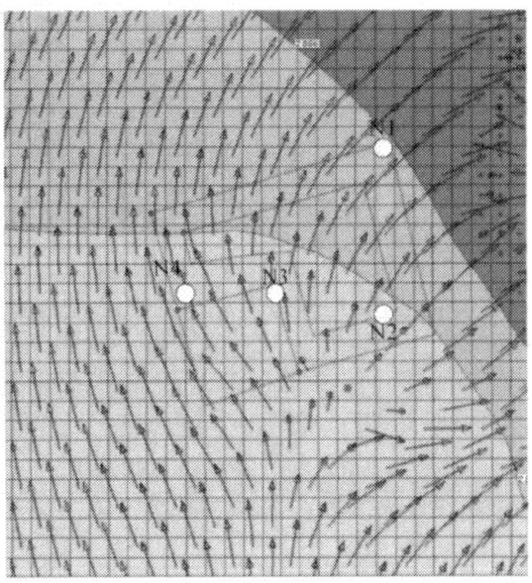

(b) The MODFLOW simulation

Figure 5. Groundwater flow model calibration.

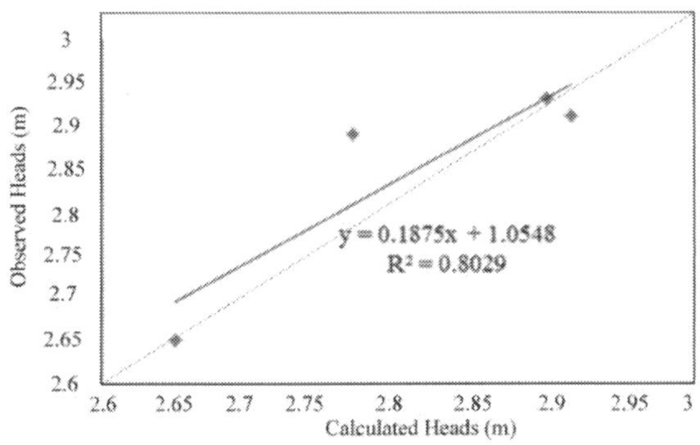

Figure 6. Correlation of calculated vs. observed heads.

In order to investigate the extent of migration of landfill leachate into the sand aquifer and to determine the role of the clay/silt layers, a chloride transport model was constructed. Although groundwater flow in the aquifer is essentially horizontal from the study area, the zone of leachate contamination beneath the landfill has penetrated downward through the entire sand aquifer thickness to a depth of 3.8-30.66 m. Chloride was selected as the conservative tracers because it is generally not likely to undergo any physico-chemical reactions (e.g., adsorption, precipitation, and biological degradation) in the aquifers, and usually it is characterized by high concentrations in leachate-contaminated groundwater near the landfill. Therefore, chloride can be used to study the dispersion and dilution of a contaminant plume. The landfill zone is the main pollution source for groundwater contamination in the study area. A plot of the correlation between concentration calculated and observed indicates that the Contaminant transport model calibration is representative of the physical conditions at the landfill site, as shown in Figure 7.

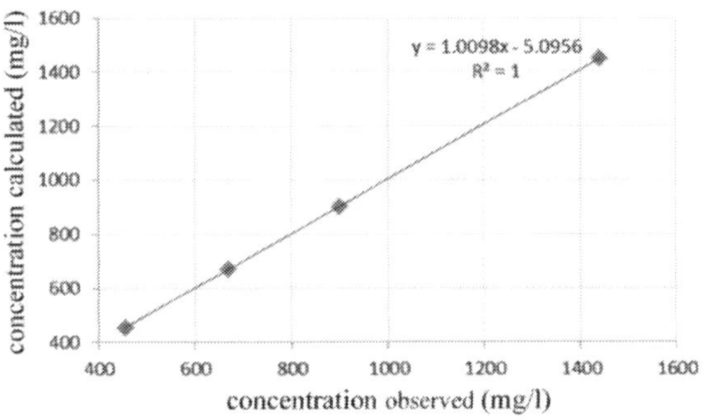

Figure 7. Correlation of calculated vs. observed concentration.

3 Numerical results

Results of chloride measurement in the study area show that a concentration of chloride in most parts is low, yet in the adjacent Hsu-Hsian creek of the Wang-Tien landfill concentration of chloride is 782 mg/l. The authors used the MODFLOW and the MT3DMS model that was run to steady-state for 1 and 10 years. As a matter of fact, chloride cannot leach and move with groundwater in the short time because of the high absorption capacity soil has for the absorption of chloride particles, but the ability to move chloride is resumed as the adsorption capacity of the soil is reached. This enables the groundwater carried chloride to go around the soil particles, which in turn allows for the transportation of high concentrations of chloride through the soil. The results illustrate that chloride did not migrate widely along the groundwater flow (Figure 8). But after 10 years, the contamination plume is moving in the direction of the northeast (into the creek which is located in east of map) instead of moving along with groundwater. As shown in Figure 8, the contamination plume is moving toward the creek and is seeping into surface water and is transported to other places.

Due to the placement of the decommissioned landfill (which is located at a higher ground level and the creek is beside the landfill, which is located at a lower ground level), movement of the groundwater in the study area is effective by the Hsu-Hsian creek that is located in east of the study area, and the movement of the contamination in the direction of the northeast (into the creek) is shown in Figure 9.

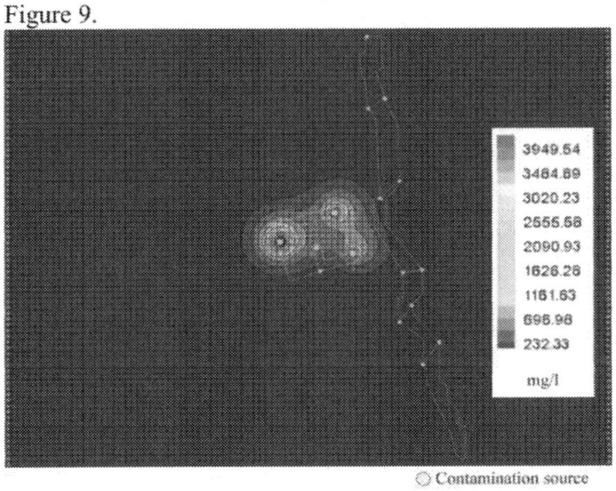

Figure 8. Migration of contamination after 1 year

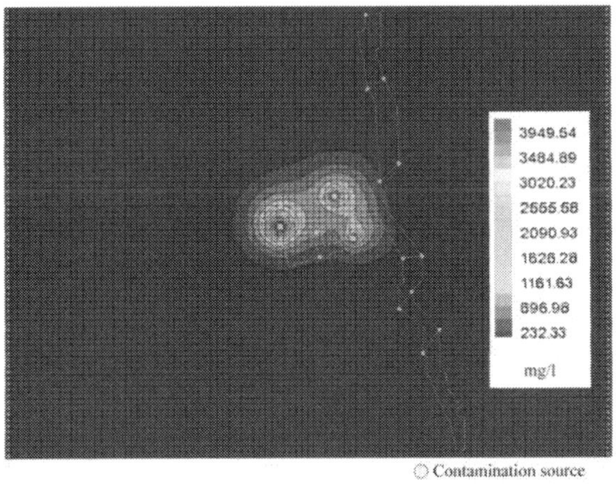

Figure 9. Migration of contamination after 10 year

4 Conclusion

An advective-dispersive transport model was established using MODFLOW and MT3DMS to simulate the chloride plume in the upper sand unit. A steady state was simulated and discussed. Chloride transport simulation shows that after 10 years operation of the closed landfill, the contamination plume is still constrained to the upper aquifer. The seasonal dynamics of the hydraulic condition, such as recharge and water level temporal variations, should be investigated further to evaluate the migration of the plume under the landfill. Groundwater flow and concentration transport modelling was valid for assessing and predicting chloride concentrations in the subsurface zone in the landfill zone. The prediction of chloride migration in groundwater might be a preliminary research, and further study must be implemented to investigate the migration of chloride better to prevent the spreading of chloride in study area. The present research confirms that the unconfined aquifer is highly contaminated by chloride, which is 782 mg/l when standard of the Monitoring Standard Value for Taiwan is declared by Environmental Protection Administration (EPA) 675 mg/l[10]. Generally, groundwater around the closed landfill is not suitable for drinking water, and some remedial measures should be considered to improve the groundwater quality around the closed landfill, particularly given the high levels of ecological and agricultural groundwater demand in this area.

Acknowledgment

This work was supported by the Ministry of Science and Technology of Taiwan (104-2221-E-006-268-) and the Environmental Protection Bureau of Tainan City.

References

1. Y. Abu-Rukah, O. Al-Kofahi. J. Arid. ENVIRON 49, 615 (2001)
2. A.F. Al-Yaqout, M.F. Hamoda. ENVIRON INT. 29, 593 (2003)
3. I.S. Babiker, A.A.M. Mohamed, H. Terao, K. Kato, K. Ohta. ENVIRON INT. 29, 1009 (2004)
4. T.H. Christensen, P. Kjeldsen, P.L. Bjerg, D.L. Jensen, J.B. Christensen, A. Baun, H.J. Albrechtsen, G. Heron. Appl. Geochem. 16, 659 (2001)
5. D. Rapti-Caputo, C. Vaccaro. Eng. Geol. 85, 111 (2006)
6. K. Kim, K.S. Ko, Y. Kim, K.S. Lee. CHEMOSPHERE 87, 851 (2012)
7. H. Bouwer, R.C. Rice. WATER RESOUR RES 12, 423 (1976)
8. A.W. Harbaugh, US Geol. Surv. (Tech. Methods 2005)
9. C. Zheng,. Department of Geological Sciences, University of Alabama, (2006) http://hydro.geo.ua.edu/mt3d/.
10. EPA. Regulations for Groundwater Monitoring. Environmental Protection Administration, R.O.C., Taiwan.http://wq.epa.gov.tw/Code/Business/Statutory.aspx?Tabs=3&Languages=en (accessed on 10 October 2014)

Internet of Things for Survey of Renewable Energy Potential (SREP) as the Basis for Hybrid Power Plant Development

Muslikhin[1], Eko Marpanaji[2], Muh. IzzuddinMahali[3]

[1,2]*Electronics Engineering Education Department, Yogyakarta State University, Karangmalang, Yogyakarta 55281, Indonesia*
[3]*Informatics Engineering Education Department, Yogyakarta State University, Karangmalang, Yogyakarta 55281, Indonesia*

Abstract. In the future, microhybrid power plant for the household scale will be something that the people want. The obstacle that is faced by the developers in micro hybrid plant development is the lack of a preliminary study. If a preliminary study is not possible, at least the developer should conduct a potential survey to determine the place where the plant will be installed. Optimal position for the hybrid power plant can be obtained through a continuous mapping at some point during the season span. This study uses research and development procedures of the internet of things (*IoT*) for its purpose of survey of renewable energy potential. The aim of this study is to obtain the real time data of the potential of renewable energy with the help of the *IoT*. Internet of things which is developed using Airlink FXT009 EDGE is operated via serial communication. Modem Airlink FXT009 is controlled by ATmega328 microcontroller which received its data from the sensor network. The database which is processed by *IoT* consists of wind velocity, wind direction, output voltage, and stationidentity. The research shows that the average of wind velocity during April-September is 3.2 m/s and the output voltage is 13.3V.On the other hand, during October-March the wind velocity is 5.3m/s, and the average output voltage at 10:00-14:00 is 10.2V with the average of speed and data updates is per 75.34 seconds. This study is limited to the ATmega328 microcontroller as the processor and need to be developed towards a mini PC.*Keywords—internet of things (IoT), energy potential survey, hybrid power plant, ATMega328*

1 Background

Our concentration needs to be focus on the major threat to the depletion of fossil fuel reserves that the world is facing. Today most of the energy demand is met by fossil. A small portion met by renewable energy technologies such as wind, solar, biomass, geothermal etc. While the projected needs of energy production in Indonesia in 2016 is amounted 52.369Gw or 3,1Gw, increased from the previous year [1].

Energy needs in Indonesia is very dependent on geographical conditions. National Energy Councilspecifically states hybrid plants, the government policy is to develop a centralized PV to electrify many remote communities far from the grid in disadvantaged areas, the islands of the state border and the other outer islands. This was driven by the spirit of the government to provide access to electricity workforce faster to people in remote areas.

Hybrid power plants centralized location was chosen after considering the techno-economic factors such as the cost of transporting the fuel to the location. PV operatesas a hybrid with existing diesel, thereby reducing fuel consumption.

According to government policy, for optimal results in the development of hybrid power plants,potential mapped is needed. Survey of renewable energy potential continue to record the energy generated in one place. This mapping results determine the type of technology, location, transmission and this process needs to adapt information technology.Errors in the mapping causes the value of energy efficiency resulting smaller[2].

Internet of Things (IoT) is a development of field of applied informatics and proved in providing ease in solving the survey, monitoring and control device. Survey hybrid energy potential domestic scale is a matter that is necessary to determine the energy generated per month [3]. Control and related technologies will be essential to solve complex problems in the hybrid plants [4]. IoT provide more responsive and reliable services; a significant cost saving features can be developed based on the needs, accurate and real-time [5]. In simple words IoT will play a major role in energy saving [6].

This paper aims to describe the design ofIoTand explain the energy generated per month since December

2014 using IoT technology. Accurate data provides optimal results in selecting the location of the placedplant.

2 Theoretical Basis

2.1. Survey of Renewable Energy

Energy sources which can be replenished by nature and its supply isnot affected by the level of consumption is often referred to as a renewable energy. The need to find the renewable, alternative and non-polluting energy has become the top priority for the national energy supply autonomy. This requires the estimationof available energy resources accurately [7]. There are various ways to conduct survey of renewable energy such as; solar, wind, geothermal heat, ocean waves, biomass, etc.As for the wind power plants, as refers to equation 1, there are several things that need to be observed.

$$Power = 0.5 \times swept\ area \times air\ density \times velocity^3 \quad (1)$$

Since the air density is affected by the location while the speed is affected by the wind velocity, it makes these variables need to be observed. In addition, the wind direction needs to be observed in each period. Survey of solar power plant potential can be done by measuring the voltage which is generated in real-time within a month. The 100Wp monocrystalline solar panel is one of the instruments that can be used to measure the voltage because this panel has an output which is range between 0.1 up to 17.6 volts.

2.2. Web Service

Web services are a set of database applications, software or any part of the software that can be accessed remotely by various devices with certain intermediaries. In general, the web service can be identified by its URL just like other webs [8]. However, the feature that distinguishes web service from other websis the interaction provided by the web service. Unlike the URL of other webs, the URL of service web only contains a set of information, commands, configuration or syntax which is used for constructing certain functions of the application, as shown in Figure 1.

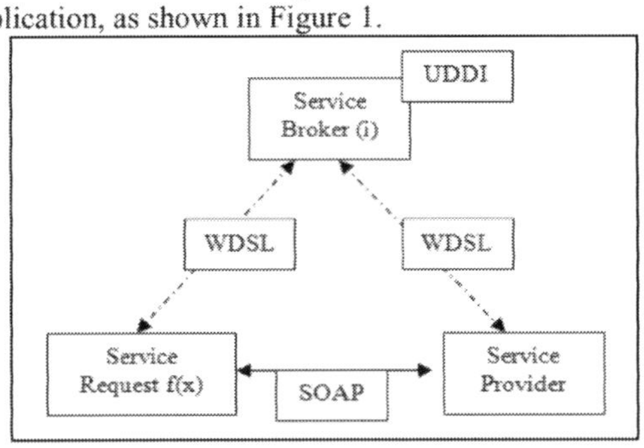

Figure1. Webservices architecture

According to Web services Architecture Working Group, the exact definition of Web service is a software system designed to support the interoperability machine-to-machine interaction over a network. Web service interfaces is described to be able using a format that can be processed by a machine (especially WSDL) [9]. Other systems that interact with the web service only requires SOAP, usually is written with HTTP and XML so that those systems have a correlation with the web standards.

2.3. Machine toMachine (M2M)

M2M terminology is derived from the term machine-to-machine, namely the integration of communication technology, computers and power which eventually enabling the long haul communication between machines [10]. M2M is a data computing application which is flowed to and from the physical interaction and biological process through sensors which are attached and processed by a microcontroller.

M2M refers to technologies that allow both wireless and wired systems to communicate with other devices of the same type. Machine to machine is a broad term because it does not determine the specific wireless or wired network, information and communication technology. This broad term is mainly used by business executives. M2M is an integral part of the Internet of Things (IoT) and give benefits for the industry and business in general [11] because it has a wide range of applications such as industrial automation, logistics, Smart Grid, Smart City, health, defense, etc which is not only for monitoring but also for controlling purposes.

2.4. Internet of Things (IoT)

Internet of Things or IoT, is a concept that aims to extend the benefits of Internet connectivity which is constantly connected. The capabilities such as data sharing, remote control, and so on, as well as on the objects in the real world. Basically, IoT refers to objects which can be identified uniquely as virtual representations in an Internet-based structure [12].

Smart connectivity with the existing network is the basic context of the integral part of IoT. With the existence of wireless Internet, Wi-Fi and 4G-LTE, the opportunities of technology development will be faster [13]. Based on its concept, IoT consists of six main elements as shown in Figure 2.

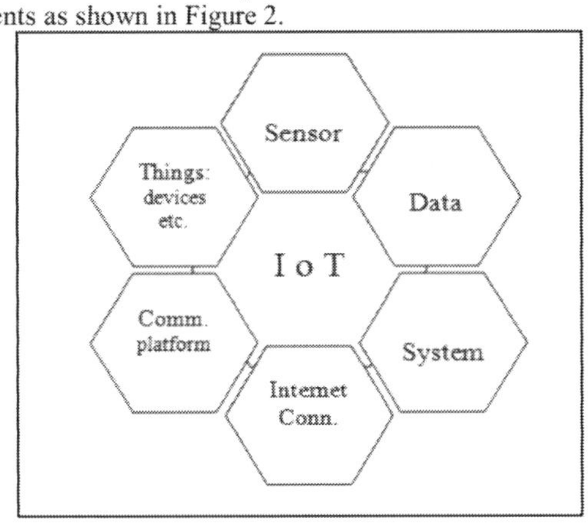

Figure 2. Part of the IoT

There are six IoT elements that emerge; sensors, devices, communication platform, internet connection, system, and data [14]. Computing paradigm will exceed the low-tech mobile computing scenarios and switch to Smartphone, portable gadgets and continue to develop into daily link. IoT refers to the identification of an object that is represented virtually in cyberspace or internet [15]. Thus, it can be said that IoT is how a real object communicated in cyberspace (internet) [16].

3 ResearchMethod

This IoT SREP research uses R&D procedure offered by Lee and Owens (2004) since it is a product-oriented research.It is divided into five phases; assessment/analysis, design, development, implementation, and evaluation phase [17].

3.1 Station Number

A total of 2 stations were implemented for this survey. The survey was conducted machine-to-machine due to limitations of sensor installed we surveyed. We would like to thank the engineers of PandansimoBantul Hybrid Power Plant and PUSDATEK (Center of Empowering Technology) Engineering Faculty, Yogyakarta State University in conducting this survey.

3.2 Location of Station

This survey was conducted in Bantul and Sleman District State, in the areas around Yogyakarta Special Region.Station 1 is locatedat coast Villages of Pandansimo, Srandakan, Bantul Regency and the station 2 is located atFaculty of Engineering, Yogyakarta State University, Sleman Regency. The distance between station 1 to station 2 about 43 km.

3.3 Classification of Station

Station 1 is located at the coastal area with the surface elevation of 3 meters above the sea level, on the ordinate of -7,9836423 and 110,2171283. Station 2 is located at an urban area with the surface elevation of 80 meters above the sea level, on the ordinate of -7,7717588 and 110,3867712. Both stations are given the same device and sensor, wind sensor WM30 and polycrystalline 100Wp, which are placed at the height of 15 meters from the ground. All the stations work in standalone for 24 hours with an independent supply.

3.4 Designing of IoT SREP

IoTSREP system development is supported by sub-fields of science that are continuous or mutually support each other i.e., mobile application, telemetry, web server, sensing, electrical and mechanical. The general overview of the system which will be developed in this research can be seen in Figure 3.

On the station area, wind speed sensor, anemometer sensor, and PV cell are installed to acquire the data of wind and solar conditions. These sensors will be installed as a standalone at selected locations, data will be sent through microcontroller ATmega328 based on asynchronous communication. The sensing results will be sent to server via EDGE-HTTP communication to data processing center. After that, the results of data reading in the field will be processed on the server, and displayed online so it can be accessed through the website monitoring or smartphone apps.

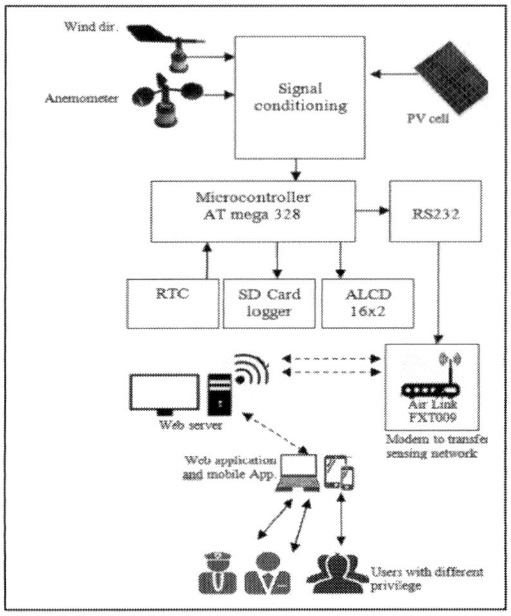

Figure 3. Working Mechanism of IoTSREP

On the station area, wind speed sensor, anemometer sensor, and PV cell are installed to acquire the data of wind and solar conditions. These sensors will be installed as a standalone at selected locations, data will be sent through microcontroller ATmega328 based on asynchronous communication. The sensing results will be sent to server via EDGE-HTTP communication to data processing center. After that, the results of data reading in the field will be processed on the server, and displayed online so it can be accessed through the website monitoring or smartphone apps.

3.5 Coding (Hardware)

This paper only provides the script code for the hardware of sensors network, ATmega328 microcontrollerand device data logger. IDE used in this programming is Arduino programming language. To facilitate the understanding of the code, then the code is divided into blocks of programs that are numbered.

```
//[1]======prog utama===========
void loop() {
  proses();
}
//[2]======rutin proses==========
void proses(){
  val=analogRead(1);
  if(val<sens)
    stat=LOW;
  else
    stat=HIGH;
  digitalWrite(13,stat);
  if(stat2!=stat){
     conter++;stat2=stat;
   }
if(millis()-last>=milisekon){
doublerps=((double)conter/encoder)/16.0*1000/mil
isekon;
double
rpm=((double)conter/encoder)/16.0*60000/miliseko
n;
```

DOI: 10.1051/matecconf/20165406003

```
//[3]=====arah====================
nilaiArah = analogRead(0);
arah=(nilaiArah-144)/2.44;
delay(500);
//[4]=====teg PV==================
bat=analogRead(2);
double vbat =bat/204.8;
delay(500);
//[5]=============================
Serial.print("AT+WIPCFG=1\r\n");
delay(500);
Serial.print("AT+WIPBR=1,6\r\n");
delay(500);
Serial.print("AT+WIPBR=2,6,11,\"internet\"\r\n");
delay(1000);
Serial.print("AT+WIPBR=4,6,0\r\n";
delay(5000);
Serial.print("AT+WIPCREATE=5,1,\"pusdatek.ft.uny
.ac.id\",80,\"\",\"\"\r\n");
delay(5000);
Serial.print("AT+WIPFILE=5,1,1,\"http://pusdatek
.ft.uny.ac.id/telemetri/tambahdata.php?arah=");S
erial.print(arah);Serial.print("&kecepatan=");Se
rial.print(rpm);
Serial.print("&station=");Serial.print(station);
Serial.print("&vbat=");
Serial.print(vbat);
Serial.print("\"\r\n");
delay(1000);
Serial.println("AT+WIPCFG=0\r\n");
delay(1000);
//[6]====SD Card parsing data ke .csv=====
String dataKirim="";
int data1 = arah;
int data2 = rpm;
float data3 = vbat;
intdata[]=
{data1,data2,data3};
  int i;
for (int i =0; i < 6; i++){
    dataKirim += data[i];
        if (i < 5) {dataKirim += ",";}
    }
File dataFile =SD.open("datalog.csv",
FILE_WRITE);
     if (dataFile) {
     dataFile.println(dataKirim);
     dataFile.close();
     Serial.println(dataKirim);
     }
     else {
Serial.println("Tidakdapat membuka SDcard");
     }
```

The explanation of the code is given in the form of numbered blocks. Blocks number 1 till 2 are in the form of main program and routin program. The essential elements of the program is on number 5, commands sent via the AT+ Command extended. The process of transfer protocol over TCP socket is started from AT+ WIPCFG= 1 and ended with AT+ WIPCFG= 0. The codes for sensor network are on block number 2, 3, and 4 which are processed through the ADC, while for the process of data logger in the SD card is on number 6.

3.6 Implementation

The realization of IoT design, SREP, has been implemented at two stations in Bantul and Sleman. The results can be seen in http://pusdatek.ft.uny.ac.id/telemetri as shown in Figure 4.

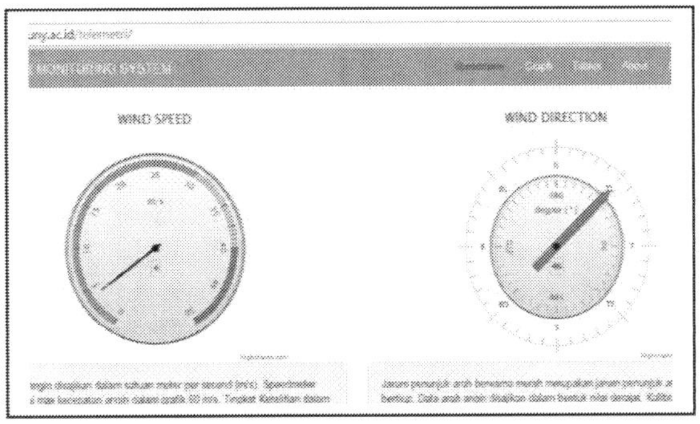

Figure 4.GUI of IoT SREP

There are three data display modes; gauge meter, polygon graphics, and tables. In the graphic, the data is presented in three different graphs for survey result of the wind velocity, direction and the results of PV output. As for graphics data, it is presented in a comprehensive manner.

Mobile application features (APK) are also available on the web in the download menu. Simple Viewer Telemetry application has yet to be available in PlayStore and AppStore. If you want to install this application, it can be done by downloaded this application in advance via website. The visualization results of APK IoT design applications, SREP, can be seen in Figure 5.

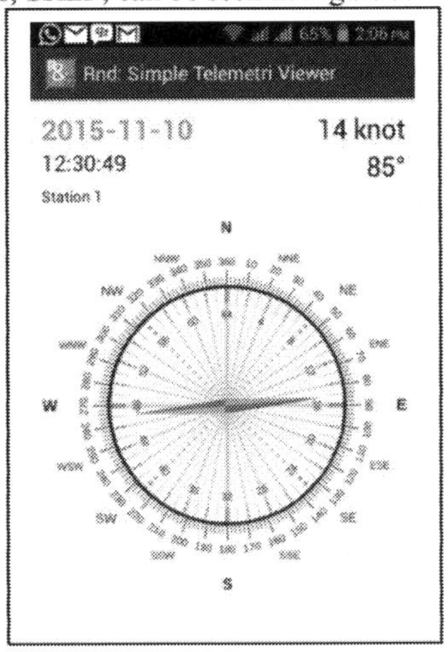

Figure 5.GUI Simple TelemetriViewerofIoT SREP

The schematic implementation from Figure 3, is designed as a compact using the ATmega328 in the minimum system.ATmega328 was selected since the chip has 32 KB of flash, 1KB EEPROM, 6 pin ADC and sufficient for the needs ofIoT SREP.

The key of the schematic realization into a hardware form,is clock 16 MHz and capacitor 22 pF as the resonator. Besides the selection of PCB materials is preferably using fiber materials.The empirical facts have shownthe neglect of those two things will cause asynchronous serial communication with baudrate of 115200 bps is disturbed. If it happens, then there will be an interference and the data received will be damaged/

corrupt. The realization of IoT SREPhardware is shown in Figure 6.

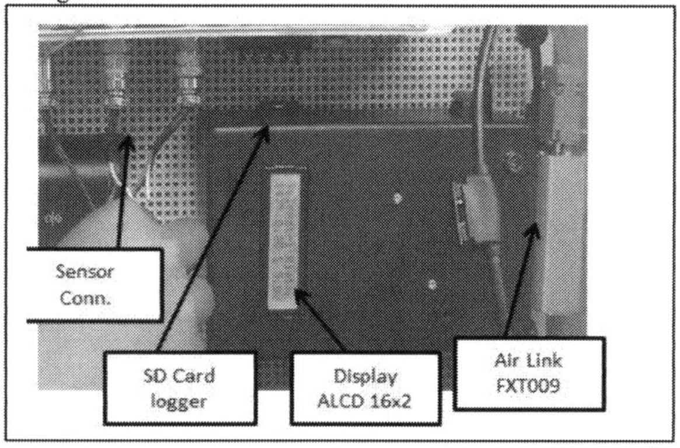

Figure 6. Main module of IoT SREP hardware

Figure 6 shows that the IoT hardware, SREP, is composed of SD card logger, ALCD 16x2 display, sensor networks, and modems AirLink FXT009. The device is packed in a waterproof box and can withstand direct sunlight. The outer part of the box should not be made of metal to reduce signal attenuation.

4 Result and Discussion

4.1 Result

The survey data of renewable energy potential using IoT, SREP is presented on the website in the form of tables, graphs polygon, and gauge meters. The data is displayed with the fastest updated time capture for every 76 seconds, with a capacity data transmission of 27 bytes.

Table 1. Survey Results of Renewable through IoT SREP[2]

Month	Wind speed (m/s)	Wind dir. (degree)	PV output (volt)
December, 2014	6.1	160	9.2
January, 2015	6.2	174	9.9
February, 2015	5.3	183	11.4
March, 2015	4.5	190	11.6
April, 2015	3.8	220	11.2
May, 2015	3.6	298	13.2
June, 2015	2.9	310	14.4
July, 2015	2.5	355	14.1
August, 2015	3.0	35	13.4
September, 2015	3.3	69	13.5
October, 2015	4.4	132	10.4
November, 2015	5.2	157	9.6
Ave. Apr. – Sept.	3.2	-	13.3
Ave. Oct. –Mar.	5.3	-	10.2

[2]data record during 10.00-14.00 (GMT+7)

IoT SREPdeveloped using EDGE FXT009 Airlink has been able to survey the potential of renewable energy.The research shows that the average of wind velocity during April-September is 3.2m/s, and the output voltage is 13.3V.On the other hand, during October-March the wind velocity is 5.3m/sand the average output voltage at 10:00-14:00 (GMT+07:00) is 10.2V with the average of speed and data updates is per 75.34 seconds.

This research presents data managed as in Table 1. The decision on the construction of a hybrid power plant can pay attention to the findings of the above data.

4.2 Discussion

This study is limited to the ATmega328 microcontroller as the processor and need to be developed towards a mini PC.In addition, based on data and many stations the opportunity to specify the type of turbine, PV, and the system becomes more efficient.

References

1. Dewan Energi Nasional. "Rencana UsahaPenyediaan Tenaga Listrik(RUPTL)PT PLN (Persero)2015-2024".http://prokum.esdm.go.id/renstra/Retrieved10 October 2014.
2. R. Rama. "Ini Alasan Pembangkit Listrik Matahari Sulit Berkembang di RI". http://finance.detik.com/read/2013/02/22/192004/2177424/1034/ini-alasan-pembangkit-listrik-matahari-sulit-berkembang-di-riRetrieved 14November 2015.
3. M. Tamvakologos.Hybrid energy system in power plant. Department of Mechanical Engineering. University of Strathclyde Engineering.Thesis.(2010).
4. S. Chen, H. Xu, D. Liu, B. Hu, and H. Wang. *A vision of IoT:Applications, challenges, and opportunities with china perspective.IEEEInternet Things J.*, **vol. 1**, no. 4, Aug. (2014), pp. 349–359.
5. R. Pozza, M. Nati, S. Georgoulas, K. Moessner, and A. Gluhak. *Neighbor Discovery for Opportunistic Networkingin Internet of Things Scenarios: A Survey. IEEE Access,* **vol. 3**, (2015), pp.1101-1131.
6. L. Atzori, A. Iera, and G. Morabito. *The Internet of Things: A survey,Comput. Netw.*, **vol. 54**, no. 15, Oct. (2010), pp. 2787-2805.
7. T.V. Ramachandra. B.V. Shruthi. *Spatial mapping of renewable energy potential.ScienceDirect Renewable and Sustainable Energy Reviews,* **vol. 11**pp. 1460–1480, (2007).
8. Technical Architecture Group. "Relationship to the World Wide Web and REST Architectures". http://www.w3.org/standards/webarch/. Retrieved 20 November 2015.
9. D.Benslimane; Dustdar, S.; A. Sheth. *Services Mashups: The New Generation of Web Applications. IEEE Internet Computing*, **vol. 12**, no, 5, (2008), pp.13–15.
10. J. Höller, V. *et.al. From Machine-to-Machine to the Internet of Things: Introduction to a New Age of Intelligence*., New York: Elsevier, (2014).
11. G. Wu, S. Talwar, K. Johnsson, N. Himayat, and K. D. Johnson, "M2M:From mobile to embedded internet,*IEEE Commun. Mag.*, **vol. 49**, no. 4,Apr. (2011), pp. 36–43.
12. D. Giusto, A. Iera, G. Morabito, L. Atzori. *The Internet of Things: 20th Tyrrhenian Workshop on Digital Communications.*, New York: Springer-Verlag, (2010).
13. G. Duncan. (2014). "You can't avoid the 'Internet of things'hype, so you might as well understand it".

http://www.digitaltrends.com/home/heck-internet-things-dont-yet/ Retrieved 25 October 2015.
14. M. Nitti, Luigi Atzori, and Irena Pletikosa Cvijikj"Friendship Selection in the Social Internet of Things:Challenges and Possible Strategies"*IEEE Internet of Things Journal*, **vol. 2**, No. 3, (2015), pp. 240-247.
15. M. Compton *et al.*, *The SSN ontology of the W3C semantic sensor network incubator group.Web Semant. Sci. Serv. Agents World Wide Web*, **vol. 17**, (2012), pp. 25–32.
16. L. Atzori, A. Iera, G. Morabito, and M. Nitti, *The Social Internet of Things (SIoT)—When social networks meet the Internet of Things: Concept, architecture, and network characterization. Comput. Netw.*,**vol. 56**, no. 16, pp. 3594–3608, (2012).
17. W Lee,. & Owens, D. L. *Multimedia based instructional design: Computer based training web based training distance broadcast training, 2nd ed.* San Fransisco: Pfeiffer.(2004).

MATEC Web of Conferences 54, 06004 (2016) DOI: 10.1051/matecconf/20165406004
MIMT 2016

The Influence of Central Regulation Methods upon Annual Heat Loss in Heating Network

Xiaofang Shan, Peng Wang, Panhong Ren and Hua Zhao

School of Municipal & Environmental Engineering, Harbin Institute of Technology, Harbin, China

Abstract: For directly buried hot water heating network, heat loss per unit length can be calculated by the thermal theory of heating network, then the calculating method of annual heat loss would be deduced. By comparing the heat loss of three different regulation schemes, it is detected that different regulation scheme has different supply and return water temperature, resulting in different heat loss. The heat loss of variable flow regulation of equivalent temperature difference is the largest, while the heat loss of constant flow regulation is the smallest, which has significant guiding to the operation of district heating system. Finally, based on the calculation of the annual heat loss of the heating network in Jiu Tai, Ji Lin Province, it is found that the distribution heat losses take up 2.4%~2.7% of the total supplying heat, which are equivalent to the heat consumptions of the heating area from 30.7×10^4 m^2 to 34.4×10^4 m^2. Therefore reducing the heat losses of large scale heating networks is imminent.

1 Introduction

With the popularization of district heating in industrial and civil buildings, heating industry is becoming a large energy consumer as well as the object of energy saving and emission reduction. Therefore how to decrease heat loss that includes distribution heat loss and hot water loss is one of important parts of energy saving in heating system. The researches of heat loss in developed Europe start quite early, they mainly study the influence factors of heat loss. Frieder Schmitt etc[1].studied the influence of pipe codes, heat insulation materials, methods to pipeline installation , supply and return water temperature, pipe buried depth and so on to heat loss systematically, thus proposing efficient ways to reduce heat loss. In China, the research of heat loss is mainly focused on theory. For example, Wang Zide[2] studied and analyzed thermal calculation of directly buried installation heating pipeline in theory. However, the method of heat loss calculation is relatively rare. Since directly buried heating pipeline has great advantage in water proof, anti-corrosion and decreasing heat loss, and directly buried pipes are usually prefabricate by manufacturer, which reduces construction difficulties, thus prefabricated directly buried insulating pipes are widely used. At present, the insulating material of prefabricated pipes is hard polyurethane foam, and the outer protective pipe is high density polyethylene with high performance of thermal insulation and water proof [3]. Therefore, the study of heat loss of widely used directly buried installation heating network has practical sense. This paper systemically introduces the calculation of heat loss and annual heat loss of directly buried heating network, and furtherly analyzes the primary influence factor of heat loss that is supply and return water temperature, which has active guidance to the design and operation of heating system. Then, taking heating network in Jiu Tai, Ji Lin Province as an example, annual hear loss is calculated as well as the percentage of heat loss in total heating quantity, which provides complete thoughts and methods for calculation of heat losses.

2 Heat loss calculations

2.1 Calculation for Heat Loss per Unit Length

Because heat loss is affected by many factors, such as warm fluid temperature inside the pipe, thickness of insulating layer, heat conductivity for soil, distance between pipe centres and ground surface, heat flow between pipes and so on, which makes it quite complex to build models and solve them. In engineering application, in order to shorten calculating time and reduce complexity of large scale heating network, there are some rational hypotheses to make calculate easier within a reasonable error range. This paper has the following hypotheses for heat loss calculation [4]:

(1) Soils are isotropic medium, and heat transfer between pipes and soils are homogenized.

(2) Neglecting temperature drop in respective directions of flow, heat transfer could be simplified to be two-dimensional.

(3) Insulating materials are constant-property that does not change with the temperature.

(4) The heat resistance of pipe wall is always neglected since it is very small. It can be approximatively thought that the temperature of pipe wall is the same as the temperature inside the pipe.

By simplifying boundary conditions and calculation methods, when heating pipes are buried in parallel, heat losses of supply pipe can be calculated with the following formula:

$$q_g = \frac{(t_g - t_w)(R_b + R_w + R_t) - (t_h - t_w)R_c}{(R_b + R_w + R_t)^2 - R_c^2} \tag{1}$$

Heat losses of return pipe can be calculated with the following formula:

$$q_h = \frac{(t_h - t_w)(R_b + R_w + R_t) - (t_g - t_w)R_c}{(R_b + R_w + R_t)^2 - R_c^2} \tag{2}$$

With the following definitions:
t_g=supply water temperature [℃]
t_h=return water temperature [℃]
t_w=ambient air temperature[℃]
R_b=insulation resistance [m^2K/W]
R_w=outer pipe resistance [m^2K/W]
R_t=ground resistance [m^2K/W]
R_c=coinciding resistance [m^2K/W]

Firstly, according to the heat conduction formula of the cylinder [5], insulation resistance can be calculated with the following formula:

$$R_b = \frac{1}{2\pi\lambda_b}\ln\frac{D_b}{D_o} \tag{3}$$

With the following definition:
λ_b=heat conductivity for insulation [W/mK]
D_b=outer insulation diameter [m]
D_o= outer pipe diameter [m]

Outer pipe resistance can be calculated with the following formula [5]:

$$R_w = \frac{1}{2\pi\lambda_w}\ln\frac{D_w}{D_b} \tag{4}$$

With the following definition:
λ_w=heat conductivity for outer pipe [W/mK]
D_w= outer pipe diameter with protective pipe [m]

The ground resistance is generally calculated by virtual heat source method. Virtual heat source method assumes there a line heat source at some point in the middle of the pipe, and its heat productivity q_l is equal to the heat productivity of pipe surface. At the same time, the method also assumes that there is line heat congruence, a negative line heat source, whose heat productivity is $-q_l$ at its symmetric point regarded the ground surface as the symmetry axis. By superimposing the thermal fields of two heat sources, then we can use the final thermal field to simulate thermal field of directly buried pipes to detect the ground resistance. When the buried depth of pipes is deeper, the ground resistance can be approximately calculated with the following formula [5]:

$$R_t = \frac{1}{2\pi\lambda_t}\ln\frac{4H}{D_w} \tag{5}$$

With the following definition:
λ_t=heat conductivity for ground [W/mK]
H=converted buried depth of pipes [m]

Because convective heat transfer should be considered while calculating heat loss of pipes, converted buried depth of pipes can be calculated with the following formula:

$$H = h + \frac{\lambda_t}{a_k} \tag{6}$$

With the following definition:
h=the distance from the ground surface to the central of pipes [m]
a_k=heat transfer coefficient of ground, 12~15W/mk,

When the pipes of heating network are installed in parallel, the heat transfer between two pipes should be considered. According to the method proposes by Szubin, the Soviet Union Scholar, the influence between two pipes could be simplified as a coinciding resistance R_c [6]. So the coinciding resistance can be calculated with the following formula:

$$R_c = \frac{1}{2\pi\lambda_t}\sqrt{(\frac{2H}{b})^2 + 1} \tag{7}$$

With the following definition:
b=distance between centers of two pipes [m]

2.2 Annual Heat Losses Calculation

For traditional heating system, supply and return water temperature as requested regulations varies with the outdoor temperature, and heat loss varied as well during heating period. So annual heat losses are affected by outdoor temperature and corresponding heating hours, and annual heat losses can be calculated with the following formula:

$$Q_l = \sum_{j=1}^{k}\int_{120}^{N_p}(q_g + q_h)l_j dn \tag{8}$$

With the following definition:
Q_l=annual heat losses [J]
l_j= the length of number j pipe [m]
k=the total number of calculated pipes
N_p=the total heating hours [h]

While calculating annual heat losses, corresponding continuous heating hours n of different outdoor temperature should be defined, and the parameter can be calculated by non-dimensional formula. When design outdoor temperature for heating t_w', outdoor mean temperature during heating period t_{pj}, and heating days N'

are known, based on the non-dimensional formula corresponding heating hours of some outdoor temperature can be calculated with the following formula:

$$n=\begin{cases}120 & t_w \le t'_w \\ \left[\left(\dfrac{1-(t_n-t_w)/(t_n-t'_w)}{\beta_o}\right)^{1/b}\times N'+5\right]\times 24 & t'_w \le t_w \le 5\end{cases} \quad (9)$$

In the formula, $\mu=\dfrac{N'-5}{N'}$, $b=\dfrac{5-\mu t_{pj}}{\mu t_{pj}-t'_w}$ and $\beta_o=\dfrac{5-t'_w}{t_n-t'_w}$ are non-dimensional parameters.

3 Analysises of influence factors of heat loss

The influence factors of heat loss include buried depth of pipes, supply and return temperature, pipe diameter and so on. For a real heating system, some parameters like buried depth of pipes and pipe diameter, are fixed. So this text analyses the main factor, supply and return temperature. Since supply and return temperature varies with different regulation ways, heat losses of three different regulation ways, constant flow regulation, centralized regulation with flow varied by steps, variable flow regulation of equivalent temperature difference, are discussed, which in fact analyses the effect of supply and return temperature on heat losses.

It is assumed that the heating pipes are prefabricated insulating pipes, steel pipes are Φ630x8, outer protective pipes are Φ760x11, insulating materials are polyurethane foam, thickness of insulating layer of supply and return pipes is δ=54mm, heat conductivity for insulation is λ_b=0.033 W/mK, outer protective pipes are high density polyethylene, heat conductivity for outer protective pipes is λ_w=0.26 W/mK, pipes are installed in parallel, the distance between two pipes is b=0.2m, mean buried depth from the top of pies to the ground surface is h=1.3m, and that design supply and return water temperature of heating network is τ_g=120℃, τ_h=60℃ respectively.

During heating period, three different regulation ways are applied, and they are constant flow regulation, centralized regulation with flow varied by steps, and variable flow regulation of equivalent temperature difference. When variable flow regulation of equivalent temperature difference is adopted, temperature difference between supply and return would main at the biggest design temperature difference, and its corresponding supply water temperature is higher than other two regulation ways. Because of higher supply water temperature, the heat losses of supply pipes are bigger than return pipes, leading to bigger heat losses under variable flow regulation of equivalent temperature difference. While constant flow regulation is adopted, the flow of heating system mains constant, supply and return varies with outdoor temperature. For its supply water temperature is lower than supply temperature under variable flow regulation of equivalent temperature difference, its corresponding heat losses are smaller as well. The reason why heat losses of centralized regulation with flow varied by steps is bigger than heat losses of constant flow regulation is that when the flow is decreased, temperature difference between supply and return pipes will increase with supply water temperature increasing. Through the calculation, heat losses of three regulation ways are shown as the following Figure 1.

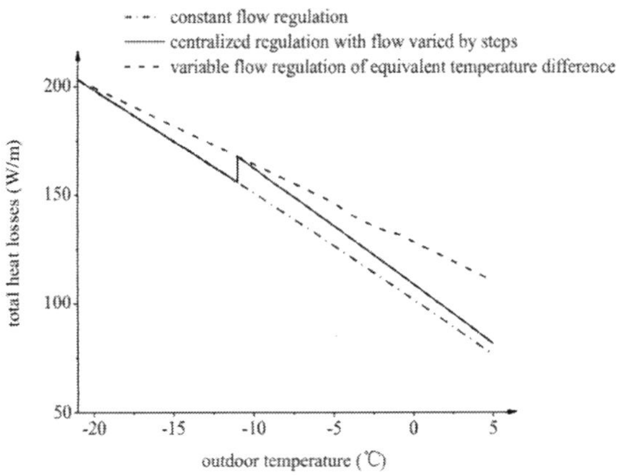

Figure 1. Heat Losses per Unit Length under Different Regulation Ways

As shown in the figure, different regulation ways have different heat losses, and heat losses of constant flow regulation are the smallest; when relative flow coefficient is 1, heat losses of centralized regulation with flow varied by steps are the same as heat losses of constant flow regulation, but when the relative flow coefficient is 0.75, in order to main indoor temperature, supply water temperature has to be increased, so in this step heat losses obviously are bigger than heat losses of constant flow regulation; compared to other two regulation ways, heat losses of variable flow regulation of equivalent temperature difference are the biggest.

4 Annual heat losses of heating network in Jiu Tai

Long term planning heating area of Jiu Tai heating network is 1250×10^4 m^2, heat load is 676.25M, and design supply and return water temperature is 120/60℃. Heating pipes are prefabricated and directly installed with hard polyurethane foam as insulation layer and high density polyethylene as outer protective pipes, and the buried depth from the top pf pipes to the ground surface is 1.3m. In Jiu Tai, the design outdoor tempreature for heating is -21℃, the number of heating days is 169, and mean outdoor temperature is -7.6 ℃ during heating peroiod.

4.1 Annual Heating Consumption of Users

According to the heating areas and design heat load, space-heating load index per unit floor is 51W/m^2

without regarding heat losses of primary network. Based on heating load duration graph of Jiu Tai, heating consumption is $Q_y=\sum Q_h$=6176735.60GJ.

4.2 Annual Heat Losses for Primary Network under Different Regulation Methods

For heating system of Jiu Tai, secondary network is applied with constant flow regulation, and primary network is applied with constant flow regulation, centralized regulation with flow varied by steps(relative flow coefficient of second step is 0.75), and variable flow regulation of equivalent temperature difference respectively. The results for three different regulation ways are shown in Figure 2..

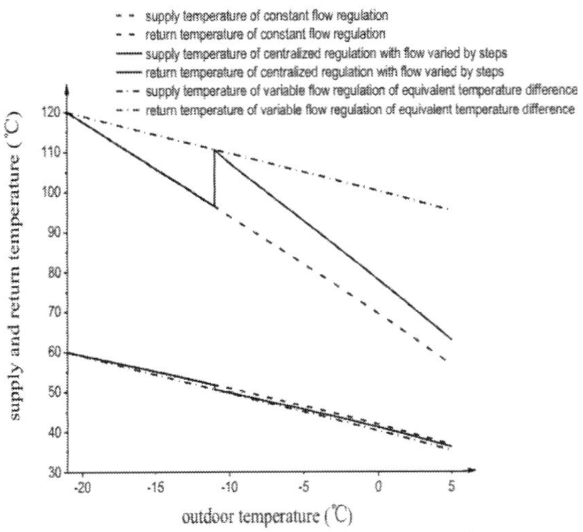

Figure 2 Supply and Return Temperature under Different Regulation Ways

According to Formula 8, annual heat losses under three different regulation ways could be got: annual heat losses for constant flow regulation are Q_f=159906.67GJ; annual heat losses for centralized regulation with flow varied by steps are Q_f=165569.92GJ; annual heat losses for variable flow regulation of equivalent temperature difference are Q_f=178919.67GJ.

4.3 Added Heating Area of Annual Heat Losses

In order to calculate added heating area of annual heat losses, according to the definition we should firstly calculate annual heating quantity per square meter (q), and then annual heat losses divided by q equals added heating area because of heat losses. While heating supply quantity includes annual heat consumption for users, heat losses for pipes and heat losses because of lost hot water (assuming that water loss ratio is 2% of system flow).Therefore, the sum of annual heat consumption for users, annual heat losses and heat losses because of lost hot water equals to total heating supply quantity. Heat transfer coefficient is the ratio of annual consumption for users and total heating supply quantity. The results are shown in Table 1.

Table 1 Added Heating Area of Heat Losses

Regulation Ways	Constant Flow Regulation	Centralized Regulation With Flow Varied By Steps	Variable Flow Regulation Of Equivalent Temperature Difference
Annual Heat Consumption for Users Q_y(GJ)	6176735.60	6176735.60	6176735.60
Annual Heat Losses Q_f(GJ)	159906.67	165569.92	178919.67
Total Heating Supply Quantity Q(GJ)	6517672.99	6523336.18	6536.685.93
Percentage of Heat Losses (%)	2.4	2.5	2.7
Added Heating Area (m^2)	30.7×10^4	31.8×10^4	34.4×10^4
Heat Transfer Coefficient (%)	94.8	94.7	94.5

From the table, we can find that the percentage of heat losses under three different regulation ways varies from 2.4% to 2.7%. The added heating area varies from $30.7\times10^4 m^2$ to 34.4×10^4 m^2. The added heating area under variable flow regulation of equivalent temperature difference is the largest, centralized regulation with flow varied by steps follows, and constant flow regulation is the smallest, which is in consistence with prior analysis.

5 Conclusion

From the above analysis, we can know that for the directly buried hot water heating network (prefabricated pipes):

(1) Different regulation ways have different heat loss per unit length. Heat loss of variable flow regulation of equivalent temperature is the biggest, centralized regulation with flow varied by steps follows, and constant flow regulation is the smallest.

(2) According to the calculation of annual heat losses of heating network in Jiu Tai, different regulation ways have different annual heat losses. During heating period, when applied with variable flow regulation of equivalent difference, supply water temperature is the highest with the biggest heat losses. While supply water temperature of constant flow regulation is the lowest with the smallest heat losses, and heat losses of centralized regulation with flow varied by steps are in the middle.

(3) Through the calculation of heating network in Jiu Tai, annual heat losses of primary network take up 2.4%~2.7% of total heating supply quantity. The added heating area under different regulation ways varies from $30.7\times10^4 m^2$ to 34.4×10^4 m^2. While the scale of heating

system is bigger, this part of heat losses is becoming bigger as well.

This paper discusses the proportions of heat loss of primary network in total heating supply quantity based on heating network in Jiu Tai. Because the difference of installation ways, heating system scale, routing, and distributions of heat loads, operation temperature and so on, the proportions of heat losses in heating supply quantity are different as well. Heat losses in heating system should analyse case by case. This paper provides calculation methods and research ideas for heat losses of prefabricated and directly buried heating pipes, and proposes some reference for study of heat losses of other types of heating network at the same time.

Acknowledgements

The authors gratefully acknowledge supports by the National Science & Technology Pillar Program in the Twelfth Five-year Plan Period (No. 2011BAJ05B04).

References

1. F. Schmitt, H.W. Hoffmann, T. Gohler. Strategies to Manage Heat Losses. IEA. **V Ⅱ** (2005)
2. Zide W. Calculation and Design of Directly Buried Heating Network. G&T. J. **22**, 457 (2002)
3. C.G. Potts, L.M. Draaijer.Developments in Rigid Polyurethane Foams for the Insulation of District Heating Pipelines. J. Cell. Plas. J. **21**, 51 (1985)
4. Ping H, Guang S. *District Heating Engineering*. (Architecture&Building, China, 2009)
5. Dexing S. *Advanced Heat Transfer*. (Architecture&Building, China, 2005)
6. A.A. Lonin. *District Heating*. (Architecture&Building, China, 1986)

An Approach of Conformance Verification between Design Models and Code Based on Abstract Syntax Tree

Zhao Liu, Yang Tian, Haihua Yan

Department of Computer Science and Engineering, Beihang University, Beijing, China

Abstract. Design models and code are products of different stages in the software development process. The conformance of design models and code plays an important role in software development process, also is a key principle to improve the maintainability of the software. Currently, the conformance on different level of abstraction is verified mostly by manual inspection. In order to avoid the errors and omissions, we propose a conformance verification approach between design models and code based on abstract syntax tree. The approach verifies the conformance by comparing the class diagram, sequence diagram with the code. Firstly, we discuss how to extract and recover model information from the abstract syntax tree, then give the conformance verification approach between class diagram, sequence diagram and code, finally implement a semi-automatic verification tool based on the approach.

1 Introduction

Traditional object-oriented software development process is divided into several stages including requirement analysis, software design and implementation. As a result, corresponding software products are requirement model, design models and code. Ensuring conformance of these products is an important principle in software development life. However, there are many factors which may reduce the conformance between design models and the code. On one hand, the code is changed while the corresponding design model is not updated. On the other hand, different understanding of the system leads to inconsistency, because developers do not refer to the design models when implementing the system.

There have been some tools on consistent maintenance between design models and code. These tools are mostly based on formal language [1]. At present, conformance verification between UML design models and source code is accomplished by manual inspection. Therefore it will be better to carry out the conformance verification depends on the support of the computer aided tools.

2 Conformance verification based on abstract syntax tree

A complete software system includes two aspects of information, static structure information and dynamic behavior information. UML(Unified Modeling Language) has become a standard to describe the structure and behavior of the software system. In UML models, class diagram describes the static structure of a system by showing the classes and the relationships between them [2,3]. Sequence diagram in UML describes dynamic behavior of a system by showing objects interaction arranged in time sequence [4]. In this paper, the class model corresponding to code is obtained by analyzing the abstract syntax tree of the code, and then the interaction information is organized into a tree structure, called message tree, and finally the conformance verification between class diagram, sequence diagram and code is carried out by comparing the class model, the tree structure with the source code. The process is shown in Figure 1.

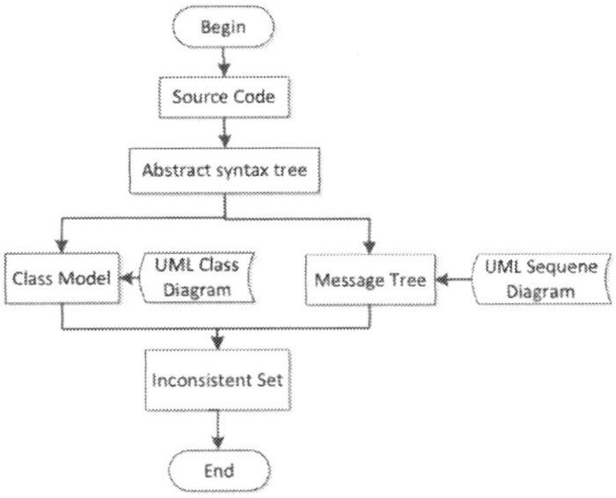

Figure 1. Conformance verification between design models and code

2.1 Abstract sytax tree of code

An abstract syntax tree is a tree of the abstract syntactic structure of source code written in a programming language. Each node of the tree denotes a construct occurring in the source code. It can be generated after lexical analysis and grammar analysis. The information in abstract syntax tree is complete for fine-grained analyzing, at the same time, it is easier to establish rules to map from the abstract syntax tree to design model.

In this paper, changing of interface of the class is our focus. Therefore four types of nodes are processed in abstract syntax tree: attribute, operation, attribute assignment and operation calls. Although the structure of abstract syntax tree depends on the grammatical structure of language and compilation tools, for most object-oriented language, source code can be transformed into an abstract syntax tree with the unified structure, as shown in Figure 2.

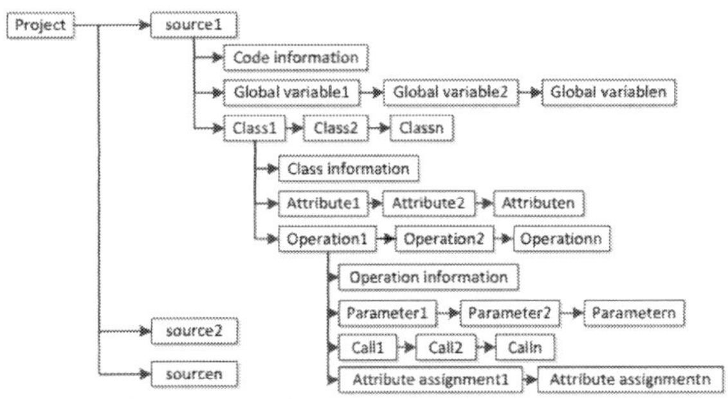

Figure 2. The structure of abstract syntax tree.

In the general case, a project is consisted of several source codes. Each source code contains the code information, global variables and several classes. Each class contains a number of attributes and operations. We can get the abstract syntax tree of each source code with common compiler tools, and then we analyze the abstract syntax tree to organize a project into the structure above. The work is for the next stage of comparison.

2.2 Conformance verification between class diagram and code

Conformance verification between class diagram and code includes two steps: model extraction and model comparison. In the step of model extraction, class model is extracted from the code by analyzing the abstract syntax tree, called implementation class model, and then in the step of model comparison, the generated implementation class model is compared with design class diagram, as a result, the program records and displays the inconsistency.

Step1: (model extraction) There are no direct mappings from UML class diagram to code for some elements in the object oriented programming language, and the implementation of some UML elements are not unique. These problems reduce the traceability in model refinement and implementation. However, we can still extract some model features by analyzing the basic grammar elements of the code. There is a good comparability between the design class diagram and the implementation class model which reverse from the code [5].

Class, attribute and operation in class diagram can directly map to the class, member variables and functions in the code and they can be obtained directly in the abstract syntax tree [6].

To obtain the relationships between classes, we have to analyze the usage and dependency relationships between classes. Three situations below can be extracted as the relationships in UML.

Class A inherits another class B. This situation describes the relationship between a subclass and a superclass. The relationship should be abstracted as generalization between two classes. Most object-oriented programming languages provide relevant keywords to implement generalization and implementation relationships, such as java using "extends" show inheritance relationship between two classes, using "implements" shows a class implements an interface, and the C++ using colon operator shows class inheritance relationship.

A member function of class A uses an instance of class B as its parameters, return values or local variables. This situation describes usage relationship between two classes. In the UML, the relationship should be defined as dependency. Dependency often refers to an occasional, weak usage relationship. Every member function of class A in the abstract syntax tree is traversed to extract the relationships. If its type of parameters, return value or local variable is class B, A dependency is established between class A and class B.

Class A uses an instance of class B as a member attribute. This situation is also a kind of usage relationship. When an instance of class B is a member attribute, to some extent, class B is a part of class A. It is a genus relationship and has a concept of "partial – whole", so the relationship is defined as association. Every member variable in abstract syntax tree is traversed to extract the relationship. If the type of member attribute is class B, then an association is established between class A and class B.

Although variable types in most object-oriented programming languages are the same with the data types in UML, there are still some exceptions, for example, C++ provides a pointer or reference type, and the types of modifiers such as const, mutable, volatile, while there are no such types in UML, when generating the implementation class model reversely, removing all pointers, references and modifiers does not affect the comparison with UML class diagram.

Step2: (model comparison) After the implementation class model is extracted from the source code, it is compared with the design class diagram. During the process, the compared elements include classes and their relationships.

The comparison is a two-way matching process. On one hand, design class diagram is compared with implementation class model. For a class in design class diagram, if there is a corresponding class in the implementation class model, then it is consistent for the two classes, if not, there may be a lack of the implementation of this class in source code. On the other

DOI: 10.1051/matecconf/20165407001

hand, implementation class model is compared with design class diagram, and we can find the elements which are in the implementation class model while not in the design class diagram.

There are four kinds of comparison results. They are "Same", "Lack", "Append" and "Inconsistent". "Same" indicates an element has the same name, visibility, and other attributes in the design class and the implementation class diagram. "Lack" indicates an element exists in the design class diagram while not exists in the implementation class model. "Append" indicates the elements appended into the implementation class model based on the design class diagram. "Inconsistent" indicates an element has the same name but different type, visibility or other attributes.

The comparison of relationships is relatively complex, so the comparison result is classified with five kinds of situations: "Lack", "Append", "Consistent", "Compatible", and "Inconsistent". Firstly, if there is a relationship exactly same in both design class diagram and implementation class model, that is to say the relationship has the same begin class and the same end class, at the same time, its type is also the same, then the comparison result is consistent. Secondly, there are four kinds of relationships in design class diagram: dependency, association, aggregation and composition while there are two kinds of relationships in our implementation class model: dependency and association. These relationships themselves have the characteristics, namely dependency < association < aggregation < composition, so if a stronger relationship in design class diagram is implemented with a weaker relationship in implementation class model, we consider the comparison result is compatible.

2.3 Conformance verification between sequence diagram and code

In object-oriented program language, different variables maybe point or refer to the same object and it is hard to recognize polymorphic objects only by analyzing the source code, so our conformance verification approach between sequence diagram and code just verify whether the message sequence in sequence diagram is implemented by order [8,9] in class level.

To illustrate our approch, we define a data structure called message tree, Messagetree=(V, E), where V is the tree nodes, V=(receiveObject, receive Class, Message Name, MessageType, ParameterList, ReturnValue), E is the parent-child relationships between nodes, E=(V, V). Messagetype includes synchronous message, asynchronous message, creating message, destroying message, return message. Asynchronous message are not considered. Our conformance verification between secuence diagram and code includes two steps: tree extraction and tree comparison. In the step of tree extraction, sequence diagram and abstract syntax tree are transformed into two message trees, then in the step of tree comparison, generated message trees are compared to verify whether the sequence diagram is implemented in the source code in class level.

Step1: (tree extraction) The message tree nodes can be obtained by messages in the sequence diagram, so a sequence diagram can be transformed into a message tree by resolve the message information. Firstly, Information from sequence diagram described in XML is extracted, these information includes the message sequence, composite fragment and its interaction domain. Secondly, a message node Vm is established and the corresponding message information is recorded in the node. At last, for each message a in the sequence diagram, we find the message b before message a, which is a synchronous message or creating message and the receive objects of message b is the send object of message a, then a parent-child relationship is added between the node a and node b. The Figure 3 shows a sequence diagram and corresponding message tree.

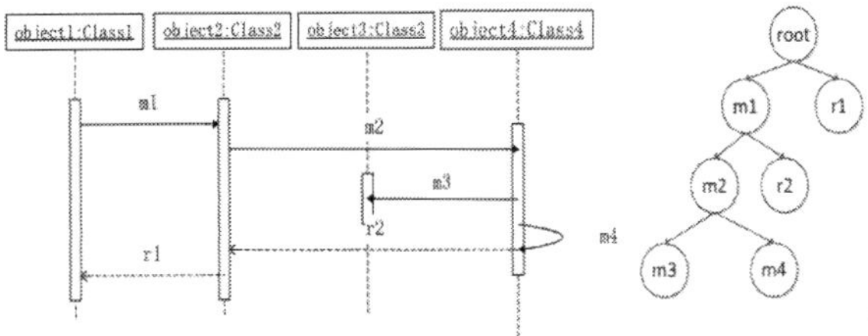

Figure 3. Design sequence diagram and corresponding message tree.

On the other hand, message tree nodes can also be obtained by analyzing the function calling, object creating and return statements in abstract syntax tree. These statements are called message statement. The algorithm below describes the process how to generate the message tree starting with a function based on abstract syntax tree.

Algorithm generate the message tree based on abstract syntax tree

```
visit(AST ast, Function function, MessageNode V)
{
Search the node corresponding to the function in ast
for(every message statement in node)
        if(statement does not contain function)
                make node V'
                add parent-child relationship(V,V')
                visit(ast, function, V')
}
```

Step2: (tree comparisons) Comparing two message trees is the second step. During the process, the compared elements include tree nodes and their order in the level. For a node in message tree of sequence diagram, if there is a corresponding node in message tree of source code and their relative orders are the same, then it is consistent for the two nodes. That is to say, the message corresponding to the node in sequence diagram is implemented in source code in class level, or its comparison result is inconsistency.

3 A prototype tool

In this section we put an example library management system to verify this method. The example is a typical information management system, readers can borrow books, return books, search books in the system, the design models of the system include a class diagram and several sequence diagram. Figure 4 are the interface of the prototype tool. Developer can draw class diagram and sequence diagram as design model in the tool.

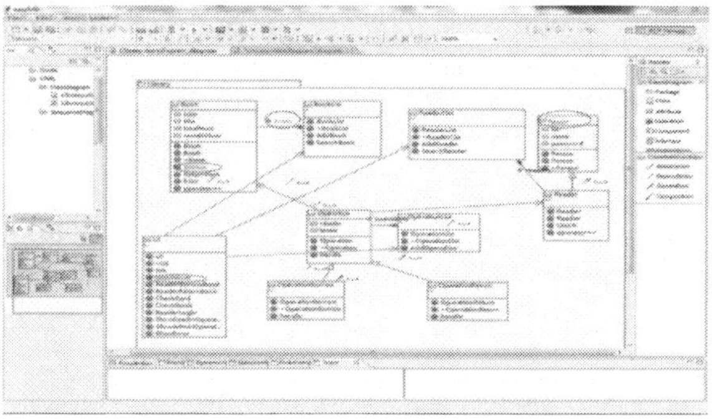

Figure 4. Drawing class diagram.with the tool.

The comparison results are displayed in the result bar, and different comparison results are represented in different marks. Figure 5 is a comparison results. There are some labels in the result bar. Elements marked with "D" are deleted in the design model. Elements marked with "A" are added into the implementation model, in another way to say, they are added into the code. Elements marked with "×" are different between design model and code.

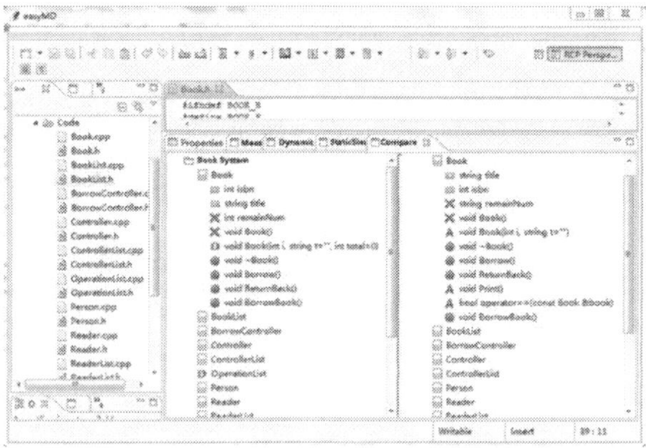

Figure 5. The comparison results.

4 Conclusions

This paper presents an approach which extracts and recovers system structure and behavior characteristics based on the abstract syntax tree, then verify the conformance between design models and code including conformance between class diagram and code, conformance between sequence diagram and code. The verification process can be accomplished by computer, and sometimes it needs a further manual inspection, so it is a semi-automation approach. The method has been embedded in conformance verification tool. Its results can locate the inconsistency between design models and code. It is an important reference for the project development and maintenance personnel.

At present, the comparison of the design model and the code is only based on element name. In the process of program, developers have to consider the specific program language, algorithm, and framework to refine the elements in design model such as adding auxiliary class, or implementing the relationship between classes with a different style. These behaviors lead that design model and code seem different but they perform the same function. In the next stage, we should classify and summarize the different implementation ways of the design model in order to compare the design model and the code more fine-grained.

At the same time, extraction and comparison rules are still relatively simple, therefore in the next step, we will summarize more detailed extraction and comparison rules to detect the consistency. If possible, we can execute the source code to obtain the object information and verify the conformance between sequence diagram and code in object level.

References

1. David Evans,John Guttag,James Horning,and Yang meng Tao. LCLint:A Tool for Using Specifications to CheckCode.
2. Booch G,Rumbaugh J,Jacobson I. The Unified Modeling Language User Guide. Addison-Wesley Publishing Company, 2009.
3. OMG Unified Modeling Language (OMG UML), Superstructure 2011.
4. Sharp R, Rountev A. Interactive exploration of UML sequence diagrams[C]//Visualizing Software for Understanding and Analysis, 2005. VISSOFT 2005. 3rd IEEE International Workshop on. IEEE, 2005: 1-6.
5. Liu Chao, Li Jian, Shen Haihua. Reversed Automatic Generation of Visualized Class Diagram of Object-Oriented Program. Journal of Beijing University of Aeronautics and Astronautics. 1998.
6. Sutton A, Maletic J I. Recovering UML class models from C++: A detailed explanation[J]. information and software technology, 2007, 49(3): 212-229.
7. Egyed A. Automated abstraction of class diagrams[J]. ACM Transactions on Software Engineering and Methodology (TOSEM), 2002, 11(4): 449-491.
8. Kung D, Hsia P. A reverse engineering approach for software testing of object-oriented programs[C]//Application-Specific Systems and Software Engineering and Technology, 1999. ASSET'99. Proceedings. 1999 IEEE Symposium on. IEEE, 1999: 42-49.
9. Kollmann R, Gogolla M. Capturing dynamic program behaviour with UML collaboration diagrams[C]//Software Maintenance and Reengineering, 2001. Fifth European Conference on. IEEE, 2001: 58-67.

MATEC Web of Conferences **54**, 07002 (2016) DOI: 10.1051/matecconf/20165407002
MIMT 2016

Variability based Approach for Minimizing over Design and under Design

Yucong Duan[1], Nanjangud C Narendra[2], Honghao Gao[3], Mingdong Tang[4], Abdelrahman Osman Elfaki[5], Shixiang Wan[1], Junxing Lu[1]

[1]*Hainan University, Haikou, China*
[2]*Cognizant Technology Solutions, Bangalore, India*
[3]*Computing Center, Shanghai University, 200444 Shanghai, P.R. China*
[4]*School of Computer Science and Engineering, Hunan University of Science and Technology, Xiangtan, China*
[5]*Department of Computer Science, University of Tabuk*

Abstract. In this paper, we have considered Over Design (OD) and Under Design (UD) as two basic forms of negative variability construction that lead to both functional incompleteness and unexpected software quality degradation during design time and run time. To make our approach economically practical to the quality assurance at the whole life cycle of software production process, we aim to ensure that OD and UD are minimized in terms of values under economical goals. These requirements are concretized with the functionality implementation and the quality management of a quality lifecycle.

1 Introduction

In [1], Boehm et al. identified that software development is a value creation process. Usually, in a software development lifecycle, there are gaps between technical decisions and targets of value creation processes. They summarized related topics as an interdisciplinary subject: Software Economics. Software economics seeks to enable significant improvements in software design and engineering through economic reasoning about product, process, program, portfolio and policy issues.

The economical influences of Software Economics on construction and evaluation of a software project can be summarized in four dimensions:

- Ultimate implementation: The composition of quality models determined by economic goals defines the business success of a software project. Quality models are built to reflect the business interests with quality properties and represent the business values with quality values.
- Integrated evaluation: The evaluation of fulfillment of quality properties shall usually be considered as a well-organized whole as considering quality properties independently might potentially impede the process of pursuing another quality target. Thus, a tradeoff between over pursuing and under pursuing a quality target should be identified under goals of Software Economics.
- Dynamic management: During software construction, the configuration of information flow and control flow may influence both efficiency and outcome of value creation process[1]. This demands a dynamic management to flexibly cater the maximization of the created value under the available resources constraints.
- Lifecycle accumulation: Software development lifecycles outline completeness for value modeling and calculation in the temporal dimension. Values accumulated alongside a complete development lifecycle or business lifecycle form the value identities for further comprehensive measurement.

Most software quality implementation serves the ultimate economic goal of software projects on the customer side. Also integrated quality evaluation is subordinated with accomplishment of economic goal towards which a software project is proposed. For example, if designing a chatting software including video chatting function, it will be unnecessary for African to use it because of bandwidth limited. Hence, a simple text chatting software is enough to ensure optimal quality and economic goal.

1.1. Integrating values of process and product quality

In software lifecycle, mismatch between technical decisions and value creation originates usually as they are considered separately in a project management perspective and a classic economic analysis. The value of process quality of a software project is calculated dividedly from values of the software product which is measured partially by its quality fulfillment. This is a practical solution from an engineering perspective.

To proceed towards harvesting the full benefit proposed in software economics, we need to consider the interrelationship among technical decisions and value creations. For example, a design decision leading to well managed late binding[2] can potentially contribute to a project with improvement on both quality properties and business values. These include higher scalability of deployment, enhanced flexibility for extension, and reduced cost for composition according to E-Service contract[3] during service composition[4] through E-Service value brokers [5]. More comprehensive solutions can be designed to integrate values of both process and product quality after exposing association between technical decision and value creation.

1.2. Variability modeling

In the past several years great achievement have been made in the area of design and implementation of quality driven or quality aware [6] for both specific software products and software product families [7]. Prominent success has been attained in practice of software product line and feature modeling [7] , MDA [6]. In such systems, a focus has been given to variability modeling that considers various superficial quality issues presented as features. In this paper, we detail the variability modeling of design variation from stage to the whole process to reveal the implementation of the economical business goal in terms of quality properties connected to technical decisions. This involves identification of quality properties related to technical implementation and leverage evaluation from stages based on temporary quality properties to the entire life cycle following process based quality model reflecting economical goal of a software project.

In a waterfall development process, a quality model is assumed to exist completely at the end of requirement specification. However, in the reality this experience does not fit for iteration based process models such as RUP or agile in which the software may suffer from constant modification. To cope with these kinds of situations and improve software quality implementation, there is a need to consider quality model evolution and identify variability management.

We propose an approach called "Problem-Quality-Constraint" (PQC), as a solution strategy for quality driven/aware software modeling. PQC is a constraint based modeling solution to manage variability towards satisfying the requirement of quality model that bears a process or lifecycle affiliated to the development process. Our approach inculcates stakeholders' assessment, agreement and economic values in quality life cycle.

The rest of the paper is organized as follows: Section 2 analyzes and models scenarios related to quality from the perspective of introduction of information, quality lifecycle and economics. Section 3 presents our strategy "Problem-Quality-Constraint". Section 4 shows an initial case study on a scenario of Over-design and Under-design and experimental results. This is followed by related works in Section 5 and in Section 6 conclusions and future directions.

2 Running example

In this section, we describe the ideal design, actual design, over design and under design in detail by an example of a bank in system.

As per Figure 1, we have demonstrated a banking system to illustrate variability based approach. The whole product process is divided into three processes including basic process, middle process and final process. Empirically, x-axis is identified as product process of development. We have confirmed an ideal line, in which seven targets belong to three processes representing different parts of system implementation. Two different colored lines (the orange line and purple line) separately shows two designs that are away from the ideal line. The parameter "cost" is chosen to calculate the deviation between actual and ideal lines, which is reflected in y-axis. In order to measure the cost of each process better, we assume that some numerical values represent the different levels of each target. Besides, there are two dotted lines on both sides of ideal line, and the variability space comes into being between them.

ID: means Ideal Design

AD: means Actual Design

ND: means Negative Design

VUB: means Variable upper bound

VLB: means Variable lower bound

SPL: means Software Product Line

Table I. The value of all targets

Factor / SP	Security	Stability	Authority distribution	Emergency capability	Response time	Throughput	Satisfaction
ID	10	11	8	9	10	10.5	9.5
AD	10	10.5	9	8	11	10	10
ND	10	13	5	11	8	11	9
VUB	11.5	12.5	9.5	10.5	11.5	12	11
VLB	8.5	9.5	6.5	7.5	8.5	9	8

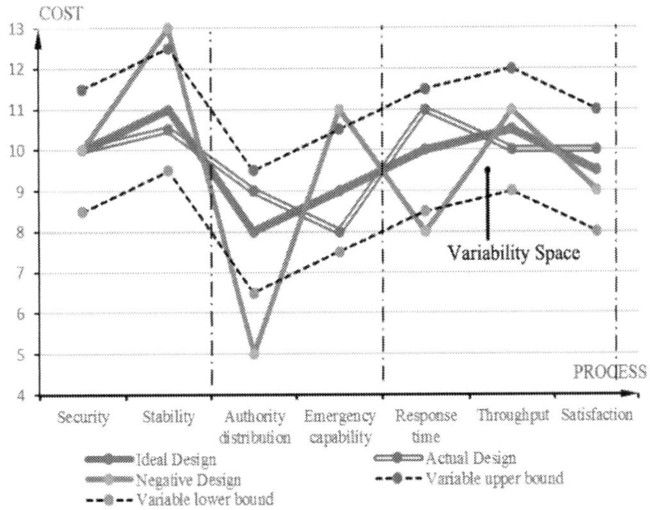

Figure 1. Idea design and actual design

Hence, the quality S_{PR-QU} of a software process line can be formulated as:

$$S_{PR-QU} := \sqrt{\frac{\sum_{i=1}^{n}(C_i - C_{ID})^2}{n}}. \quad (1)$$

Where:

- C_i: actual cost of the target
- C_{ID}: ideal cost of the target
- n: number of all targets

Table II. The values of S_{PR-QU}

Target	Ideal Design	Actual Design	Negative Design	Variable upper bound	Variable lower bound
S_{PR-QU}	0	0.732	1.75	1.5	1.5

The numerical value of SPR-QU shows degree of deviation during the software process line. The larger it is, lower the overall product quality, and vice versa. By this quality index SPR-QU, a comprehensive solutions can be designed to maximize both process and product quality based on association between technical decision and value creation.

3 Analysis of fundamental issues

3.1. Criteria of Information Introduction

In software product lines, it is not easy to set priority among several modeling decisions. But, the use of constraint based approach helps in such cases. The enforcement of creating a priority among difference modeling decisions without concrete proof will bear the risk of Over-design (OD) and Under-design (UD).

Definition 1: [Over Design]. The design introduces unnecessary information for directly promoting the intended goal. This surplus information bears the risk of refraining quality improvement in subsequent stage and may reduce the variability space of subsequent modeling.

Definition 2: [Under Design]. The design lacks some necessary information for the demand. This lost information extends the variability space of software modeling decision, which adds the software product cost.

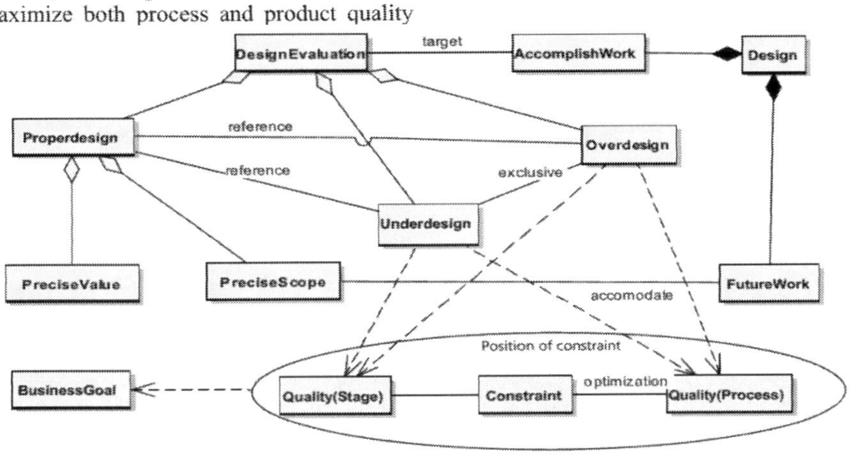

Figure 2. Constraint driven design

Over Design/OD is a form of Negative Information Introduction (NPII) [8]. Considering system design from perspective of dichotomy, the deviation (ΔD) in design is the discrepancy between the ideal design (IDD) and the actual design (ACD). It is denoted as:

$$\Delta D = IDD - ACD$$

There are two situations associated with ΔD:

- If ΔD>0, then ΔD belongs to IDD, which implies that ΔD is a space of Under-design (UD). i.e., the actual design is quite below the ideal design of system.
- If ΔD>0, then ΔD belongs to ACD, which implies that ΔD is a space of OD. i.e., the actual design of the system is beyond the expectation of ideal system design.

Then the problem facing a design/implementation is trying to reduce the space of ΔD denoted as ΔPD which is the combination of space of under and over design as formalized below

$$\Delta P_D = \Delta P_{OD} \cup \Delta P_{UD}. \quad (2)$$

Where

- UD space ΔP_{UD} represents the unaccomplished design space that can be filled later on.
- ΔP_{OD} is an OD space.

Figure 2 illustrates the relationships between IDD, OD and UD.

3.2. Essence of quality: quality lifecycle

3.2.1. Hypothesis on quality

The quality properties considered in this work are assumed to be orthogonal, i.e., they do not affect each other.

3.2.2. Quality lifecycle

Software Quality is defined mostly in terms of functional, quality attributes/property and resource specifications. The last two specifications are related to the non-functional requirement of system. Quality assessment has to be done throughout the software life cycle and considered as quality lifecycle (QL):

- Design stage
- Execution stage
- Maintenance stage
- Economic analysis stage

At each stage (STG) of QL, quality properties (QP) are of interest to various Stakeholders (SK) and are evaluated using various criteria (CR). Table I shows the quality properties QP and associated evaluation criteria CR. An agreement (AG) is usually a compromise after a series of adaptation/change among stakeholders. During a QL, the agreement of a specific quality property may be changed from stage to stage. The agreement on quality properties will be the reference of quality driven modification on the implementation of software system. Figure 2 shows that UD and OD influence both quality of process and quality of final output. In this work, we propose to integrate these as per the principles software economics to bridge the implementation of UD and OD at individual stages. And UD and OD are evaluated according to the quality model of the whole process as presented next.

Table III. Software quality criteria

Quality property (QP)	Software Quality Criteria (CR)
Correctness	Traceability, consistency, completeness
Reliability	Error tolerance, consistency, accuracy, simplicity
Efficiency	Execution efficiency, storage efficiency
Integrity	Access control, access audit
Usability	Operability, training, communicativeness, input/output volume, input/output rate
Maintainability	Consistency, simplicity, conciseness, modularity, self-descriptiveness
Testability	Simplicity, modularity, instrumentation, self-descriptiveness
Flexibility	Simplicity, modularity, instrumentation, self-descriptiveness
Portability	Modularity, self-descriptiveness, machine independency, software system independence
Reusability	Generality, modularity, software system independency, machine independency, self-descriptiveness
Interoperability	Modularity, communications commonality, data commonality

3.2.3. Quality Economics

In the section II, we create a running example about ideal design and actual designs. From the value analysis perspective and based on the running example, we derive the goal of quality driven design in three steps:

1. Tradeoff for each stage - at the tradeoff point of quality negotiation, stakeholders reach agreement on quality properties of a stage. A general tradeoff point is the state where the sum of the satisfaction of all stakeholders reaches maximum. Thus, the tradeoff point of a quality property among stakeholders in a stage is formulated as:

$$AG(qp_i)|_{STG} ::= \max\left(\sum_{j=1}^{n} qp_i \times w_{SK_j}\right). \quad (3)$$

Where:

- $qp_i \in N$ – the ith quality property
- w_{SK_j} : weight given to the jth stakeholder

To reach the $AG(qp_i)|_{STG}$, a change has to be made on the design, implementation or the processing itself. This kind of change is denoted as CG-STG and its variability space is denoted as $CP_{CG\text{-}STG}$.

2. Tradeoff on evolving stages - Demands on a quality property from a stakeholder may be evolved from stage to stage in a QL, as well as the agreement of quality properties among stakeholders. During a QL, the aim is to maximize overall quality, and this can be formulated as:

$$AG(qp_i)|_{QL} ::= \max\left(\sum_{k=1}^{m}\left(\sum_{j=1}^{n} qp_i \times f_k\left(w_{SK_j}\right)\right)\right) \\ == \max\left(\sum_{k=1}^{m} AG(qp_i)|_{STG=k}\right). \quad (4)$$

Where

- $qp_i \in N$ –a quality property
- $SK_j \in N$ – a stakeholder
- $k \in N$ –a stage
- w_{SK_j} : weight given to the jth stakeholder
- $f_k(\ )$ is the modification function at the kth stage on the weight given to the jth stakeholder

To reach the $AG(qp_i)|_{QL}$, a change need to be made in a holistic manner on the distribution of design, implementation or processing decisions as interrelated across stages. This kind of change is denoted as *CG-QL*. The variability space of this kind of change is denoted as $CP_{CG\text{-}QL}$.

3.2.4. Transformation from short run to long run

From the problem solution perspective, the optimization action (OA) of design/implementation decision on quality properties is a form of change which belongs to either CPCG-STG or CPCG-QL. CPCG-QL is not a simple sum of individual CPCG-STG: $\sum(\mathrm{CP}_{CG-STG})$, instead it is expected to be an optimization of $\sum(\mathrm{CP}_{CG-STG})$ based on the global business strategy. From the long run vs. short run point of view [9], the ultimate goal of a design process is maximizing the global business value related to the project (GBV) which can be implemented through adaptation on CPCG-STG towards forming CPCG-QL. This process is denoted as: $OA(CP_{CG-STG}) \Rightarrow CP_{CG-QL}$. Existing quality driven approaches have been introduced to handle individual CPCG-STG [26].We would like to expand it from individual CPCG to a whole CPCG-QL.

4 Problem-quality-constraint

Our solution of lifecycle quality assurance assistance for software projects which is called Problem-Quality-Constraint (PQC) is composed of three main components:

4.1 Problem modeling

The problem model associated to a software project can be modeled using function or process modeling.

- Functional modeling [7] at an abstract level - Actions in a system construction are expected to contribute positively to fill the gap between project schedule and final ideal system. We model the reference to modeling directions as deviations from the ideal model which bears the maximization of satisfaction of quality requirements. The reference is distinguished as over-design and under-design. Reducing the spaces of over-design and under-design will aid in attaining the satisfaction of quality goals.
- Process modeling - we model the optimal variability space to make it correspond to stages and lifecycle from the perspective of economic quality. For example, we shall close to ideal design line when designing actual line by the process including technical decision and value creation.

4.2 Quality model

Keeping the intended space of variability is of interest to some stakeholders who work to meet non-functional requirements (NFR) of a system or its design process. These represent a form of quality requirement and denoted as NFRV. In addition to explicitly describe NFRV the most important goal of design is to reduce the variability space of quality lifecycle by reducing the space of under-design and avoiding increment of the space of over-design. We identify this as a quality requirement in the development process by transforming design knowledge/ information from stakeholders to model. It is denoted as: NFRP. A simple quality model (QMS) is adopted following works in [10], which use the aggregation of NFRV and NFRP as precondition to meet system implementation. It is formulated as.

$$QM_S := NFR_V \cup NFR_P. \quad (5)$$

In Figure 3, quality stages are modeled as three common targets towards minimizing the quality loss of modeling practice before the accomplishment of a software project:

- Maximize the scope of known information in the models. The excess in the boundary of proper design will turn into OD practice.
- Minimize the scope of unknown information in the models. The excess in the boundary of proper design will turn into UD practice.
- Maintain the intended variability or scopes in models which might be used for satisfying various quality properties.

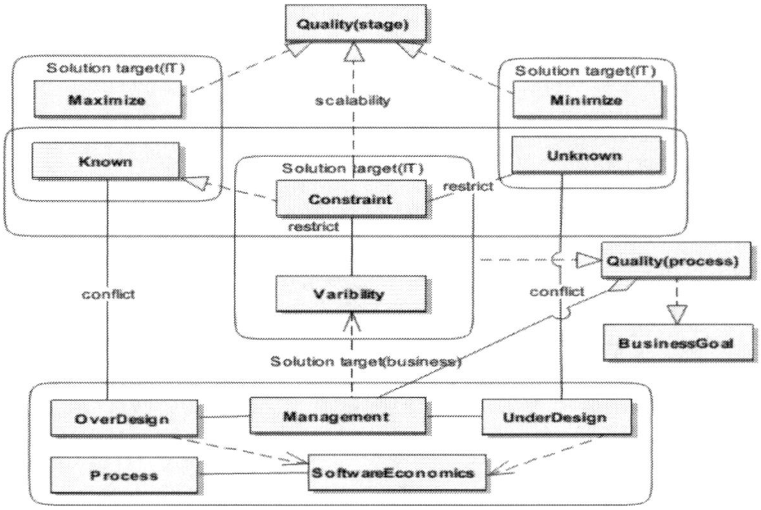

Figure 3. Constraint based solution framework

4.3 Constraint Based Solution

There are various understandings of the concept of constraint for design:

In [11] constraint is defined as the relationship between set of design variables that are related to each other. A relation between independent design variables represents a simple constraint whereas a relation between dependent design variables represents a complex constrain. This can be implemented by using constraint modeling language – OCL in UML. It is defined as expressions typically specifying invariant conditions that must hold for the system being modeled or queries over objects described in a model[1].

In this work, constraints are defined as expressions of relationships among modeling elements for restricting the unknown space in a constraint space (CRP) to form a product which minimizes the chances of OD and UD. The OD and UD are measured according to an evolutionary quality lifecycle which concretizes the goals of software economics.

From a general sense, relationships among modeling elements can be viewed as implicit form of constraints. For example, among UML relationships [12] - *association*, *aggregation* and *composition*, without a specific project context, *association* accommodates a CRP which contains both *aggregation* and *composition*. Using *association* instead of either *aggregation* or *composition* leaves space for choosing a fitting implementation with desired quality without incurring OD.

Constraint used to construct a variability space or a CRP avoids the drawback of increasing ΔP_{OD} with a prudence decision.

CRP accommodate multiple choices which might make a difference in terms of influencing AG(qpi)|STG. However, if the choices have different influences on CPCG-QL, CRP contributes to following two aspects:

- The package of choices alleviates the risk of increasing ΔP_{OD}.
- Postponing of a choice effectively reduces the risk of increasing ΔP_{OD}

Technically CRP is the source of many design patterns [5], [13] and can be used to plan quality aware later binding [2]. To assure that CRP contributes positively, the benefit of reducing risk of ΔPOD shall surpasses the loss of performance such as in later binding. The concrete planning relies on the specific tradeoff among quality properties.

5 Case study

In this section, we share our intermediate results through practicing PQC and experimental value.

5.1. Feature based modeling of e-commerce

We adapted the feature model of [15] to create Figure 3 in which service is implemented from up to down. There are two types of structures in Figure 3: Static and Dynamic. In Static, all components are static service choices. On the contrast, we propose the new concept of Dynamic structure, which is a sequential services bundle. Obviously, the sum of all cost and value-added includes all selected components cost and value-added in the variability construction.

Variant of "Validation" in design (B) with optional and is OD relative to design (A) with mandatory as validation is required in every E-commerce process.

Variant point of "Translation" in design (B) represented with "alternative" is UD relative to "Or" in design (A) as several of languages may be selected at a time.

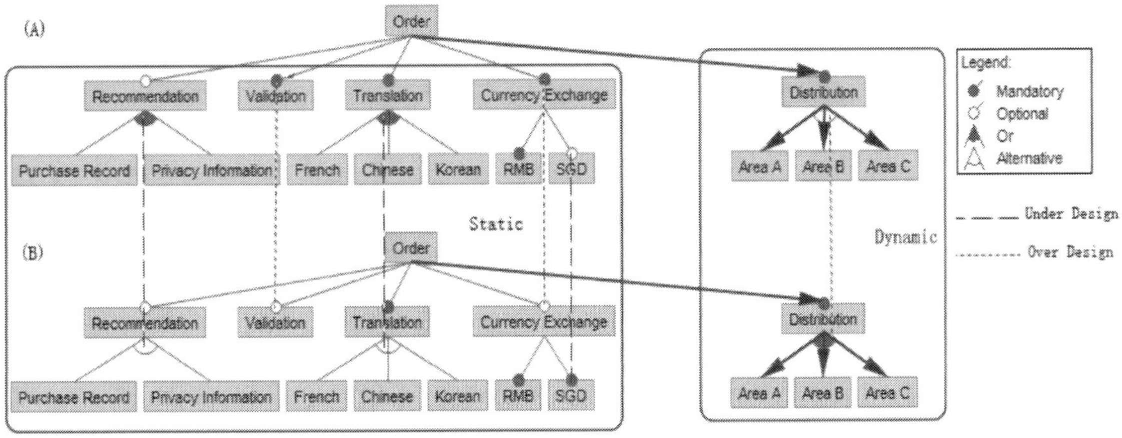

Figure 4. Ideal-design (A) vs. under-design and over-design (B) in a feature model

Table IV. Value added to the design vs. cost tradeoffs

Component	Cost	Value Added
Order	200	2
Recommendation	100	1
Validation	200	2
Translation	300	3
Currency Exchange	400	4
Distribution	100	1
Purchase Record	200	2
Privacy Information	200	2
French	100	1
Chinese	100	1
Korean	100	1
RMB	300	3
SGD	200	2
Area A	300	3
Area B	300	3
Area C	300	3

Variant point of "Distribution" in design (B) represented with "Or" is OD relative to "Alternative" in design (A) as only one distribution broker will be selected at a time.

The above analysis is based on quality criterion of precision of design from the perspective of a design stage. From the real project perspective, precision decisions are expected. However it is not practical or feasible for many situations. The coping strategy is "design for change" i.e., there is a need to leave spaces to positively accommodate "unknown" to avoid easily fall into the category of either OD or UD. We can identify the case that seeking a precision decision is very expensive, "Optional" in design (A) is better than "Alternative" in design (B). This also is in line with the commonsense knowledge "unknown" is more precise than "wrong" or "improper".

To elaborate the difference between ideal design and actual design including OD and UD, an experiment is demonstrated by building a calculating model about getting the sum of all component values. We utilize the following tradeoff for quality metrics: Cost vs. Value Added. Cost refers to the general expense of design. Value added refers to the gains of price or value brought by a design implementation. An increase in cost produces more values added features to the design. This is provided for the design elements in Table IV.

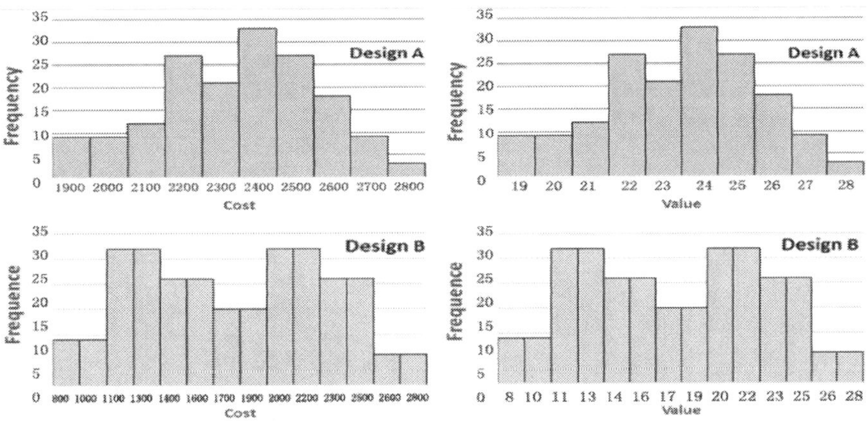

Figure 5. Histograms of cost and value provided by exhaustive combinations

As shown in Figure 4, we see there are many difference about the cost and value added between design (A) and design (B). We assume that design (A) is the ideal design, then we get design (B) in the case of UD and OD. Obviously, the design (B) has a greater volatility than the design (A). Besides, the interval of design (B) is bigger than design (A), which shows the actual design including UD and OD is less instable than the ideal design. In other words, UD and OD severely affects quality of the software system. Hence, UD and OD are nagative design models, we should do our best to make the reachment: $\Delta P_D \to 0$.

6 Related work

Dao et al. [17] presented a problem solution framework for relating features to properties of non-functional requirement(NFR). It supports architecting specific systems with focused features binding to properties of NFR. Alebrahim and Heisel [18] proposed a UML profile to model NFR embedded problem description and bridges the transfer to solution description with the problem to solution mapping of feature modeling. Most of existing approaches are at the methodological level and based on post-active analysis which does not guide the proactive design activities in a system development process. However, our approach provides systemic guide for designing directly by leveraging a constraint driven design approach.

Eiffel language implemented contract based implementation [19]. Boogie language [20] allows specifying pre- and post- conditions to accommodate nondeterministic choices in program flows. However, both of them are restricted to code level and do not fully cover software design activities that span across processes.

In [15], Czarnecki et al. identified gaps in mapping Functional Model to Decision Model. The authors discussed the gap in binding time. This is partially because of the missing of an explicit cognition link between the implementation of binding strategies and quality properties. In addition, the authors ignored the avoidance of the practical situation of over-design vs. under-design at functional level, which are main sources of quality issues from a constructive perspective. Our approach can be applied to fill such gaps as it models the quality assurance based on the quality properties derived from functional level analysis. In particular, QPC supports the bridging of this gap with quality property identification and quality model construction steps.

Demuth et al. [21] proposed constraint based approach to accommodate the choice facing model transformation in a side-effect free manner. Our approach leverages the constraint based modeling from aiding a design to driving the main process of a design.

7 Conclusion and future work

Enlightened by quality mapping and modeling stated in [17], [18], this paper presents a strategy for assuring quality in the software development process, called "Problem-Quality-Constraint" (PQC), where business value constrains are defined. We aim to fill the gap between the schedule and final ideal system using well defined quality goals throughout the software lifecycle. The proposed approach will be usable in incremental iterations process model, such as RUP. Thus, PQC runs in iterations taking into account stakeholder's participation in assessing quality attribute. Stakeholder participation is important to reach agreement both at given stage and evolving stages which are more difficult to quantify.

The key contribution of this paper can be summarized as follows.

- We introduced software economics driven approach that adopt a holistic view to integrate achievements on quality considering criteria on quality tradeoff, quality modeling, quality evaluation and tradeoff, during a quality lifecycle.
- Instead of mapping quality properties to functions at higher level in a retrospective and top-down manner, we propose to start from model the basic form of design quality analysis at the level of over-design vs. under-design.

Over-design (OD) should be avoided by postponing some decisions to later stages as "unknown" which is more precise than "wrong" or "improper".

The introduction of software economics perspective helps relating quality properties that are considered as unrelated and categorized as process quality and product quality. This fits for both proceeding towards quality modeling and retrospective modification of a software product. From the basic form of quality analysis in the form of UD and OD, we can obtain the clue to make quality aware design decisions.

We have identified that OD deserves more attention. OD is a result of rigidly fulfilling precision design ideology by making design decisions which introduces potentially more loss of quality than gains in a quality lifecycle. From the perspective of process quality assurance, we have identified that constraint based modeling [22], [23] manage to accommodate abstract or temporarily not distinguishable information of design decisions. It can postpone the decision to a later stage while balancing the loss of performance. This will contribute directly inevitable choices which are invulnerable to either over-design or sub-design of functions which initiates the deficiency of software quality. Thereafter we propose that a corresponding solution lies in employing constraint as a major form of design activity which is called constraint driven design [24].

In the future, we would like to see PQC contributes directly to the design, composition, runtime, maintenance and evaluation of corresponding systems integrating both functional and non-functional aspects of a system under the constraints of expected business value.

Since PQC is based on modeling related to change of quality directly, it can be used to supplement the implementation of higher level decisions as presented in [17], [18]. PQC will also fit the emerging trend of system design, such as MASHUP [25] in which systems are built on top of existing services with independent functions and corresponding QoS documented in their e-Contracts, and service broker patterns (SVB) [5] which embodies the maximization of business value of a system design.

This paper has considered quality assurance that spans over the entire life cycle of a project focusing on software economics and variability techniques. At the next stage, we will explore the open question of evaluating the level of OD and UD through measurement related to design.

Acknowledgment

This paper was supported in part by CNSF grant 61363007 and by HNU Research program grant KYQD1242 and HDSF201310.

References

1. B. W. Boehm and K. J. Sullivan, "Software economics: a roadmap," in Proceedings of the Conference on The Future of Software Engineering, pp. 319–343, 2000.
2. C. Pautasso and G. Alonso, "Flexible binding for reusable composition of web services," in Proceedings of the 4th international conference on Software Composition, pp. 151–166, 2005.
3. Y. Duan, "A Survey on Service Contract," in Proceedings of the 2012 13th ACIS International Conference on Software Engineering, Artificial Intelligence, Networking and Parallel/Distributed Computing, pp. 805–810, 2012.
4. A. Kattepur, A. Benveniste, and C. Jard, "Negotiation Strategies for Probabilistic Contracts in Web Services Orchestrations," in Proceedings of ICWS2012, pp. 106–113.
5. Y. Duan, A. Kattepur, H. Zhou, Y. Chang, M. Huang, and W. Du, " Service Value Broker Patterns: An Empirical Collection," in SNPD2013, 675-682.
6. M. L. Drago, C. Ghezzi, and R. Mirandola, "A quality driven extension to the QVT-relations transformation language," Computer Science, vol. 27, no. 2, 2012.
7. K. Lee, K. C. Kang, and J. Lee, "Concepts and Guidelines of Feature Modeling for Product Line Software Engineering," Springer Berlin Heidelberg, pp. 62-77, 2002.
8. Y. Duan and R. Lee, "Knowledge Management for Model Driven Data Cleaning of Very Large Database," Ed. Springer-Verlag, pp. 143–158, 2012.
9. M. Feldstein, "Domestic saving and international capital movements in the long run and the short run," European Economic Review, pp. 129-151, 1983.
10. L. Chung, J. Cesar, and S. P. Leite, "Non-functional requirements in software engineering," Springer Berlin Heidelberg, pp. 363-379, 2009.
11. A. Nassaj and J. Lienig, "A New Methodology for Constraint-Driven Layout Design of Analog Circuits," pp. 996–999, December 2009.
12. O. M. Group, "OMG Unified Modelling Language (OMG UML), Infrastrucgture,".
13. E. Gamma, R. Helm, R. E. Johnson, and J. M. Vlissides, "Design Patterns: Abstraction and Reuse of Object-Oriented Design," in Proceedings of the 7th European Conference on Object-Oriented Programming, pp. 406–431, 1993.
14. D. Garlan "Software architecture: a raodmap," in Proceedings of the Conference on the Fureture of Software Engineering, pp. 91-101, 2000.
15. K. Czarnecki, P. Grünbacher, R. Rabiser, and K. Schmid, "Cool Features and Tough Decisions : A Comparison of Variability Modeling Approaches," in Proceedings of the sixth international workshop on variability modeling of software-intensive systems, pp. 173-182, 2012.
16. P. Y. Schobbens, P. Heymans, and J. C. Trigaux, "Feature Diagrams: A Survey and a Formal Semantics," in Proceedings of the 14th IEEE International Requirements Engineering Conference, pp. 136–145, 2006.
17. T. M. Dao, H. Lee, and K. C. Kang, "Problem Frames-Based Approach to Achieving Quality Attributes in Software Product Line Engineering," Proceedings of the 2011 15th International Software Product Line Conference, pp. 175–180, 2011.
18. A. Alebrahim and M. Heisel, "Supporting quality-driven design decisions by modeling variability," Proceedings of the 8th international ACM SIGSOFT conference on Quality of Software Architectures, pp. 43–48, 2012.
19. M.B. Eiffel,"A language and environment for software engineering," Journal of Systems and Software, pp. 199-246, 1988.
20. M. Barnett, B. E. Chang, R. Deline, B. Jacobs, and K. R. Leino, "Boogie: A modular reusable verifier for object-oriented programs," in Formal Methods for Components and Objects: 4th International Symposium, FMCO2005, LNCS, pp. 364–387, 2006.
21. A. Demuth, R. E. Lopez-herrejon, and A. Egyed, "Constraint-driven Modeling through Transformation," Springer Berlin Heidelberg, pp. 248-263,2012.
22. A. Demuth, R. E. Lopez-Herrejon, and A. Egyed, "Constraint-Driven modeling through transformation," in Proceedings of the 5th TPMP, pp. 248–263, 2012.
23. A. Egyed and D. S. Wile, "Support for Managing Design-Time Decisions," IEEE Trans Softw Eng, vol. 32, no. 5, pp. 299–314, May 2006.
24. K. Lano, "Constraint-driven development," Inf Softw Technol, vol. 50, no. 5, pp. 406–423, April 2008.

25. M. Ogrinz, "Mashup Patterns: Designs and Examples for the Modern Enterprise," 1st ed. Addison-Wesley Professional, 2009.
26. T. Al-Naeem, I. Gorton,M. Babar, F.A. Rabhi, and B. Benatallah, "A quality-driven systematic approach for architecting distributed software applications," ICSE2005, pp. 244-253.
27. Y. Duan, A. Kattepury, F. Getahun, A. Elfaki, and W. Du, "Releasing the Power of Varibility: Towards Constraint Driven Quality Assurance," AAI2013. pp. 15-20.

MATEC Web of Conferences **54**, 08001 (2016)
DOI: 10.1051/matecconf/20165408001
MIMT 2016

A two-stage approach for front-view vehicle detection and counting at nighttime

Van-Huy Pham[1], Duc-Hau Le[2]

[1]*Faculty of Information Technology, Ton Duc Thang University, Ho Chi Minh City, Vietnam*
[2]*School of Computer Science and Engineering, Water Resources University, , Hanoi, Vietnam*

Abstract. In this paper, we introduce an approach to car detection and counting using a two-stage method of car dectection and counting. For car hypothesis, we propose a method of headlight candidate extraction using k-means clustering based segmentation which is used as a multi-thresholding method to segment the gray-image of the traffic scene with the lowest level and highest level of intensities included in seed configuration. In verification stage, both individual and paired headlights are tracked during their existence in the ROI. Kalman filter is used to track the detected headlights and counting decision is given when the pairs of headlights follow specified counting rules. The experiments are evaluated on real world traffic videos at the resolution of 640x480, 15fps. The method is robust with various situations of illumination at nighttime.

1 Introduction

Although vehicle detection systems are reported to produce really high accuracy in detecting and tracking vehicles at daytime [1], its inherent characteristic of using motion detection heavily relies on the assumption that the shape of the vehicles are always good so that the neighboring moving pixels can be classified as one vehicle. This assumption, however fails badly when the illumination becomes weaker, especially at nighttime, where the shapes of the vehicles are hardly visible to normal CCTV cameras. In addition, the reflection of vehicle headlights on the road do appear very similar in motion perspective to its actual vehicle shape. Most of motion detection algorithms will easily detect the whole big blob including vehicle, vehicle's reflection on the road, and any occluded vehicles right behind as one big vehicle. Those disadvantages of Daytime Traffic Surveillance Systems have motivated a different research stream which approaches Nighttime Traffic Surveillance with a separate methodology.

As methods applied for detecting and counting cars daytime are generally quite different for that at nighttime [2-4], together with the proposed front-view car detection and counting at daytime based on car's windshields described in [1], we develop a system for car counting at nighttime to cover a more complete range of solving the car counting problem using a fixed camera capturing car images at front-view.

At nighttime, the only promising cue is the headlights which can be easily extracted with reasonable threshold of brightness. The straightforward and feasible approach is using the headlight appearance in pairs and tracking the pairs during their appearance in traffic scenes to help counting decision.

The headlight candidate extraction process is the hypothesis generation and the pairing process works as the verification stage. We propose a method of headlight candidate extraction using k-means clustering based segmentation [5]. The method is much like an adaptive multi-threshold segmentation that is robust with various situations of illumination (darkness) at nighttime. Similar to the method of daytime car counting, Kalman filters are used to track the detected headlights and counting decision is given when the pairs of headlights follow specified counting rules. The experiments are evaluated on real world traffic videos at the resolution of 640x480, 15fps.

The rest of this paper is organized as follows. Car hypothesis is introduced in Section 2, including the perspective transformation and headlight candidate detection. In Section 3, we describe method of car verification by pairing detected headlight candidates that satisfy the size and distance criteria. In the Section 4, our proposed counting system is described. The experimental results for the 2 stages mentioned are shown in Section 5 and concluded in Section 6.

2 Car hypothesis based on headlight detection

2.1 Perspective transformation

Our proposed system uses a fixed camera looking down the traffic lanes to get car images. Because of the perspective view, when the human eye looks at a scene, objects in the distance appear smaller than objects close by. This causes the difference in distance estimation between pairs of car's headlights with respect to the distance from the camera. To suppress the horizontal and vertical constrain errors when dealing with headlight pair matching, we take into account a simple perspective transformation, bird's eye view transformation, a common task in navigation to convert the camera view of the scene into a top-down "bird's-eye" view. Figure 1 shows an example of a transformation of a scene and its bird's eye view.

For simplicity, our system requires user input of four points for the transformation at the beginning of the detection process. The corresponding minimum rectangle including the four input points is automatically calculated and used as the region of interest where the algorithm is applied to.

Figure 1. User ROI and the transformed region

In our method, the transformation is only applied to the center points of the blobs of headlight candidates obtained from the detection stage by using the homography matrix remaping the input trapezoid to the region of interest. It is no need to remap the original image piece into a bird's-eye view because it can reduce the image quality by interpolation during the transformation process.

Figure 1 shows an example of a user input trapezoid, the derived ROI and the transformed region.

2.2 Car headlight detection

The traffic scene illumination condition varies at different time at nighttime and depends on the traffic infrastructure itself. In the early nighttime the background is lit brighter, but in later time, the background is darker and only car headlights are clearly visible. Another problem is to deal with light reflection from headlights.

We propose a more robust method to extract car headlights based on k-means clustering. K-means clustering work as an adaptive multi-thresholding segmentation that clusters different levels of brightness in the gray image.

Let consider a traffic scene at night time. The highest level of brightness is of the car headlights, the lower levels are of the reflection of headlights on the background, the lit background and the dark background in decrease order of brightness, respectively. Headlight regions can be easily extracted using thresholding method with a specified high threshold. But using a static threshold make the system not robust with various lighting condition of traffic scenes. The idea here is to partition the image into difference regions of the brightness levels corresponding to the four object regions mentioned. *K*-means method works well in this situation which the number of region k is equal to the number of object regions.

Headlight candidates are regions of the highest level of intensity from *k*-means segmentation. The reflection of lights on background also appears in a clear pattern that can be detected by considering the shape and brightness transition. Some morphological operations such as opening are used to reduce noise and filter out small regions.

Figure 2 shows a segmentation result in different levels of gray levels of a traffic scene (upper right) and the extracted light candidates (lower right).

Using the projective transformation, it is easy to find the appropriate size of headlights, those white blobs that match the criteria will be considered as positions headlights on the vehicles, and the center weight of the blob will be recorded as the headlight position with a corresponding size.

Headlights detection works well as long as headlight brightness is high enough that all headlight regions are in the highest level of intensity of the image. In that case, the segmentation results are usually over-segmentations which mean that there are more regions of headlights than needed but also other regions such as light reflection, and other highly lit objects. Over-segmentation is what we are expecting for the next steps because noise and other objects out of interest can be filter out in the pairing step, and we don't expect for missed regions of headlight candidates.

 DOI: 10.1051/matecconf/20165408001

Figure 2. K-means clustering for headlight extraction (k=6): K-means clustering result (top) on a traffic scene (left) and the extracted headlight regions.

3 Headlight pairing for car verification

Once individual headlights are detected, car positions can be estimated by relevant pairs of headlights thanks to that cars and lights are rigid of objects and each pair of lights with some constrains potentially represents a position of car. Finding relevant pairs is matching individual lights that hold for specific criteria. Perspective transformation is applied at first to the estimated headlight positions to suppress distance errors. Then every pair of headlights is judge for matching using the following constrains:

3.1 Size constrain

Size constrain is apply to filter out non-related regions such as over-sized regions of light reflection or of other object rather than car headlights. The size of headlight region should be in a specified range:

$$A_{min} < A_{Blob} < A_{max}$$

where A_{min}, A_{max} are minimum and maximum area for a headlight region; A_{Blob} is blob area of the headlight. The lower and upper bound area thresholds depend on the input image size of the traffic scene.

The relative size of the two blobs of headlights is not taken into account because experimentally, two headlights in each pair can have different intensities of brightness causing various sizes of blobs. In this case, the blob's centers still work as headlight positions as shown in example in Figure 3.

Figure 3. An example of different headline appearance in a pair

3.2 Relative position constrains

With a given input image size (traffic scene) and after the projective transformation, we assume that the two headlight positions belonging to the same car are at a relative distance with some small errors.

The two headlights of the same car are an appropriate distance away from each other and are aligned horizontally on the image plane:

$$\left|P_X^i - P_X^j - P_X^0\right| < \varepsilon_X$$

$$\left|P_Y^i - P_Y^j\right| < \varepsilon_Y$$

where P_X^i and P_Y^i are coordinates of the ith headlight position ($1<i<n$), P_X^0 is the pre-defined distance between headlights of a car that we expected. ε_X , ε_Y are allowable errors in pixels of the positions in horizontal and vertical direction, respectively.

Every matching pair of headlight position is consider as a car position assumedly having the two corresponding headlights at the positions. Recall that at the headlight extraction stage, we aim at extract all the possible headlight positions with allowing some more noise positions. Together with some special cases such as two side headlights in two cars are too close that they can be matched, we decide to allow redundancy pairings for counting purpose. It means that there can be many false positive regions where cars are falsely detected. By tracking individual headlights and pairs of headlights, the false positives can be easily eliminated. Figure 4 shows some common appearance situations of headlights.

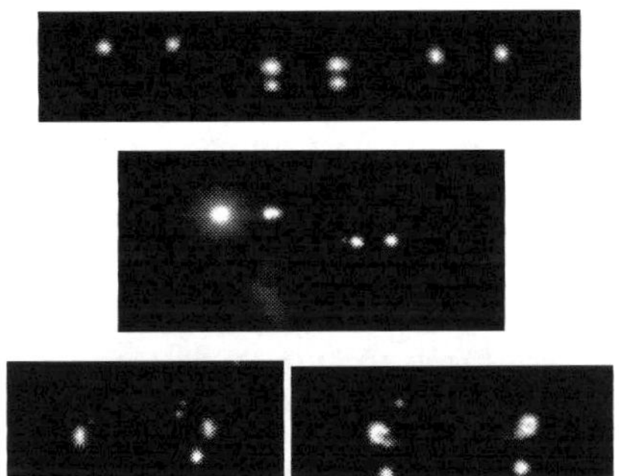

Figure 4. Some situations of headlight appearance

DOI: 10.1051/matecconf/20165408001

4 Car tracking and counting

4.1 Car Headlight Tracking

Individual headlights are tracked using Kalman filter with the same behaviors as car tracking in daytime. Each detected headlight position is associated with a Kalman filter and predicted position is used when missed detection occurs (change in brightness or occlusion). Number of detected frame and continuous missing frame are also recorded to feed the pair tracking.

Tracking pairs of headlights is on the basis of tracked individual headlights. The existence of each pair relies on the existence of both its composition. Number of detected frame and continuous missing frame are also recorded to help the counting decision

4.2 Headlight-based Car Counting

Both tracking objects are tracked and judged at the exit of the ROI for the counting decision. The number of relevant pairs of headlights are counted as the number of cars.

Rule-based judgment for counting is similar to the daytime car counting. The overall counting process is show in Figure 5.

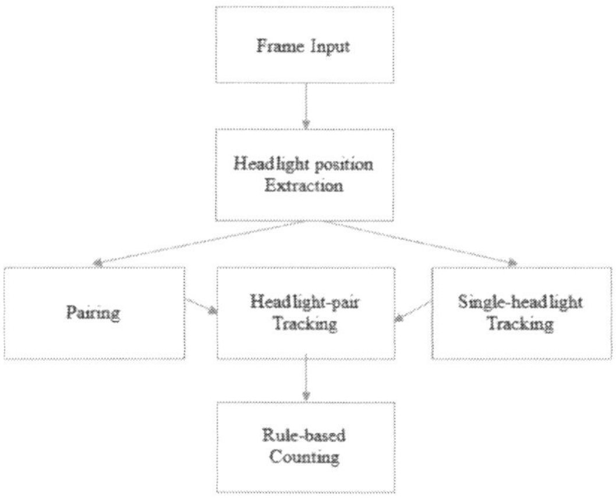

Figure 5. The proposed car counting flowchart in nighttime

4.3 Occlusion handling

By the tracking process using Kalman filter, temporarily occlusion can be overcome by using predictions. In the case of our system setup, most of occlusions occur when a car is occluded by another car ahead with one or two headlights occluded. Another case but not very often is some cars with one-operating light only.

5 Experimental results

The experiments perform on 640x 480 input frames. K-means clustering based method of headlight extraction give quite good results. Experimental results are shown below with different k(s). It is shown that $k=6$ is good for the segmentation. The seeds for k-means are chosen with the bias to the right of the gray level axis so that the dark background and the headlights are well clustered. Experiments show that including 0 and 255 gray level in the initial seeds gives better results. For example, with k=5, the seed configuration can be [0, 120, 200, 220, 255]. Figure 6 shows some results of the segmentation results with different k.

Figure 6. K-means clustering headlight extraction results with $k = 3, 4, 5$

Some results of the extraction process are shown in Figure 7 and Figure 8. A result of the pairing process is shown in Figure 7. Figure 8 shows a car detection result for a night scene in a region of interest. The result is quite good with the existence of remarkably reflection on the background. Figure 9 shows one more detection result with false positives.

Figure 7. A result of car headlight pairing. Paired lights are connected by lines.

Figure 8. A headlight based car detection result: car position are estimated by pairs of headlights.

Figure 9. An example of redundancy of tracking of headlight pairs. A traffic scene (left), headlight extraction result (right) and pairing result (below).

Experiments on some real-world traffic videos at the resolution of 640x480, 15fps, our system can obtain the accuracy at about 98%. The accuracy varies with respect to the illumination condition, occlusion density …

6 Conclusion

In comparison with car detection in daytime, detection of car in nighttime is considered independently because the only promising cue is the headlights which can be easily extracted with reasonable threshold of brightness. The extraction process is the hypothesis generation and the pairing process works as the verification stage.

References

1. Van Pham, H. and B.-R. Lee, *Front-view car detection and counting with occlusion in dense traffic flow.* International Journal of Control, Automation and Systems. **13**(5): p. 1150-1160 (2015).
2. Taktak, R., M. Dufaut, and R. Husson. *Vehicle detection at night using image processing and pattern recognition.* in *Image Processing, 1994. Proceedings. ICIP-94., IEEE International Conference.* 1994.
3. Wang, J., X. Sun, and J. Guo, *A region tracking-based vehicle detection algorithm in nighttime traffic scenes.* Sensors. **13**(12): p. 16474-16493 (2013).
4. Cucchiara, R., M. Piccardi, and P. Mello, *Image analysis and rule-based reasoning for a traffic monitoring system.* Intelligent Transportation Systems, IEEE Transactions on. **1**(2): p. 119-130 (2000).
5. Pham, V.H. and B.R. Lee, *An image segmentation approach for fruit defect detection using k-means clustering and graph-based algorithm.* Vietnam Journal of Computer Science. **2**(1): p. 25-33 (2014).

MATEC Web of Conferences **54**, 08002 (2016) DOI: 10.1051/matecconf/20165408002
MIMT 2016

Lossy Image Compression Using PCA and Contourlet Transform

Yaxiong Chen, Zhangcan HUANG,Hao Sun, Mengying Chen, Hua Tan

School of Science, Wuhan University of Technology, Wuhan 430070, China

Abstract. With the rapid development of Internet, image information is growing. It requires a lot of image storage and transmission. In order to reduce the storage and get better image quality, image compression algorithm is studied. The paper proposes a new image compression algorithm that combines principal component analysis (PCA) and Contourlet Transform (CT). Because PCA has good image quality, but the compression ratio is low, and CT compression algorithm has high compression ratio and good PNSR value. The image is decomposed by PCA. The image data is divided into blocks, and each block is used as a sample vector, then select covariance matrix of k larger eigenvalues corresponding eigenvector to realize image compression. Then the image is compressed again using CT compression algorithm. Compared with the results of JEPG2000 and CT compression algorithm, the results show that the proposed algorithm has better performance than JEPG2000 and CT compression algorithm. In the same compression ratio, PNSR value of proposed algorithm is about 3dB higher than that of JEPG2000, and 2dB higher than that of CT compression algorithm.

1 Introduction

With the rapid development of Internet, image information is growing. It requires a lot of image storage and transmission [1]. The amount of information in the digital image is very large, and it occupies a large amount of resources. So it is important to use appropriate method to compress the image in order to store and transmit. Image compression is divided into lossless compression and lossy compression [2-3]. Principal component analysis algorithm is a kind of feature vector based lossy compression algorithm, which has many applications in image compression and digital watermarking [4-8]. Principal component analysis converts the data set into a low dimensional feature space, and retains most of the original data information content [9]. Contourlet transform are used in the image denoising [10], image fusion [11] and image enhancement [12], and the effect is good. CT compression algorithm is to make full use of contourlet transform to decompose the image with multi-scale and multidirection [13], despite contourlet transform can produce 4 / 3 redundancy, but the use of hard threshold of coefficients processing and run length coding with redundancy is eliminated and realize image compression [14]. JPEG2000 is a new generation of still image compression standard, it uses the wavelet transform, the optimized truncation embedded block encoding and efficient code stream organization form [15-16].In this paper, a new image compression algorithm is proposed, which combines PCA and contourlet transform. CT compression algorithm is to first convert by contourlet transform, using hard threshold for contourlet transform coefficients, treatment with run length coding for image compression. Because PCA has good image quality, but the compression ratio is low, and CT compression algorithm has high compression ratio and good PNSR value. The proposed algorithm has been tested in the famous images such as airfield, Lena, etc., and has been compared with JEPG2000 and CT compression algorithm.

2 Image compression based on principal component analysis and Contourlet transform

2.1 Principal component analysis of image compression

Principal component analysis is a kind of exploratory statistical analysis method, which will be distributed on a set of variables to focus on a few principal components. Principal component analysis is a kind of non-supervised method, because it does not need to output information.

Assume a sample set X, contains m samples, the dimension of each sample is n,

$$X = \{x_1, x_2, \ldots, x_m\}$$
$$x_i = (x_{i1}, x_{i2}, \ldots, x_{in}) \in R^n \tag{1}$$

The sample set X is organized into a sample matrix, each row of which is a sample, and each column is a dimension, S is m*n matrix.

$$\hat{x}_i = x_i - \overline{x} \tag{2}$$

$$\overline{x} = \frac{1}{m}\sum_{i=1}^{m} x_i$$

The sample matrix of $\hat{x}_i$ is denoted as $\hat{S} \in R^{m\times n}$, then the covariance matrix of S is C, which is known by the formula (1).

$$C = \frac{1}{m}\hat{S}^T\hat{S} = \frac{1}{m}\sum_{i}^{m}\hat{x}_i^T\hat{x}_i \tag{3}$$

According to matrix theory, we know that there exists an orthogonal matrix P, $P^TCP=A$. For the specific operation of the orthogonal matrix P: The C eigenvalue decomposition, get the eigenvalue matrix as a feature vector, and the orthogonal matrix is denoted as P. Obviously, P, $A \in R^{n*n}$. To take the maximum of the former K eigenvalues, then the K feature values of the A_1 feature vectors are composed of new feature vector matrix, which is the principal component of PCA transform.

Because of $P_1^TCP_1=A_1$, it can be obtained

$$P^T(\frac{\hat{S}^T\hat{S}}{m})P = \frac{(\hat{S}P)^T\hat{S}P}{m} = A \tag{4}$$

$$\frac{(\hat{S}P_1)^T\hat{S}P_1}{m} = A_1 \tag{5}$$

Then $S_1 = \hat{S}P_1$, the formula (4) deformation is

$$\frac{1}{m}S_1^TS_1 = A_1 \tag{6}$$

$$\hat{S} = S_1P_1^T \tag{7}$$

PCA is used to image compression and block image data, each block as a sample vector, and then calculate corresponding feature vector based on the k eigenvalues of covariance matrix C, which is main component of PCA, and give up the corresponding feature vector of the smaller $n\text{-}k$ eigenvalues, which constitute the transformation matrix P_1. The center of the sample matrix $\hat{S}$ after transformation to get S_1, the compression ratio is $n{:}k$, The formula (6) can reconstruct the original image, and the mean square error is equal to the sum of $n\text{-}k$ eigenvalues.

2.2 Contourlet transform of image compression

Contourlet Transform is the use of Laplacian pyramid decomposition(LP)and directional filter banks(DBF) to achieve a multiresolution, local and multi direction image representation method. Firstly, the image is decomposed by using LP to capture the singular points, and then the DFB is used to connect the singular points of the similar direction. LP is a low pass filter for the original image, and then the low frequency image is sampled and filtered, The difference between the original image and the predicted image is Laplasse's decomposition. The low pass image continues to carry on the Laplasse decomposition, and finally completes the multi scale decomposition of the image. In the reconstruction, the low pass image is sampled and filtered with a band-pass image to complete the reconstruction, the transformation process is performed on the low frequency image, and the image is reconstructed. After the image is decomposed by Laplasse, the singular points of the close direction of the image are connected to the linear structure, so that the image can be decomposed by using a binary tree decomposition. The band pass image is divided into 2^L wedge frequency sub band. The wedge frequency spectrum is obtained by using the combination of Quincunx Bank and DFB.

Because of the structure of the contourlet transformation, the support set is the "long form" structure which is transformed with the scale, so that the image can be decomposed completely. Image Peppers can be carried out by contourlet transform.

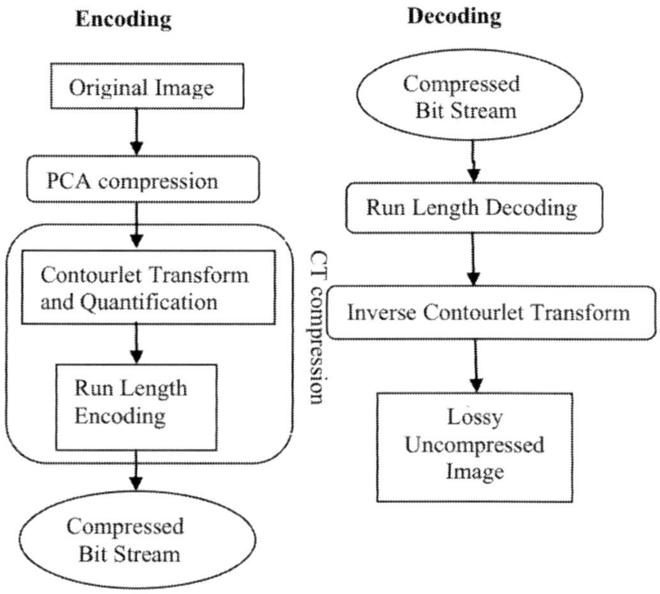

Figure 1. Block diagram of the proposed image compression algorithm

Secondly, Contourlet transform has a large number of coefficients, which represent the direction information and texture information. Large coefficients contain more information, more important than small coefficients. With the decomposition of the scale, the Contourlet coefficient decays rapidly. By the decomposition, the fine scale contains a large number of small coefficients, these small coefficients around 0 floating. This paper uses a simple hard threshold processing to remove the small coefficient of the decomposition. While retaining the high frequency sub bands in the larger coefficient, so that the image quality can be guaranteed for the follow-up of the compression process. As shown below:

$$g(x, y) = \begin{cases} f(x, y), |f(x, y)| > \beta \\ 0, |f(x, y)| \le \beta \end{cases} \tag{8}$$

Among them, $g(x, y)$ indicates that the coefficients are processed by threshold, $f(x, y)$ represents the change coefficient before processing, β represents setting the threshold, the greater the β, the greater the compression ratio.

Finally, using the run-length coding of quantitative coefficients are encoded.

2.3 Algorithm of image compression

In order to ensure the compression quality and improve the compression ratio, this paper combines PCA and Contourlet transform, and the image is compressed by two levels. Specific algorithms are shown in figure 1.

The image data is divided into blocks, and each block is used as a sample vector, then select covariance matrix of k larger eigenvalues corresponding eigenvector to realize the first-stage compression of the image. Reconstruction of image uses contourlet transform. Contourlet transform coefficients are hard threshold processing. To achieve image compression using run length encoding in entropy coding.

Reconstruction of image data based on PCA and contourlet transform. Run length decoding of compressed results are run in contourlet inverse transformation. The image is reconstructed.

3 Experiments and discussions

In order to verify the effectiveness of the proposed method, 5 images of the compression experiments were carried out. All images are 512*512 gray-scale images. According to the literature, contourlet transform using '9-7' pyramid decomposition and 'pkva' direction filter 3 layer decomposition, several layers direction is [1, 2, 3]. By changing the contribution of the singular value of the image signal and the contourlet coefficient quantization threshold, the reconstructed images are obtained under different compression ratios, and the PNSR and SSIM are calculated.

Table 1. PNSR values in dB for 20:1 compression.

Image/Method	CT	JEPG2000	Proposed
airfield	30.46	27.32	31.63
Goldhill	31.96	32.18	34.19
Lena	34.16	35.99	36.64
Peppers	33.89	35.07	36.03
Boats	32.57	33.18	35.13

Table 2. PNSR values in dB for 40:1 compression.

Image/Method	CT	JEPG2000	Proposed
airfield	30.17	24.88	31.06
Goldhill	31.52	29.72	33.33
Lena	33.12	32.75	35.88
Peppers	32.95	32.4	35.5
Boats	31.87	29.76	33.5

Table 1 shows the comparison PNSR value of the proposed algorithm, JEPG2000 and CT compression algorithm in the 20:1 compression ratio. And the Table 2 and Table 3 respectively present the PNSR values which the compression ratios are 40:1 and 80:1. According to the Table 1, Table 2 and Table 3, it can be found that the proposed algorithm is the best one of the three algorithms.

Table 3. PNSR values in dB for 80:1 compression.

Image/Method	CT	JEPG2000	Proposed
airfield	27.95	22.64	30.87
Goldhill	28.53	27.69	32.67
Lena	29.01	29.62	34.87
Peppers	29.33	29.54	34.6
Boats	28.7978	26.76	33.31

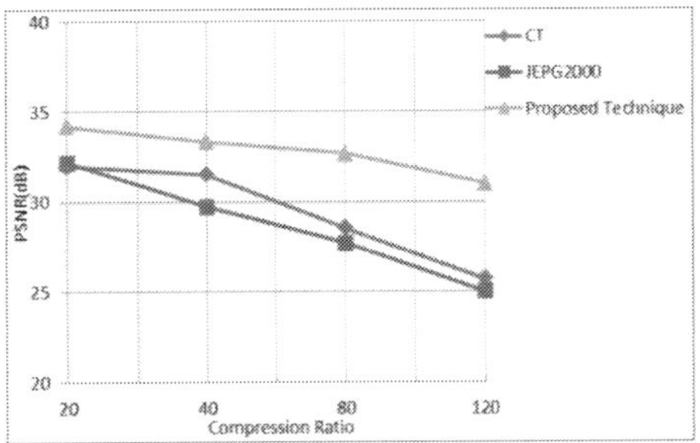

Figure 2. Comparison of PNSR values of peppers image for different compression ratios for the proposed algorithm, JEPG2000 and Contourlet transform.

Under the same compression ratio, the PNSR value of proposed algorithm is 4dB higher than that of JEPG2000, and 2dB higher than that of CT compression algorithm. Besides, Table 4, Table 5 and Table 6 respectively show the SSIM values in the compression ratio of 20:1, 40:1 and 80:1.

Table 4. SSIM comparison for 20:1 compression.

Image/Method	CT	JEPG2000	Proposed
airfield	0.5546	0.698	0.74
Goldhill	0.6541	0.839	0.8312
Lena	0.8057	0.91	0.9057
Peppers	0.7933	0.875	0.876
Boats	0.7515	0.898	0.894

Table 5. SSIM comparison for 40:1 compression.

Image/Method	CT	JEPG2000	Proposed
airfield	0.4758	0.606	0.6409
Goldhill	0.598	0.752	0.7733
Lena	0.738	0.868	0.8838
Peppers	0.7399	0.839	0.8581
Boats	0.672	0.828	0.812

Obviously, the proposed algorithm is similar to JEPG2000 under the compression ratio of 20:1. it can be seen that the proposed algorithm is much better than the others under the compression ratio of 40:1 and 80:1. With the increasing of the compression ratio, the proposed algorithm is better.

 DOI: 10.1051/matecconf/20165408002

Table 6. SSIM comparison for 80:1 compression.

Image/Method	CT	JEPG2000	Proposed
airfield	0.3392	0.52	0.6117
Goldhill	0.4442	0.67	0.7089
Lena	0.5322	0.808	0.8375
Peppers	0.5642	0.783	0.823
Boats	0.5145	0.74	0.8021

compression ratio. Because of the large eigenvectors contribution to the image signal, PCA has good image quality, so it can be combined with CT compression algorithm to improve the image quality and SSIM value in the same compression ratio.

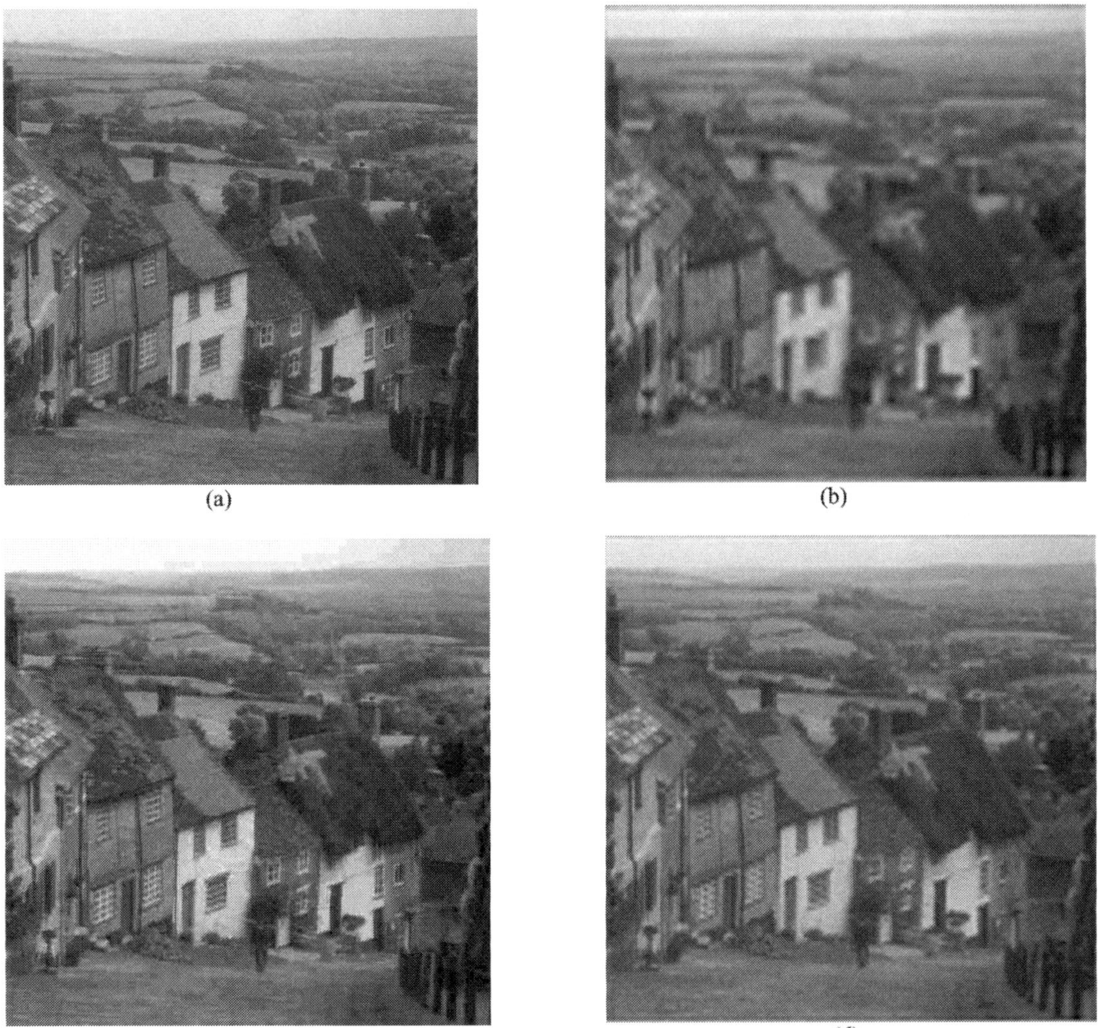

(a) (b) (c) (d)

Figure 3. Original Goldhill image, and compressed image by using(b)Contourlet transform, (c)JEPG2000, and(d) proposed algorithm at 40:1 compression ratio.

Figure 2 shows that the PNSR value of the proposed algorithm is higher than the other algorithms, when the compression ratio become more higher, the proposed algorithm is better than the other two algorithms.

Figure 3 shows that image Goldhill is compressed by three methods in the compression ratio of 40:1. Overall, quantitative and qualitative results show that the proposed algorithm has better image quality and

4 Conclusions

In this paper, a new lossy compression algorithm is proposed, which combines PCA and Contourlet transform. The proposed algorithm uses PCA to ignore the smaller eigenvectors to reconstruct the compressed image. This image was compressed again by using CT compression algorithm. The advantages of PCA and

Contourlet transform are effectively combined. The compression ratio of proposed algorithm is the product of the compression ratio of PCA and the compression ratio of CT compression algorithm. The result of proposed algorithm is obviously better than JEPG2000 and CT compression algorithm, which has a certain application prospect.

References

1. An H, Meng L, Zhao L, et al, Long-distance Transmission and High-speed and Real-time Storage Technology of Image Data, *J. Video Engineering.* **37**(3) (2013) 175-178.
2. Groach M, Garg D A, DCSPIHT: Image compression algorithm, *J. International Journal of Engineering Research and Applications.* **18**(2) (2012) 560-567.
3. Rufai A M, Anbarjafari G, Demirel H, *Lossy medical image compression using Human coding and singular value decomposition* (Signal Processing and Communications Applications Conference, 2013).
4. Du, Qian, and James E. Fowler. "Hyperspectral image compression using JPEG2000 and principal component analysis." *IEEE Geoscience and Remote Sensing Letters,* **4**.2 (2007): 201-205.
5. Ting, Lim Sin, David Yap Fook Weng, and Nurulfajar Bin Abdul Manap. "A Novel Approach for Arbitrary-Shape ROI Compression of Medical Images Using Principal Component Analysis (PCA)." *Trends in Applied Sciences Research,* **10**.1 (2015): 68.
6. Du, Qian, et al. "Hyperspectral image compression and target detection using nonlinear principal component analysis." in Proc. *Satell. Data Compression Commun. Process. IX*, 2013, vol. 8871
7. Shi, Qiuyan, Xingsong Hou, and Xueming Qian. "Hyperspectral image compression based on DLWT and PCA." *Proceedings of the 7th* ACM *International Conference on Internet Multimedia Computing and Service*, 2015.
8. Saboori, Arash, and S. Abolfazl Hosseini. "A new method for digital watermarking based on combination of DCT and PCA." IEEE *Telecommunications Forum Telfor (TELFOR)*, 2014.
9. Wang, Chih-Wen, and Jyh-Horng Jeng. "Image compression using PCA with clustering." IEEE *International Symposium on Intelligent Signal Processing and Communications Systems (ISPACS),* 2012.
10. Min, Dong, Zhang Jiuwen, and Ma Yide. "Image denoising via bivariate shrinkage function based on a new structure of dual contourlet transform."*Signal Processing,* **109** (2015): 25-37.
11. Hou C, Yang M S, Zhang G D, Image Enhancement Method Based on Wavelet-Contourlet Transform Image Fusion of Intelligent Monitoring, *J. Control Engineering of China.* **21**(1) (2014) 62.
12. BAI, Chun-xia, and Wen-zhong ZHAO. "Treatment of multi-dimensional signal based on complex wavelet-contourlet transform [J]." *Automation & Instrumentation,* **1** (2011): 044.
13. Do, Minh N., and Martin Vetterli. "The contourlet transform: an efficient directional multiresolution image representation." *IEEE Transactions on Image Processing,* **14**.12 (2005): 2091-2106.
14. Jian Liu,Yinglei Cheng, Qiang Xu. SAR image compression algorithm based on Contourlet transform[J]. *Science and technology and Engineering*, **12** (2012):7252-7255.
15. Grosbois, Raphael, Diego Santa-Cruz, and Touradj Ebrahimi. "New approach to JPEG 2000 compliant region of interest coding." *SPIE International Symposium on Optical Science and Technology*,2001.
16. Christopoulos, Charilaos, Athanassios Skodras, and Touradj Ebrahimi. "The JPEG2000 still image coding system: an overview." *IEEE Transactions on Consumer Electronics,* **46**.4 (2000): 1103-1127.

A novel object detection technique for dynamic scene and static object

Xiu Li, Liansheng Chen , Zhixiong Yang and Huimin Wang

Shenzhen Key Laboratory of Information Science and Technology Graduate School at Shenzhen, Tsinghua University Shenzhen, China

Abstract. The object detection in video streams plays an important role in computer vision applications. The background subtraction, comparing each new frame to a model of the background, is one of the most popular method. However, the static background is practically impossible, and dynamic background makes the perfect object detection difficult. In addition, the problem of static object significantly affects the performance. In the paper, based on the visual background extraction (ViBe) algorithm, we presented a new method to deal with all problems. For the dynamic scene, we presented a new update mechanism to obtain the more robust model in dynamic regions. To address the static object issue, we cancelled the propagation mechanism in ViBe, designed an algorithm to detect region where it is always detected as foreground, and distinguished the static object from ghost with a self-designed measurement, which combined the knowledge of region contrast and edge. We described our method in full details, and compared it with other background subtraction techniques. Our experimental results show that the proposed method outperforms several state-of-the-art object detection approaches. In addition, it can process 60 frames per second on an Intel i5 3.1 GHz CPU with C++ code.

1 Introduction

In image processing and computer vision applications, accurate object extraction from the video stream is an important precursor for consecutive tracking, recognition and behaviour analysis. Background subtraction, comparing an observed image with the preserved background model, has become the mainstream technique for accurate foreground extraction.

In the simplest case, a static background frame is compared with the current frame, and pixels with high deviation are classified as foreground. However, it is rarely the case, and the stationary camera does not mean a static background because of complicated environments, such as water wave, spring and swaying leaves, et al. [1], which trigger many false alarms. The camera jitter [2] makes the situation worse.

In addition, there are problems which have not been paid enough attention but significantly affect the performance, intermittent object [1], ghost and bootstrapping [2]. When the intermittent object rests, and the model is updated blindly [3], the foreground object will be gradually incorporated into model and removed from the detection result. If the object moves again, both it and the newly revealed parts of the background, called ghost, are detected. If the model is updated conservatively [3], everything seems to be OK. Actually, it is not the case. The problem of bootstrapping makes the initial background model polluted by some foreground information inevitably. The conservative update mechanism will lead to ghost when an object initially in the background model moves. The problem of bootstrapping is a special case for problems of intermittent object and ghost. No existing method can remove the ghost, detect the intermittent object and solve the problem of bootstrapping simultaneously. We call these three problems as static object because all negative effects result from the static behaviour of the object.

Both the phenomena of dynamic scene and static object are common. Dynamic scene, such as water wave, spring and swaying leaves, et al., is inevitable in the natural environment. People and cars at the traffic lights are typically static object. Both of them introduce negative effects for the accurate object detection. Based on visual background extraction (ViBe) [3], we presented a new method, which can deal with all problems mentioned above.

The movement of dynamic background leads to the pixel value failing to be compared with the corresponding model, which results in false alarms. However, the displacement in consecutive frames is small, and thus corresponding model can be found in a small neighbourhood, which named as the neighbourhood match. If there exists a successful match, we will replace two samples in the model with the value, which called sharp update. If these two operations are conducted for all pixels, there will be many false negatives and foreground information will pollute the background model. Thus, these operations just worked in the dynamic background region.

To solve the problem of intermittent object, ghost and bootstrapping, we present a unified framework. First, we cancel the propagation mechanism [3], and update the model conservatively. In this way, ghost and static object will be detected as foreground all the time. As a result there will be some regions, where are always detected as foreground. Then, we design an algorithm to detect these regions. Finally, a new measurement is designed to distinguish ghost from the static object, based on which the model in the region of ghost is modified and the previous results are corrected. Thus, the ghost will be removed, and the static object will be detected successfully.

The rest of this paper is organized as follows. In section 2, we present related background subtraction methods. In section 3, a detailed description of our algorithm is provided. Then, the experimental results of the proposed method in several videos are given in Section 4. Finally, we summarize this paper and suggest possible future work in Section 5.

2 Related Work

Background subtraction has become the mainstream technique in object detection. The main idea is to create and maintain a model of the scene without objects, and detect the foreground by comparing the current frame with the estimated background. The advantage of this concept is that no prior knowledge is required to detect the object as long as their appearance differs enough from the background (i.e. they are not camouflaged [2]). A multitude of algorithms and methods for background modelling have been developed, such as the Gaussian mixture model (GMM) [4], kernel density estimate (KDE) [5], Codebook (CB) [6], Visual background Extraction (ViBe) [3] and many other improved versions [7][8][9][10] based on these methods. Excellent survey papers can be found in [2][11][12][13].

Because of the dynamic nature of real-world scenes, there are inevitably many false alarms. Numerous methods have been presented to handle with this issue. Some statistical models [4][5] are used to represent the multi-modal essence of dynamic background. In fact, these models are only able to represent very small background movement. Many people hold the idea the model converges too slowly, and designed some algorithms [9][14] to speed up the update process in the dynamic region. However, there is still some delay in the model. Some leave the problem to be solved in regularization step [7].

Static object is another problem in background subtraction. If the model is updated blindly, ghost and static object will be incorporated in the model. If the model is updated conservatively, ghost and static object will be detected forever. It is a difficult task to detect static object all the time and remove ghost in no time. Some works have been done to tackle with this problem. Taycher [15] firstly proposed the statistically consistent method for incorporating feedback from the high-level motion model to modify adaptation behaviour, which depends heavily on an accurate tracker. Wang and Bunyak [16] fused flux tensor and split Gaussian based on detection results to get the region of static object and ghost, and exploited edge knowledge in these regions to distinguish them. It detects static foreground and removes ghost successfully, but fails to run in real time.

3 The Proposed Method

Based on ViBe [3], we propose a new approach to handle the problem of dynamic scene and static object. For the dynamic scene, we present a mechanism of the neighbourhood match and sharp update to remove false alarms and speed up update process in the dynamic region, which is described in Section 3.1. For static object, we design an algorithm to find out the regions, where is always detected as foreground. Then, we figure out the objects covering the regions by clustering the foreground pixels in these regions with the sequential leader clustering (SLC) algorithm [17]. Finally, we distinguish static object from ghost by a self-designed measurement with the knowledge of region contrast [18] and edge. Detailed description is provided in Section 3.2. An overall framework is shown in Figure 1.

3.1. Dynamic Scene Noise Removal

In outdoor environments with fluctuating background, the false alarms are derived from two sources. First, there are false alarms because of random noise, which should be homogeneous over the entire image. Second, there are false alarms because of small movements in the scene background that are not represented in the background model. For example, a tree branch moves further than it did during model generation. In addition, small camera displacements because of wind load are common in outdoor surveillance and trigger many false alarms. This false detection is usually spatially clustered in the image, which is not easy to eliminate using morphology or noise filtering. Because these operations might also affect small and occluded targets. This method aims to suppress the false detections due to small and unmodelled movements in the scene background.

In ViBe, the pixels are compared with their own models and the ones similar with at least two samples in the model considered as background, having a chance of one sixteenth to make its model and the model of neighbourhood updated with its current value. In this way, only pixels sharing similar appearance with the samples in the model can be incorporated into the model. Thus, the model just contains samples from one object in most cases. Although there is a probability the model contains samples from different objects because two objects are similar with the same object may be different, it takes a long time to collect enough samples and requires special condition the pixel values vary slowly, such as slow illumination. In addition, with the mechanism of propagation, updating the model of its neighbourhood with its own pixel value, the model in the edge of objects may have the chance of containing samples from different objects, but collecting enough samples takes a long time. The models in the region of dynamic background should always be multi-modal. Thus, the algorithm of ViBe fails to handle the problem of the dynamic scene. The false alarms because of the small

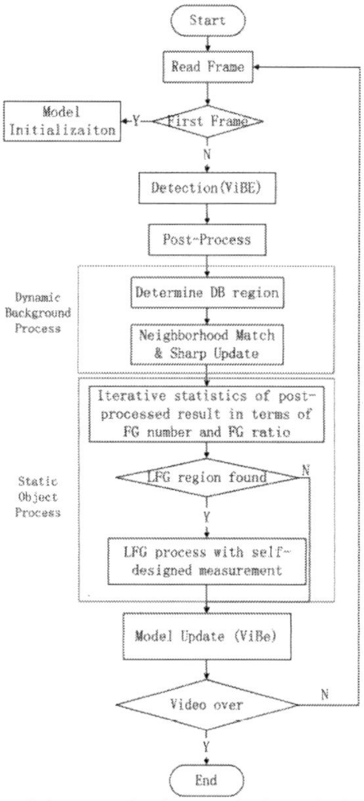

Figure 1. Overall framework of our method. DB is short for Dynamic Background and LFG is short for Long-time Foreground.

movements of background objects prevent the moved background objects incorporating into the models of new pixel locations, and the unchanged models will cause similar false alarms, which is a drop-dead halt.

Parts of the background move to occupy a new pixel, but it was not part of that pixel, then it will be detected as a foreground object. However, this object will have a high probability to be a part of the background distribution at its original pixel. If only a small displacement occurs between consecutive frames, we judge whether a detected pixel is caused by a background object that has moved by comparing the detected pixel value with models in the neighbourhood and finding out some matched models. We name the procedure of finding matched models in the neighbourhood as neighbourhood match, which can break the drop-dead halt.

The object of dynamic background does not stay too long on one spot. Collecting enough samples (typically 2 samples) to represent the moving object needs about the time of 32 frames in the situation that the object is there. If the object is always moving, it will take more time to collect enough samples. Moreover, if enough samples fail to be collected immediately, some false alarms of background objects will also fail to be found out and removed, which further prevent the background objects incorporating into the models of new pixel locations. Although the mechanism of neighbourhood match breaks the drop-dead halt, the slow update mechanism leads to a vicious circle. To break the vicious circle for the false alarms of moving background object, we replace two samples in the model of new pixel location with two copies of the current pixel value, which is named as sharp update. In this way, the moving background objects can be incorporated into the models of new pixel locations immediately, and further movements can also be detected. Thus, a more robust and reliable model is obtained, and fewer false alarms will be detected, which is beneficial to post process, such as median filtering and morphological operations.

In addition, operations of the neighbourhood match and sharp update are only conducted in the region of dynamic background, which is indicated in a 2D map of pixel level, named as blink map. In [19], for each pixel, the previous updating mask (prior to any modification) and a map with the blinking level are stored. This level is determined as follows. For a pixel, if the current updating label is different from the previous one, then the blinking level is increased by 15 (the blinking level being kept within the [0,150] interval), while the level is decreased by 1. A pixel is considered as blinking if its level is larger or equal to 30. Directly detecting blinking pixels following the steps above is inconvenient, since the borders of moving foreground objects would also be included in the result. Thus, we find blinking pixels just in the region of background, which is indicated in the post-processed and dilated updating mask. As it is shown in Figure 2, we presented several results from ViBe and our method, and much fewer false alarms are detected in our results, which demonstrates the advantage of our model.

3.2 Static Object Detection

The static objects (true positives) and revealed background by removed objects (false positives) called as ghost are detected as foreground by traditional method, such as ViBe. If the model is updated blindly, ghost and static object will be incorporated into the model. If the model is updated conservatively, ghost and static object will be detected forever. In fact, the ghost should be included into model, and the static object should be detected as foreground. It is hard to distinguish them from each other just with the temporal knowledge. Thus, we should exploit some spatial knowledge. The static objects are different from the surrounding background while the revealed background by removed objects is similar with the surrounding background, and our method is designed based on it. Detailed descriptions are given as follows.

First, we choose the conservative update mechanism and cancel the spatial propagation mechanism [3] in ViBe in order to make the static objects, and revealed background by removed objects detected as foreground all the time. Then, we find out them by searching for regions, it is always detected as foreground. Considering the factors of noise, illumination change, occlusion, et al, the static objects and the revealed background by removed objects are not detected as foreground continuously. Thus, a more reliable approach is designed.

The main idea is where that if a pixel is detected as

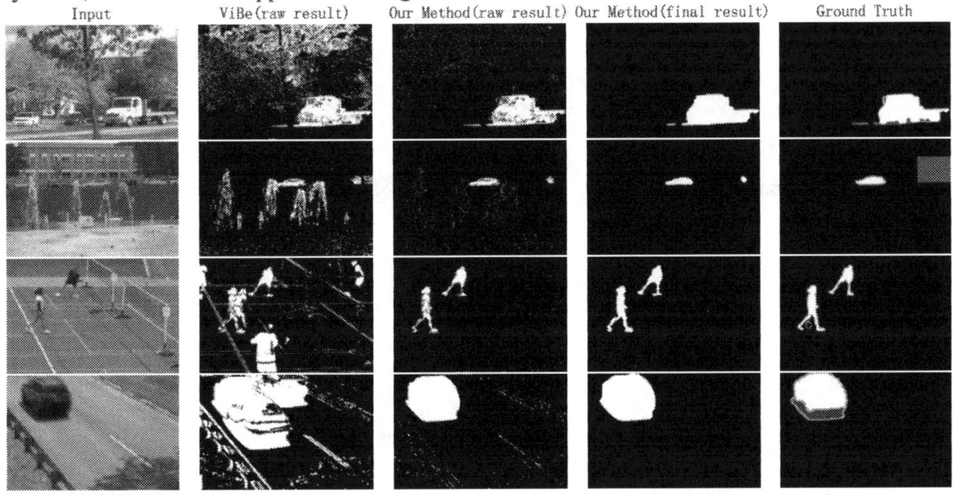

Figure 2. Comparison with ViBe (the threshold R = 10)

foreground frequently enough in a certain period, it is considered as a long-time foreground (LFG, for short) pixel. Specifically, if in a period, a pixel is detected as foreground more than 500 times and the ratio of foreground is more than 80%, we consider it as a LFG pixel. In addition, the pixels in a LFG object are detected at different time, but the time interval is not big. In order to detect the complete LFG, we stop detecting new LFG pixels when no new LFG pixel is detected for 100 frames. Furthermore, these foreground pixels are stored and will be used to figure out the object covering the region of long-time object with the algorithm of SLC [17].

Finally, we design a measurement to distinguish the static object and ghost. The concept of region contrast given in [11] can be used to measure the similarity between the central and surrounding region. However, it is difficult to set a global threshold for all cases. Thus, we compute the region contrast between the central and surrounding model, RC_M, and the region contrast between the central cluster result and surrounding model, RC_C, respectively.

$$RC\_M = (\sum_{p_i \in P_{MC}} \sum_{p_j \in P_{MS}} D(p_i, p_j)) \div (\# P_{MC} \times \# P_{MS}) \quad (1)$$

$$RC\_C = (\sum_{p_i \in P_{CC}} \sum_{p_j \in P_{MS}} D(p_i, p_j)) \div (\# P_{CC} \times \# P_{MS}) \quad (2)$$

where $D(p_i, p_j)$ is the color Euclidean distance metric between pixels p_i and p_j in the R*G*B space, P_{MC} stands for the pixels in the model of the center region of a long-time foreground object, the white region in Fig 3.b, P_{MS} stands for the pixels in the model of the surrounding region of a long-time foreground object, the gray region in Fig 3.b, P_{CC} stands for the pixels in the cluster result of the center region of a long-time foreground object, $\# P_{MC}$, $\# P_{MS}$, $\# P_{CC}$ are the number of pixels in P_{MC}, P_{MS}, P_{CC}.

If RC_M > 1.5*RC_C, the model is modified with the cluster result. Illustrations are given in Fig 3.

However, it only works when the background is flat. To solve this problem, we further exploit the knowledge of edge inspired by the idea given in [16]. Specifically, we measure the degree of flatness with variance and compute the variance for center model (VAR_M) and center cluster (VAR_C) in the LFG region, respectively.

$$VAR\_M = (\sum_{p_i \in P_{MC}} D(p_i, \overline{p_{MC}})) \div (\# P_{MC}) \quad (3)$$

$$VAR\_C = (\sum_{p_i \in P_{CC}} D(p_i, \overline{p_{CC}})) \div (\# P_{CC}) \quad (4)$$

where $\overline{p_{MC}}$ and $\overline{p_{CC}}$ are the average pixels in P_{MC} and P_{CC}.

And if 0.66*RC_C<RC_M<1.5*RC_C and max(VAR_M, VAR_C)>8000, we follow the steps in [16] to respectively figure out the edges in model image, cluster image and binary image indicating the LFG object, and modify the model when the edges in binary image are more similar with the ones in the model image. A visual and intuitionistic flow chat is given is Fig 4.

In addition, when some LFG objects are spatially close, these objects are detected as a combined LFG object due to some morphological operations in our method, which is not beneficial for the distinguishing procedure. In order to handle the problem, we further exploit the knowledge of the starting time of LFG object. In details, we split the combined object into several parts, each of which just belongs to a real LFG object, with analysis of histogram of the starting time of LFG object and deal with each part as mentioned above. The main idea of the histogram analysis is to find some isolated intervals, whose histogram value is larger than certain threshold. Specifically, we first generate a histogram of ts with the bin width being about 100. To find the left border of each interval, we look for the first bin, whose histogram value is larger than 0.005, from the right border of the previous interval and set the left border of the found bin as the left border of a new interval. To find the corresponding right border, we look for the first bin, whose histogram value is less than 0.005 from the corresponding left border and set the left border of the found bin as the corresponding right border of the new interval. We present a concrete case in Figure 5.

The modified model will be used for the correction of previous wrong result and object detection in following frames. Furthermore, we find that our method can deal with the problem of bootstrapping in Figure 2.

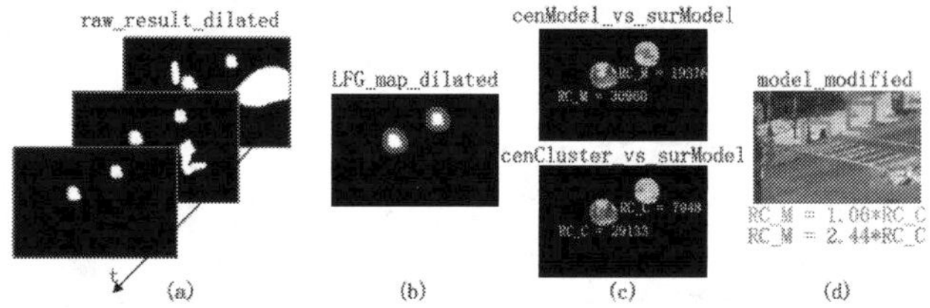

Figure 3. These images are all chosen from the video of abandonedBox (a) dilated version of raw result, (b) obtained map of LFG region (the center white region is LFG region, and the surrounding gray region is obtained with dilation operation) (c) comparative image (in the up image the center region is model and the surrounding region is model as well, and in the down image the center region is the cluster result using SLC and the surrounding region is model) used to compute the measurement of region contrast, (d) modified model image

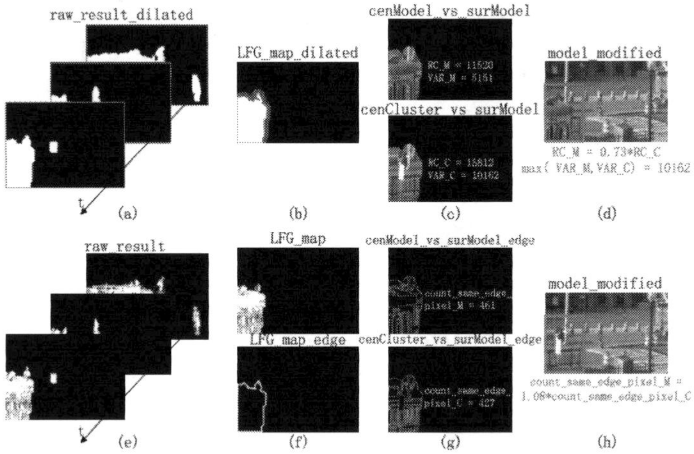

Figure 4. These images are all chosen from the video of tramstop (a) dilated version of raw result, (b) obtained map of LFG region from dilated results, (c) comparative image (VAR_X is the variance of the center region), (d) modified model image (the model is not modified successfully with the knowledge of region contrast but 0.66*RC_C<RC_M<1.5*RC_C and max(VAR_M, VAR_C)>8000), (e) raw result, (f) obtained map of LFG region from raw result and obtained edge from it, (g) obtained edge from (c), (h) modified model image

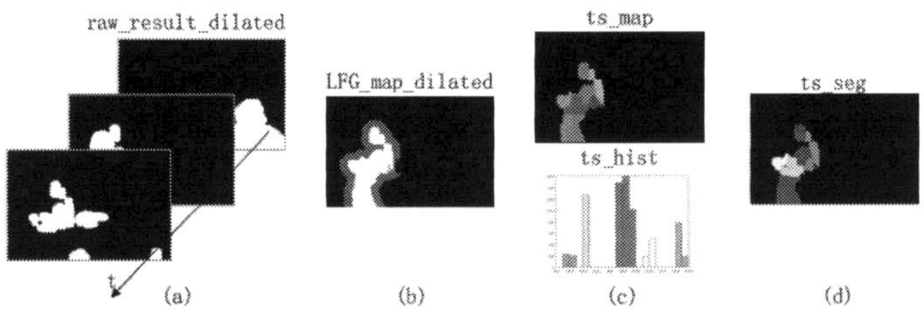

Figure 5. These images are all chosen from the video of diningRoom (a) dilated version of raw result, (b) obtained map of LFG region from dilated results, (c) map of the starting time of LFG object and the histogram of it, (d) the segmentation of ts_map from the analysis of its histogram

4 Experimental Results

The proposed method is evaluated using the dataset and evaluation metrics in CVPR 2012 Change Detection challenge (CDnet 2012) [1].A fixed set of parameters is used for all sequences, and some of them inherit from [3][19]. All experiments run on an Intel i5-3450 processor with 3.10GHz, 16GB RAM and Win7 OS. The algorithm is implemented by C++ codes, and the average processing speed is about 60 fps for the image size of 320x240.

4.1. Performance for dynamic scene and static object

In the CDnet 2012 dataset, the video sequences of dynamic background (DB) and camera jitter (CJ) share the problem of dynamic scene, and the ones of intermittent object (IO) share the problem of static object. Some of these video sequences share the problems of shadow, highlight, illumination change and camouflage. To demonstrate the contributions of our method, we compare it with several state-of-the-art methods in the video sequences of badminton in CJ, canoe, fall, fountain01 and overpass in DB, and tramstop in IO, which do not share the problems of shadow, highlight, illumination change and camouflage. These state-of-the-art methods include ViBe [3] on which our method is based; SuBSENSE [9] is the best one in all improved versions for ViBe; CDet [20] performs best for the video sequences in DB; GPRMF [20] performs best for the video sequences in CJ; and FTSG [16] performs best for the video sequences in IO. Detailed statistical data is provided in Table I and the data in bold is the best in each row. Obviously, our method performs better than the other five methods for dealing with the problem of dynamic scene and static object. For the dynamic scene, our method is better than the other methods on most of the video sequences, and a little worse than the best one is in a few of video sequences. The overall measurement of our method is better than that of the other methods. Although our performance is not the best for static object, the gap between the best one and our method is small. In addition, FTSG fails to run in real time, while our method succeeds and performs much better than FTSG for the dynamic scene. Finally, for both dynamic scene and static object, the overall performance of our method is better than that of the other methods.

4.2. Overall performance and discussion

We compared our method with methods mentioned in section 4.1 using the metrics of recall, precision, F-Measure and time cost. Table II shows the statistical data. Our method performs much better than ViBe, and increases the first three metrics by more than 10%. Although SuBSENSE, CDet and FTSG obtain better performance, the gaps are not large. These three methods design some tricks to deal with the problem of shadow, highlight, illumination change and camouflage, which are the reasons of these gaps. However, these tricks introduce some extra time cost. Thus, our method runs faster than the other three methods.

5 Conclusions

In this paper, we presented a ViBe-based background subtraction algorithm, which deals with the problem of dynamic scene and static object well. Neighbourhood match can remove the false alarms of dynamic scene, and sharp update can speed up the update process, which make the model more robust and reliable. A novel algorithm is presented to detect the LFG region, and a new measurement is designed to distinguish static object from ghost. Experiments show they work well together. As mentioned above, no specific mechanisms are designed to deal with the shadow, highlight, illumination change and camouflage in our algorithm. We believe a good representation of the pixel will be the answer, and try to find out the appropriate representation to deal with these issues in the future.

Acknowledgment

This research was partly supported by National Natural Science Foundation of China (NSFC, Project No.: 71171121/61033005) and National "863" High Technology Research and Development Program of China. (863 Project No.: 2012AA09A408).

Table 1. Method Comparison on performance for dynamic scene and static object

category	video name	method / measure	ViBe	SuBSENSE	CDet	GPRMF	FTSG	Our Method
dynamic scene	badminton	recall	0.799	0.922	**0.971**	0.823	0.917	0.965
		precision	0.576	0.843	0.851	**0.964**	0.916	0.937
		F-Measure	0.669	0.881	0.907	0.888	0.916	**0.951**
	canoe	recall	0.858	0.659	**0.966**	0.945	0.913	0.950
		precision	0.855	**0.993**	0.974	0.978	0.985	0.957
		F-Measure	0.856	0.792	**0.970**	0.961	0.948	0.953
	fall	recall	0.789	0.857	**0.998**	0.969	0.988	0.988
		precision	0.267	0.876	0.887	0.199	0.878	**0.942**
		F-Measure	0.399	0.866	0.926	0.331	0.930	**0.964**
	fountain02	recall	0.802	0.923	**0.950**	0.884	0.947	0.945
		precision	0.862	**0.966**	0.948	0.937	0.955	0.965
		F-Measure	0.831	0.944	0.949	0.910	0.951	**0.955**
	overpass	recall	0.763	0.785	0.882	**0.980**	0.944	**0.980**
		precision	0.851	0.943	0.919	0.891	0.941	**0.970**
		F-Measure	0.805	0.857	0.900	0.934	0.943	**0.975**
static object	tramstop	recall	0.405	0.417	0.972	0.734	**0.987**	0.981
		precision	0.272	0.831	**0.976**	0.1934	0.967	0.964
		F-Measure	0.325	0.555	0.974	0.306	**0.977**	0.972
overall	overall	recall	0.736	0.761	0.957	0.889	0.949	**0.968**
		precision	0.614	0.909	0.926	0.694	0.940	**0.956**
		F-Measure	0.648	0.816	0.937	0.722	0.944	**0.962**

Table 2. Method comparison on overall performance

	ViBe	SuBSENSE	CDet	GPRMF	FTSG	Our Method
recall	0.682	0.828	**0.903**	0.837	0.838	0.846
precision	0.736	0.857	0.840	0.814	**0.868**	0.854
F-Measure	0.668	0.826	**0.861**	0.794	0.835	0.818
time cost (ms/frame)	**2 (C++) i5 3.1GHz**	22 (C++) i5 3.3GHz	25 (C++) 2.8GHz	unknown	100 (Matlab) Intel 2.4GHz	16 (C++) i5 3.1GHz

References

1. Goyette, Nil, Pierre-Marc Jodoin, Fatih Porikli, Janusz Konrad, and Prakash Ishwar. "Changedetection. net: A new change detection benchmark dataset." In *Computer Vision and Pattern Recognition Workshops (CVPRW), 2012 IEEE Computer Society Conference on*, pp. 1-8. IEEE, 2012.
2. Bouwmans, Thierry. "Traditional and recent approaches in background modeling for foreground detection: An overview." *Computer Science Review*11 (2014): 31-66.
3. Barnich, Olivier, and Marc Van Droogenbroeck. "ViBe: A universal background subtraction algorithm for video sequences." *Image Processing, IEEE Transactions on* 20, no. 6 (2011): 1709-1724.
4. Stauffer, Chris, and W. Eric L. Grimson. "Adaptive background mixture models for real-time tracking." In *Computer Vision and Pattern Recognition, 1999. IEEE Computer Society Conference on.*, vol. 2. IEEE, 1999.
5. Elgammal, Ahmed, David Harwood, and Larry Davis. "Non-parametric model for background subtraction." In *Computer Vision—ECCV 2000*, pp. 751-767. Springer Berlin Heidelberg, 2000.
6. Kim, Kyungnam, Thanarat H. Chalidabhongse, David Harwood, and Larry Davis. "Real-time foreground–background segmentation using codebook model." *Real-time imaging* 11, no. 3 (2005): 172-185.
7. Sheikh, Yaser, and Mubarak Shah. "Bayesian modeling of dynamic scenes for object detection." *Pattern Analysis and Machine Intelligence, IEEE Transactions on* 27, no. 11 (2005): 1778-1792.
8. St-Charles, Pierre-Luc, Guillaume-Alexandre Bilodeau, and Robert Bergevin. "A Self-Adjusting Approach to Change Detection Based on Background Word Consensus." In *Applications of Computer Vision (WACV), 2015 IEEE Winter Conference on*, pp. 990-997. IEEE, 2015.
9. St-Charles, Pierre-Luc, Guillaume-Alexandre Bilodeau, and Robert Bergevin. "SuBSENSE: A Universal Change Detection Method With Local Adaptive Sensitivity." *Image Processing, IEEE Transactions on* 24, no. 1 (2015): 359-373.
10. Zivkovic, Zoran. "Improved adaptive Gaussian mixture model for background subtraction." In *Pattern Recognition, 2004. ICPR 2004. Proceedings of the 17th International Conference on*, vol. 2, pp. 28-31. IEEE, 2004.
11. Brutzer, Sebastian, Benjamin Höferlin, and Gunther Heidemann. "Evaluation of background subtraction techniques for video surveillance." In *Computer Vision and Pattern Recognition (CVPR), 2011 IEEE Conference on*, pp. 1937-1944. IEEE, 2011.
12. Parks, Donovan H., and Sidney S. Fels. "Evaluation of background subtraction algorithms with post-processing." In *Advanced Video and Signal Based Surveillance, 2008. AVSS'08. IEEE Fifth International Conference on*, pp. 192-199. IEEE, 2008.
13. Radke, Richard J., Srinivas Andra, Omar Al-Kofahi, and Badrinath Roysam. "Image change detection algorithms: a systematic survey." *Image Processing, IEEE Transactions on* 14, no. 3 (2005): 294-307.
14. Hofmann, Martin, Philipp Tiefenbacher, and Gerhard Rigoll. "Background segmentation with feedback: The pixel-based adaptive segmenter." In*Computer Vision and Pattern Recognition Workshops (CVPRW), 2012 IEEE Computer Society Conference on*, pp. 38-43. IEEE, 2012.
15. Taycher, Leonid, John W. Fisher III, and Trevor Darrell. "Incorporating object tracking feedback into background maintenance framework." In *Application of Computer Vision, 2005. WACV/MOTIONS'05 Volume 1. Seventh IEEE Workshops on*, vol. 2, pp. 120-125. IEEE, 2005.
16. Wang, Rui, Filiz Bunyak, Guna Seetharaman, and Kannappan Palaniappan. "Static and moving object detection using flux tensor with split Gaussian models." In *Computer Vision and Pattern Recognition Workshops (CVPRW), 2014 IEEE Conference on*, pp. 420-424. IEEE, 2014.
17. Hartigan, John A. *Clustering algorithms*. John Wiley & Sons, Inc., 1975.
18. Cheng, Ming, Niloy J. Mitra, Xumin Huang, Philip HS Torr, and Song Hu. "Global contrast based salient region detection." *Pattern Analysis and Machine Intelligence, IEEE Transactions on* 37, no. 3 (2015): 569-582.
19. Van Droogenbroeck, Marc, and Olivier Paquot. "Background subtraction: Experiments and improvements for ViBe." In *Computer Vision and Pattern Recognition Workshops (CVPRW), 2012 IEEE Computer Society Conference on*, pp. 32-37. IEEE, 2012.
20. changedetection website. http://changedetection.net/.

MATEC Web of Conferences 54, 08004 (2016) DOI: 10.1051/matecconf/20165408004
MIMT 2016

An Efficient Image Co-segmentation Algorithm based on Active Contour and Image Saliency

Zhizhi Zhang[1], Xiabi Liu[1], Nouman Qadeer Soomro[2]

[1]*Beijing Laboratory of Intelligent Information Technology, School of Computer, Beijing Institute of Technology, Beijing 100081, China*
[2]*Department of Computer Science, DHA Suffa University, Karachi Pakistan*

Abstract. Image co-segmentation is the problem of extracting common objects from multiple images and it is a very challenging task. In this paper we try to address the co-segmentation problem by embedding image saliency into active contour model. Active contour is a very famous and effective image segmentation method but performs poor results if applied directly to co-segmentation. Therefore, we can introduce additional information to improve the segmentation results, such as saliency which can show the region of interest. In order to optimize the model, we propose an efficient level-set optimization method based on super-pixels, hierarchical computation and convergence judgment. We evaluated the proposed method on iCoseg and MSRC datasets. Compared with other methods, our method yielded better results and demonstrated the significance of using image saliency in active contour.

1 Introduction

Co-segmentation is a relatively new and fast emerging research field in computer vision. Different from the traditional segmentation, which separates the foreground objects from the background, co-segmentation is defined as the simultaneous segmentation of similar regions from two (or more) images. The necessity of co-segmentation is increasing in recent years as many applications require desired-objects to be carved-out from bulk of images or videos, stored in scattered and haphazard manner. Co-segmentation is difficult due to similar backgrounds and variations in targeted objects.

The existing co-segmentation methods mainly include Markov Random Field (MRF) based, active contour based, clustering based, user interaction based and graph based. MRF based co-segmentation methods include Rother et al. [1], Mukherjee et al. [2], Hochbaum et al. [3] and Vicente et al. [4]. Rother et al. [1] is the first work of co-segmentation. They used the l_1-norm to measure the foreground consistency, and used the Trust Region Graph Cut (TRGC) algorithm to optimize the co-segmentation model. After Rother et al. [1], several other co-segmentation methods were proposed by improving the model construction and optimization. Vicente et al. [4] summarized the previous works [1-3] and presented a new MRF model based on Boykov-Jolly model. To optimize the proposed model, the Dual Decomposition (DD) optimization method was used. Co-segmentation methods based on clustering mainly include Joulin et al. [5]. They proposed a co-segmentation method based on discriminative clustering. In their paper, normalized cuts and kernel methods are used in a discriminative clustering framework: the goal is to assign foreground/background labels jointly to all images, so that a supervised classifier trained with these labels leads to maximal separation of the two classes [5]. User interaction based methods mainly include Batra et al. [6]. They proposed co-segmentation method called iCoseg based on a few user interactions. ICoseg can intelligently recommend regions to scribble on, and users following these recommendations can achieve good quality cutouts with significantly less time and energy than exhaustively examining all cutouts [6]. Vicente et al. [7] introduced graph based method. They presented the concept of object co-segmentation, which focused on segmenting "objects" rather than "stuff". In their work, they firstly got the candidate object-like segmentations and then described the relationship between the regions by the fully connected graph. Finally, Loop Belief Propagation (LBP) was used to obtain results. In 2012, Meng et al. [8] proposed a co-segmentation model based on active contour, and later in 2013 Meng et al. [9] improved the co-segmentation method. The ideas of former and later works were somehow the same. Based on these two closely related works, energy function was established on foreground similarity among the images and background consistency in each image. Then level-set formulation and the optimization are illustrated.

This paper presents a co-segmentation method by introducing image saliency into well-known active contour based segmentation method and we called ACIS. According to active contour schema, we design an energy function, in which the image saliency is considered for giving additional information of foreground. The energy function is minimized by using the level-set strategy. In order to speed up the computation, we propose an efficient level-set optimization based on hierarchical segmentation, super-pixel and better termination. We conducted the experiments on widely used iCoseg image database and MSRC image database. Experimental results demonstrated the effectiveness of the proposed co-segmentation algorithm. The contributions of this paper are summarized as follows:

- A new image co-segmentation model is constructed by introducing the saliency information into the active contour framework. Saliency information can bring useful foreground features to achieve better performance.
- An efficient algorithm is developed through combining super-pixels, hierarchical computation and convergence judgment for solving the co-segmentation problem based on the model above.

The rest of this paper is organized as follows. Section 2 introduces the proposed co-segmentation model based on active contour and image saliency. Section 3 describes our optimization algorithm. Experiments and results are discussed in Section 4. We conclude the paper in Section 5.

2 The proposed co-segmentation model

As described in Section 1, the goal of the co-segmentation is extracting common objects from a group of images. Active contour is a classical and effective method for image segmentation. It is possible to apply it on co-segmentation task. Our proposed co-segmentation model considers not only foreground consistency but also background consistency, moreover, not only consistency within a single image but also consistency between images. Besides, we introduced the saliency information into the active contour framework to achieve better performance. According to the active contour algorithm, given original images, each image should have initial contour, and each contour keeps changing according to the energy function until the energy is minimized. In this paper, the initial contour is a rectangle without artificial participation because we observed the initial contour has little effect on the final results in the experiments, and the size of rectangle is determined by the parameters.

Let $I = \{I_1, I_2, \dots, I_n\}$ be a group of images, suppose C_k is the initial contour of I_k. Based on the initial contour set $C^0 = \{C_1, C_2, \dots, C_n\}$, the evolution of the contour curve is performed to achieve common object segmentation. Let ω_{int}^k be the inner region of the k-th image curve, and ω_{ext}^k be the outer region of the k-th image curve. So the goal of the co-segmentation is to get the curve when the energy is minimized:

$$\mathrm{C}_k^* = \arg\min_{C_k} \mathrm{E}(C_k)\,, i = 1, \dots, n. \quad (1)$$

For a given curve C_k, according to the active contour Chan-Vese (C-V) model [10], our energy function $\mathrm{E}(C_k)$ of the curve is expressed as:

$$\begin{aligned}\mathrm{E}(C_k) = E_c^k - \lambda_{sf} E_{sf}^k - \lambda_{sb} E_{sb}^k \\ -\lambda_f E_f^k - \lambda_b E_b^k - \lambda_{co} E_{co}^k,\end{aligned} \quad (2)$$

where λ_{sf}, λ_{sb}, λ_f, λ_b, and λ_{co} are weights for each item. The details of each term in Eq. 2 are described as follows.

The regularization term of the curve: E_c^k:

Regularizing terms mainly describe the attribute of curve, including internal curve area and curve length:

$$E_c^k = \mu \cdot Area(\omega_{int}^k) + \nu \cdot Length(C_k)\,, \quad (3)$$

where $Area(\omega_{int}^k)$ refers to curve area, $Length(C_k)$ refers to the curve length, and μ and ν represent the weights for two factors.

The energy of foreground consistency of the k-th image: E_{sf}^k:

The foreground consistency of a single image is used to guarantee the consistency of the foreground pixels. It also ensures the differentiation of foreground from background. When the pixels in the curve are consistent with the foreground pixels, the pixels are more likely to be foreground pixels. Thus the energy can be designed as the sum of consistency between the curve pixels and the inner region pixels:

$$E_{sf}^k = \int_{(x,y)\in\omega_{int}^k} f\left(I_k(x,y), g(\omega_{int}^k)\right) dxdy. \quad (4)$$

In above Eq. 4, $g(\omega_{int}^k)$ is for region metric, $f\left(I_k(x,y), g(\omega_{int}^k)\right)$ is for measuring similarity between pixels and regions [8]. If the pixels are more similar to the inner region, the E_{sf}^k is larger, otherwise smaller.

The energy of background consistency of the k-th image: E_{sb}^k:

The background consistency of a single image is similar to foreground consistency of a single image, except it is used to ensure consistency of the background pixels. Because the background of images may be similar, we should take the energy of background consistency into account and λ_{sb} would be zero when the background of images are not similar. Thus, the energy of background consistency is designed as the consistency between the pixels inside the curve and the external region. The corresponding equation is:

$$E_{sb}^k = \int_{(x,y)\in\omega_{ext}^k} f\left(I_k(x,y), g(\omega_{ext}^k)\right) dxdy. \quad (5)$$

If the pixels are more similar to the outside region, the E_{sb}^k is larger, otherwise smaller.

The energy of foreground consistency of the k-th image and other images: E_f^k:

The foreground consistency of multiple images guarantees the consistency of the image foreground region with other images, which can be expressed as the sum of the image curve and the interior region of the rest of the images:

$$E_f^k = \sum_{j=1, j\neq i}^{n} \int_{(x,y)\in\omega_{int}^k} f\left(I_k(x,y), g(\omega_{int}^j)\right) dxdy. \quad (6)$$

If the curve of the image is consistent with most of the inner regions, the E_f^k is larger, and smaller in vice versa.

The energy of background consistency of the k-th image and the rest of the images: E_b^k:
Similar to foreground consistency of multi images, background consistency of multi-images can be expressed as the sum of the image curve and the outer region of the rest of images:

$$E_b^k = \sum_{j=1,j\neq i}^{n} \int_{(x,y)\in\omega_{ext}^k} f\left(I_k(x,y), g\left(\omega_{ext}^j\right)\right) dxdy. \quad (7)$$

If the external regions are more consistent with each other, the E_b^k is larger.

The saliency energy of the k-th image: E_{co}^k:
Saliency is used to obtain more foreground information. During this image saliency process, the pixels with greater saliency values are more likely to be foreground pixels, making the discrimination between foreground and background easier. Thus, the energy can be designed as the sum of the difference between the saliency value of pixel and the saliency value of region:

$$E_{co}^k = \int_{(x,y)\in\omega^k} h(J_k(x,y), l(\omega^k))dxdy. \quad (8)$$

In above Eq. 8, $J_k(x,y)$ refers to the saliency value for each point of the k-th image, $l(\omega^k)$ represents the saliency value of measuring the region saliency, and $h(J_k(x,y), l(\omega^k))$ refers to pixel and region saliency differences. Therefore, if the saliency value of pixel is larger, then E_{co}^k is larger, and vice versa.

3 Optimization method

This section aims at model optimization and proposes an efficient level-set optimization method which is based on super-pixel, hierarchical segmentation and better termination.

3.1 Basic level-set optimization

The level-set optimization method of Osher and Sethian [11] can be used to minimize the energy function expressed in Eq. 2. Based on the level-set function $\varphi(x,y,t)$, Eq. 2 can be expressed as:

$$\begin{aligned} E(\varphi_k) = {} & \mu \int_{\Omega_k} H(\varphi_k)dxdy + \nu \int_{\Omega_k} \delta(\varphi_k)|\nabla\varphi_k|dxdy \\ & -\lambda_{sf} \int_{\Omega_k} f\left(I_k(x,y), g\left(\omega_{int}^k\right)\right) H(\varphi_k)dxdy \\ & -\lambda_{sb} \int_{\Omega_k} f\left(I_k(x,y), g(\omega_{ext}^k)\right)(1 - H(\varphi_k))dxdy \\ & -\lambda_f \sum_{j=1,j\neq i}^{n} \int_{\Omega_k} f\left(I_k(x,y), g\left(\omega_{int}^j\right)\right) H(\varphi_k)dxdy \\ & -\lambda_b \sum_{j=1,j\neq i}^{n} \int_{\Omega_k} f\left(I_k(x,y), g\left(\omega_{ext}^j\right)\right)(1 - \\ & \quad H(\varphi_k))dxdy \\ & -\lambda_{co} \int_{\Omega_k} h\left(J_k(x,y), l\left(\omega_{int}^k\right)\right) H(\varphi_k)dxdy. \quad (9) \end{aligned}$$

Here, Ω_k is the k-th image in the image domain, φ_k is the k-th image level-set function, $\delta(x)$ is the Dirac function, $H(x)$ is the Heaviside function. Subsequently, $\delta(\varphi_k)$ stands for C_k, $H(\varphi_k)$ is for the inner region and $1 - H(\varphi_k)$ is for the outer region. The Dirac and Heaviside functions are defined as follows:

$$\begin{cases} H(x) = \frac{1}{2}\cdot\left(1 + \frac{2}{\pi} argtan\frac{x}{\varepsilon}\right) \\ \delta(x) = \frac{dH(x)}{dx} = \frac{1}{\pi}\cdot\frac{\varepsilon}{\varepsilon^2+x^2} \end{cases}. \quad (10)$$

Energy function expressed in Eq. 9 can be optimized using Euler Lagrange by keeping $f\left(I_k(x,y), g(\omega_{ext}^k)\right)$ and φ_k as independent and keeping $g\left(\omega_{int}^k\right)$ and $g(\omega_{ext}^k)$ as constant.

Set $\Delta t = 1$, we can get the iterative formula of the level-set function:

$$\begin{aligned} \varphi_k^{N+1} = {} & \varphi_k^N + \delta(\varphi_k)(-\mu + \nu\cdot div\left(\frac{\nabla\varphi_k}{|\nabla\varphi_k|}\right) \\ & +\lambda_{sf} f\left(I_k(x,y), g\left(\omega_{int}^k\right)\right) \\ & -\lambda_{sb} f\left(I_k(x,y), g(\omega_{ext}^k)\right) \\ & +\lambda_f \sum_{j=1,j\neq i}^{n} f\left(I_k(x,y), g\left(\omega_{int}^j\right)\right) \\ & -\lambda_b \sum_{j=1,j\neq i}^{n} f\left(I_k(x,y), g\left(\omega_{ext}^j\right)\right) \\ & -\lambda_{co} h\left(J_k(x,y), l\left(\omega_{int}^k\right)\right). \quad (11) \end{aligned}$$

Finally, co-segmentation algorithm based on the active contour using image saliency is obtained and the iterative equation of the level-set function is expressed in Eq. 11.

Given a group of images $I = \{I_1, I_2, \dots, I_n\}, n \geq 2$, and initial contour sets $C^0 = \{C_1, C_2, \dots, C_n\}$, we can get the initial level-set $\varphi^0 = \{\varphi_1^0, \varphi_2^0, \dots, \varphi_n^0\}$. At each step of the iteration, we calculate the energy-value of each item, then update φ_k^{N+1} values according to Eq. 11, and finally, calculate internal area $g\left(\omega_{int}^k\right)$ and external area $g(\omega_{ext}^k)$. If the convergence conditions are obtained, algorithm is terminated. When working with level-sets and Dirac delta functions, a standard procedure is to reinitialize to the signed distance function to its zero-level curve. This prevents the level-set function to become too flat, or it can be seen as a rescaling and regularization [10].

3.2 An efficient level-set optimization

If the algorithm derived above is directly calculated on pixels, the calculation will be very expensive. In order to speed up the optimization, the following strategies, including super-pixels, hierarchical computation and convergence judgment, are introduced.

3.2.1 Super-pixels

To accelerate the computation, super-pixels based computation is performed to reduce the computational burden drastically. Taking 200×200 image as an example, the calculation directly on original pixels computes the whole 40000 pixels totally, contrarily number reduced to 1000, or 1/20th of the original level, if super-pixels are used. In this paper, we use Simple Linear Iterative Clustering (SLIC) method [12] to generate super-pixels since this method is commonly used and has good results, and the default super-pixels number is 1000.

3.2.2 Hierarchical Computation

From the view of down sampling, hierarchical data processing can be used to speed up computation. Hierarchical computation includes several layers and processes layer by layer, each layer processes one size of the original image and runs co-segmentation algorithm until convergence. The hierarchical structure starts from small size of images with the quick convergence to the object boundary and then propagates the segmentation from the images with small size to the images with large size. As initial contour of the last layer is somehow close to the final results, it is possible to achieve faster convergence. This whole reducing scheme is called cascade acceleration and it significantly increases the speed of the algorithm.

3.2.3 Convergence Judgment

In general, the algorithm sets an iteration number, and continues until the number of iterations is achieved. Large iteration number guarantees the convergence but on the huge computational cost while small iteration number, for avoiding the computational cost, hinges on the uncertainty of satisfactory convergence. In this paper, to our algorithm judges whether it is converged during iterative process and terminate timely. Specifically, the algorithm is converged if φ_k^{t+1} tends to remain stable and in this way we can set a relatively large number of iterations without worrying of computational cost. In this way, we can accelerate the algorithm and ensure good segmentation results at the same time.

To sum up, our algorithm of co-segmentation is summarized in Algorithm 1.

4 Experiments

In order to evaluate the proposed co-segmentation algorithm, we conducted the experiments on commonly used datasets and compared our algorithm with other competitive methods.

4.1 Experimental Setup

4.1.1 Image dataset

The iCoseg image dataset [13] and MSRC image dataset [14] are used in our experiments, which are often used in co-segmentation field. The iCoseg image dataset is established by Batra et al. [5], which contains a total of 37 categories. Each category has several to dozens of images. The MSRC image dataset was established by Criminisi et al. [14]. This image dataset includes 23 categories, and each category has dozens of images.

The effectiveness of results obtained by our method is evaluated with ground truths. The error rates are measured, which is defined as the number of misclassified pixels divided by the number of the total pixels.

4.1.2 Parameter setting

Algorithm 1. Our co-segmentation algorithm based on active contour and image saliency (ACIS)

Step 1. Set initial parameters μ, ν, λ_{sf}, λ_{sb}, λ_f, λ_b and λ_{co}.
Step 2. Extract the information of each image I, super-pixel information and corresponding saliency information J.
Step 3. Initialize the contour of each image $C^0 = \{C_1, C_2, \dots, C_n\}$.
Step 4. The original image, the corresponding super-pixel information and saliency information are set to different sizes.
Step 5. Using super-pixels to achieve co-segmentation of different image sizes and enter Step 6 until all sizes are processed.
  Step 5.1. Given C^0, the initial level set function φ^0 is obtained according to [10].
  Step 5.2. According to φ^0, get internal area representation $g(\omega_{int}^k)$ and external regional representation $g(\omega_{ext}^k)$.
  Step 5.3. Iterative calculation.
    Step 5.3.1. Calculate each energy item according to Eq. 11.
    Step 5.3.2. The level set function of each image φ_k^{t+1} is updated according to φ_k^t and Eq. 11.
    Step 5.3.3. Reinitialize to the signed distance function to update φ_k^{t+1}.
    Step 5.3.4. Based on φ_k^{t+1}, calculate internal area $g(\omega_{int}^k)$ and external area $g(\omega_{ext}^k)$.
    Step 5.3.5. Judging convergence condition, if meet the convergence condition enter Step 5, otherwise enter Step 5.3.1.
Step 6. Output segmentation results.

The parameters of our algorithm are set up by serious experiments. The actual values leading to the reported results are listed as follows.
- For the initial contour, $\alpha = 0.2 \cdot \min\{\text{height}, \text{width}\}$, height stands for the height of an image, width stands for the width of an image.
- In Eq. 11, considering the relationship between each item, we here set $\mu = 0.01$, $\upsilon = 0.05$, $\lambda_{sf} = \lambda_f \in [0.5, 1.1]$, $\lambda_{sb} = \lambda_b = 1$, $\lambda_{co} \in [0.1, 0.3]$.
- The parameter for measuring whether pixels are similar is $\beta = 15$ [8].
- In the experiments, the level number of the cascade acceleration is three, the expansion scale is $\{0.1, 0.5, 1\}$ and the corresponding iteration numbers are $\{100, 30, 30\}$.

4.1.3 Compared Methods

GrabCut [15] is a very famous image segmentation method, we compared this method to show ACIS has better performance. We also compared our method with other co-segmentation methods including CoSB [16] and CoSand [17]. CoSB is based on active contour and CoSand is based on thermal expansion. We set default

DOI: 10.1051/matecconf/20165408004

parameter values according to the original papers [15], [16], [17].

4.2 Evaluation of saliency methods

Saliency item in Eq. 11 greatly affects the segmentation results. As the weight of saliency item is increased, the segmentation results become more close to saliency results. In this paper, we consider four saliency methods, including GS [18], MR [19], SF [20] and wCtr [21]. We compared them through careful evaluation. The results are shown in Fig. 1. The first row in Fig. 1 represents the original images and the second row represents the ground truth while the remaining rows represent the saliency images extracted by GS [18], MR [19], SF [20], and wCtr [21], respectively.

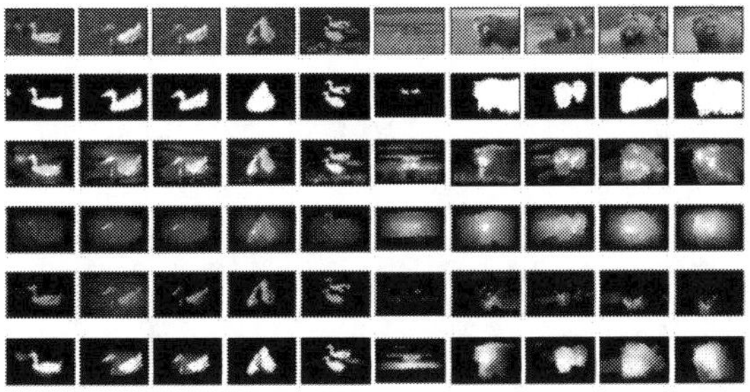

Figure 1. First row is the original image, second row is the ground truth, and other rows are the saliency extraction results using GS, MR, SF and wCtr, respectively, on iCoseg image dataset.

From the results shown in Fig. 1, we can see that extraction results of the GS method for the foreground saliency are good, but some unnecessary backgrounds also have high saliency value. MR method obtains lower value and cannot accurately extract the foreground saliency. The SF method fails to accurately extract the foreground saliency in the bear class. Meanwhile, the extraction results of the wCtr method are more accurate and robust than other methods. Table 1 shows the comparison of the different saliency detection methods on the error rate with different thresholds. We can see that in each threshold and in each category, wCtr has the lowest error rate. Therefore, we chose the wCtr method for the extraction of the saliency in this paper.

Table 1. The error rates comparison of different saliency detection methods.

Thresholds	Category	GS [18]	MR [19]	SF [20]	wCtr [21]
Mean value	Goose	0.159	0.227	0.054	**0.050**
	Bear	0.204	0.242	0.312	**0.152**
1.2*Mean value	Goose	0.120	0.200	0.054	**0.047**
	Bear	0.204	0.233	0.316	**0.159**
Fixed value(127)	Goose	0.065	0.117	0.096	**0.041**
	Bear	0.198	0.192	0.316	**0.179**
Avg		0.158	0.202	0.191	**0.105**

4.3 Results on iCoseg Image dataset

The first part is the subjective comparison of co-segmentation results. Fig. 2 shows the comparison of different segmentation methods on the iCoseg image dataset. From left to right and from top to bottom, the classes are Sox players, Pyramids, Brighton and Helicopter, respectively.

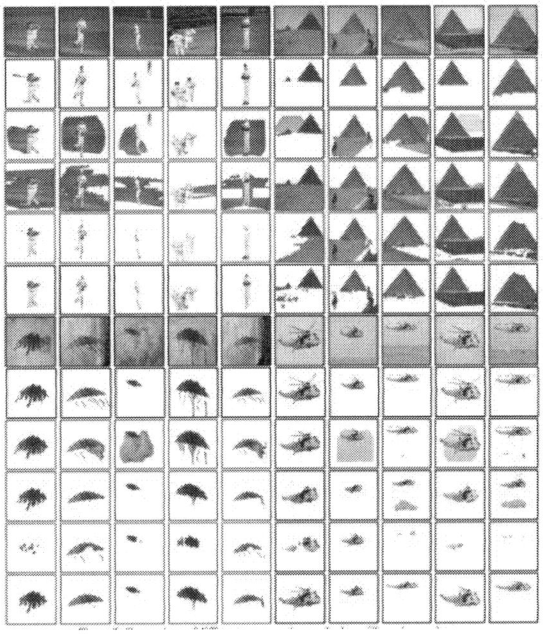

Figure 2. Comparison of different segmentation methods on iCoseg image dataset.

For each class, the first row contains the original images and the second row contains ground truth images. The third row shows segmented images using GrabCut [15]. Results shown in the fourth and fifth rows are segmented results using CoSB and CoSand, respectively. Results in the sixth row are obtained using our method.

According to Fig. 2, for the class of Sox Players, the method CoSand accurately segments the foreground image from the grass but in due process over-segment the foreground objects, resulting in loss of information. In the contrary, GrabCut and CoSB have poor results. Since foregrounds and backgrounds of Pyramids class are very similar, Grab Cut and CoSB failed to distinguish Pyramid from desert. Somehow, our method can separate them in some degree. Moreover, for the Brighton class, the results of segmentation obtained by CoSB are good, but GrabCut has bad results and results of CoSand appear over-segmented. For the Helicopter class, GrabCut and CoSB failed on some images and CoSand also appears over-segmented. In general, the effects and robustness of our method, ACIS, are more excellent compared with other methods. Table 2 shows the comparison of the different methods and our method on the error rate.

GrabCut is based on single image segmentation. So it performs well in skating and gymnastics class because the background is relatively simple, but its performance deteriorates on other classes. CoSB performs well in Brighton class. Because this method only considers the consistency of foreground and background without adding more foreground information. So in many cases of co-segmentation, especially under the complex background, it is difficult to define a suitable energy function. The CoSand, a thermal diffusion method, has good results on the classes of skating and hot balloons, but due to heavily relying on the color, the method will fail easily if the foreground has different colors. ACIS is based on active contour using saliency, active contour guarantees the curve close to the edge of the object and saliency ensures stronger robustness. Therefore, our method has good co-segmentation results in most classes.

Table 2. The comparison of error rates on iCoseg image dataset.

Category	GrabCut [15]	CoSB [16]	CoSand [17]	ACIS (Our)
Bear	0.241	0.331	0.105	**0.073**
Sox players	0.283	0.449	0.027	**0.023**
Stonehenge	0.239	0.230	0.075	**0.064**
Pyramids	0.246	0.380	0.266	**0.130**
Elephants	0.131	0.285	0.120	**0.121**
Helicopter	0.142	0.054	0.071	**0.011**
Cheetah	0.094	0.092	0.294	**0.065**
Brighton	0.104	**0.016**	0.047	**0.016**
Margate	0.218	0.111	0.264	**0.086**
Gymnastics	**0.013**	0.053	0.038	0.020
Skating	**0.019**	0.184	**0.019**	0.028
Balloons	0.050	0.054	**0.028**	0.037
Avg	0.148	0.187	0.113	**0.056**

4.4 Results on MSRC Image dataset

Fig. 3 shows the comparison of different segmentation methods on MSRC image dataset. From left to right and from top to bottom, the classes are cow, face, sign and cat, respectively.

For the cow class, GrabCut and ACIS performed well despite of different foreground colors, while CoSB and CoSand performs badly because of the different colors of foregrounds. For the face class, ACIS has the best segmentation effects, while other methods do not have satisfactory results due to many interferences and complexity of background. For the sign class, all the methods have relatively good segmentation results, but CoSand takes the sign post as the foreground in the 2nd image while the sign post actually belongs to the background. For the cat class, ACIS also has the best segmentation results. Comparison of the error rates of different segmentation methods are given in the following Table 3.

As shown in Table 3, ACIS has the best average value. GrabCut has relatively lower error rate in most classes while CoSB and CoSand did not perform well on the dataset. The reason is that the foreground and background are relatively complex and different colors in foreground make it more difficult. ACIS has the lowest error rate in the classes of the face and cat and the average error rate is lower. Thus, it is proved from aforementioned results that our method is effective.

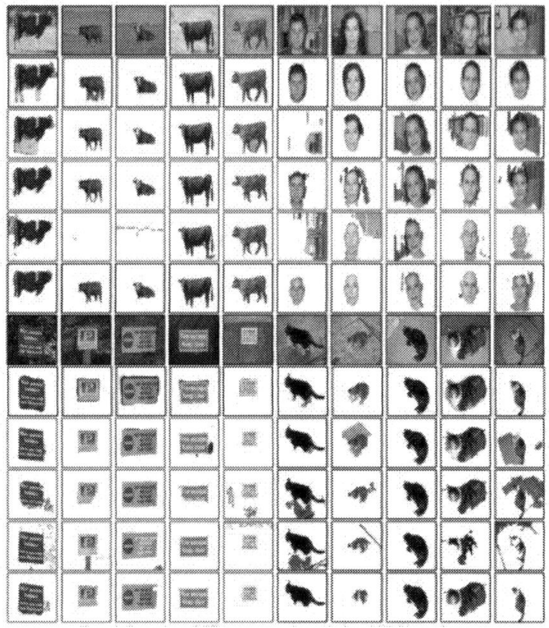

Figure 3. Comparison of different segmentation methods on MSRC image dataset.

Table 3. The comparison of error rates on MSRC image dataset.

Category	GrabCut [15]	CoSB [16]	CoSand [17]	ACIS (Our)
Cow	**0.063**	0.070	0.122	0.068
Face	0.232	0.175	0.278	**0.126**
Bicycle	**0.204**	0.235	0.316	0.239
Sign	**0.038**	0.082	0.077	0.042
Bird	**0.039**	0.151	0.137	0.072
Chair	**0.111**	0.198	0.281	0.116
Cat	0.091	0.125	0.152	**0.055**
Avg	0.111	0.148	0.195	**0.103**

The main difference between ACIS and CoSB is that CoSB is based on active contour while ACIS is based on active contour and image saliency. So the above experiments show that introducing saliency information can improve co-segmentation results. In conclusion, compared with other methods, our method yielded better results and demonstrated the significance of using image saliency in active contour.

4.5 Efficiency

Except the comparison of error rates, ACIS is also an efficient image co-segmentation algorithm because we adopt an efficient level-set optimization method. Table 4 shows the comparisons of computation cost on MSRC image dataset. From the results we can see that our optimized level-set is much faster than the original level-set.

Table 4. Computation cost comparison of level-set and optimized level-set on MSRC image dataset.

Category	Level-set [10]	Optimized level-set
Cow	>1000s	115.2s
Face	>1000s	91.6s
Bicycle	>1000s	298.6s
Sign	>1000s	87.8s
Bird	>1000s	37.0s
Chair	>1000s	37.8s
Cat	>1000s	32.4s
Avg	>1000s	100.0s

5 Conclusion

This paper has proposed an efficient co-segmentation algorithm based on active contour and image saliency. Firstly, we construct a new image co-segmentation model by introducing the saliency information into the active contour framework. Secondly, we proposed an efficient algorithm which is using level-set method with super-pixels, hierarchical segmentation, and better termination to optimize the model above. In order to let the algorithm has more robust and better segmentation effect, the choice of saliency method is discussed using rigorous evaluation. Finally, we conducted a series of experiments on iCoseg and MSRC datasets. Compared with other methods, our method yielded better co-segmentation results and computationally inexpensive. In future work, we will adopt more accurate foreground information, such as co-saliency, since saliency is just for a single image.

References

1. C. Rothe, V. Kolmogorov, T. Minka, and A. Blake, "Cosegmentation of Image Pairs by Histogram Matching - Incorporating a Global Constraint into MRFs," *Proceedings of the IEEE Computer Society Conference on Computer Vision and Pattern Recognition*, **1**, 993-1000 (2006).
2. L. Mukherjee, V. Singh, R. And Dyer C, "Half-Integrality based Algorithms for Cosegmentation of Images," *2009 IEEE Computer Society Conference on Computer Vision and Pattern Recognition Workshops*, 2028-2035 (2009).
3. S. Hochbaum D, and V. Singh. "An efficient algorithm for Co-segmentation," *IEEE International Conference on Computer Vision*, 269-276 (2009).
4. S. Vicente, V. Kolmogorov, and C. Rother, "Cosegmentation revisited: Models and optimization, " *ECCV*, 465-479 (2010).
5. A. Joulin, F. Bach, and J. Ponce, "Discriminative clustering for image co-segmentation," *IEEE Conference on Computer Vision & Pattern RecognitionIEEE Computer Society*, 1943-1950 (2010).
6. D. Batra, A. Kowdle, and D. Parikh D, "iCoseg: Interactive co-segmentation with intelligent scribble guidance," *Proceedings / CVPR, IEEE Computer Society Conference on Computer Vision and Pattern Recognition*, 3169-3176 (2010).
7. S. Vicente, C. Rother, and V. Kolmogorov, "Object cosegmentation," IEEE Conference on Computer Vision & Pattern Recognition, **42**, 2217-2224 (2011).
8. F. Meng, H. Li, and G. Liu, "Image co-segmentation via active contours," *IEEE International Symposium on Circuits & Systems*, 2773–2776 (2012).
9. F. Meng,H. Li, and G. Liu, "Image Cosegmentation by Incorporating Color Reward Strategy and Active Contour Model," Cybernetics IEEE Transactions on, **43**, 725-737 (2013).
10. F. Chan T and A. Vese L, "Active contours without edges," IEEE Transactions on Image Processing A Publication of the IEEE Signal Processing Society, **10**, 266–277 (2001).
11. S. Osher and A. Sethian J, "Fronts propagating with curvature-dependent speed: algorithms based on Hamilton-Jacobi formulations," Journal of Computational Physics, **79**, 12-49 (1988).
12. R. Achanta, A. Shaji, K. Smith, et al. "SLIC Superpixels Compared to State-of-the-art Superpixel Methods," IEEE Transactions on Pattern Analysis and Machine Intelligence, **34**, 2274–2282 (2012).
13. Available at : http://chenlab.ece.cornell.edu/projects/touch-coseg/CMU_Cornell_iCoseg_dataset.zip, last accessed: 04/11/2015.
14. Available at : http://research.microsoft.com/en-us/um/people/antcrim/data_objrec/msrc_objcategimagedatabase_v2.zip, last accessed: 04/11/2015.
15. C. Rother, V. Kolmogorov, and A. Blake, "Grabcut: Interactive Foreground Extraction Using IteratedGraph Cuts," ACM Transactions on Graphics, **23**, 309–314 (2004).
16. F. Meng, H. Li, K. Ngan, et al. "Cosegmentation from similar backgrounds," IEEE International Symposium on Circuits & Systems, 353 – 356 (2014).
17. G. Kim, P. Xing E, L. Fei-Fei, et al. "Distributed cosegmentation via submodular optimization on anisotropic diffusion," *Computer Vision (ICCV), 2011 IEEE International Conference on. IEEE*, 169-176 (2011).
18. Y. Wei, F. Wen F, W. Zhu, et al. "Geodesic Saliency Using Background Priors," Lecture Notes in Computer Science, **7574**, 29-42 (2012).
19. C. Yang C, L. Zhang, H. Lu, et al. "Saliency Detection via Graph-Based Manifold Ranking," IEEE Conference on Computer Vision & Pattern Recognition, **9**, 3166-3173 (2013).
20. A. Hornung, Y. Pritch, P. Krahenbuhl, et al. "Saliency filters: Contrast based filtering for salient

region detection," *Cvpr. IEEE Computer Society*, 733-740 (2012).
21. W. Zhu, S. Liang, Y. Wei, et al. "Saliency Optimization from Robust Background Detection," *IEEE Conference on Computer Vision & Pattern Recognition*, 2814 – 2821 (2014).

DOI: 10.1051/matecconf/20165408005

Runway Detection based on Geometric Constraints of Runways for Safe Landing

Chang-hyun Lee , Hwan-hee Lee , Myeong jin Lee

Korea Aerospace University, 76 Hanggongdaehang-ro Deogyang-gu, Goyang, Gyeonggi, 10540, Korea

Abstract. A novel runway detection algorithm is proposed for safe landing assistance for aircrafts. The shape of the runway projected into the image frame is modeled using the geographical location and size of the runway, the location of the aircraft, and the camera parameters. Candidates for the boundary lines of the runway are extracted from image frames by the geographical relationship among the real boundary lines of the runway. The performance of the proposed algorithm is evaluated with real IR video frames, and is shown to achieve more than 88% of runway detection rate within 1km distance to the runway.

1 Introduction

Runway detection in aerial video sequences is useful video analytics technology which can be applied to the automatic landing guidance for commercial aircrafts and unmanned aerial vehicles (UAV). Followed by detection algorithms of moving obstacles on the found runways, it can control flight control systems or help pilots to decide landing. With high resolution electro-optical (EO), infra (IR), or hybrid EO-IR image sensors, it can extract tiny objects which are invisible to naked human eyes, and detect objects also in nighttime or bad weather conditions.

Obstacle detection on runways has been a classical and important issue in airports. Runway detection has been studied in various aspects for military or commercial purposes. Recently, with the development of automatic flight control system and UAVs, and with the increasing demand for the airport security, automated runway detection has been studied especially for automatic landing or guidance for landing. Conventional studies for runway detection are classified into runway detection from an aerial image for military purpose [1-7] and runway detection for automatic landing or landing assistance from runway approaching video sequences [8-11].

Previous methods detecting runways full frame searching is performed every frame because they did not consider temporal relationship among extracted boundaries in adjacent frames. Furthermore, the purpose of runway detection is an early warning from a distance to pilots. However, most of the existing researches only presented the experimental results on situations where aircrafts are close enough to runways. Although some can detect runways at a distance, it is probable that interference from neighboring structures around runways may degrades their performance. Commonly, all the conventional literatures have not presented enough experimental results for various approaching sequences.

In this paper, we propose a runway detection algorithm with IR video and GPS inputs, which considers geometric shapes of a runway and spatio-temporal relationship of vertices and line candidates of the runway. Straight lines in image frame are extracted using Hough transform. After checking the adjacency among the lines based on the size and the shape of the runway within a frame and filtering out lines with large deviation from the runway detected in previous frames, final vertices and edges surrounding the runway are calculated. Geometric parameters constraining the size and shape of the runway are defined and updated adaptively based on the distance to the runway, the latitude of the aircraft, and the focal length of the IR camera.

The organization of this paper is as follows. In section 3, the size and the shape of a runway mapped into an image frame is analyzed. Then, a runway detection method is proposed, which is based on locations and their adjacency of boundary edges of runways. In section 4, we present experimental results for the proposed algorithm with real video sequences of approaching to a runway. In section 5, we present our conclusions.

2 Landing Assistance System Based on Image Processing

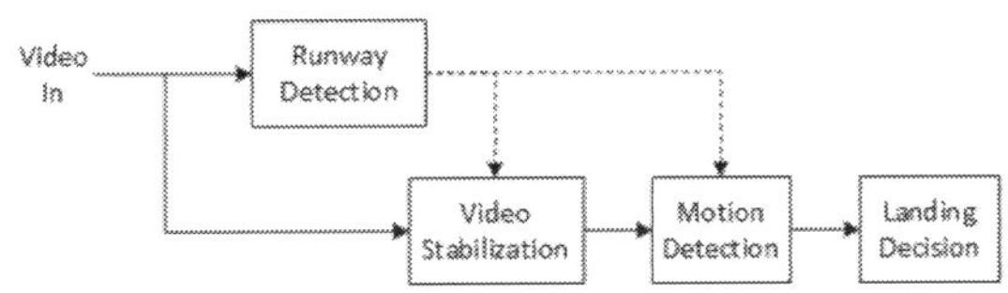

Figure 1. Landing Assistance System

DOI: 10.1051/matecconf/20165408005

For airplane safety, obstacle detection and avoidance during landing is an essential problem. For our study, a video assisted landing (VAL) using a vision sensor, shown in Fig. 1, of which concept was first introduced in [7], is considered. This system can provide real-time awareness of potential obstacles and collision, leading to prompt reaction and avoidance in critical surroundings. We use an infrared (IR) camera to operate the system even at night.

Although a camera can be fixed to an aircraft, because extrinsic parameters of an IR camera, such as the pose and the location, change over time during flight, the region of interest (ROI) and the shape of runway change every frame. A runway and obstacle region are usually very small at a distance, then get larger in the vision field when the airplane is approaching the runway. Therefore, the video stabilization module fixes the ROI to a specific location within a frame by linear transform between adjacent frames with Homography matrices found on extracted features.

Runway detection plays an important role to provide a ROI for video stabilization and moving obstacle detection. The detected runway can be used for video stabilization, navigation, and motion detection of moving objects on runways. Motion detection is performed on stabilized runway and its surrounding region to search possible obstacles.

3 Proposed Runway Detection Algorithm

In this section, a runway detection algorithm based on spatio-temporal relationship among extracted edges is proposed.

3.1. Overview of the proposed runway detection algorithm

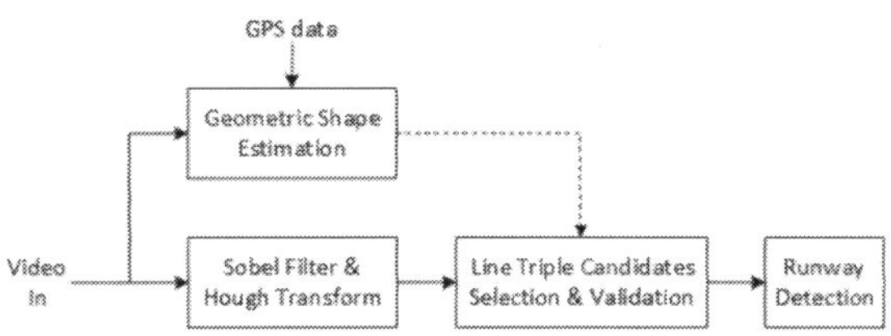

Figure 2. Proposed Runway Detection Algorithm

Fig. 2 describes the proposed runway detection algorithm. Because the size and the shape of a runway captured by IR cameras change over time as an aircraft approaches to the runway, both parameters and their constraints reflecting their spatio-temporal relationship should be defined. Also, during detection process, they should be updated every frame considering the distance to the runway and the current pose of the aircraft. In our study, only GPS sensor is used to capture the distance to the runway. If sensors to capture current pose of an aircraft are used, the accuracy and the processing speed of runway detection are expected to be enhanced further.

Line candidates for a runway are found by applying Hough transform on binary edge images extracted by Sobel filter. By assuming the shape of the runway on image frames to be a trapezoid, the line candidates are filtered by spatial constraint and construct line triple candidates enclosing a runway. Finally, they are filtered again by the temporal constraint; a line triple candidate should be close to the runway region in the previous frame. Detailed explanation on parameters and their adjustment will be done in the following section.

3.2 Mapping of a runway on an image plane

For runway detection, the imaging process of three-dimensional objects can be modeled as the viewing pipeline of conventional imaging systems. Firstly, a projection is applied which maps the runway in 3-D world to a 2-D object on an image sensor. Secondly, a coordinate system in the image sensor is defined by specifying a point as an origin, and two perpendicular vectors. And the image sensor coordinate mapping is applied to express the runway mapped on the image sensor in terms of the chosen two-dimensional image coordinate system. Finally, the runway on the image sensor is mapped on the image frame based on the given horizontal and vertical resolutions of the image frame and the size of the image sensor by means of a two-dimensional device coordinate transformation which utilizes camera calibration matrix.

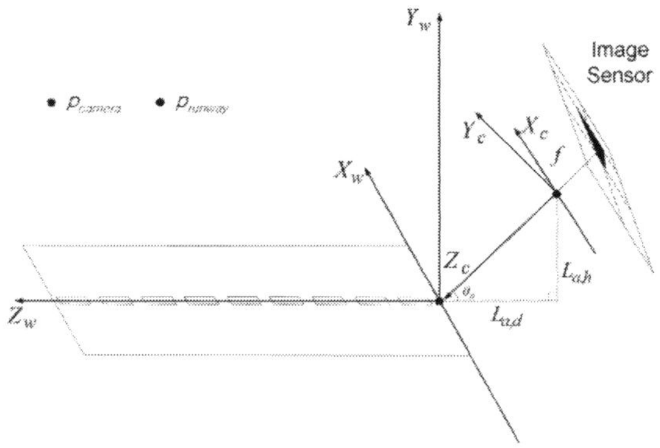

Figure 3. Runway on the camera coordinate

Fig. 3 shows the process of perspective projection of the real-world runway to the runway on the image sensor. $L_{a,d}$, $L_{a,h}$ represent the distance between the camera and the runway, the height of the camera. While the airplane approaches to the runway, the roll and yaw angles are less than the pitch angle, and assumed to be zero for the simplicity of the analysis. Then, the pitch angle is the dominant parameter reflecting current pose of an aircraft and determines the shape of the runway on an image plane. If an approaching direction of an aircraft is parallel to the running direction of the runway, the shape of the runway on an image sensor can be approximated to a trapezoid due to the perspective projection. Shape of the runway and relationship among edges of the runway, can be used to find real edges of the runway among edges found by image processing techniques.

The translation vector for an image frame can be calculated from the GPS coordinates of the camera and the end of the runway as follows.

$$T_s = \begin{bmatrix} 1 & 0 & 0 & 0 \\ 0 & 1 & 0 & -L_{a,h} \\ 0 & 0 & 1 & -L_{a,d} \end{bmatrix}.$$

(1)

For the pitch angle Θ_p the rotation matrix is given as follows.

$$R_s = \begin{bmatrix} 1 & 0 & 0 \\ 0 & \cos\theta_p & \sin\theta_p \\ 0 & \sin\theta_p & \cos\theta_p \end{bmatrix},$$

(2)

where θ_p is the angle between the heading direction of the aircraft and the running direction of the runway.

The camera calibration matrix is given as follows.

$$\mathrm{K} = \begin{bmatrix} f\frac{R_w}{2S_x} & 0 & \frac{R_w}{2} \\ 0 & f\frac{R_h}{2S_y} & \frac{R_h}{2} \\ 0 & 0 & 1 \end{bmatrix},$$

(3)

where f, S_x, S_y, R_h, and R_w represent the focal length of the camera lens and the width and the height of an image sensor, and the horizontal and vertical resolutions of the image frame, respectively.

The runway position in the camera coordinate can be transformed from its position in world coordinate as follows.

$$\begin{bmatrix} X_c \\ Y_c \\ Z_c \end{bmatrix} = KR_sT_s \begin{bmatrix} X_w \\ Y_w \\ Z_w \\ 1 \end{bmatrix}.$$

(4)

3.3 Line candidates and geometric constraints

3.3.1 line candidates and geometric constraints

Images captured during the approaching phase of the aircraft to the runway include horizontal and vertical lines on the trapezoid shape of the runway as shown in Fig. 4. Location of the surrounding lines of the runway, their distances, and angles among the lines can provide information for runway detection.

For incoming images, Sobel filter is applied to extract horizontal and vertical edges. Hough transform is applied to the binary edges to find lines which possibly enclose the runway. These lines are validated with respect to their lengths and angles, and finally grouped into line candidates.

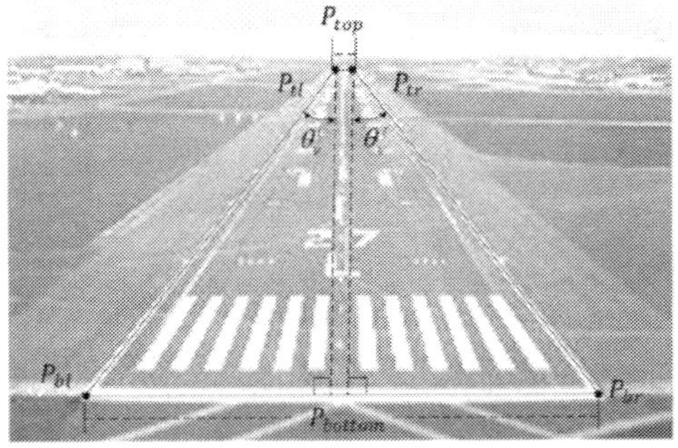

Figure 4. Runway shape on an image frame

3.3.2 geometric constraints on the runway

From the geometric constraints of the lines enclosing the runway, their lengths and the angles between them should be fallen into certain ranges depending on the distance to the runway, the altitude, and the approaching angle of the aircraft. Therefore, the geometric constraints on the lines can be interpreted as the threshold values on the lengths of the horizontal lines of the runway and the angle between the vertical lines of the runway and can be used for runway detection.

The bottom horizontal side and the left and the right lateral sides of the trapezoid matches the bottom, the left, and the right edge lines of the runway, respectively. However, the length of the line mapped from the top edge line of the runway is nearly zero from a distance or quite small compared to those of the other lines. Therefore, the top edge line is set to the line which is parallel to the bottom horizontal edge line and matches the largest number of edge pixels above the lateral lines. Two angles Θ_v^l and Θ_v^r represent the angles between the vertical line to the bottom edge line and the lateral edge lines, respectively. The lengths of the bottom and top edge lines in the video frame captured at time t, P_{bottom} and P_{top} are calculated using positions of the vertices of the trapezoid.

To identify the edge lines of the runway in input video frames, margins on the lengths of the horizontal lines $P_{top,t}$ and P_{bottom} should be given to compensate the shape distortion of the runway trapezoid especially from the error of the pitch angle. Therefore, lines detected in each video frame satisfying the following constraints on the lengths of $P_{top,t}$ and $P_{bot,t}$ are selected as line candidates.

$$P_{top}^{max} = P_{bottom}^{min} = (P_{top} + P_{bottom})/2, \quad (5)$$

$$P_{top}^{min} = (1-\alpha)P_{top}, \quad (6)$$

$$P_{bottom}^{min} = (1+\alpha)P_{bottom}, \quad (7)$$

where variables α is a constant, 0.3 in our experiment, reflecting the ratios of changes in edges of a runway.

3.4 Validation of line candidates based on geometric constraints

DOI: 10.1051/matecconf/20165408005

3.4.1 validation of vertical line candidates of the runway

There exist two long parallel line segments around the runway in the direction of aircraft approaching. Once they are projected on the image plane, they are not parallel but make a certain angle each other. Hereafter, these two lines on the image plane will be called as the vertical lines of a runway. Vertical line candidates are taken from extracted lines in the video frame by checking whether the angle between vertical line candidate pair satisfies the range of the vertical angle θ_v. The vertical line candidate pairs of which angles are validated construct the following set.

$$V_{\theta,t} = \left\{(v_i, v_j): \theta_v^{min} < \frac{\theta_{v_i}-\theta_{v_j}}{2} < \theta_v^{max}, \forall v_i, \forall v_j \in V_H\right\}, \quad (8)$$

where V_H, v_i, θ_{v_i}, and θ_v^{min} represent the set of the vertical lines after Hough transform, the ith vertical line in set V_H, the angle of v_i and the minimum angle between two vertical lines given by 5°in our experiment, respectively.

The distance between top or bottom end-points of each vertical line pair has to be within the range of minimum and maximum of Eq. 5-7. Although the y-coordinates of center positions of a vertical line pair may not be same due to incomplete line detection or the shape distortion of the runway trapezoid, if the center positions in a vertical pair are close enough, the vertical line pair is considered as set of runway candidates. Fig. 5 shows the validation result of vertical line pair.

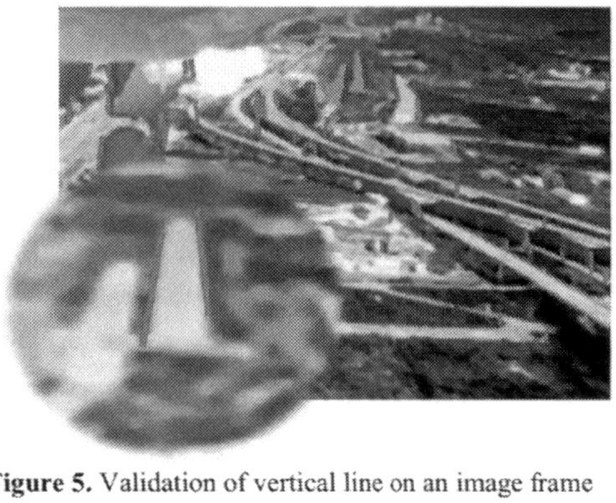

Figure 5. Validation of vertical line on an image frame

3.4.2 validation of line triple candidates of the runway

Geometrically, the bottom horizontal line of the runway is located below the vertical line pair of the runway. Also, the center position of the bottom horizontal line should be located between the vertical line pair as shown in Fig. 6. By using these geometric properties, the bottom horizontal line candidate can be chosen from the horizontal lines extracted by Hough transform. Then, the vertical line pair and the bottom horizontal line construct a new line triple candidate for the runway.

The bottom horizontal line should be below the vertical line pair, and close enough to them. All the elements in set of line triple after checking whether the bottom horizontal line is below the vertical line are checked under this distance constraint, which limits the distance between one

Figure 6. Validation of line triple on an image frame

of the center points of vertical lines and another center point of the bottom horizontal line within a certain threshold. The distance is calculated only by using y-coordinates of the center points.

If the x-coordinate of the center position of the bottom horizontal line is located between the x-coordinates of the bottom end-points of the vertical line pair, the line triple is finally considered as the element of the line triple candidate for the runway.

3.5 Runway detection

The runway in approaching video sequences appears as a group of long vertical and horizontal lines other than other ground structures. Finally, the line triple with maximum line length sum is selected as the runway region. Fig. 7 is result of final runway detection on the image frame.

In the landing assistance system, image stabilization is performed after runway detection. Thus, the accuracy of the vertices of the runway has the great effect on the image stabilization performance. To extract the final vertices of the runway, three lines in the final line triple are extended to construct an initial trapezoid of which top width is 3 pixels wide, which is empirically determined to cover the whole runway at 1.2km distance to the runway, and the histogram based refinement of the runway region is performed.

Figure 7. Final runway detection

4 Experimental Result

4.1 Experimental environments

To evaluate the performance of the proposed runway detection algorithm, ten landing IR sequences are used for simulation. These sequences were captured in summer season by an onboard IR camera in QVGA(320x240) resolution and 15 Hz frame rate. The focal length of the camera is 19mm and the size of the IR image sensor is 8.1×6.4 mm^2.

For safe landing of an aircraft, in our study, the minimum distance between the aircraft and the runway which a pilot should determine landing convincing no obstacles on the runway is set to 400m. Then, the runway detection algorithm should be able to detect the runway and to search for obstacles on the runway long before the minimum distance. In our video collection system, video sequences are collected during the approaching phase of the aircraft; 1.2km to 0.4km from the runway.

The ground truth for runway detection is manually picked for all video frames in each sequence. Table 1 summarizes the information of the video sequences used for simulation.

Table 1. IR Sequences for simulation

Seq	Dis.(km)	Runway area(pixel)	# of frames
0	1.19 – 0.38	180 – 641	316
1	1.19 – 0.39	160 – 623	285
2	1.20 – 0.37	142 – 474	331
3	1.17 – 0.39	125 – 520	315
4	1.19 – 0.37	156 - 569	328

4.2 Experimental results

Table 2 shows the runway detection rate. For the segment of 1.00-0.96km to the runway, the detection rate is 86%. For all the segments with the distance less than 700m, the detection rate is 100% for all IR video sequences. The low detection rate 67% of the segment of 0.8-0.76km in Seq2 is caused by the occlusion of the runway by the exhaust pipe of the aircraft. This kind of occlusion depends on several factors such as viewing angle adjustment of the IR camera attached on the aircraft, air turbulence around the aircraft, and the pilot's flight control style.

The proposed runway detection algorithm can be used for the detection of the initial runway position for safe landing decision controller. If the resolution of the camera is increased and sensors measuring the pose of the aircraft can be used, more accurate constraints on the shape of the runway can be estimated for more accurate runway detection.

Table 2. Runway detection rate(%)

Dist.(km)	Seq0	Seq1	Seq2	Seq3	Seq4	Avg.
1.20-1.16	0.45	0.73	0.20	0.67	0.64	0.54
1.16-1.12	0.40	0.33	0.33	0.44	0.73	0.45
1.12-1.08	0.53	0.71	0.80	0.33	0.60	0.60
1.08-1.04	0.80	0.79	0.87	0.67	0.87	0.80
1.04-1.00	0.60	0.71	1.00	0.94	0.61	0.77
1.00-0.96	0.84	0.73	1.00	0.78	0.93	0.86
0.96-0.92	0.87	0.93	0.87	1.00	0.94	0.92
0.92-0.88	1.00	1.00	1.00	1.00	0.95	0.99
0.88-0.84	1.00	1.00	1.00	1.00	1.00	1.00
0.84-0.80	1.00	1.00	1.00	1.00	1.00	1.00
0.80-0.76	1.00	1.00	0.67	1.00	1.00	0.93
0.76-0.72	1.00	1.00	1.00	1.00	0.93	0.99
0.72-0.68	1.00	1.00	1.00	1.00	1.00	1.00
0.68-0.64	1.00	1.00	1.00	1.00	1.00	1.00
0.64-0.60	1.00	1.00	1.00	1.00	1.00	1.00
0.60-0.56	1.00	1.00	1.00	1.00	1.00	1.00
0.56-0.52	1.00	1.00	1.00	1.00	1.00	1.00
0.52-0.48	1.00	1.00	1.00	1.00	1.00	1.00
0.48-0.44	1.00	1.00	1.00	1.00	1.00	1.00
0.44-0.40	1.00	1.00	1.00	1.00	1.00	1.00
0.40-0.36	1.00	1.00	1.00	1.00	1.00	1.00

5 Conclusions

A novel runway detection algorithm is proposed for safe landing assistance. Candidates for the boundary lines of the runway are extracted from image frames by the geographical relationship of the runway, and the temporal adjacency of the centroids of the runway is used for final detection. It is expected that the exact orientation and velocity of the aircraft as well as high resolution of the IR camera can enhance the runway detection rate.

Acknowledgment

This work was supported by the Gyeonggi Regional Research Center (GRRC) support program supervised by Gyeonggi Province. The first and the second authors were supported by Expert Education Program of Maritime Transportation Technology (GNSS Area), the Ministry of Land, Transport and Maritime Affairs (MLTM) of the Korean government.

References

1. Y. Ma, J. Wang, D. He, and J. Pang, *Accurate line detection by adjusting hough transform threshold adaptively*, WiCOM, 1-6 (2010)
2. N. Di, M. Zhu, and Y. Wang, *Real time method for airport runway detection in aerial images*, ICALIP, 563-567 (2008)
3. Y. Dong, B. Yuan, H. Wang, and Z. shi, *A runway recognition algorithm based on heuristic line extraction*, IASP, 292-296 (2011)
4. X. Wang, B. Li, and Q. Geng, *Runway detection and tracking for unmanned aerial vehicle based on an improved canny edgy detection algorithm*, IHMSC, **2**, 149-152 (2012)
5. S. Cao, J. Jiang, Y. Yuan, *Runway detection Using Line Segment Statistical Model*, ICICIP (2012)
6. F. Zhang, W. Sun, Y. Guo, *Airport Runway Extraction Method in Aviation reconnaissance Image*, IMSNA (2013)

7. O. Aytekin, U. Zongur, and U.Halici, *Texture-based airport runway detection*, IEEE, **10**, 471-475 (2013)
8. Y. Shanmin, F. Ziliang, Z. Xiuqiong, and Y. Hongyu, *A multi-scale morphologic runway detector*, ICCET, **4**, 524-528 (2010)
9. C.-H. Pai, Y.-P. Lin, G. Medioni, and R. Hamza, *Moving object detection on a runway prior to landing using an onboard infrared camera*, CVPR, 1-8 (2007)
10. M. abdul Rahiman, Mahalekshmi, C.Monisha, S. Remya, G. Manoj Kumar, M. Rajasree, *Automated object recognition and pattern matching analysis of runways using surface track data*, ICECT, **3**, 46-51 (2011)
11. K. Abu-Jbara, W. Alheadary, G. Sundaramorthi, C. Claudel, *A Robust Vision-based Runway Detection and Tracking Algorithm for Automatic UAV Landing*, ICUAS (2015)

MATEC Web of Conferences **54**, 08006 (2016) DOI: 10.1051/matecconf/20165408006
MIMT 2016

Scaled-neighborhood Patches Fusion for Multi-view Stereopsis

Ning An , Yicong He, Hang Dong and Fei Wang

Institute of Artificial Intelligence and Robotics, Xi'an Jiaotong University, 710049, P. R. China

Abstract. In this paper, we present a multi-view stereo reconstruction approach which fuses scaled-neighborhood information. PMVS proposed by Furukawa is one of the most excellent algorithms, and it has a good performance on many datasets both the accuracy and the completeness. However, there are still further improvements on this algorithm. PMVS cannot perform well in the presence of slanted surfaces, which are usually imaged at oblique angles. According to these aspects, on the one hand we propose to estimate the initial normal of every seed patch via fitting quadrics with scaled-neighborhood patches, which greatly improves the accuracy of the normal. On the other hand, we present to compute scaled-window for the further optimization based on texture. And it has been tested that employing the scaled-window will dramatically smooth the surfaces and enhance the reconstruction precision.

1 Introduction

Multi-view stereo (MVS) is to reconstruct a complete 3D model from a collection of images taken from a scene or an object. Nowadays, huge numbers of images could be easily available with rapidly development of the modern digital camera and the Internet. It has seen a surge of interest in computer vision that how to exploit the largest and most diverse images collection ever assembled to reconstruct the 3D model about a scene [1, 2]. Much progress has been made, both in terms of precision and performance. Snavely et al. [2] proposed a method to automatically rebuild 3D large-scale scenes based on massive photographs from the Internet imagery, which realizes sparse 3D scene modeling from multi-view.

In the research of dense reconstruction, Furukawa[3] presented a novel algorithm, PMVS, based on patches which has been tested on various datasets. It has been proved to perform well on completeness and accuracy. In this paper, we revisit PMVS and show that it appears some limitations when the 3D modeling encounters slanted surfaces. Particularly, PMVS cannot achieve good result where the scene has both plain regions and curved regions. We will show that, with special design, this classical approach can yield some of the higher quality reconstructions.

The rest of this paper is organized as follows: we will first review related work (Section 2), and provide a more detailed overview of our method (Section 3). Section 3.1 briefly describes Furukawa's patch-based MVS algorithm. We then present the individual stages of our method, including initial estimation of patch normal via fitting quadrics with scaled-neighborhood patches (Section 3.2) and scaled-window for further optimization based on texture (Section 3.3). Experimental results and discussions are given in Section 4. We conclude with results in Section 5.

2 Related work

In the process of MVS reconstruction, there are many factors influencing results such as slanted surfaces, low-texture regions, large concavities, which increase the difficulty for features matching across images. And, it is more difficult to reconstruct a 3D model where the local curvature is too high. And these factors have been always affecting the research of MVS.

Recently, the MVS problem has achieved a great development, yielding a variety of reconstruction algorithms. Most of methods need some prior information as auxiliary to get more precise results. According to the taxonomy of Seitz et al. [4], MVS algorithms can be divided into four categories: 3D volumetric approaches [5, 6], surface evolution techniques [7, 8], algorithms that compute and merge depth maps [9, 10], and techniques that grow regions or surfaces starting from a set of extracted features or seed points [3, 11], our algorithm apparently falls into the last category.

A multi-view framework for computing dense depth estimations was first proposed by Szeliski [12]. Esteban et al. [7] proposed a method based on texture and silhouette information, and fused the silhouette force into the snake framework. Based on these methods, Furukawa et al. [3] presented quite an accurate Patch-based MVS (PMVS) approach that starts from a sparse set of matched key points, and repeatedly expands these till visibility constraints are invoked to filter away false matches. The authors tested their method on the datasets provided by Strecha et al. [13], the available evaluation, and their results were significantly more accurate and complete than the few other submitted ones.

Hiep et al. [14] proposed a minimum s-t cut based global optimization that transforms a dense point cloud into a visibility consistent mesh, followed by a mesh-based variational refinement that captures small details, smartly handling photo-consistency, regularization and adaptive resolution. In Bleyer's [15] approach, a 3D scene is represented as a collection of visually distinct and spatially coherent objects. Inspired by Markov Random Field models of image segmentation, they employed object-level color models as a soft constraint, which can improve depth estimation in powerful ways. Jancosek et al. [16] augmented the existing Labatut CGF 2009 method with the ability to cope with these difficult surfaces just by changing the t-edge weights in the construction of surfaces by a minimal s-t cut. Their method uses Visual-Hull to reconstruct the difficult surfaces which are not sampled densely enough by the input 3D point cloud. They proved that the method can considerably better reconstruct difficult surfaces while preserving thin structures and details in the same quality. Qi Shan [17] leveraged occluding contours to improve the performance of multi-view stereo methods. This proposed approach outperforms state of the art MVS techniques for challenging Internet datasets, yielding dramatic quality improvements both around object contours and in surface detail.

We focus attentions on the research of improving 3D reconstruction on high curvature regions and thin structures. We look back on the patch-based method, PMVS, and on the framework of iterative optimization for every new seed patch, we propose an improvement approach that is inspired by the concept of multi-scale. It has been showed that, with careful design, this classical method can yield some of the higher quality reconstructions of any multi-view stereo algorithm, remaining highly scalable and offering excellent performance.

3 Algorithm Overview

This paper considers the problem of reconstructing the dense geometry of a 3D model from a number of images, and the camera poses and intrinsic parameters have been previously obtained. Methods about sparse multi-view stereo reconstruction and how to obtain the intrinsic parameters please refer to [1]. The dense MVS is a classic computer vision problem that has been extensively studied and a number of solutions have been published. Our multi-view reconstruction algorithm takes a set of calibrated images as input, which are captured from different viewpoints around the object to be reconstructed.

Decipted in Figure 1, similar to the process of PMVS, features being extracted are first matched across multiple images, yielding a sparse set of patches associated with salient image regions. The pairs of features are constrained to lie the corresponding epipolar lines, and triangulate the 3D points. We compute the normalized cross correlation (NCC) score based on scaled-window whereas PMVS adopts fixed-window. The seed patches reconstructed by our approach are proved to be more steady, which reduces the impact of outliers in some degree.

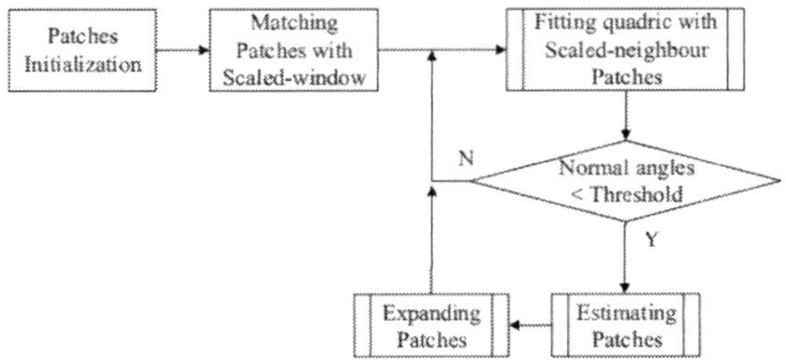

Figure 1. Algorithm framework.

PMVS estimates the surface orientation while enforcing the local photometric consistency, which is significant in practice to obtain accurate models for diversiform datasets. We employ this idea to compute scaled-window for the further optimization based on texture in the next iterations. In the step of optimization, we propose a design to fuse neighborhood priors to initially estimate the patch normal, that multi-scale quadrics are fitted according to *K*-neighborhood patches for acquiring the precise normal. It can be obviously improper in PMVS that the patch normal is assumed to parallel with the viewing ray passing through the center intersects the plane containing the patch. The application of the theory of multi-scale spatial is effectively improved the reconstruction accuracy with fine surface detail, especially on high curvature parts and thin structure regions.

3.1 Briefly Description of PMVS

PMVS is one of the most classical multi-view stereopsis algorithms. A patch p is essentially a local tangent plane approximation of a surface (Figure 2). Its geometry is fully determined by its center $\boldsymbol{c}(p)$, unit normal vector $\boldsymbol{n}(p)$ oriented toward the cameras observing it, a reference image $R(p)$ where p is visible and visible images $V^*(p)$.

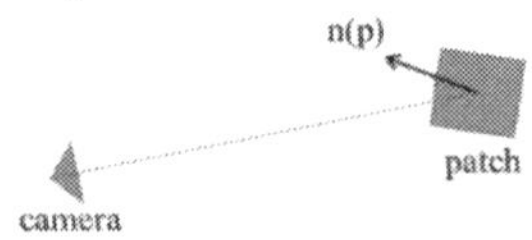

Figure 2. Patch model

It is implemented as a match, expand, and filter procedure. Features found by Harris and difference-of-Gaussians (DoG) operators are first matched across multiple images, yielding a sparse set of seed patches. Expansion is used to spread the initial matches to nearby pixels and obtain a dense set of patches. The geometric parameters, $\boldsymbol{c}(p)$ and $\boldsymbol{n}(p)$, are optimized by minimizing the photometric discrepancy score $g^*(p)$. It is repeatedly expanded before using visibility consistency and regularization constraint to filter away false matches.

$$c(p) \leftarrow \{Triangulation \ \ from \ \ f \ \ and \ \ f'\} \quad (1)$$

$$n(p) \leftarrow \overrightarrow{c(p)O(I_i)} / \left|\overrightarrow{c(p)O(I_i)}\right| \quad (2)$$

$$R(p) \leftarrow I_i \quad (3)$$

$$V(p) \leftarrow \left\{I \mid n(p) \cdot \overrightarrow{c(p)O(I)} / \left|\overrightarrow{c(p)O(I)}\right| > \cos(\theta)\right\} \quad (4)$$

$$V^*(p) = \{I \mid I \in V(p), h(p, I, R(p)) \leq \alpha\} \quad (5)$$

$$g^*(p) = \frac{1}{\left|V^*(p) \setminus R(p)\right|} \sum_{I \in V^*(p) \setminus R(p)} h(p, I, R(p)) \quad (6)$$

To simplify computations, PMVS constrains $c(p)$ to lie on a ray such that its image projection in one of the visible images and $n(p)$ is assumed to parallel with the viewing ray passing through the center which intersects the plane containing the patch. Considering the reconstruction of high curvature and slanted surfaces, the initial estimation of patch normal determined by any visible image diverges from the true normal. It cannot make any sense by minimizing the photometric discrepancy score for the optimization of the geometric parameters. In addition, there are both plain regions and fold regions existing, and PMVS employs the fixed-window to compute photometric discrepancy scores for the further optimization, which is unable to perform well.

3.2 Normal Estimation by Scaled-Quadric

Our algorithm follows the framework of patch-based PMVS, however, some key designs are proposed. We compensate for estimating normal of seed patch by employing fitting quadrics with scaled-neighborhood patches, which improves the quality especially in high curvature regions and slanted surfaces. From the Figure 3, the dashed lines are the initial estimation of patch normal in PMVS. It can be assumed almost true when the reconstructed surface exactly faces towards cameras, while some high curvature regions exist unavoidable.

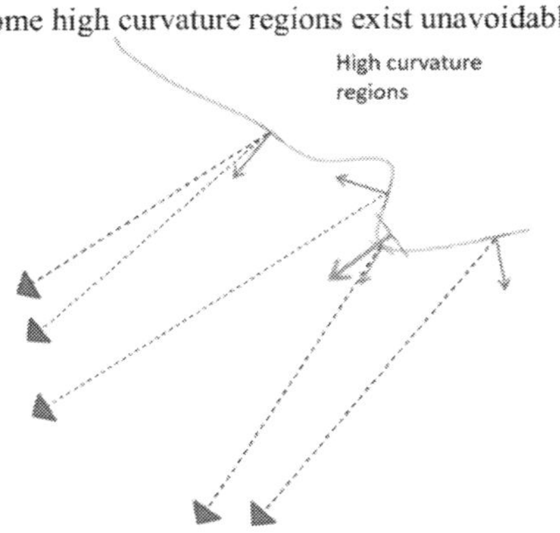

Figure 3. Cameras and the surface of 3D scenes

Shi Li-Min[18] has found the deficiency in PMVS, and proposed a method that K neighborhoods around the seed patch are selected to fit quadrics. The normals of fitted quadrics are computed to be as the initial estimation normal of the seed patch. This method rectifies the estimation of patch normal but still has some limitations. The value of K is set beforehand, and if the reconstructed region is plain, some outliers will lead to the deviation when the quadrics are fitted without enough patches. The large K neighborhood patches has certain endurance to fault tolerance when the plain regions are reconstructed. From the Figure 3, the normal computed with the large K neighborhood patches has some deviations while the smaller K will be more accurate.

We propose an approach to search multi-scale K neighborhoods around the seed patch that more effective patches are acquired to fit quadrics. It can be found more appropriate quadrics to estimate the normal when the high curvatures are reconstructed. Firstly, we search $\sigma \times K$ neighborhood patches to fit a quadric and search $\lambda\sigma \times K$ neighborhood patches to fit another quadric in a lower scale (Figure 4). Then we get two normals V1, V2. When the two normals satisfy the formula (7), the normals are regarded as the same direction, and the normal computed in higher scale is set to be the initial estimation of the seed patch. Otherwise, $\lambda^2\sigma \times K$ neighborhood patches are searched to compute the normal V3, checking whether V2 and V3 meet the threshold. And repeat the above process. If the two normals still have comparatively large deviation, the normal computed in smaller scale is used to as the initial estimation of the seed patch. Right now the new seed patch already has a relatively accurate initial normal, which can dramatically improves the further optimization of the geometric parameters, $c(p)$ and $n(p)$.

$$\cos^{-1}(normal1, normal2) < AngleThreshold \quad (7)$$

3.3 Scaled-Window for Further Optimization

As mentioned above section, it is imperative that the idea of adaptive multi-scale is introduced in the complex 3D reconstruction. During the process of optimization based on texture, it obtains amazing results when we employ this concept to rectify the texture window. $c(p)$ is constrained to lie on a ray such that its image projection in one of visible images does not change, reducing its number of degrees of freedom to one and solving only for a depth. $n(p)$ is parameterized by Euler angles, yaw and pitch. We use normalized cross correlation (NCC) on the square texture region as a metric for minimizing the photometric discrepancy.

$$NCC(p_0, p_1) = \frac{\sum_i (p_0(i) - \overline{p_0}) \cdot (p_1(i) - \overline{p_1})}{\sqrt{\sum_i (p_0(i) - \overline{p_0})^2 \cdot \sum_i (p_1(i) - \overline{p_1})^2}} \quad (8)$$

It has a certain limitation that PMVS algorithm employs the fixed-window for optimization. The fixed-window is not usually large, or some holes will appear on the reconstructed surfaces because of high curvature regions. However, some plain regions (building wall, ground) exist, and the reconstructed surface will be extremely corrugate if the fixed-window is not large enough. The 3D point cloud looks just like fish scale, which has seriously affected the accuracy of stereo reconstruction.

According to every patch, the scaled-window is selected independently. We choose large texture windows for the plain regions, whose NCC scores could maintain the stability and the 3D model gets more smooth. During the reconstruction of fold regions, our approach adaptively selects smaller texture window for optimization. It maintains the sensibility to high curvature regions, which helps to reconstruct thin structures well. The algorithm is intuitively described in Figure 4.

```
For each new patch candidate p'
   Search   σ×K neighborhood Patches;
   Search  λσ×K neighborhood Patches;
   Fit quadratic and compute the normal V1, V2;
   If acos(V1,V2) < AngleThreshold1
      n(p') = V2, window_size = WZ;
   else if acos(V1,V2) < AngleThreshold2
      n(p') = V1, window_size = WZ×δ;
   else
      Search  λ²σ×K neighborhood Patches
      Fit quadratic and compute the normal V3;
      If acos(V1,V3) < AngleThreshold3
         n(p') = V1, window_size = WZ×δ;
      else
         n(p') = V3, window_size = WZ×δ²;
```

Figure 4. Estimation of the normal via fitting quadrics with scaled-neighborhood patches and optimization based on scaled-window. $K \in [30,60]$, $\sigma=2$, $\lambda=0.6$, $\delta=0.8$

4 Results

Our algorithm is implemented by VC++ with the CGAL library. We demonstrate our multi-view stereo algorithm with a number of reconstructions. Images are acquired from the real-world scenes during the experimental process, and we have some particular comparison with the result of PMVS. It is mainly analyzed on three 3D models from the reconstruction. The first set of data is taken from a sculpture, and the other two are the open datasets from Strecha [13]. When we reconstruct a scene, it is shot from various angles firstly, yielding a set of pictures. Then structure-from-motion software [19] is applied to restore the position of every camera and get the P matrix for per visible image. We take the P matrix as input of the further dense 3D reconstruction.

The first experiment is carried on a sculpture of typography with nine images (3264×2448 resolution), called "Typography-P9". Typography-P9's surface is composed of many slanted planes, and parts of them have characters carved with thin structures. PMVS builds 355142 patches while our algorithm obtains 423148 patches with ten percent more. The denseness reconstructed has improved a lot (Table 1). At the same time, from the comparison of 3D models, our approach performs obviously better than PMVS on both accuracy and completeness on the local region. The regions with red circles (see Figure 5, 6) appear large holes because these areas are slanted and close to the edge of visibility, with small texture for further optimization. And our method conducts an estimation fusing multi-scale neighborhood patches to rectify the normal of a new seed patch, with adaptive scaled-window for optimization, regarding some new patches supported by small size of texture. It apparently shows that our 3D reconstruction of fold regions can be seen very smooth, with more dense point cloud, in contrast (Figure 5).

Figure 5. First row : some example images of Typography-P9. 3D point cloud in the last row are computed using PMVS and our algorithm.

The second experiment is on the dataset Herz-Jesu-P8 [13]. The eight images have been precisely calibrated corresponding cameras for their intrinsic and extrinsic parameters. We can directly obtain the P matrix for every image. Our method produces 1565694 patches with 25.7 percent more on the denseness of 3D point cloud (Figure 6, Table 1). This scene has a large flat building wall, which leads to the fish-like scale of the 3D model reconstructed by PMVS with small fixed-window, while large window may lead to some holes on the thin structure surface. Our adaptive approach will choose the large texture window for optimization when the scene has large plain regions. It is the careful design that helps to achieve the fine rebuilding result.

We adopt the dataset of Castle-P10 [13] to make another experiment. This scene has an obvious character of bending building wall, which is more apparently proved the limitation of PMVS. From the 3D modeling, the PMVS's reconstruction of the front building wall facing towards cameras is just similar to ours, only deficient on some regions with thin structures. However, the bending building wall is almost impossible to rebuild by PMVS. By comparison, our method can not only build the bending building wall, but also have an accurate recovery of carvings, embossments and windows on details (Figure 6).

5 Conclusion

The experiments are proved that the method proposed by us has a better performance on the high curvature regions and slanted surfaces. In comparison to PMVS, the 3D models reconstructed by our algorithm have a great improvement on the accuracy, denseness and thin

Figure 6. Datasets from Strecha [13]. Top row: from left to right, an image from Herz-Jesu-P8, the 3D model reconstructed by PMVS, by Ours, the bending building wall and the 3D model reconstructed by PMVS, by Ours. Bottom row: from left to right, an image from Castle-P10. the 3D model reconstructed by PMVS. by Ours. the embossment and the 3D model reconstructed by PMVS. by Ours.

structures. The estimation of patch normal via fitting quadrics with scaled-neighborhood patches enhances the sensibility to the high curvature regions. And the application of multi-scale also improves the stability of the normal estimation. It is presented to be more efficient to rebuild a scene existing both plain regions and fold regions with scaled-window for the further optimization based on texture. The adaptiveness guarantees the patches to be optimal during iterations, which leads to a more accurate and complete 3D model.

Table 1. Comparisons of denseness

Datasets	Resolution	Number of patches (PMVS)	Number of patches (Ours)	Percent
Typography-P9	3264×2448	355142	423148	19.2%
Herz-Jesu-P8	3072×2048	1246046	1565694	25.7%
Castle-P10	3072×2048	1540772	1863400	20.9%

Acknowledgment

This work was supported by the Natural Science Foundation of China (No61273366), National High Technology Research and Development Program (No2015BAH31F01).

References

1. N. Snavely, S.M. Seitz, R. Szeliski S. Modeling the world from internet photo collections. *International Journal of Computer Vision* **80**(2) :189-210. (2008).
2. S. Agarwal, Y. Furukawa, N. Snavely, et al. Building rome in a day. *Communications of the ACM* **54**(10). (2011).
3. Y. Furukawa, and J. Ponce. Accurate, dense, and robust multiview stereopsis. *Pattern Analysis and Machine Intelligence.* **32**(8) :1362-1376. (2010).
4. S.M. Seitz, B. Curless, et al. A comparison and evaluation of multi-view stereo reconstruction algorithms. *Computer Vision and Pattern Recognition.* Vol. **1** :519-528. (2006).
5. M. Sormann, C. Zach, J. Bauer, et al. Watertight multi-view reconstruction based on volumetric graph-cuts. *Image analysis*. 393-402. (2007).
6. G Vogiatzis, et al. Multiview stereo via volumetric graph-cuts and occlusion robust photo-consistency. *Pattern Analysis and Machine Intelligence,* **29**(12) :2241-2246. (2007).
7. C.H. Esteban, F. Schmitt. Silhouette and stereo fusion for 3D object modeling. *Computer Vision and Image Understanding.* **96**(3) :367-392. (2004).
8. A. Zaharescu, E. Boyer, R. Horaud. Transformesh: a topology-adaptive mesh-based approach to surface evolution. *Computer Vision-ACCV* 166-175. (2007).
9. C. Strecha, W. von Hansen, L. Van Gool, P. Fua, U. Thoennessen Combined depth and outlier estimation in multi-view stereo. *Computer Vision and Pattern Recognition.* Vol. **2** :2394-2401. (2006).
10. Shen, Shuhan, and Z.Y. Hu. How to Select Good Neighboring Images in Depth-Map Merging Based 3D Modeling. *Image Processing.* **23**(1). (2014).
11. D. Bradley, T. Boubekeur, and W. Heidrich. Accurate multi-view reconstruction using robust binocular stereo and surface meshing. *Computer Vision and Pattern Recognition.* 1-8. (2008).
12. R. Szeliski. A multi-view approach to motion and stereo. *Computer Vision and Pattern Recognition.* Vol. **1**.(1999).
13. C. Strecha, W. von Hansen, L. Van Gool. On benchmarking camera calibration and multi-view stereo for high resolution imagery. *Computer Vision and Pattern Recognition.* 1-8. (2008).
14. V.H. Hiep, R. Keriven, P. Labatut. Towards high-resolution large-scale multi-view stereo. *Computer Vision and Pattern Recognition.* 1430-1437. (2009).
15. M. Bleyer, C. Rother, P. Kohli. Object stereo-joint stereo matching and object segmentation. *Computer Vision and Pattern Recognition.* 3081-3088. (2011).
16. J. Michal, and T. Pajdla. Multi-view reconstruction preserving weakly-supported surfaces. *Computer Vision and Pattern Recognition.* 3121-3128. (2011).
17. Q Shan, B. Curless, Y. Furukawa, C. Hernandez, S.M. Seitz. Occluding contours for multi-view stereo. *Computer Vision and Pattern Recognition.* 4002-4009. (2014).

18. L.M. Shi, F.S. Guo, Z.Y. Hu. An Improved PMVS through Scene Geometric Information. *Acta Automatica Sinica*, **37**.5(2011).
19. C. Wu. Towards linear-time incremental structure from motion. *3D Vision,* 127-134. (2013).

Micro Satellite Thermal Balance Testing: Orbit Heat Flux Simulation Method and Verification

Liu Jia [1] ,Li Ming[1] and Gao Qinghua [2]

[1]*DFH satellite Co., LTD, Beijing 100094, China*
[2] *Beijing Institute of Spacecraft Environment Engineering, Beijing 100094, China;*

Abstract. Comparing to conventional satellites, micro satellite features smaller envelop size, lower cost, and shorter R&D period. At the same time, while micro satellite generally runs within the LEO, it is more susceptible to the earth IR radiation. To simulate the orbital heat flux efficiently for the micro satellite, in this paper the thermal balance testing respectively using infrared cage and solar simulator are performed. The resulting differences between the two simulating approaches are examined with care. Conclusions can be drawn from analysis: in thermal balance testing of micro satellite, in terms of simulating orbital heat flux, the results of two simulators are close enough to consider that it is feasible to choose infrared cage as orbital heat flux simulator, given thermal coating material of typical IR emittance and solar absorptance ratio. Results demonstrate that by using infrared cage simulator, the actual temperature is about 5℃ higher than theoretical value. The background heat flux interference stays stable at 17% of intensity of simulated orbital heat flux, and the value could be referred to in the thermal mathematical model modification.

1 Introduction

Driven by rapid growth of technical innovation and demand for astronautic task, researches and developments of satellites are required to be both low-cost and short-period, which are met by the emergence of micro satellite. However, design of thermal control subsystem in spacecraft is challenged by complexity and uncertainty introduced by mission complexity. In order to verify both scheme and assembly of thermal control subsystem, it is crucial to perform the ground thermal balance testing and obtain on-board temperature data [1, 2, and 3].

In contrast to conventional satellites, micro satellite features smaller envelop size, lower cost, and shorter R&D period. At the same time, while micro satellite generally runs within the LEO, it is more susceptible to the earth IR radiation. The contribution of this paper are: a) studies on orbit heat flux simulation method for ground thermal balance testing of micro satellite; b) performing two different types of ground thermal balance testing based on infrared cage and solar simulator respectively; c) analysing the difference between the results of two simulations.

2 Options for ground simulation of orbit heat flux

Methods for ground simulation of orbit heat flux can be sorted into two categories: incident and absorptive. Specifically, incident simulator (such as solar simulator) mimics parameters of orbit heat flux including irradiance, directions and spectral characteristics. Meanwhile, absorptive simulator uses heating devices (such as infrared cage, heater, etc.) to imitate orbital heat absorbed by the surface of satellite. In contrast to conventional satellites, micro satellite features smaller envelop size, lower cost, and shorter R&D period. At the same time, while micro satellite generally runs within the LEO, it is more susceptible to the earth IR radiation. To simulate orbit heat flux efficiently, the following approaches are studied [4, 5]:

Solar Simulator. For small-sized spacecraft or units, solar simulator is the most appropriate option. Nevertheless, in micro satellite's case, the system is susceptible to earth IR radiation, the solar simulator fails to simulate earth IR, whereas it can only provide solar spectrum. In addition, the lower budget of micro satellite prohibits solar simulator due to the high running expenditure.

Foil Heater. This method simulates orbit heat flux by sticking foil heaters onto exterior surface panel. The advantage is to enable transient simulation, and to facilitate the modification of thermal mathematic model. But concerning micro satellite's short R&D period, the thermal balance testing is only performed during the flight-ready phrase (rather than the missing CDR phase), thus incurs two problems below: 1) limited by practical situation, the mocked surface panels are usually used for the testing, therefore radiative and conductive characteristics of the real panels can only be theoretically assumed, instead of being verified by the testing; 2) after the thermal balance testing, the mocked panels need to be replaced by the real ones, which raises the R&D cost as well as reliability risk.

Infrared Cage. The cage is an enveloped frame with several belts twisted on it. The energy is generated by heated stainless steel band, and the absorptive flux density is measured with radiometer on the satellite surface. The cage is split to several areas as heat fluent. The belt on an area is one chain and will be electrified during test. Whenever it powered on, heat will emit as infrared energy. Different area will be put different current as heat fluent in orbit. The power can be modified in each case we want to simulate during test.

Cold Plate. Unlike the infrared cage, surface of cold plate can be refrigerated or be heated, depending on the demand. By surrounding the satellite with cold plate, this simulation method is able to eliminate interference caused by background heat flux, which helps improving the thermal mathematic model. Additionally, the approach allows using all flight-ready parts and devices, hence leads to the most accurate simulation results. The shortcomings with cold plate simulation are requirements for larger vacuum simulator and plenty of assembly instruments. Moreover, to achieve ideal thermal uniformity, the simulator demand for larger sized cold plates and satellite, compare to the IR cage method.

To sum up, as to micro satellite, the infrared cage simulator turns out to be the most feasible and economical option. In this paper, infrared cage simulation is used to modify the thermal mathematical model, and the model is taken to predict the on orbit status. To verify the validity of infrared cage simulator, the difference between the results of both IR cage and solar simulator is examined.

3 Testing procedure

3.1. Testing Objects

The objects are two pieces of 200mm×200mm×1mm aluminum alloy plates, each weighs approximately 111.2 gram. The faced sides are painted with SR107-ZK white thermal coating (a=0.17, ε =0.87) and aluminum Kapton film (a=0.36, ε=0.69) respectively. The back sides are both covered with 25μm aluminum Mylar, and optical parameters are taken as beginning of life. The objects are as illustrated in Fig. 1.

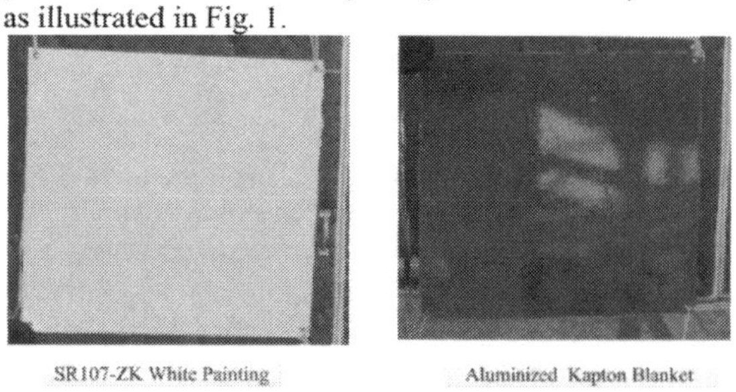

SR107-ZK White Painting   Aluminized Kapton Blanket

Figure 1. Tested Objects

3.2 Testing process

To analyze different effects from infrared cage and solar simulator, the thermal balance testing is performed in two phases.

In the solar simulator testing phase, the optical system of solar simulator and other parts would generate infrared radiation to the testing objects, thus necessitates radiometer to measure actual flux density that arrives at the objects.

Heat flux that absorbed by radiometer is written as:

$$q_0 = \alpha_s S + \varepsilon_H q_1 = \varepsilon_H \sigma T^4 \quad (1)$$

Heat flux that arrives at radiometer, i.e. arrives at the object, is written as:

$$q_H = \frac{q_0}{\varepsilon_H} = \frac{\alpha_s}{\varepsilon_H} S + q_1 = \sigma T^4 \quad (2)$$

where q_0 denotes the absorbed heat flux of radiometer (W/m^2); q_H is heat flux that arrives at radiometer; q_I is infrared radiation caused by the system (W/m^2); S is solar irradiation(1S$_0$ is 1410W/m^2); a_S is the absorptance of radiometer black surface, as 0.93; ε_H the emittance of radiometer black surface, as 0.88; T is the temperature of radiometer (K).

In contrast, in the infrared cage phase, the arrived heat flux to be simulated consists of not only radiation from solar simulator, but also from the combination of the optical system of solar simulator and others, which is denoted by q in above Eq. 2. In this phase, it is necessary to adjust the currency through the heating band, to match the target temperature on radiometer with the temperature T in solar simulator phase.

Figure2. Technical Status of solar simulator

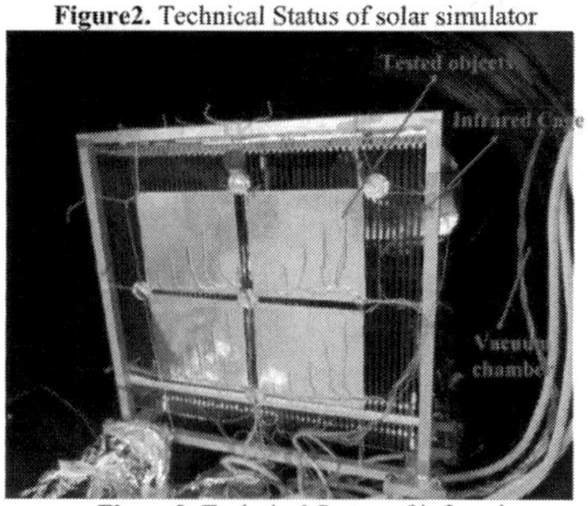

Figure3. Technical Status of infrared cage

Testing status of both phases is demonstrated in Fig.2 and Fig.3, and the technical statuses are set to be as consistent as possible.

3.3 Test results and analysis

3.3.1 The first phase thermal balance test

The first phase thermal balance test was carried out in the vacuum chamber with solar simulator, assuming there are 3 sorts of external heating to effect on the tested objects:

a) Solar irradiation(S):$1S_0$ is 1410W/m^2, so simulated orbital heat flux are $0.5S_0$, $0.8S_0$, $1.0S_0$;
b) The infrared radiation of faced side (q_1): Besides the solar irradiation, the infrared radiation from the chamber or optical system of solar simulator is considered which measured by 4 radiometers outside the facula.
c) The infrared radiation of revised side (q_2): To measure the background infrared radiation to test objects, 1 radiometer was installed on the revised side of test objects.

The energy balance for computing the temperature of surface is given by:

$$C_s \frac{dT_s}{dt} = \alpha_s S + \varepsilon_1 q_1 + \varepsilon_2 q_2 - (\varepsilon_1 + \varepsilon_2)\sigma T_s^4 \tag{3}$$

Where C_s is the equivalent heat capacity of the tested object; T_s is the temperature of the tested object; ε_1, ε_2 are the emittance of SR107-ZK white painting and aluminum Kapton film respectively; σ is Stefan-Boltzmann constant.

According to the sampling of the first phase thermal balance test and calculation by Eq.3, the results are shown as Table 1 and Table 2. Note that Tested objects temperature (T_s') means the average values of 4 sampling thermocouples; radiometer temperature (T_1) in the faced side represents the average values of 5 sampling radiometers; radiometer temperature (T_2) in the reversed side represents the average values of one sampling radiometer; and totally absorbed heat flux of the faced side denotes q_0You are free to use colour illustrations for the online version of the proceedings but any print version will be printed in black and white unless special arrangements have been made with the conference organiser. Please check with the conference organiser whether or not this is the case. If the print version will be black and white only, you should check your figure captions carefully and remove any reference to colour in the illustration and text. In addition, some colour figures will degrade or suffer loss of information when converted to black and white, and this should be taken into account when preparing them.

3.3.2 The second phase thermal balance test

The second phase thermal balance test was carried out in the vacuum chamber with infrared cage; there are just 2 sorts of external heating to effect on the tested objects.

The energy balance for computing the temperature of surface is given by:

$$C_s \frac{dT_s}{dt} = \varepsilon_1 q_1 + \varepsilon_2 q_2 - (\varepsilon_1 + \varepsilon_2)\sigma T_s^4 \tag{4}$$

According to the sampling and calculation of the first phase thermal balance test, the results are shown as Table 3 and Table 4.

3.3.3 Analysis on results of two phases.

Compare actual temperatures obtained from infrared cage phase to the ones from solar simulator, together with approximated calculated temperatures, conclusions can be drawn:

1. Speaking of micro satellite thermal balance testing with stable heat flux, in thermal balance testing of micro satellite, in terms of simulating orbital heat flux, the results of two simulators are close enough to consider that it is feasible to choose infrared cage as orbital heat flux simulator, given coating material of typical a/ε ratio. Results demonstrate that by using infrared cage simulator, the actual temperature is about 5℃ higher than theoretical value.
2. Background heat flux interference generated by infrared cage is stronger than its counterpart from solar simulator. However, as to the same testing object, the background interference decreases while simulated heat flux goes up, which implies solar simulator is sensitive to optical characteristics of object's surface.
3. Comparing to background interference intensity of solar simulator increases from 5% to 15% of simulated heat flux while the latter falls from 1.0 S_0 to 0.5 S_0, the interference stays around 17% while infrared cage through the same process. Thus would facilitate the model modification.

Table 1 Test conditions of SR107-ZK using solar simulator (a=0.17, ε=0.87)

	Faced side				Reverse side		T_s' [℃]	T_s [℃]	ΔT
	S	T_1 [℃]	q_1 [W/m^2]	q_0 [W/m^2]	T_2 [℃]	q_2 [W/m^2]			
Case1-w	$0.5S_0$	-86.44	59.95	179.8	-112.84	25.84	-47.62	-53.57	*5.95*
Case2-w	$0.8S_0$	-77.45	72.35	264.11	-109.12	28.32	-27.57	-33.36	*5.79*
Case3-w	$1.0S_0$	-73.2	78.85	318.55	-107.25	29.64	-18	-22.67	*4.67*

Table 2 Test conditions of Aluminum Kapton film using solar simulator (a=0.36, ε=0.69)

	Faced side				Reverse side		T_s' [℃]	T_s [℃]	ΔT
	S	T_1 [℃]	q_1 [W/m^2]	q_0 [W/m^2]	T_2 [℃]	q_2 [W/m^2]			
Case1-k	$0.5S_0$	-86.44	47.54	301.34	-112.84	25.84	-13.47	-18.86	*5.39*
Case2-k	$0.8S_0$	-77.45	57.38	463.46	-109.12	28.32	13.79	8.41	*5.38*
Case3-k	$1.0S_0$	-73.2	62.53	570.13	-107.25	29.64	25.03	22.74	*2.29*

Table 3 Test conditions of SR107-ZK using infrared cage (a=0.17, ε=0.87)

	Faced side				Reverse side		T_s' [℃]	T_s [℃]	ΔT
	S	q_0 [W/m^2]	Refer as	q_H [W/m^2]	T_2 [℃]	q_2 [W/m^2]			
Case4	$0.5S_0$	179.8	Case1-w	206.66	-97.74	37.03	-46.58	-50.64	4.06
Case5	$0.8S_0$	264.11	Case1-w	303.58	-90.94	43.12	-25.82	-30.38	4.56
Case6	$1.0S_0$	318.55	Case2-w	366.15	-86.73	47.25	-14.91	-19.56	4.65

Table 4 Test conditions of Aluminum Kapton film using infrared cage (a=0.36, ε=0.69)

	Faced side				Reverse side		T_s' [℃]	T_s [℃]	ΔT
	S	q_0 [W/m^2]	Refer as	q_H [W/m^2]	T_2 [℃]	q_2 [W/m^2]			
Case7	$0.5S_0$	253.8	Case1-k	436.73	-84.2	49.87	-8.98	-14.31	5.33
Case8	$0.8S_0$	406.08	Case2-k	671.69	-75.39	59.84	17.68	12.82	4.86
Case9	$1.0S_0$	507.6	Case3-k	826.28	-70.42	66.08	31.41	27.14	4.27

Conclusion

Simulation of orbital heat flux plays a key role in the satellite thermal balance testing. Conclusions can be drawn from analysis: in thermal balance testing of micro satellite, in terms of simulating orbital heat flux, the results of two simulators are close enough to consider that it is feasible to choose infrared cage as orbital heat flux simulator, given coating material of typical a/ε ratio. Results demonstrate that by using infrared cage simulator, the actual temperature is about 5 ℃ higher than theoretical value. The background heat flux interference stays stable at 17% of intensity of simulated orbital heat flux, and the value could be referred to in the model modification.

References

1. Gilimore D. Satellite Thermal Control. Second Edition. California: The Aerospace Corporation Press, 2002
2. Hou Zengqi and Hu Jingang, Spacecraft thermal control—theory and applications, Beijing: Chinese Technology Press, 2007, (in Chinese)
3. Baturkin Volodymyr. Micro-satellites thermal control--concepts and components. Acta Astronautica. 2005. 56(1-2): 161-170
4. Westheimer David T. Active Thermal Control System Considerations for the Next Generation of Human Rated Space Vehicles. AIAA-2005-342
5. Liu Jia. Analysis of Orbital Thermal Environment for Inclined-LEO Small Satellite, 2014 5th international conference on mechanical and aerospace engineering (ICMAE 2014). Madrid, Spain, 2014.

MATEC Web of Conferences **54**, 09002 (2016) DOI: 10.1051/matecconf/20165409002
MIMT 2016

Optimal Rendezvous Guidance Using Linear Quadratic Control

Gun-Hee Moon, Byung-Yoon Lee, Min-Jea Tahk, and David Hyunchul Shim

Korea Advanced Institute Science and Technology, Daejeon, Rep. of. Korea

Abstract. This paper handles with an energy optimal guidance law for rendezvous mission, based on linear quadratic control (LQC) problem. Rendezvous of two satellites are expressed by a nonlinear relative orbit dynamic model. The LQC problem minimizes integral of control input quadratic term with given final time and terminal states. A linear relative orbit dynamic, also called as the Clohessy-Wiltshire equation, is utilized as governing equation for optimal rendezvous guidance law. It is proven that renewing costates like an initial time is identical to propagating it from initial time to current time. Thus optimal guidance law can be formulated in state feedback form. To enhance computation efficiency, this work uses Taylor series expansion for the exponential of system matrix. The proposed algorithm is verified through nonlinear relative orbit simulations.

1 Introduction

Recently, needs on satellite proximity operation are growing faster. For decades, global major countries in the space industry have built the International Space Station (ISS), and its size is now up to a soccer field. Its enormous size makes the ISS requires many maintenance crews to cooperate by turns. In addition, fuel and food need to be supplied almost from the ground until now. The space-X have made Falcon-9, a nine tandem busted space cargo ship to do commercial transport service missions on the ISS. Sooner or later, the satellite constellations or satellite online maintaining missions will open new markets in the space industries.

The proximity operation in space is considered as two of two particles problems (earth-sat1, earth-sat2). The key modeling on the relative orbit dynamics is done by Clohessy and Wiltshire [1]. They derived a linear relative orbit dynamic model, so named after as the Clohessy-Wiltshire Equation (CW Equation). It is utilized for many proximity operation conceptual studies; In [2], [3], Kim have studied about relative orbit navigations based on mono image sensor and image processing technique. Kim utilized the CW equation for simulating the relative orbit to get a virtual images of satellite. In [4], Owis used an approximating sequence riccati equations for satellite formation flying guidance. It transforms the nonlinear relative orbit dynamic into a state affined form, then solved it sequentially as if it is a linear quadratic regulation (LQR) problem. Luo has used the CW equation as the governing equation of the rendezvous trajectory and formulated the guidance problem as an LQR problem [5]. Luo transformed the problem into a parameter optimization problem finding full state feedback gain K, and solved it through simulated annealing technique to compare with the gain K from the algebraic Riccati equation solution. For more detail description about LQR, referred to read [6-7].

Spencer used a potential field guidance for satellite proximity operation guidance in his Ph. D. dissertation [8]. Spencer also utilized the CW equation as governing equation, and he introduced advanced relative orbit elements to describe the chaser orbit.

For a real-time trajectory optimization of missile or launch vehicle, Padhi has developed *Model Predictive Static Programming* (MPSP) algorithm [9-10]. It is inspired from the approximate dynamic programming and the model predictive control.

In this work, we derived the nonlinear relative orbit dynamic model and the chaser satellite elementary orbit model, then introduced details of the CW equation. The CW equation is employed as a dynamic constraint equation for Linear Quadratic Control (LQC) Problem. The LQC is distinguished from LQR in two aspects. 1. The LQC specifies terminal time and terminal state, so those are constraints that must be satisfied. 2. The LQC has no penalty on the integration of the state quadratic term. The LQC solution can be derived into a state feedback form, because the propagated costate variables in each time is same as the costate by renewal from current state. And the suggested algorithm is verified through cases of simulations.

The rest of this paper is organized as follows; In chapter 2, nonlinear relative orbit dynamic equation is introduced, and related target satellite orbit element is remarked. Then linear relative orbit dynamic equation, the Clohessy-Wiltshire equation, is derived from the nonlinear model. In chapter 3, an optimal rendezvous guidance method, which is based on the linear quadratic control problem, is proposed. In chapter 4, suggested method is utilized on the nonlinear relative orbit dynamic model for several cases of scenarios, and we discuss about results of the nonlinear simulation. After then some brief closing session is at the end of the paper

2 Relative Orbit Dynamic

2.1 Nonlinear Relative Orbit Dynamic

Consider two satellites traveling in a similar orbit, one is a chaser and the other is a target. Position vector to the target is $\boldsymbol{r}_T$, position vector to the chaser is $\boldsymbol{r}_C$, and relative position vector from the target to the chaser is $\boldsymbol{\rho}$. Those vectors are related as follow,

$$\boldsymbol{r}_C = \boldsymbol{r}_T + \boldsymbol{\rho}. \tag{1}$$

Fig. 1 shows two coordinates, one is inertia frame (I-frame) and the other is local vertical local horizontal frame (L-frame). I-frame is fixed on the center of earth, x-axis, and y-axis is respectively aligned to the North Pole and the vernal equinox, remaining z-axis is properly aligned according to the right hand sign convention. L-frame is attached to the target ship, its x is a radial direction from the origin of I-frame to the origin of L-frame, z axis is oriented to angular momentum direction of the target, y axis is determined by the right hand sign convention again.

The second order derivative of (1) with respect to I-frame leads to next,

$${}^I\ddot{\boldsymbol{r}}_C - {}^I\ddot{\boldsymbol{r}}_T = {}^I\ddot{\boldsymbol{\rho}} \tag{2}$$

where the superscription on the left of a vector indicates the frame which change of the vector is measured on. The external forces acting on the target are solely the gravity, and those on the chaser are the gravity and the control acceleration $\boldsymbol{a}_T$.

$$\ddot{\boldsymbol{r}}_T = -\frac{\mu_e}{|r_T|^3}\boldsymbol{r}_T \tag{3}$$

$$\ddot{\boldsymbol{r}}_C = -\frac{\mu_e}{|r_T+\rho|^3}(\boldsymbol{r}_T + \boldsymbol{\rho}) + \boldsymbol{a}_T \tag{4}$$

Here we construct the earth gravitational parameter μ_e of 3.98601E5 (km^3/s^2) to express the gravity force. Now we change the reference frame of time derivatives to observe changes of the relative distance in the L-frame. The time derivatives of $\boldsymbol{\rho}$ in I-frame decomposed to 4 separated vectors; time derivative term of $\boldsymbol{\rho}$ in L-frame, tangential acceleration term, Coriolis term, and centripetal acceleration term.

$${}^I\ddot{\boldsymbol{\rho}} = {}^L\ddot{\boldsymbol{\rho}} + (\boldsymbol{\alpha}\times\boldsymbol{\rho}) + 2(\boldsymbol{\omega}\times\dot{\boldsymbol{\rho}}) + \boldsymbol{\omega}\times(\boldsymbol{\omega}\times\boldsymbol{\rho}) \tag{5}$$

where $\boldsymbol{\omega}$ is angular velocity vector of L-frame respect to I-frame, and $\boldsymbol{\alpha}$ is angular velocity acceleration vector of L-frame respect to I-frame.

Substituting (3), (4) and (5) to (2), then rearranging it gives following nonlinear relative orbit vector dynamic equation.

$$\begin{aligned}{}^L\ddot{\boldsymbol{\rho}} = &-\frac{\mu_e}{|r_T+\rho|^3}(\boldsymbol{r}_T+\boldsymbol{\rho}) + \frac{\mu_e}{|r_T|^3}\boldsymbol{r}_T + \boldsymbol{a}_T \\ &-(\boldsymbol{\alpha}\times\boldsymbol{\rho}) - 2(\boldsymbol{\omega}\times\dot{\boldsymbol{\rho}}) - \boldsymbol{\omega}\times(\boldsymbol{\omega}\times\boldsymbol{\rho})\end{aligned} \tag{6}$$

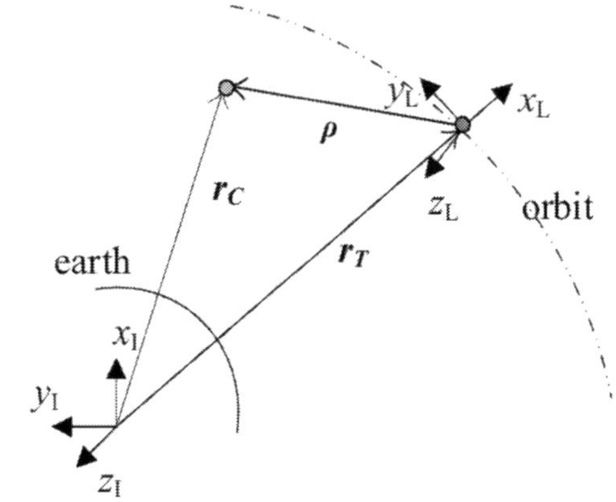

Figure 1. Relative Orbit Geometry. (Target and Chaser)

Each vectors in (6) ought to be described in L-frame.

$${}^L\boldsymbol{\rho} = x\boldsymbol{e}_x + y\boldsymbol{e}_y + z\boldsymbol{e}_z \tag{7}$$

$$\boldsymbol{r}_T = r_T\boldsymbol{e}_x \tag{8}$$

$$\boldsymbol{a}_T = a_x\boldsymbol{e}_x + a_y\boldsymbol{e}_y + a_z\boldsymbol{e}_z \tag{9}$$

where $\boldsymbol{e}_x$, $\boldsymbol{e}_y$, and $\boldsymbol{e}_z$ represent the orthonormal reference vector of each axis of L-frame.

As z-axis of L-frame and angular momentum per unit mass of the target are in same direction, the angular velocity vector has only z-component as,

$$\boldsymbol{\omega} = \omega_T\boldsymbol{e}_z \tag{10}$$

$$\dot{\boldsymbol{\omega}} = \dot{\omega}_T\boldsymbol{e}_z. \tag{11}$$

Here, ω_T is rate of transportation or rate of true anomaly, which is related to the angular velocity of L-frame. As the target motion is independent from relative motion of the chaser, it needs dynamic of the target orbit to describe r_T, ω and $\dot{\omega}_T$.

2.2 Target Satellite Elementary Orbit

The rate of the position vector $\boldsymbol{r}_T$ is describable in L-frame as follows,

$${}^I\dot{\boldsymbol{r}}_T = {}^L\dot{\boldsymbol{r}}_T + \boldsymbol{\omega}_T\times\boldsymbol{r}_T = \dot{r}_T\boldsymbol{e}_x + r_T\omega_T\boldsymbol{e}_y. \tag{12}$$

Here, $\dot{r}_T$ indicates velocity of target in radial direction. Then, it is trivial to,

$$\begin{aligned}{}^I\ddot{\boldsymbol{r}}_T &= -\frac{\mu_e}{r_T^3}\boldsymbol{r}_T = -\frac{\mu_e}{r_T^2}\boldsymbol{e}_x \\ &= {}^L\ddot{\boldsymbol{r}}_T + \dot{\boldsymbol{\omega}}_T\times\boldsymbol{r}_T + 2(\boldsymbol{\omega}_T\times\dot{\boldsymbol{r}}_T) + \boldsymbol{\omega}_T\times(\boldsymbol{\omega}_T\times\boldsymbol{r}_T) \\ &= (\ddot{r}_T - r_T\omega_T^2)\boldsymbol{e}_x + (r_T\dot{\omega}_T - 2\dot{r}_T\omega_T)\boldsymbol{e}_y\end{aligned} \tag{13}$$

Then, by integrating (12) and (13) with respect to I-frame, one can obtain the orbit of the target satellite.

As y-component of the acceleration vector described on L-frame should equal to zero, it is rearranged about changes of rate of the transportation as follows,

$$\dot{\omega}_T = -2\dot{r}_T\omega_T / r_T. \tag{14}$$

From x-component of the acceleration vector, the rate of transportation is calculated as,

$$\omega_T = \sqrt{\frac{1}{r_T}\left(\frac{\mu_e}{r_T^2} + \ddot{r}_T\right)}. \tag{15}$$

However this form contains an unknown, second order derivative of r_T, thus it cannot gives ω_T. Rather than from (12), the rate of transportation can be computed.

The velocity component of the target ship in I-frame can be described in L-frame by a direction cosine matrix C_I^L which translating a vector in I-frame to L-fame vector. Thus the radial component and the rate of transportation is calculated as,

$$\begin{aligned} \dot{r}_T &= C_I^L \, {}^I\dot{\boldsymbol{r}}_T \cdot \boldsymbol{e}_x \\ \omega_T &= C_I^L \, {}^I\dot{\boldsymbol{r}}_T \cdot \boldsymbol{e}_y / r_T \end{aligned} \tag{16}$$

The C_I^L changes by the true anomaly, which is a basic orbit element. It indicates in which direction a satellite exists with respect to the origin of I-frame. Then true anomaly of the target satellite is calculated by atan2 function as,

$$\nu = \text{atan2}\,({}^I\boldsymbol{r}_{Ty}, {}^I\boldsymbol{r}_{Tx}). \tag{17}$$

Since the orbit plane is perpendicular to the z-axis of L-frame, the C_I^L is computed as follows,

$$C_I^L = R_3(\nu)C_I^{L0} = \begin{bmatrix} \cos\nu & \sin\nu & 0 \\ -\sin\nu & \cos\nu & 0 \\ 0 & 0 & 1 \end{bmatrix} C_I^{L0} \tag{18}$$

where C_I^{L0} is the DCM from I-frame to L-frame at initial time, and R_3 indicates the DCM rotating in z-axis with a mount of ν.

2.3 The Clohessy-Wiltshire Equation

In order to design a rendezvous guidance law, a linear relative orbit dynamic equation is extracted from the nonlinear equation. At first, consider following approximation which is the reciprocal of the gravity acceleration term in (6).

$$\begin{aligned} |r_T + \rho|^{-3} &= (r_T^2 + 2r_T \cdot \rho + \rho_T^2)^{-3/2} \\ &= |r_T|^{-3}\left(1 + \frac{2r_T \cdot \rho}{r_T^2} + \frac{\rho^2}{r_T^2}\right)^{-3/2} \\ &\simeq |r_T|^{-3}\left(1 + \frac{2r_T \cdot \rho}{r_T^2}\right)^{-3/2} \quad (\frac{\rho^2}{r_T^2} \ll 1) \\ &\simeq |r_T|^{-3}\left(1 - \frac{3r_T \cdot \rho}{r_T^2}\right) \quad (1+x)^n \approx 1 + nx \end{aligned} \tag{19}$$

Substituting (19) to (6), gives next.

$$\begin{aligned} {}^L\ddot{\boldsymbol{\rho}} = \frac{\mu_e}{|r_T|^3}\frac{3r_T \cdot \rho}{r_T^2}(\boldsymbol{r}_T + \boldsymbol{\rho}) - \frac{\mu_e}{|r_T|^3}\boldsymbol{\rho} + \boldsymbol{a}_T \\ - (\boldsymbol{\alpha} \times \boldsymbol{\rho}) - 2(\boldsymbol{\omega} \times \dot{\boldsymbol{\rho}}) - \boldsymbol{\omega} \times (\boldsymbol{\omega} \times \boldsymbol{\rho}) \end{aligned} \tag{20}$$

When an object flies a circular orbit, it undergoes centripetal accelerations. In this case this acceleration should be equal to the gravity, then frequency of the satellite orbit becomes like,

$$a_n = \frac{\mu_e}{r_T^2} = r_T \omega_T^{*2} \rightarrow \omega_T^* = n \triangleq \sqrt{\frac{\mu_e}{r_T^3}}. \tag{21}$$

Here we, by construction, defined the mean motion frequency n, which is related to a concept, called mean motion frequency in space dynamic.

As you see from (15), when the radial component of acceleration is zero, that is, a satellites flies a perfect circular orbit, the rate of transportation is same as the mean motion frequency.

Now, substituting from (7) to (11) and (21), to (20), arrange it component-wise, then it becomes next.

$$\begin{aligned} &\ddot{x} - 2\omega_T\dot{y} - (\omega_T + 2n^2)x - \dot{\omega}_T y - \frac{3n^2x^2}{r_T} = a_x \\ &\ddot{y} + 2\omega_T\dot{x} - (\omega_T^2 - n^2)y + \dot{\omega}_T x - \frac{3n^2xy}{r_T} = a_y \\ &\ddot{z} + n^2 z - \frac{3n^2xz}{r_T} = a_z \end{aligned} \tag{22}$$

In general, the relative position components x, y and z are quite small relative to r_T, thus the last terms on the left hand side of each equation in (22) are negligible. If the orbit is quite similar to a circular orbit, the rate of ω_T is ignorable too. Then we have following linear relative orbit dynamic equation.

$$\begin{aligned} &\ddot{x} - 2n\dot{y} - 3n^2x = a_x \\ &\ddot{y} + 2n\dot{x} = a_y \\ &\ddot{z} + n^2z = a_z \end{aligned} \tag{23}$$

This equation is called as the *Clohessy-Wiltshire equation*. It is a linear state space model of the relative distance and speed between the chaser and the target on L-frame.

3 Optimal Rendezvous Guidance

3.1. Linear Quadratic Control

Let's define a linear quadratic control problem (LQC), with fixed terminal time and terminal states. For a given system, it supposed to find the control input u that minimizes following cost function while satisfies the terminal states.

$$\min_u J = \frac{1}{2}\int_0^{t_f} u^T R_{uu} u dt \tag{24}$$

Subject to

$$x(t_f) = x_f \quad x(t_0) = x_0 \quad \dot{x} = Ax + Bu \tag{25}$$

where $x \in \mathbb{R}^n$ is state, $u \in \mathbb{R}^m$ is control input, $R_{uu} \in \mathbb{R}^{n\times n}$ is cost weighting. This LQC problem can be solved by the calculus of variation. First, define the Hamiltonian of given object function as follows

$$\mathcal{H} \triangleq \frac{1}{2}u^T R_{uu} u + \lambda^T (Ax + Bu) \tag{26}$$

$\lambda \in \mathbb{R}^n$ is costates. The necessary condition of critical point is the Karush-Kuhn-Tucker condition (KKT),

$$\mathcal{H}_u = R_{uu}u + B^T\lambda = 0 \tag{27}$$

$$\mathcal{H}_\lambda = Ax + Bu = \dot{x} \tag{28}$$

$$-\mathcal{H}_x = -A^T\lambda = \dot{\lambda} \tag{29}$$

where subscriptions on the Hamiltonian means a partial derivatives of it with the subscribed variables. For the solution to have the optimality, R_{uu} should be positive definite. ($R_{uu} > 0$) Rearranging (27) gives,

$$u^* = -R_{uu}^{-1}B^T\lambda \tag{30}$$

Merging (28) and (29) gives the augment of costate-sate equation as,

$$\begin{bmatrix} \dot{x} \\ \dot{\lambda} \end{bmatrix} = \begin{bmatrix} A & -BR_{uu}^{-1}B^T \\ 0 & -A^T \end{bmatrix} \begin{bmatrix} x \\ \lambda \end{bmatrix} \triangleq F \begin{bmatrix} x \\ \lambda \end{bmatrix} \tag{31}$$

The blocked transition matrix is nominated as system matrix F. This is an unforced linear system state space model, whose well-known solution is,

$$\begin{gathered} \begin{bmatrix} x(t) \\ \lambda(t) \end{bmatrix} = e^{F(t-t_0)} \begin{bmatrix} x_0 \\ \lambda(t_0) \end{bmatrix} \\ e^{F(t-t_0)} = \Phi(t,t_0) = \begin{bmatrix} \Phi_{xx} & \Phi_{x\lambda} \\ \Phi_{\lambda x} & \Phi_{\lambda\lambda} \end{bmatrix}. \end{gathered} \tag{32}$$

The exponential matrix of F is nominated as a state transition matrix Φ (STM), that has information of state propagation from initial time t_0 to a given time t, and it is useful to express STM with blocks of matrices like above. Thus the final state is simply described as,

$$\begin{bmatrix} x_f \\ \lambda(t_f) \end{bmatrix} = \Phi(t_f,t_0) \begin{bmatrix} x_0 \\ \lambda(t_0) \end{bmatrix} \tag{33}$$

Here, we already know the initial and final states, x_0 and x_f, as they are given boundary conditions. On the other hand, the initial and final costates, $\lambda(t_0)$ and $\lambda(t_f)$, are unknown variable. Therefore, we have 2n system of equations with 2n unknown variables. The remaining problem is just how to solve this system of equations.

If we manipulated the STM like a blocked matrices, (33) is turn into,

$$\begin{aligned} x_f &= \Phi_{xx}(t_f,t_0)x_0 + \Phi_{x\lambda}(t_f,t_0)\lambda(t_0) \\ \lambda(t_f) &= \Phi_{\lambda x}(t_f,t_0)x_0 + \Phi_{\lambda\lambda}(t_f,t_0)\lambda(t_0) \end{aligned}. \tag{34}$$

where the subscription, $x\lambda$, indicates the transition from initial λ to final x, and so on. By rearranging it, the initial and final costates are computed as follows,

$$\begin{aligned} \lambda(t_0) &= \Phi_{x\lambda}^{-1}(t_f,t_0)(x_f - \Phi_{xx}(t_f,t_0)x_0) \\ \lambda(t_f) &= \Phi_{\lambda x}(t_f,t_0)x_0 + \Phi_{\lambda\lambda}(t_f,t_0)\lambda(t_0) \end{aligned}. \tag{35}$$

Then, with (32) and (35),

$$\begin{aligned} x(t) &= \Phi_{xx}x_0 + \\ &\quad \Phi_{x\lambda}\Phi_{x\lambda}^{-1}(t_f,t_0)(x_f - \Phi_{xx}(t_f,t_0)x_0) \\ \lambda(t) &= \Phi_{\lambda x}x_0 + \\ &\quad \Phi_{\lambda\lambda}\Phi_{x\lambda}^{-1}(t_f,t_0)(x_f - \Phi_{xx}(t_f,t_0)x_0) \end{aligned} \tag{36}$$

Now by substituting costate value at a given time, $\lambda(t)$, to (30), one can obtain an energy optimal control history on LQC problem.

3.2 Optimal Rendezvous Guidance

Now we adjust the LQC problem to rendezvous mission of two satellites. The satellites dynamic is expressed by the Clohessy-Wiltshire equation,

$$\dot{\mathbf{x}} = A\mathbf{x} + B\mathbf{u} \tag{37}$$

$$\mathbf{x} = \begin{bmatrix} x & y & z & \dot{x} & \dot{y} & \dot{z} \end{bmatrix}^T \tag{38}$$

$$A = \begin{bmatrix} \{0\}_{3\times3} & & & I_{3\times3} & & \\ 3n^2 & 0 & 0 & 0 & 2n & 0 \\ 0 & 0 & 0 & -2n & 0 & 0 \\ 0 & 0 & -n^2 & 0 & 0 & 0 \end{bmatrix}, \quad B = \begin{bmatrix} \{0\}_{3\times3} \\ I_{3\times3} \end{bmatrix} \tag{39}$$

where the mean motion angular rate, n, is updated by time. For this linear system, a modification of the LQC is utilized to feedback current states by time.

The time left until rendezvous is defined as time to go as follows.

$$t_f - t = t_{go} \tag{40}$$

The system matrix is computed by (31), then the STM is approximated as follows,

$$e^{F(t_f-t)} = \Phi(t_{go}) \simeq I + Ft_{go} + \frac{1}{2!}F^2t_{go}^2 + \frac{1}{3!}F^3t_{go}^3 \tag{41}$$

This imitation form of STM is quite accurate when the eigen values of F exist on the left half plane of imaginary plane. By substituting t_{go}, it indicates state transition from current t to the final t_f and the STM is function of the t_{go}. The STM from t_0 to t_f can be separated with two STMs; STM from t to t_f, and STM from t_0 to t. The STM from t to t_f is named as the STM-t_{go}. Then the block of matrices are expressible as follows,

$$\begin{aligned} \Phi(t_f,t_0) &= \Phi(t_{go})\Phi(t,t_0) \\ \Phi_{xx}(t_f,t_0) &= \Phi_{xx}(t_{go})\Phi_{xx} + \Phi_{x\lambda}(t_{go})\Phi_{\lambda x} \\ \Phi_{x\lambda}(t_f,t_0) &= \Phi_{xx}(t_{go})\Phi_{x\lambda} + \Phi_{x\lambda}(t_{go})\Phi_{\lambda\lambda} \end{aligned}. \tag{42}$$

Now substitute (42) to (32), rearrange current costate $\lambda(t)$ to be expressed with the current states $x(t)$ and the STM-t_{go}.

$$\begin{aligned} &\lambda(t) \\ &= \Phi_{\lambda x}x_0 + \Phi_{\lambda\lambda}\lambda_0 \\ &= \Phi_{x\lambda}^{-1}(t_{go})[\Phi_{xx}(t_f,t_0) - \Phi_{xx}(t_{go})\Phi_{xx}]x_0 \\ &\quad + \Phi_{x\lambda}^{-1}(t_{go})[\Phi_{x\lambda}(t_f,t_0) - \Phi_{xx}(t_{go})\Phi_{x\lambda}]\lambda_0 \\ &= \Phi_{x\lambda}^{-1}(t_{go})[(\Phi_{xx}(t_f,t_0) - \Phi_{xx}(t_{go})\Phi_{xx})x_0 \\ &\quad + (\Phi_{x\lambda}(t_f,t_0) - \Phi_{xx}(t_{go})\Phi_{x\lambda})\lambda_0] \\ &= \Phi_{x\lambda}^{-1}(t_{go})([\Phi_{xx}(t_f,t_0)x_0 + \Phi_{x\lambda}(t_f,t_0)\lambda_0] \\ &\quad - \Phi_{xx}(t_{go})[\Phi_{xx}x_0 + \Phi_{x\lambda}\lambda_0]) \\ &= \Phi_{x\lambda}^{-1}(t_{go})(x_f - \Phi_{xx}(t_{go})x(t)) \end{aligned} \tag{43}$$

Now, substitute current costate to (30), in order to get a state feedback form of suboptimal guidance law.

$$u^* = -R_{uu}^{-1}B^T\Phi_{x\lambda}^{-1}(t_{go})(x_f - \Phi_{xx}(t_{go})x(t)) \tag{44}$$

As command u^* is computed on L-frame, and the dynamic equation of the nonlinear relative orbit is expressed on L-frame too, we can directly substitute the u^* to a_T.

In the rest of paper, this optimal rendezvous guidance is utilized on the nonlinear relative orbit dynamic model for several cases of initial and final condition sets.

4 Simulation

In this section, we implement the suggested optimal guidance law on some cases of virtual rendezvous scenarios, in order to verify the method on the nonlinear orbit dynamic models. Here comes operation introduction and simulation results. Then some discussions concerned with the results are attached below.

4.1 Satellite Proximity Operation Scenarios

Consider a maintenance mission on a communication satellite (COMSAT), which is skidding the earth in a circular orbital. To extend design life of the COMSAT, a service satellite takes rendezvous with the COMSAT and do some maintenance job.

The servicing satellite will have a mobile robot for mission and it can dock to the COMSAT somehow. It is launched from ground and released near the COMSAT within the visible distance. It approaches to the COMSAT from the below, and has to be attached very slowly, not to crash or damage the COMSAT.

KOREASAT 6, also called *Olleh 1*, is a commercial COMSAT launched by the KT Corporation in 2010. It services direct broadcasting and fixed satellite service to the users of Republic of Korea within 15 years. It weighs

Table. 1. Nonlinear Rendezvous Simulation Scenarios
(x, y, z, v_x, v_y, v_z) relative orbit state in L-frame,
(r_{Tx}, r_{Ty}, v_{Tx}, v_{Ty}) target satellite state in I-frame

	CASE			
	#1	**#2**	**#3**	**#4**
x_0 **(km)**	-1	-1	-0.5	-0.5
y_0 **(km)**	-0.5	1	0.1	0.1
z_0 **(km)**	0.2	0	0	0
v_{x0} **(m/s)**	0	5	5	5
v_{y0} **(m/s)**	5	0	-5	-5
v_{z0} **(m/s)**	-5	10	0	0
r_{Tx} **(km)**	42169	42169	42169	42169
r_{Ty} **(km)**	0	0	0	0
v_{Tx} **(m/s)**	0	0	0	0
v_{Ty} **(m/s)**	4074.5	4074.5	4074.5	4074.5
t_f **(sec)**	1000	1000	1000	1500
x_f **(km)**	0	-1	0	-10
y_f **(km)**	0	0	0	0
z_f **(km)**	0	0	0	0
v_{xf} **(m/s)**	0	0.1	0	1
v_{yf} **(m/s)**	0	0	0	0
v_{zf} **(m/s)**	0	0	0	0

about 2.6 ton, and satellite period is 1437 min/rotate. It is flying at an altitude of 35791 km. For some virtual repairing missions on the KOREASAT 6, we launch a virtual satellite and release it in a near orbit. Table 1 lists each case of simulation initial conditions. The initial and final states of relative orbits are expressed in L-frame, on the other hand, the initial states of target states are described in I-frame.

4.2 Simulation Results

Fig. 2 depicts 3-dimensional trajectories of the chaser satellites of each case. It presents relative position of the chaser in L-frame. Red star marker indicates the target satellite, and green ring means initial position of the chaser. Even though the chaser moves in wrong direction at first due to initial velocity, it eventually settles down on the specified terminal position.

Each component of relative positions is shown in Fig. 3 by time. The end time of simulation is different case by case, as specified in the simulation scenarios. For case 1 to 3, it terminates on 1000 sec, on the other hand, Case 4 terminates at 1500 sec.

The suggested guidance law exactly satisfied the terminal position condition, as shown in Fig. 4. The figure displays normalized relative position errors of the

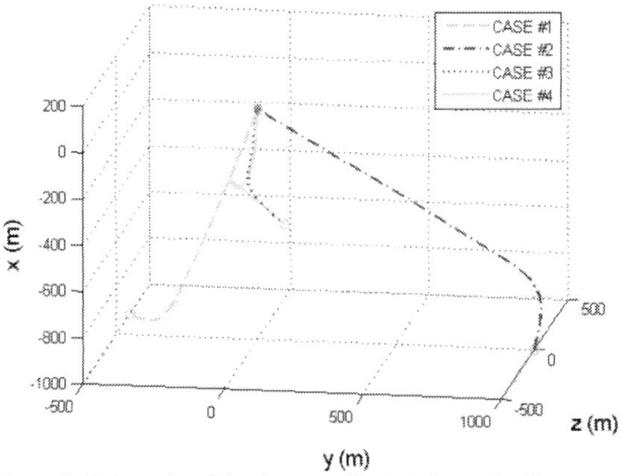

Figure 2. Trajectories of the chaser satellite in L-frame. (nonlinear relative orbit dynamic simulation)

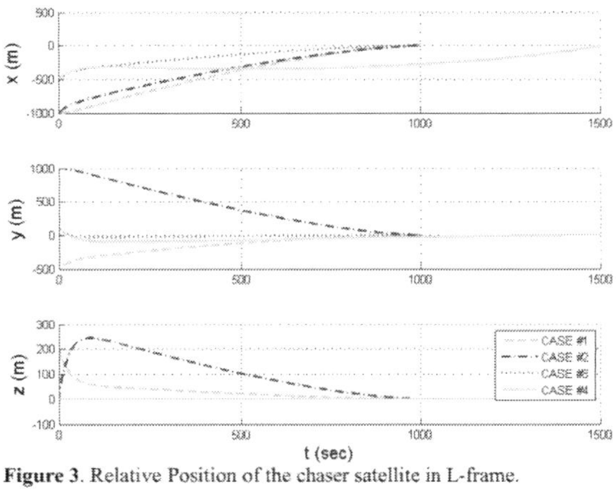

Figure 3. Relative Position of the chaser satellite in L-frame.

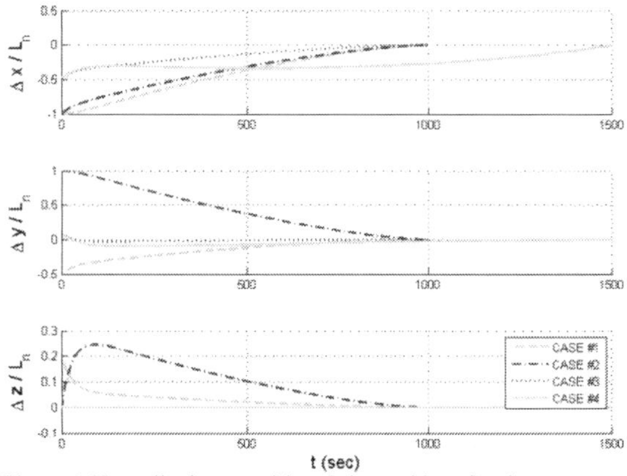

Figure 4. Normalized errors of the relative position of each component in L-frame. Presented values are $(\rho-\rho_f)/L_n$, where L_n is 1000 m.

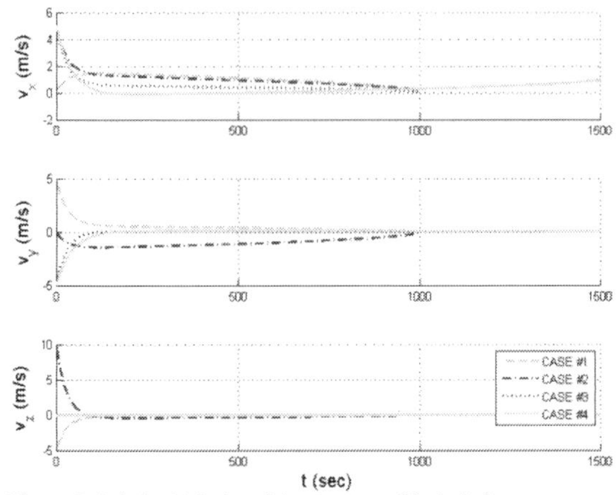

Figure 5. Relative Velocity of the chaser satellite in L-frame.

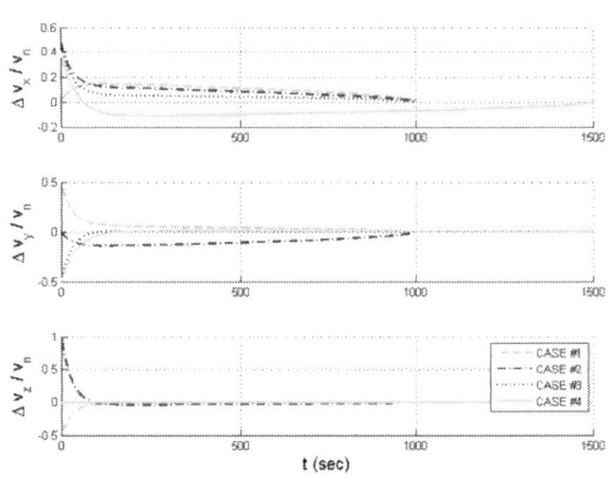

Figure 6. Normalized error of the relative velocity of each component in L-frame. Presented values are ΔV/Vn, where Vn is 10m/s.

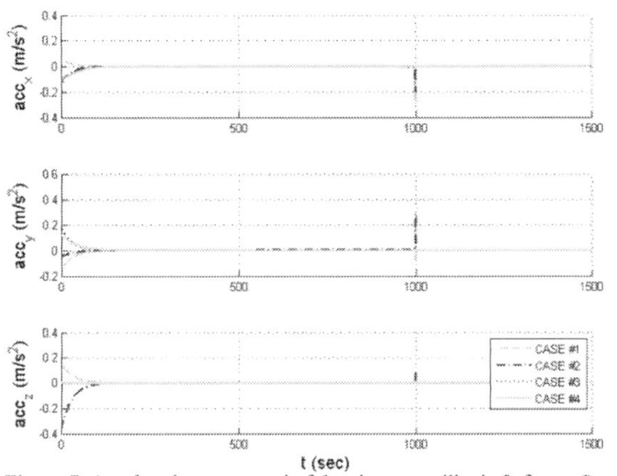

Figure 7. Acceleration command of the chaser satellite in L-frame. It shows little diverge in terminal phase.

chaser. Those errors are normalized with length normalizer L_n of 1000 m for each component. The errors all become zero on the final, therefore the chaser reaches to the specified terminal position at the given time.

Fig. 5 presents changes of the relative velocity components. In all cases, those converge to the specified terminal velocity at the nominated terminal time from the given initial values. Especially x-component velocity of CASE. 4 converges to 1 m/s, which is the specified value on Table. 1.

In order to check precisely whether the terminal velocity constraints are satisfied or not, we attach the normalized errors of the relative velocity on Fig. 6. The relative velocity error is divided by velocity normalizer V_n of 10 m/s. For all cases, the normalized errors are converged to zero, therefore the terminal velocity constraint are satisfied.

Acceleration command is displayed in Fig. 7 for each component of L-frame. It varies smoothly in most time, but it is about to be diverged at the end of simulation. However, the divergence is quite small and short, thus it does not affect the terminal state value that much. And if we consider with a thrust actuator, it may not respond to such a short and strong command due to system lag and bandwidth.

4.3 Discussion

Comparison with Other Method.

Glide Slope method is a common way of manned rendezvous. It is usually used for space shuttle rendezvous on the International Space Station. It fires sequences of jet on the local vertical or the local horizontal axis. It is quite insightful but takes long time, and impulse on the jet has to be controlled precisely in short time interval.

Linear Quadratic Regulation (LQR) problem has a cost function that has penalty not only on the input quadratic term, but also on the state quadratic term. As it doesn't have the terminal time and state constraints, the KKT condition leads it to the differential Riccati equation (DRE) of $2n+1$ variables; n final states, n initial costates and 1 terminal time. However, we alternatively solve algebraic Riccati equation (ARE), because the DRE is unable to solve analytically. The solution of ARE is identical to the optimal solution, only when $t_f \rightarrow +\infty$, thus it is always suboptimal.

Model Predictive Sequential Programming (MPSP) is a discretized nonlinear real-time optimization algorithm.

It is inspired from both the Approximate Dynamic Programming and the Model Predictive Control, so it calculates sensitivity on the terminal state of each input sequence and adjust the input sequence repeatedly. As it needs terminal states estimation and sensitivities of each states, it internally has to simulate for each input sequence updates. Additionally, the augmented sensitivity matrix getting larger as the time steps are increased, which often lead the matrix be singular. Meanwhile, the CW equation is reliable in most of rendezvous scenarios, thus the problem is suitable to be handled by the LQC. It is much faster and cheaper way to the optimal rendezvous.

Real-Time Implementation and Expectation

The guidance command was generated at every 0.01 sec interval during the simulation, which was conducted on Intel PC (i7-4770 3.4 GHz) with MATLAB 2013b. Computing 1500 sec simulation takes 135 sec at the slowest. Thus we can assume onboard flight computers of satellites may have a chance to calculate the command just in time also.

The major computation load of the algorithm occurs at matrix inversion and exponential matrix calculation. Here we approximated the exponential matrix to the third order of Taylor series expansion. However, if we can analytically solve the exponential of the system matrix and the inversion of $\Phi_{x\lambda}$, it will much fasten the algorithm.

In the view of system integration, measurement errors, system uncertainties, and control system lag may cause little performance degradation. To ensure the system safety, further studies on stability margin are required.

5 Conclusion

A linear quadratic control problem is used to design optimal rendezvous guidance law. The LQC minimizes the integration of input sequence energy from t_0 to t_f. The LQC is solved by the KKT condition. In this work, we proved that the current costate obtained by the feedback of current state results in identical costate of the optimal trajectory, so that, we can compute the guidance command in state feedback form. Proposed algorithm utilized the Clohessy-Wiltshire equation as a governing equation, which is a linear system equation. It is quite similar compared to the nonlinear orbit dynamic model, as the nonlinear effect is negligible. By solving the LQC for CW equation, this paper made an optimal rendezvous guidance law. It is shown to be useful to satisfy the terminal time and states through the four cases of simulations.

Acknowledgment

This work was supported by a grant (No. NRF-2012 M1A3A3A02033464) from the National Research Foundation of Korea (NRF) funded by the Korean Ministry of Science, ICT and Future Planning.

References

1. W. H. Clohessy, R. S. Wiltshire, "Terminal guidance system for satellite rendezvous," J. Aerospace Sci. vol. 27, no. 9, 1960, pp. 653-658.
2. S. H. Kim, H. L. Choi, and D. H. Shim, "Mono-vission based satellite to satellite relative navigation using region-based active contour method," 2014 Korea Society for Aeronautical & Space Science Autumn Conference, Proceeding, pp. 333-336, Jeju, Korea, 2014
3. S. H. Kim, H. L. Choi, and D. H. Shim, "A study of mono-vision based relative navigation for satellite proximity operation," 2013 Korea Society for Aeronautical & Space Science Autumn Conference, Proceeding, pp. 714-717, Jeju, Korea, 2013
4. A. H. Owis, M. A. Amer, "Satellite formation control using the approximating sequence riccati equations," Theory and Application of Mathematics & Computer Science, vo. 3, no. 2, pp 103-113, 2013.
5. Y. Z. Luo, and G. J. Tang, "Spacecraft optimal rendezvous controller design using simulated annealing," Aerospace Science and Technology, vol. 9, 2005, pp. 732-737.
6. K. Morris, *Introduction to Feedback Control,* London, UK: Harcourt Academic Press, 2001, pp. 173-181.
7. D. E. Kirk, *Optimal Control Theory: An Introduction*, New York, NY: Dover Publication Inc., 1970, pp. 200-201.
8. D. A. Spencer, "Automated trajectory control for proximity operations using relative orbital elements," Ph. D. Dissertation, Daniel Guggenheim School of Aerospace Engineering, Georgia Institute of Technology, Atlanta, GA, May 2015.
9. R. Padhi, and M. Kothari, "Model predictive static programming: A computationally efficient technique for suboptimal control design," International Journal of Innovative Computing, Information and Control, vol 5, no. 2, pp. 399-411, Feb. 2009.
10. M. Kothari, R. Padhi, "A nonlinear suboptimal robust guidance scheme for long range flight vehicle with solid motors," Automatic Control in Aerospace, vol. 3, no. 1, 2010.

MATEC Web of Conferences **54**, 09003 (2016) DOI: 10.1051/matecconf/20165409003
MIMT 2016

Lunar Lander Landing Site Decision in Low-Fuel Case

Min-Guk Seo, Seong-Min Hong and Min-Jea Tahk

KAIST, Department of Aerospace Engineering, Daejeon, Korea

Abstract. This paper deals with lunar landing site decision algorithms for the case when the lunar lander is not able to reach the original landing site. A new landing site which minimizes the fuel consumption of the lunar rover is determined from landing site candidates located within the maximum reachable downrange of the lunar lander. A path planning algorithm is introduced to determine the new landing site which has the shortest path to the original landing site among the candidates. Numerical simulations are conducted to verify the performance of the proposed algorithm.

1 Introduction

In a lunar landing mission, the lander may need a new landing site if the landing on the original landing site is not possible due to various disturbances during the operation. A good candidate for the new landing site is an area from which the rover is able to travel to the original landing site, where the mission should take place, with minimum energy consumption. A new landing site decision algorithm is designed with full information of obstacles to satisfy this requirement. The first step of the algorithm is to obtain the safe reachable region of a lunar lander by solving the landing trajectory optimization problem for maximum downrange. The new landing site is chosen among all the landing site candidates as the one whose path to the original landing site is determined to be the shortest one by applying path planning algorithm.

Landing guidance is an important issue in the planetary exploration field. The common objective of planetary landing guidance algorithms of previous studies is to precisely land on the predetermined landing site with minimum fuel consumption[1-5]. This implies that those algorithms are not suitable to figure out the maximum downrange of the lunar lander under fuel limits. In this paper, the optimal landing guidance problem is reformulated as a downrange maximization problem. The dynamics of the lunar lander is simply modeled in a 2-dimensional local Cartesian coordinate system. The optimal solution is obtained using Gauss Pseudospectral Optimization Solver (GPOPS). The outer-most contour points of the reachable landing region, which are new landing site candidates, are determined from the maximum reachable downrange and the obstacle information.

Path planning algorithms with obstacle avoidance studied in many previous articles[6-13] do not focus on guaranteeing shortest path. This paper suggests a new path planning algorithm especially for the lunar rover to minimize path length. The obstacles on the lunar surface are assumed to be given as circles and polygons. The resulting path map consists of straight line segments and circular arcs. The shortest path starting from the landing site is obtained by applying Dijkstra's algorithm[14]. The landing site candidate with the minimum path length is chosen as the new landing site.

This paper is organized as follows. The landing trajectory optimization for downrange maximization is handled in Section 2. The path planning algorithm for lunar lander is introduced in Section 3. Simulation results are given in Section 4 for the performance analysis of the algorithm. The overall conclusion of this paper is addressed in Section 5.

2 Landing Trajectory Optimization for Downrange Maximization

In the case when a lunar lander cannot reach the original landing site where exploration missions are supposed to be performed, a landing site decision algorithm is needed to find a new landing site. One requirement of the new landing site is that the lunar lander should be able to land on it safely with limited amount of fuel of the thruster and the attitude control system. This implies that the landing site candidates should be the points where the lunar lander can make a landing and not on obstacle regions. The lunar rover has to travel to the original landing site to perform its missions after the lunar lander reaches the new landing site. In order to minimize the energy and the time consumption, the point which has the shortest path to the original landing site is chosen as the new landing site among candidate points.

As mentioned previously, the points in the reachable region for lunar lander on the lunar surface are candidates of the new landing site. Since the original landing site is supposed to be outside of the reachable region in this paper, all the paths from the candidates should cross the outer-most contour of the reachable landing region. This means that the landing site candidates are limited to the points on the outer-most contour. The landing trajectory optimization for downrange maximization is utilized to figure out the contour.

Several assumptions are considered to define landing trajectory optimization problem.

1) The motion of the lunar lander is described in inertial Cartesian coordinate system with its origin on the lunar surface, which is assumed to be a flat plane.

2) The altitude axis is normal to the lunar surface and the other axes are on the surface.

3) The effect due to the rotation of the moon is neglected and the uniform gravitational acceleration is considered.

4) The initial velocity component normal to the altitude axis and the attitude angle rate are 0.

5) The major axis of the lunar lander, the axis parallel to the main thruster direction, is parallel to the altitude axis at the initial time.

6) The rotation with respect to any axis normal to the major axis is possible by the combination of the attitude control system.

Under those assumptions, 1) ~ 6), the trajectory optimization problem is defined in the 2-dimensional inertial Cartesian coordinate system, and the outer-most contour is defined as a circle centered at the original landing site with the maximum downrange as the radius. The effect of altitude control thrusters on the lunar lander is velocity change, and the attitude control is conducted by attitude control system equipped on the lander. The control inputs of the lunar lander are defined as main thruster and attitude thruster in this model The dynamic model of the lunar lander is defined from all the assumptions given above.

$$\begin{aligned}
\dot{x} &= V_x \\
\dot{h} &= V_h \\
\dot{V}_x &= \frac{T_D \cos\theta}{m} \\
\dot{V}_h &= \frac{T_D \sin\theta}{m} - g \\
\dot{m} &= -\frac{(T_D + T_{att})}{I_{sp} g} \\
\dot{\theta} &= q \\
\dot{q} &= \frac{T_{att} \cdot L_T}{I_{zz}}
\end{aligned} \tag{1}$$

where x and h are downrange and altitude of lunar lander. V_x and V_h denote velocity components in each axis direction. Mass, moment of inertia, moment arm of the attitude thruster, attitude angle, and attitude angle rate of lunar lander are symbolized as m, I_{zz}, L_T, θ, and q. The attitude angle is defined as the angle between the major axis and the downrange axis. I_{sp} is specific impulse of lunar lander engine, which is used at the main thruster and attitude control thruster, and g is the gravitational acceleration on the lunar surface. The control input variables are main thrust, T_D, and attitude control thrust, T_{att}.

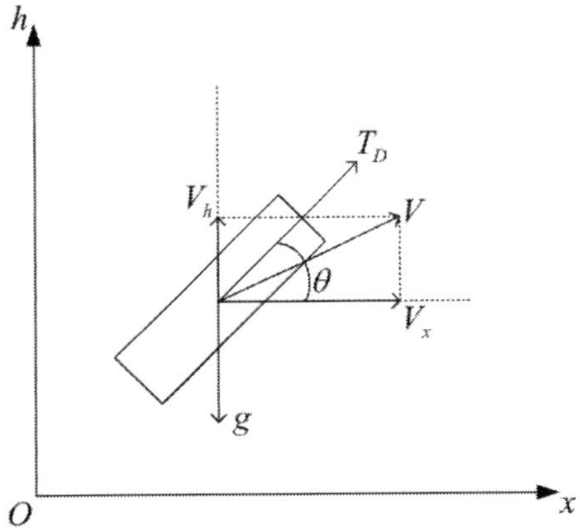

Figure 1. Coordinate System and State Variable Definitions.

The objective function is defined to maximize the magnitude of x.

$$\min_{T_D, T_{att}} -|x| \tag{2}$$

The initial conditions of state variables are given from the assumptions addressed before.

$$\begin{aligned}
&x(t_0) = x_0 \quad h(t_0) = h_0 \quad V_x(t_0) = 0 \quad V_h(t_0) = 0 \\
&m(t_0) = m_0 \quad \theta(t_0) = 90^\circ \quad q(t_0) = 0
\end{aligned} \tag{3}$$

The altitude of the lunar lander is bounded to be larger than 0 to protect it from crashing on the lunar surface. The sensors like cameras on the lunar lander measure the information of the lunar surface for safe landing and hazard avoidance. The attitude angle and its rate are limited to consider the operation performance of those sensors.

$$\begin{aligned}
&h(t_f) = 0 \quad V_x(t_f) = 0 \quad V_h(t_f) = 0 \\
&\theta(t_f) = 90^\circ \quad q(t_f) = 0
\end{aligned} \tag{4}$$

$$\begin{aligned}
&h \geq 0 \\
&\theta_{min} \leq \theta \leq \theta_{max} \\
&q_{min} \leq q \leq q_{max}
\end{aligned} \tag{5}$$

The main thrust is limited by the performance of the actuator. The attitude control thruster is also limited under actuator performance, but it can be a negative value considering both direction of the thruster.

$$\begin{aligned}
&T_{D_{min}} < T_D < T_{D_{max}} \\
&T_{att_{min}} < T_{att} < T_{att_{max}}
\end{aligned} \tag{6}$$

The restriction on the amount of fuel is expressed as a mass limit of the lunar lander.

$$m_{min} \leq m \leq m_{max} \tag{7}$$

The solution of the optimization problem defined in (1)-(7) is obtained by applying a proper optimization algorithms. GPOPS is the optimization tool utilized in this paper. This tool is based on the variable-order Gaussian quadrature method and deals with continuous-time optimal control problems. The outer-most contour of the reachable region is decided from the optimal solution as the circle on the lunar surface centered at the initial position on the surface with the maximum downrange as radius. Assuming that the information of obstacles are given, the landing site candidates are defined as the points dividing this contour into certain small and equal intervals and not located on the obstacles.

3 Path Planning Algorithm for Lunar Rover

As addressed before, the new landing site is the one among candidates with the minimum path length to the original landing site. The path planning algorithm between two points is required for this process. The designed algorithm should not only ensure the minimum path length but also guarantee the safety of lander by planning the path not to cross the obstacles on the lunar surface.

The path planning algorithm introduced in this paper starts from constructing path candidates from a landing site candidate to the original landing site. The obstacles on the lunar surface are assumed to be given as geometrical figures as stated below. Each of the original landing site and a landing site candidate is able to be expressed as a point on the lunar surface. The craters are able to be approximated as circles[15]. The circular obstacle centered at (x_c, y_c) with the radius of r_c is expressed as $C(x_c, y_c, r_c)$ in this paper. The information of other obstacles which are not round, like rocks, are able to be detected from the sensor measurements[16]. Each of those obstacles is able to be approximated as a polygon, which is figure consists of straight segments, surrounding it[17].

In the case when the obstacles are circles, the minimum path is consists of some of following path segment candidates: tangent lines from two end points to the circles, common tangent lines of two arbitrary selected circles, and circular arc connecting points of contact on each circle[18]. When the obstacles are polygons, path segment candidates are as follows: straight segments between two end points and the vertexes of the polygons, straight segments between two vertexes from different polygons, and the sides of the polygons[19]. Since the obstacles on the lunar surface includes both circles and polygons, the group of path segment candidates consists of all the sorts of them stated above and the tangent lines from the vertexes of the polygons to the circular obstacles.

The straight line connecting two points, (x_1, y_1) and (x_2, y_2), are given as follows.

$$a_{PP}x + b_{PP}y + c_{PP} = 0 \tag{8}$$

where

$$\begin{aligned} a_{PP} &= y_2 - y_1 \\ b_{PP} &= -(x_2 - x_1) \\ c_{PP} &= -x_1(y_2 - y_1) + y_1(x_2 - x_1) \end{aligned} \tag{9}$$

The path segment is defined by (8) and two end points, (x_1, y_1) and (x_2, y_2).

The two tangent lines to the circle $C(x_c, y_c, r_c)$ from the point (x_p, y_p) outside of the circle are obtained as below.

$$a_{CP}x - y + c_{CP} = 0 \tag{10}$$

where

$$\begin{aligned} a_{CP} &= \frac{f_{CP} \pm \sqrt{f_{CP}^2 - g_{CP}}}{(x_c - x_p)^2 - r_c^2} \\ c_{CP} &= y_p - ax_p \\ f_{CP} &= (x_c - x_p)(y_c - y_p) \\ g_{CP} &= \left[(x_c - x_p)^2 - r_c^2\right]\left[(y_c - y_p)^2 - r_c^2\right] \end{aligned} \tag{11}$$

The points of contact on the tangent lines, (x_t, y_t), obtained as (10) and (11) are calculated as,

$$\begin{aligned} x_t &= \frac{f_{x_t} \pm \sqrt{f_{x_t}^2 - g_{x_t}}}{a_{CP}^2 + 1} \\ y_t &= a_{CP}x_t + c_{CP} \end{aligned} \tag{12}$$

where

$$\begin{aligned} f_{x_t} &= \left[a_{CP}(c_{CP} - y_c) - x_c\right] \\ g_{x_t} &= (a_{CP}^2 + 1)\left[x_c^2 + (c_{CP} - y_c)^2 - r_c^2\right] \end{aligned} \tag{13}$$

The equations of path segment candidates are given as (10), and the end points of them are (x_p, y_p) and (x_t, y_t).

The common tangent lines between two circles, $C(x_{c_1}, y_{c_1}, r_{c_1})$ and $C(x_{c_2}, y_{c_2}, r_{c_2})$, are given as follows.

$$a_{CC}x + b_{CC}y + c_{CC} = 0 \tag{14}$$

where

$$\begin{aligned} a_{CC} &= \frac{f_{x_c} f_{r_c} + f_{y_c}\sqrt{f_{x_c}^2 + f_{y_c}^2 - f_{r_c}^2}}{f_{x_c}^2 + f_{y_c}^2} \\ b_{CC} &= \frac{f_{y_c} f_{r_c} - f_{x_c}\sqrt{f_{x_c}^2 + f_{y_c}^2 - f_{r_c}^2}}{f_{x_c}^2 + f_{y_c}^2} \\ c_{CC} &= -\left(x_{c_2}a_{CC} + y_{c_2}b_{CC} \pm r_{c_2}\right) \end{aligned} \tag{15}$$

$$f_{x_c} = (x_{c_1} - x_{c_2}) \qquad f_{y_c} = (y_{c_1} - y_{c_2}) \qquad f_{r_c} = (\pm r_{c_1} \pm r_{c_2}) \tag{16}$$

The point of contact on $C(x_{c_1}, y_{c_1}, r_{c_1})$, (x_{t_1}, y_{t_1}), and one on $C(x_{c_2}, y_{c_2}, r_{c_2})$, (x_{t_2}, y_{t_2}), are calculated as,

$$\begin{aligned} x_{t_1} &= x_{c_1} - (\pm r_{c_1}a_{CC}) \qquad & y_{t_1} &= y_{c_1} - (\pm r_{c_1}a_{CC}) \\ x_{t_2} &= x_{c_2} + (\pm r_{c_2}a_{CC}) \qquad & y_{t_2} &= x_{c_2} + (\pm r_{c_2}a_{CC}) \end{aligned} \tag{17}$$

The mathematical signs in front of the same variable are equal. This results in four common tangent lines except for $(x_{c_1} - x_{c_2})^2 + (y_{c_1} - y_{c_2})^2 < (r_{c_1} + r_{c_2})^2$ case. Two tangent lines, obtained when the mathematical signs in front of r_{c_1} and r_{c_2} of (15) are different from each other, and the points of contact on them are possible in this case. The equations of path segment candidates are expressed as (14). (x_{t_1}, y_{t_1}) and (x_{t_2}, y_{t_2}) are the end points of them.

The path candidates should be designed not to pass through any of obstacles for safety of the lunar rover. This implies that the path segment candidates, whose part is included in any of the obstacles, should be excluded. The acceptability test is designed for an arbitrary path segment candidate, denoted by two end points, (x_1, y_1) and (x_2, y_2) , and the equation of straight line, $ax+by+c=0$.

The first process of this test deals with circular obstacles. The distance between an arbitrary path segment candidate and the center of an arbitrary circular obstacle, $C(x_c, y_c, r_c)$, is calculated as below.

$$d = \frac{ax_c + by_c + c}{\sqrt{a^2+b^2}} \tag{18}$$

It is obvious that the path segment candidate does not pass through the circular obstacle in $d \geq r_c$ case. Two intersection points of $ax+by+c=0$ and $C(x_c, y_c, r_c)$, $(x_{i_{LC}}, y_{i_{LC}})$, are calculated when $d < r_c$ in order to figure out the feasibility of the segment.

$$x_{i_{LC}} = \frac{f_{i_{LC}} \pm \sqrt{f_{i_{LC}}^2 - h_{i_{LC}} g_{i_{LC}}}}{h_{i_{LC}}}$$
$$y_{i_{LC}} = -\frac{a}{b}x_{i_{LC}} - \frac{c}{b} \tag{19}$$

where

$$f_{i_{LC}} = -\frac{a}{b}\left(\frac{c}{b} + y_c\right) + x_c$$
$$g_{i_{LC}} = x_c^2 + \left(\frac{c}{b} + y_c\right)^2 - r_c^2 \tag{20}$$
$$h_{i_{LC}} = \left(\frac{a}{b}\right)^2 + 1$$

The path segment candidate is excluded when any of two $x_{i_{LC}}$ satisfies $x_1 < x_{i_{LC}} < x_2$ since this means that some part of it is located inside of $C(x_c, y_c, r_c)$. This process is applied to all the possible combinations of a path segment candidate and a circular obstacle.

The second process, which is applied to the path segment candidates not excluded in the first process, handles polygon obstacles. In the case when some part of a path segment candidate is laid inside of a polygon obstacle, it should intersect with one or more sides of this polygon. The intersection point of an straight line, which arbitrary path segment is included in, and a straight line passing through two vertexes, (x_{P_1}, y_{P_1}) and (x_{P_2}, y_{P_2}), is expressed as follows.

$$x_{i_{LP}} = \frac{(x_1y_2 - y_1x_2)(x_{P_1} - x_{P_2}) - (x_1 - x_2)(x_{P_1}y_{P_2} - y_{P_1}x_{P_2})}{(x_1 - x_2)(y_{P_1} - y_{P_2}) - (y_1 - y_2)(x_{P_1} - x_{P_2})}$$
$$y_{i_{LP}} = -\frac{a}{b}x_{i_{LP}} - \frac{c}{b} \tag{21}$$

The path segment candidate is excluded when this intersection point is laid on it, $x_1 < x_{i_{LP}} < x_2$, and the side of a polygon obstacle with those vertexes as the end points, $x_{P_1} < x_{i_{LP}} < x_{P_2}$, at the same time. This process is applied to all the arbitrary combinations of a path segment candidate and a side of polygon obstacle. The safe path segment candidates are obtained from the acceptability test.

The construction of path candidates is completed by connecting the end points of the safe path segment candidates, which are called as nodes in the later part of this paper, on obstacles. The nodes on polygon obstacles are obviously their vertexes. This implies that all the sides of every polygon obstacles should be included in the group of path segment candidates. The path segment candidates on circular obstacles are obtained as the circular arcs between two nearby nodes on the same circular obstacle. When a part of a circular obstacle is overlapped by another circular obstacle, the path segment candidates on it and containing the overlapped part of its contour should be excluded. Two circular obstacles, $C(x_{c_1}, y_{c_1}, r_{c_1})$ and $C(x_{c_2}, y_{c_2}, r_{c_2})$, are defined to be overlapped when the following condition is satisfied.

$$\sqrt{(x_{c_1} - x_{c_2})^2 + (y_{c_1} - y_{c_2})^2} < (r_{c_1} + r_{c_2}) \tag{22}$$

The intersection points of those two circles, $(x_{i_{CC}}, y_{i_{CC}})$, are obtained as below.

$$x_{i_{CC}} = \frac{f_{i_{CC}} \pm \sqrt{f_{i_{CC}}^2 - h_{i_{CC}} g_{i_{CC}}}}{h_{i_{CC}}}$$
$$y_{i_{CC}} = -\frac{a_{i_{CC}}}{b_{i_{CC}}}x_{i_{CC}} - \frac{c_{i_{CC}}}{b_{i_{CC}}} \tag{23}$$

where

$$a_{i_{CC}} = -2(x_{c_1} - x_{c_2})$$
$$b_{i_{CC}} = -2(y_{c_1} - y_{c_2})$$
$$c_{i_{CC}} = x_{c_1}^2 - x_{c_2}^2 + y_{c_1}^2 - y_{c_2}^2 - r_{c_1}^2 + r_{c_2}^2$$
$$f_{i_{CC}} = x_{c_1} - \frac{a_{i_{CC}}}{b_{i_{CC}}}\left(\frac{c_{i_{CC}}}{b_{i_{CC}}} + y_{c_1}\right) \tag{24}$$
$$g_{i_{CC}} = x_{c_1}^2 + \left(\frac{c_{i_{CC}}}{b_{i_{CC}}} + y_{c_1}\right)^2 - r_{c_1}^2$$
$$h_{i_{CC}} = 1 + \left(\frac{a_{i_{CC}}}{b_{i_{CC}}}\right)^2$$

Two circular arcs on $C(x_{c_1}, y_{c_1}, r_{c_1})$, called 'arc A' and 'arc B' connecting two intersection points given in (20) exist; one of them is located inside of $C(x_{c_2}, y_{c_2}, r_{c_2})$, and the other one is not. The overlapped arc is figured out from the center of arc A, (x_{ARC_A}, y_{ARC_A}), and the center of arc B, (x_{ARC_B}, y_{ARC_B}).

$$x_{ARC_A} = x_{c_1} + r_{c_1}\frac{\lambda_x}{\sqrt{\lambda_x^2+\lambda_y^2}}$$
$$y_{ARC_A} = y_{c_1} + r_{c_1}\frac{\lambda_y}{\sqrt{\lambda_x^2+\lambda_y^2}}$$
$$x_{ARC_B} = x_{c_1} - r_{c_1}\frac{\lambda_x}{\sqrt{\lambda_x^2+\lambda_y^2}} \tag{25}$$
$$y_{ARC_B} = y_{c_1} - r_{c_1}\frac{\lambda_y}{\sqrt{\lambda_x^2+\lambda_y^2}}$$

where

$$\lambda_x = \frac{x_{c_1} - \frac{a_{i_{CC}}}{b_{i_{CC}}}\left(\frac{c_{i_{CC}}}{b_{i_{CC}}} + y_{c_1}\right)}{1+\left(\frac{a_{i_{CC}}}{b_{i_{CC}}}\right)^2} - x_{c_1}$$
$$\lambda_y = -\frac{\frac{a_{i_{CC}}}{b_{i_{CC}}}\left(x_{c_1} - \frac{a_{i_{CC}}}{b_{i_{CC}}}\left(\frac{c_{i_{CC}}}{b_{i_{CC}}} + y_{c_1}\right)\right)}{1+\left(\frac{a_{i_{CC}}}{b_{i_{CC}}}\right)^2} - \frac{c_{i_{CC}}}{b_{i_{CC}}} - y_{c_1} \tag{26}$$

If $\sqrt{\left(x_{ARC_A}-x_{c_2}\right)^2+\left(y_{ARC_A}-y_{c_2}\right)^2} < r_{c_2}$, the path segment candidates containing 'arc A' should be excluded. The ones containing 'arc B' are excluded when $\sqrt{\left(x_{ARC_B}-x_{c_2}\right)^2+\left(y_{ARC_B}-y_{c_2}\right)^2} < r_{c_2}$. The same process is needed to be applied to $C\left(x_{c_2}, y_{c_2}, r_{c_2}\right)$ for the verification of an unacceptable circular arc.

The optimal path from a landing site candidate to the original landing site is determined by applying Dijkstra's algorithm[14]. The visibility adjacency matrix is constructed from the lengths of the path segment candidates and the coordinates of the nodes.

The optimal path planning algorithm explained in this section is applied to all the landing site candidates obtained in Section 2. This provides the lengths of the path starting from all the landing site candidates. The one with the minimum path length is decided as the optimal landing site.

4 Simulation

Simulations are performed to verify the performance of the proposed landing site decision algorithm. The lunar lander is assumed to be placed stationary on the point (x_0, y_0, z_0).

$$\begin{bmatrix} x_0 \\ y_0 \\ h_0 \end{bmatrix} = \begin{bmatrix} 0 \\ 0 \\ 20 \end{bmatrix} (km) \tag{27}$$

The original landing site on the lunar surface (x_{OL}, y_{OL}) is defined as below.

$$\begin{bmatrix} x_{OL} \\ y_{OL} \end{bmatrix} = \begin{bmatrix} -68 \\ 50 \end{bmatrix} (km) \tag{28}$$

The map of obstacles is given as shown in the following figure.

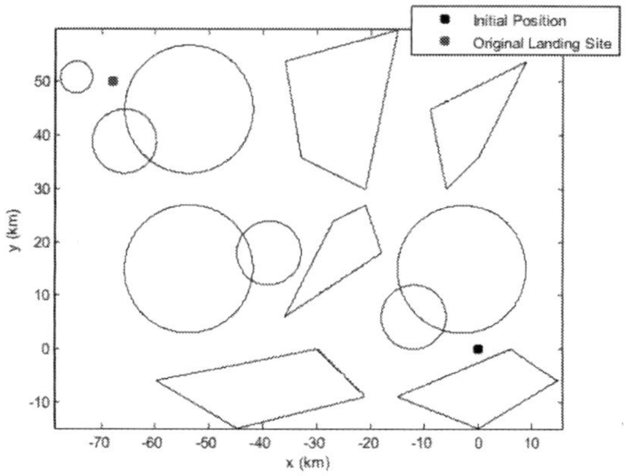

Figure 2. Coordinate System and State Variable Definitions.

4.1. Landing Trajectory Optimization for Downrange Maximization

The landing trajectory optimization is performed first to figure out the maximum downrange of the lunar lander. The initial conditions of the lunar lander are decided from (3) and (27).

$$x(t_0)=0 \quad h(t_0)=20km \quad V_x(t_0)=0 \quad V_h(t_0)=0$$
$$m(t_0)=400kg \quad \theta(t_0)=90^\circ \quad q(t_0)=0 \tag{29}$$

The terminal constraints are given in (4). The limits of attitude angle and its rate are defined as below.

$$\theta_{min}=70^\circ \quad \theta_{max}=110^\circ$$
$$q_{min}=-10^\circ/\sec \quad q_{max}=10^\circ/\sec \tag{30}$$

The control inputs are given as follows.

$$T_{D_{min}}=0kN \quad T_{D_{max}}=1kN$$
$$T_{att_{min}}=-50N \quad T_{att_{min}}=50N \tag{31}$$

The maximum and the minimum mass of the lunar lander are specified to decide the amount of available fuel during landing.

$$m_{min}=100kg \quad m_{max}=400kg \tag{32}$$

The moment of inertia and moment arm are also chosen as,

$$I_{zz}=150kg\cdot m^2 \quad L_T=1.0m \tag{33}$$

The optimal landing trajectory is obtained with GPOPS as follows.

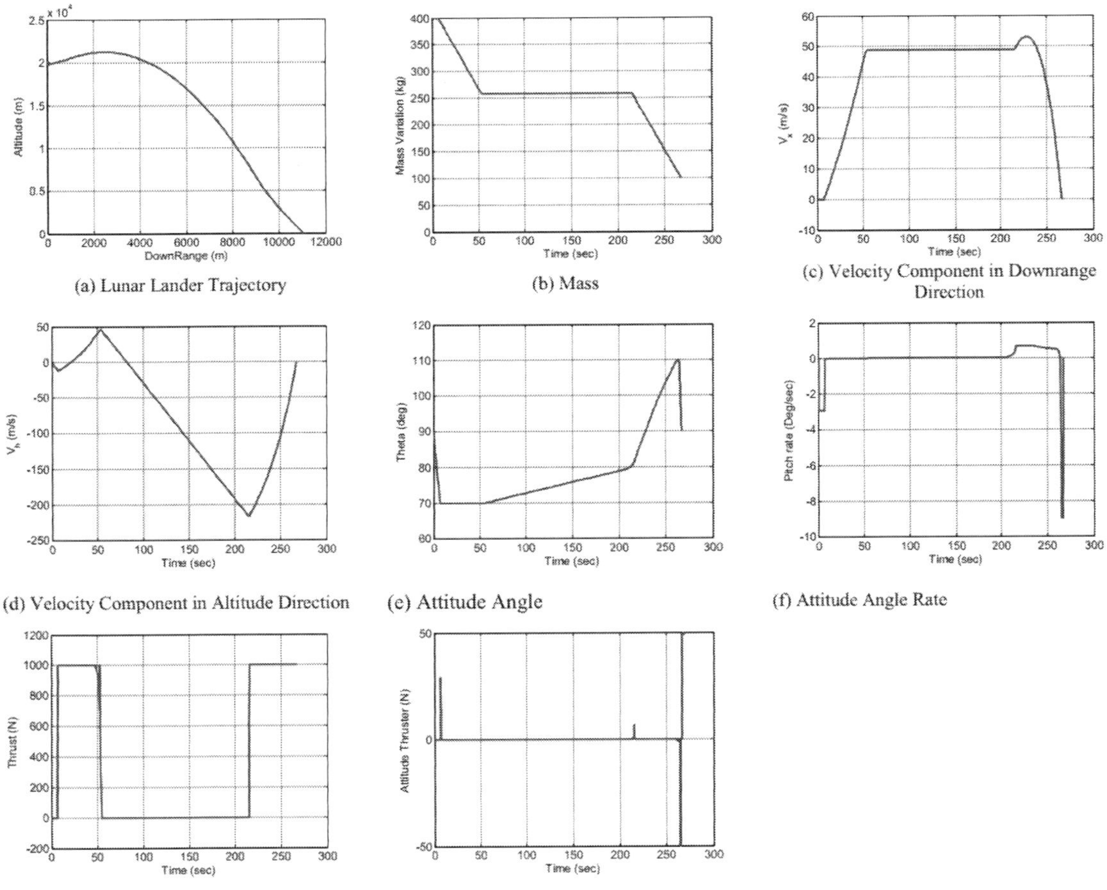

(a) Lunar Lander Trajectory (b) Mass (c) Velocity Component in Downrange Direction

(d) Velocity Component in Altitude Direction (e) Attitude Angle (f) Attitude Angle Rate

(g) Main Thrust (h) Attitude Angle Rate Command

Figure 3. Landing Trajectory Optimization for Maximum Downrange.

The maximum downrange, x_{max}, is $11.01km$. As shown in Fig. 3-(c) to (f), state variables satisfy the terminal constraints for soft landing. The control inputs, the attitude angle and its rate are bounded by their own limits as given in Fig. 3-(e) to (h). All of the available fuel is consumed as addressed in Fig. 3-(b).

4.2. Optimal Landing Site Decision with Path Planning Algorithm

The outer-most contour is defined as the circle centered at the initial position on the lunar surface, (x_0, y_0), with the radius of x_{max} figured out in the previous step. 120 points on this contour, dividing it in equal intervals, are chosen as the landing site candidates. The locations and shapes of obstacles are given as drawn in Fig. 2. The result of optimal landing site decision algorithm is obtained as Fig. 4.

Fig. 4 shows that the path segment candidates and nodes are generated from the given obstacle map. The cyan points, laid on the outer-most contour of the reachable landing region, are the feasible candidates of new landing site, and the red point located on $(-10.9949, -0.5762)$ is the optimal landing site. The green dotted lines are the feasible path segment candidates obtained with the optimal landing site, and the red circles on the obstacles are nodes. The solid magenta line represents the optimal path and its length is $92.4417km$.

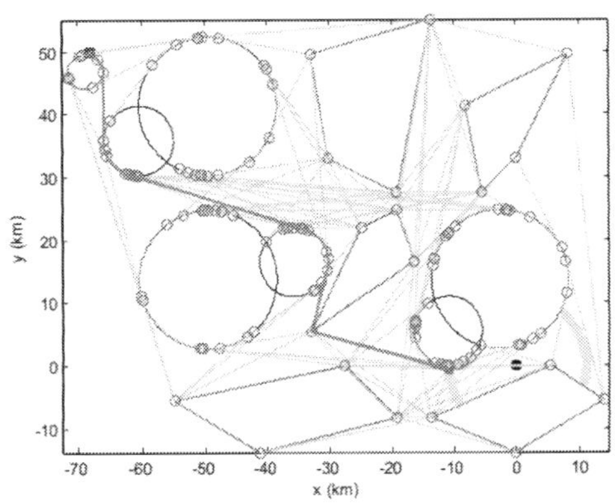

Figure 4. Optimal Path and Landing Site for Lunar Rover.

5 Conclusion

The landing site decision algorithm for lunar lander is proposed in this paper. The outer-most contour of safe reachable region is defined by solving landing trajectory optimization problem. This problem is defined to figure out the maximum downrange with the terminal constraints and the energy limits in consideration. GPOPS is utilized to figure out the solution. The algorithm to find the shortest path between a landing site candidate and an original landing site is designed to select the most suitable landing site for lunar rover energy consumption minimization. The path segment candidates and nodes are generated from a given obstacle map by applying the theories of geometry. The shortest path is derived by applying Dijkstra's algorithm.

Acknowledgment

This work was supported by the National Research Foundation of Korea(NRF) Grant funded by the Korean Government(MSIP)(No. 2014M1A3A3A03034589).

References

1. A. R. Klumpp, Automatica, **10**, 2, (1974)
2. R. R. Sostaric, J. R. Rea, *AIAA Guidance, Navigation, and Control Conference*, (2005)
3. F. Najson and K. D. Mease, *AIAA Guidance,* Navigation*, and Control Conference*, (2005)
4. B. Aç ıkmeşe, S. R. Ploen, Journal of Guidance, Control, and Dynamics, **30**, 5, (2007)
5. L. Blackmore, B. Açıkmeşe, D. P. Scharf, Journal of Guidance, Control, and Dynamics, **33**, 4, (2010)
6. B. Siciliano, L. Sciavicco, L. Villani, G. Oriolo, *Robotics: Modeling, Planning and Control,* (Springer Science & Business Media, 2009)
7. V. J. Lumelsky, A. A. Stepanov, IEEE Transactions on Automatic Control, **AC-31**, 11, (1986)
8. V. J. Lumelsky, A. A. Stepanov, Algorithmica, **2**, 1-4, (1987)
9. K. Fuzimura, H. Samet, IEEE Transactions on Robotics and Automation, **5**, 1, (1989)
10. E. Gat, M. G. Slack, D. P. Miller, R. J. Firsby, *IEEE International Conference on Robotics and Automation*, (1990)
11. M. Gerke, *Proceedings of the American Control Conference*, **4**, (1999)
12. N. A. Melchior, R. Simmons, *IEEE International Conference on Robotics and Automation*, (2007)
13. M. Tarokh, Fuzzy Sets and Systems, **159**, 21, (2008)
14. E. W. Dijkstra, Numerische Mathematik, **1**, 1, (1959)
15. B. J. Jeon, B. G. Park, M. J. Tahk, 2013 *Asia-Pacific International Symposium on Aerospace Technology*, (2013)
16. F. Junhua, C. Pingyuan, C. Hutao, *Systems and Control in Aeronautics and Astronautics (ISSCAA), 2010 3rd International Symposium on*, (2010).
17. K. S. Kim, G. H. Moon, J. W. Kim, J. W. Jeong, H. L. Choi, H. C. Shim, M. J. Tahk, *The 2nd International Conference on Robot Intelligence Technology and Application*, (2013)
18. D. S. Kim, K. Yu, Y. Cho, D. Kim, C. Yap, *Computational Science and Its Applications-ICCSA 2004*, (2004)
19. G. H. Moon, Master's Thesis, Korea Advanced Institute of Science and Technology, Daejeon, Republic of Korea (2014)

Wireless Communication and Its Effect to Tunneling Giant Magneto-Resistive (TGMR) Reader in Head Gimbal Assembly Process

Dutharuthai Na Pombejara [1,3], Chaiyan Jettanasen [2], Kraisorn Rittipad [3]

[1]College of Data Storage Innovation, King Mongkut's Institute of Technology Ladkrabang, Thailand
[2]Faculty of Engineering, King Mongkut's Institute of Technology Ladkrabang, Thailand
[3]Seagate Technology (Thailand) Ltd. Nakornratchasima, Thailand

Abstract. TGMR reader is current technology of magnetic recording read sensor which dramatically improves readout amplitude, speed and recording areal density. In advanced manufacturing environment, automated tools are increasingly used and wireless communication device is aimed to be employed to enhance the operation efficiency. However, electromagnetic fields from wireless communication devices could harm or damage the most energy sensitive devices/equipment, such as TGMR reader sensor, in head gimbal assembly process. This study surveys, therefore, electromagnetic interference from wireless communication devices using CTM048, EM Field Meter with CTS001 directional antenna and electromagnetic field module, then use Agilent 34410A to measure TGMR resistance and thermal change of TGMR head while placing in the field of wireless communication device. Build-in Differential-Ended Temperature Coefficient of Resistivity sensor is utilized for proximal TGMR temperature. Impact to electrical performance of TGMR pre and post actions simulating HGA assembly process will be confirmed using Quasi-static Tester.

1 Introduction

Nowadays wireless communication has been integrated to every activity owing to its convenience. Such interference may cause undesirable effects to sensitive equipment or sensitive devices. There were some studies about Electromagnetic Interference (EMI) effect to magnetic recording head, but none of them studied on real TGMR head. Previous studies about EMI effect from wireless communication were done on 5 Gb/in^2 PtMn GMR sensors [1] [2] [3], while current technology is up over 500 Gb/in^2 sensor. A study mentioned about TGMR head, but no technical discussion or end result of TGMR sensor impact presented in the paper [4]. Newer studies had simulated TGMR sensor impact in stand-alone stage. The consideration of connection to slider body (Key components of Slider are Reader, Writer and Advance Air Bearing Surface), Trace-Gimbal Assembly (TGA), suspension or other assembly was excluded from simulation scope [5] [6]. A study concludes that magnetic heads in production can be exposed to electrical overstress exposure of different types due to the ultra-sensitive nature of magnetic heads [7]. This study is to confirm if the electromagnetic fields from wireless communication devices will harm the most energy sensitive devices, such as TGMR recording reader sensor, in assembly process.

2 Structures and Process

2.1 The difference between Giant Magneto-Resistive (GMR) and Tunneling Giant Magneto-Resistive (TGMR) head structure

GMR Current-in-Plane (CIP) architecture has been replaced by TGMR Current-Perpendicular (CPP) architecture to allow higher data recording density

Key distinguishing element of TGMR is Magnetic Tunnel Junction (MTJ), which consists of two ferromagnets separated by an angstrom-thin insulator. Electrons can tunnel from one ferromagnet to the other if the barrier is thin enough. With this quantum mechanical phenomenon, the MTJ can be switched between two states of electrical resistance, high resistance and low resistance, then provides higher MR Ratio ($\Delta R/R_0$: change of TGMR resistance/minimum resistance of TGMR), higher readout amplitude and speed as well as lower areal resistance (*RA*: Resistance x Area) and head noise [8].

TGMR resistance is ten times higher than GMR resistance. This makes TGMR undergoing ten times higher joule heat than GMR. The angstrom-thin metal oxide barrier makes TGMR more susceptive to external energy. Fortunately, not like GMR, the TGMR head do not have insulator between buffer and shield, so TGMR can dissipate thermal directly through top and bottom

 DOI: 10.1051/matecconf/20165409004

metal shield. Example of simplified structure of GMR and TGMR recording reader head are explained in **Figure1** and **Figure2**. Actual structure and material may be difference base on design and fabrication.

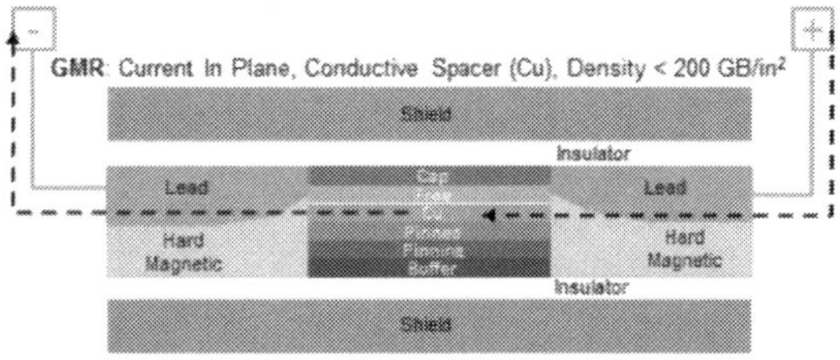

Figure1. Example of simplified GMR head structure.

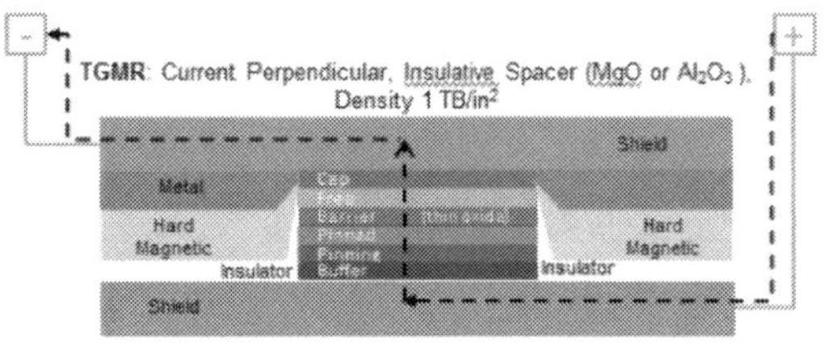

Figure2. Example of simplified TGMR head structure.

2.2 Head-Gimbal Assembly and its process

Head-Gimbal Assembly (HGA) is an important piece part of Hard Disk Drive in which slider body, which containing read-write head, located and flies above magnetic recording media.

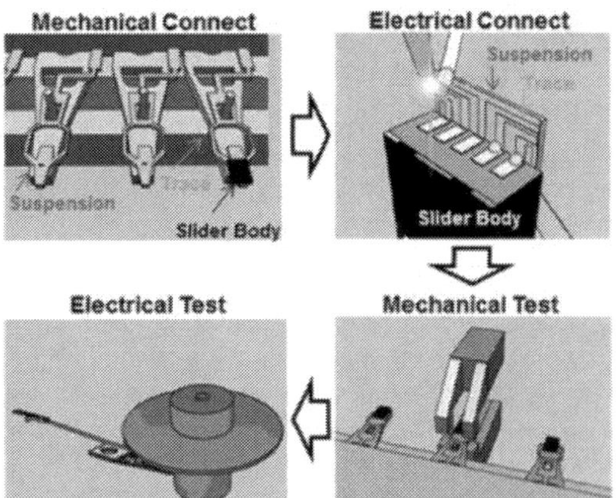

Figure3. Head-Gimbal Assembly process steps

The manufacturing process of Head-Gimbal Assembly is basically in full automation. As shown in Figure.3, robot end-effector pick-and-place each piece part, make mechanical connection and electrical connection between Trace-Gimbal Assembly (TGA) and Slider body. Then check performance via mechanical test and electrical test processes. Piece parts of HGA are contacted and in-contacted with grounded metal or static dissipative material many times along the automated process.

3 Experiment and discussion

3.1 Wireless communication devices in factory environment - Electromagnetic Interference

Electric Field from wireless communication devices is measured using CTM048, EM Field Meter with CTS001 directional antenna and electromagnetic field module. This measuring instrument has Peak and Average Measurement Range of 0.001 V/m - 20.00 V/m, Frequency Bandwidth of 1 MHz to 2.5 GHz, Antenna Factor Range of -40.0 to 40.0 dBm-1, Antenna Frequency Response of 10 MHz – 2.5 GHz. The measurement is performed in empty large shield room, at distance of 1 inches (25.4 mm) between CTS001 antenna and wireless communication devices, which are (1) Portable Two-way Radio Transceiver (walkie-talkie), (2) Wireless LAN Access Point (WiFi router), (3) GSM Cellular phone, (4) Cordless phone, (5) Laptop computer, and (6) Wireless mouse & Keyboard set.

In **Figure4**, the result shows that E-Field of higher than 5 V/m is observed from three types of wireless communication devices, Portable Two-way Radio Transceiver, GSM Cellular Phone and Cordless Phone. Whereas the E-Field of higher than 1 V/m is observed

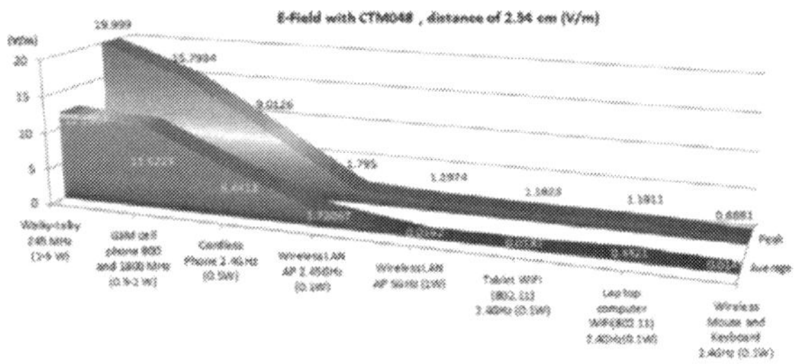

Figure4. Result of Electric field survey of wireless communication devices

3.2 Thermal change and DC voltage of TGMR head while placing in the field of wireless communication device

Build-in Differential-Ended Temperature Coefficient of Resistivity (DETCR) sensor [9] [10] has been used to observe actual thermal change at TGMR head. In this experiment, the build-in DETCR sensor was located inside the slider body, approximately 1 micron (1×10^{-6} m) from TGMR reader.

Agilent 34410A with shielded-cables and Gryphic probe was modified to measure DETCR resistance and DC voltage of TGMR while placing TGMR head in the field of wireless communication devices.

The response to local changes in temperature at the DETCR sensor is based on equation (1) and (2). The change of DETCR resistance ($\Delta R_{DETCR} = R_{DETCR} - R_{0\_DETCR}$) is proportional to the change of DETCR temperature ($\Delta T_{DETCR} = T_{DETCR} - T_{0\_DETCR}$) while R_{0_DETCR} is original resistance of the DETCR, R_{DETCR} is resistance of the DETCR at the point of time to compare, T_{0_DETCR} is original temperature of the DETCR, T_{DETCR} is temperature of the DETCR at the point of time to compare, and α is specific to DETCR material.

$$\Delta R_{DETCR} = \alpha \ \Delta T_{DETCR} \quad (1)$$

$$R_{DETCR} - R_{0\_DETCR} = \alpha * (T_{DETCR} - T_{0\_DETCR}) \quad (2)$$

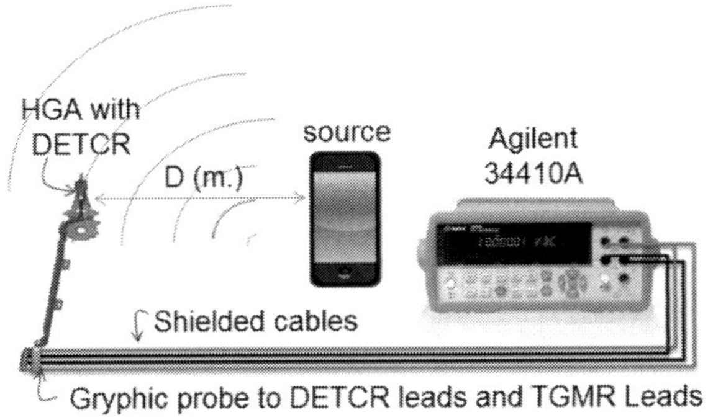

Figure5. Measurement set-up on Thermal change and DC voltage of TGMR head while placing in the field issued from wireless communication device

Result is shown in **Figure6.1**; the temperature changes when placing TGMR in the field of wireless communication device. The original temperature (T_0) is 22 degree Celsius and the change of temperature (ΔT) is less than 3 degree Celsius, so maximum temperature is less than 25 degree Celsius which is needed to be compared whether it is high enough to reach the threshold of soft level of stability failure of TGMR. Typically, the threshold should be more than 100 degree Celsius.

Figure6.2 shows measurement result of DC voltage drops on TGMR, while placing TGMR head in the field of wireless communication device(s). Overall average of DC voltage is 1.2×10^{-6} VDC and overall maximum is 11.78×10^{-6} VDC. As average TGMR resistance is about 300 Ohms, so maximum calculated power on TGMR is 4.8×10^{-13} Watts or 0.48 pico Joule/second. Next, this should be compared with threshold of TGMR failure which is typically susceptible to the energy in nano Joule range.

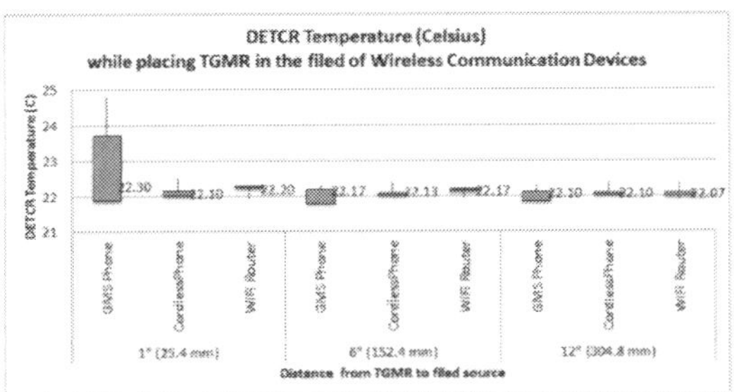

Figure6.1. Temperature of TGMR head while placing in the field of wireless communication device

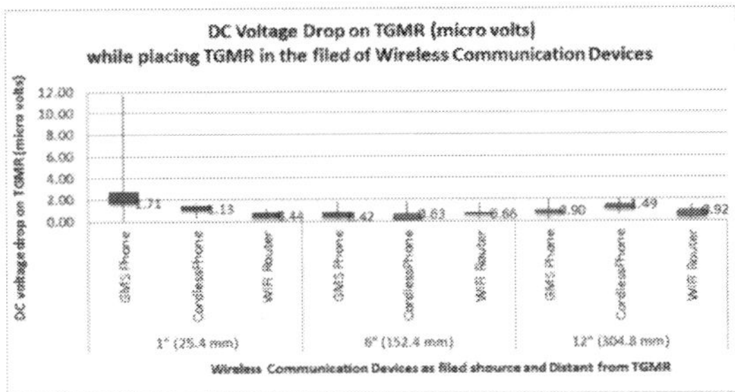

Figure6.2. DC voltage of TGMR head while placing in the field of wireless communication device

3.3 Electrical Performance of TGMR pre and post actions simulating HGA assembly process

Four types of wireless communication devices: GSM Cellular Phone, Cordless Phone, WiFi Router and Tablet Computer have been chosen to be field sources in this experiment because they are mostly used nowadays. HGAs are placed in the field generated from each device at D (cm.) distance far away as set up shown in **Figure5**.

Four actions of HGA process have been simulated during the HGAs are in-field as shown in **Figure7**. There are; (I) No contact to HGA at all (II) Grounded-Metal 10 times contact to open traces of TGMR (III) Grounded-Non-Metal 10 times contact to open traces of TGMR (IV)10 times pick-and-place on base plate with Non-Metal End-Effector

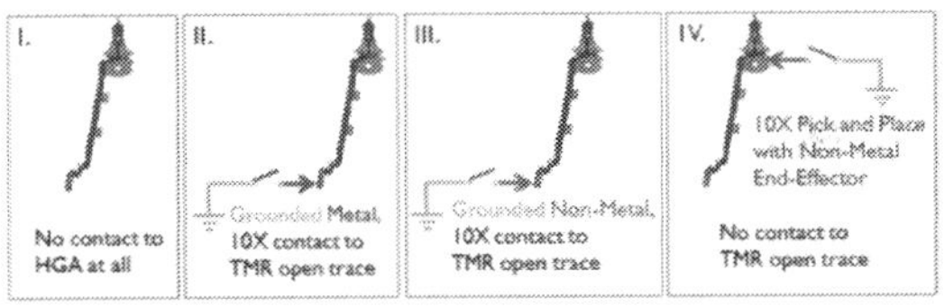

Figure7. Four actions represent HGA process

Electrical performance of HGAs has been tested before and after the actions with Quasi-static Tester (QST) to see how the experiment impacts to electrical performance of HGAs. Both catastrophic failure parameters (TGMR resistance, Amplitude, Slope) and instability parameters (Asymmetry, Barkhausen Jump, Hysteresis Loop, and Spectral Maximum Amplitude Noise/ S.M.A.N.) were observed and compared with total sample size of 180 HGAs. QST TGMR resistance measurement is static test and quite direct to see obvious physical change of TGMR sensor such as insulator or oxide layer breakdown. The other following parameters are measured and calculated from the Transfer Curve as shown in **Figure8**.

Amplitude is the key performance of reader sensor. The test measures peak amplitude in micro-volt at defined point of magnetic field. Amplitude is directly related to the change of TGMR resistance as shown in (3). While I is the bias current, V is the voltage drop on the sensor, $\Delta R/R_0$ is the intrinsic TGMR ratio, ε is the efficiency or amount of free layer rotation. External energy such as Thermal, EOS/ESD can cause Pinned Layer reversal in TGMR and Slope test can reveal the magnitude and location of reversal over the sweep test range. Asymmetry test basically shows the symmetry of the transfer curve which can indicate performance of magnetic layers in TGMR. Equation (4) shows calculation of asymmetry from absolute value of the maximum and minimum amplitude. Hysteresis test is measurement area between forward amplitude (Fi) and reverse amplitude (Ri) of the amplitude. Equation (5) shows calculation of hysteresis, in the unit of micro-volt/Oe, from amplitude and magnetic field increment (Inc). Barkhausen-jump (micro-volt) is maximum amplitude jumping between two adjacent measurement points. Percent of Barkhuasen jump (*%BJ*) can be calculated with peak-to-peak amplitude ($Amp._{pk\text{-}pk}$) as shown in (6). Spectral Maximum Amplitude Noise (S.M.A.N.) test measures noise density by sweeping through a range of DC magnetic fields. This parameter is invented to provide capability of high frequency instability identification. [11] [12] Example of S.M.A.N. test is shown in Figure9.

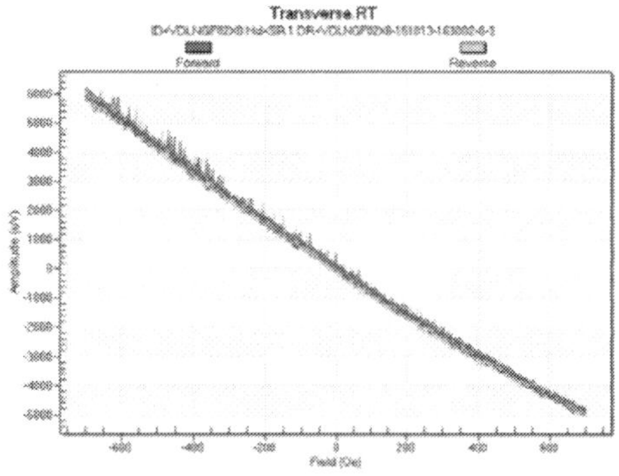

Figure8. Example of Amplitude Test by QST

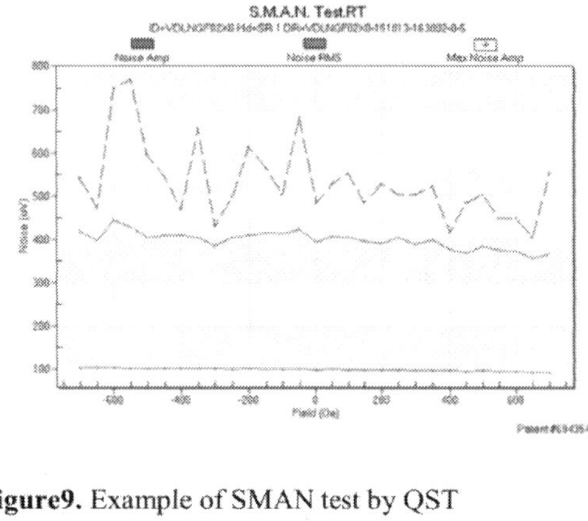

Figure9. Example of SMAN test by QST

$$\text{Amplitude} = I\times\Delta R\times\varepsilon = I\times R_0\times\frac{\Delta R}{R_0}\times\varepsilon = V\times\frac{\Delta R}{R_0}\times\varepsilon \quad (3)$$

$$\text{Asymmetry}(\%) = \frac{|MAX.amp|-|MIN.amp|}{|MAX.amp|+|MIN.amp|}\times 100 \quad (4)$$

$$Hysteresis(\mu V/Oe) = Inc\times\sum|Fi-Ri| \quad (5)$$

$$\%BJ = [BJ(\mu V)/Amp._{Pk-Pk}]\times 100 \quad (6)$$

The result from QST shows that, both catastrophic failure and instability failure parameters related to TGMR electrical performance (TGMR resistance, Amplitude, Slope, Asymmetry, Barkhausen Jump, Hysteresis-Loop, and Spectral Maximum Amplitude Noise) are not significantly changed. It is illustrated by correlation plot between pre and post actions simulating HGA assembly processes, in **Figure10.1** to **Figure10.7**. This could confirm with the result of measuring Thermal change and DC voltage of TGMR head while placing in the field of wireless communication device in the section 3.2.

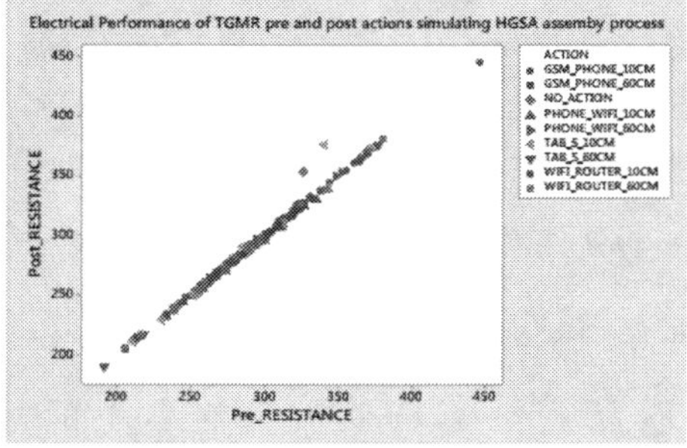

Figure10.1 Correlation of **"TGMR Resistance"**

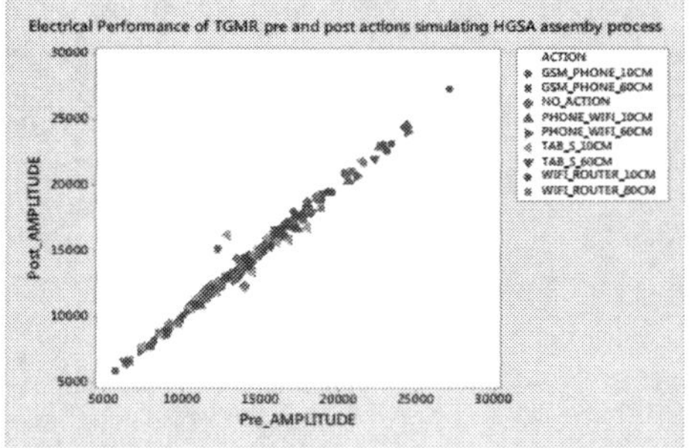

Figure10.2 Correlation of **"TGMR Amplitude"**

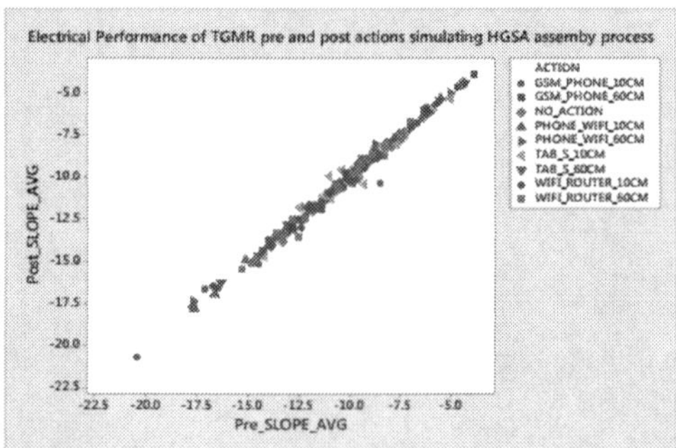

Figure10.3 Correlation of **"TGMR Slope"**

DOI: 10.1051/matecconf/20165409004

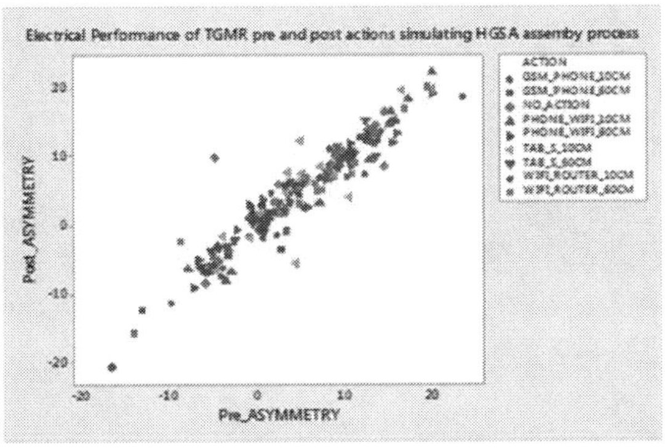

Figure10.4 Correlation of **"Asymmetry"**

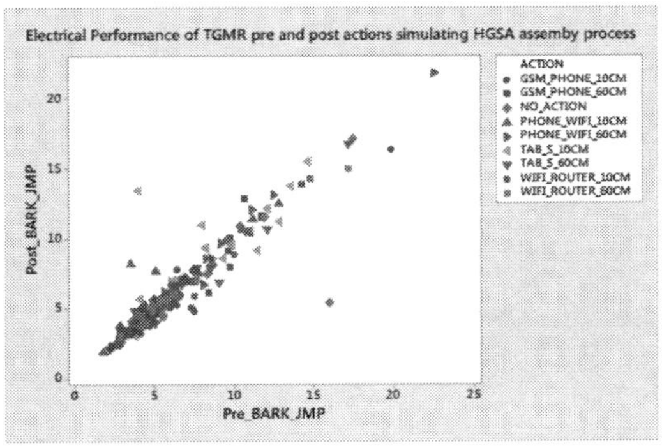

Figure10.5 Correlation of **"Barkhausen Jump"**

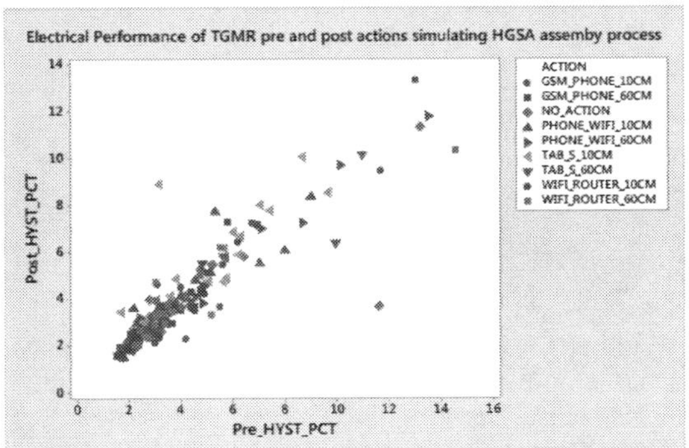

Figure.10.6 Correlation of **"Hysteresis Loop"**

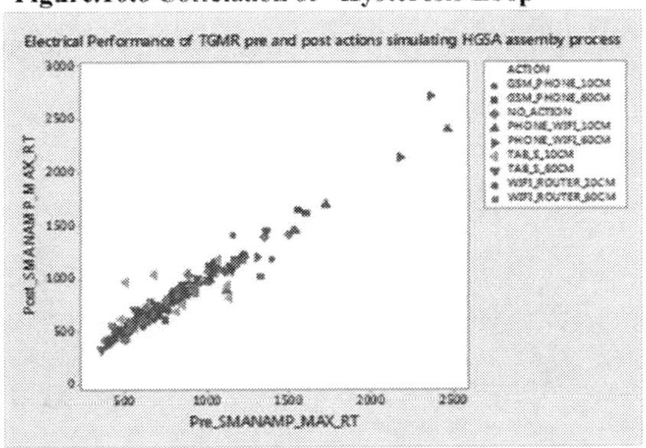

Figure10.7 Correlation of **"S.M.A.N."**

4 Conclusions

In this study, Electromagnetic Interference of 1-16 V/m had been observed from GSM cell phone, Cordless Phone, WiFi Router and Tablet Computer. However, when placing TGMR head in the electromagnetic fields issued from those wireless communication devices, the thermal change of TGMR head and the DC voltage of TGMR are not significant to create instability nor catastrophic failure to TGMR heads.

It had been proved with key QST parameters on 180 HGA samples for pre and post actions simulating HGA assembly process in the field of those wireless communication devices. There is no significant change in TGMR resistance, Amplitude, Slope, Asymmetry, Barkhausen Jump, Hysteresis Loop, and Spectral Maximum Amplitude Noise (S.M.A.N.) in all experiment conditions.

Wireless communication devices may be considered to enhance operation efficiency or operation cost saving. Nevertheless, it should be studied to make sure the effect on energy sensitive devices as well as automated production equipment and testers to be in acceptable level.

Acknowledgements

This work was supported by; [i] Seagate Technology, PLC, [ii] Seagate Technology, Thailand, Ltd, [iii] National Electronics and Computer Technology Center (NECTEC) Thailand, [iv] College of Data Storage Innovation, King Mongkut's Institute of Technology Ladkrabang, [v] Faculty of Engineering, King Mongkut's Institute of Technology Ladkrabang, Thailand.

References

1. Wallash, D. Smith, EOS/ESD Symposium 98, 4B.6.1 (1998).
2. C.F. Lam, D. Martinez, C. Chang, EOS/ESD Symposium 99, 4B.6.2 (1999).
3. V. Kraz, A. Wallash, EOS/ESD Symposium 00, 2C.10.1 (2000).
4. S. Natori, T. Wada, Y. Itou, H. Sakata, K. Kataoka, H. Tanabe, T. Ohtsu, EOS/ESD Symposium 06, 4B.3.1 (2006).
5. S. Kaengrang, A. Kaewrawang, DSTCON 4th, pp. 61 (2011).
6. S. Puapairoj, C. Sa-ngiamsak, DSTCON 4th, pp. 149 (2011).
7. V. Kraz, P. Tachamaneekorn, D. Napombejara, EOS/ESD Symposium 05, 4A.4.1 (2005).
8. T. Zhao, K. Zhang, H. C. Wang, Y.H. Chen, M. Li, US Patents, US20070111332 A1 (2007).
9. M.A. Dufresne, B.W. Karr, G.J. Kunkel, Z. Wei, US Patents, US20120120982 A1 (2012).
10. D. Liu, K. Haapala, S.E. Ryun, L. Zhou, US Patents, US8638349 B1(2014).
11. H.Patland, W.A. Ogle, US6943545 B2 (2005)
12. H.Patland, W.A. Ogle, EOS/ESD Symposium 02, pp. 130 (2002).

Detection of R-Peaks in ECG Signal by Adaptive Linear Neuron (ADALINE) Artificial Neural Network

Jeong-Hwan Kim, Sang-Eun Park, Gyeo-Wun Jeung, Kyeong-Seop Kim

School of Biomedical Engineering, Konkuk University, KOREA

Abstract. This research proposes a new method to detect R-peaks in electrocardiogram by using the prediction value from adaptive linear neuron (ADALINE) artificial neural network. With this aim, the weights of four input neurons in ADALINE are updated for each encoded ECG vector-segment and the value of an output neuron is compared with the actual ECG followed by applying finite impulse response filter. Our simulated experiments with the MIT-BIH ECG database that represents the long-term recordings from the heart disease patients show that our proposed algorithm can detect R-peaks in ECG data with the accuracy of more than 99%.

Keywords: ECG, R-peak, neural network, ADALINE, arrhythmia, Premature Ventricular Contraction, finite impulse response filter, MIT-BIT database.

1 Introduction

Electrocardiogram (ECG) that records the electric potentials resulting from a heart's activity can offer the clinical decision parameters to diagnose an abnormal heart sinus rhythm. The primary features that characterize ECG waveform include R-peak, QRS complex, S and T fiducial points in terms of time-location and morphological shape [1].

In the ECG data, the most crucial information is the time locations occurring in R-peak because this feature can be utilized to monitor especially Premature Ventricular Contraction (PVC) arrhythmia [2] and Heart Rate Variability (HRV) [3] in the long-term based heart health monitoring environment. So far, concerning the automatic detection of R-peaks in ECG data, the adaptive threshold level was sought [4] or a differential operator was applied to compute the gradient of ECG waveform [5]. Also R-peak characteristic point was estimated by encoding the ECG data in wavelet transform [6, 7, 8] or Hilbert transform domain [9]. In our study, a new algorithm to detect R-peaks is proposed by evaluating the predicting feature from ADALINE artificial neural network model [10, 11] and comparing with the actual ECG voltage level. With this aim, ECG data is encoded successively in a four dimensional vector-segment format and feed forwards into four input neurons in ADALINE to update their weights. The value of an output neuron is computed as a prediction value for each ECG segment and compared with the actual voltage level to estimate the time locations of R-peak.

2 Prediction of ECG data by adaline neural network

2.1 ADALINE artificial neural network architecture

ADALINE has single output neuron that receives the weighted sum from the multiple input neurons with an optional bias. Fig.1 shows the architecture of an ADALINE neural net consisted with n-input neurons X_1, X_2, … , X_n , an output neuron y with specifying weights of w_1, w_2, …, w_n and a bias b.

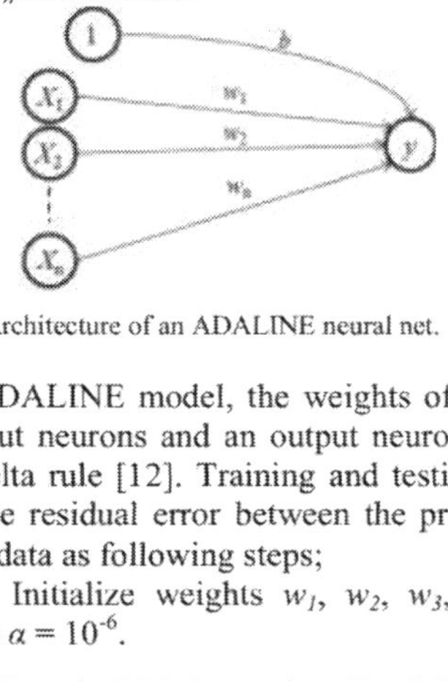

Figure 1. Architecture of an ADALINE neural net.

In the ADALINE model, the weights of connections between input neurons and an output neuron are trained using the delta rule [12]. Training and testing algorithm to extract the residual error between the predicted ECG and the real data as following steps;

Step 1: Initialize weights w_1, w_2, w_3, w_4 and set learning rate $\alpha = 10^{-6}$.

Step 2: Encode ECG input data $\boldsymbol{X} = [X_1, X_2, X_3, \ldots, X_N]$, N: the number of samples, into an ith segment $\boldsymbol{S}_i$, $i = 1, 2, \ldots .N$ in a fourth dimensional vector format:

$$S_1 = \begin{bmatrix} x_1 \\ 0 \\ 0 \\ 0 \end{bmatrix}, S_2 = \begin{bmatrix} x_2 \\ x_1 \\ 0 \\ 0 \end{bmatrix}, S_2 = \begin{bmatrix} x_3 \\ x_2 \\ x_1 \\ 0 \end{bmatrix}, \cdots, S_{N-1} = \begin{bmatrix} x_{N-1} \\ x_{N-2} \\ x_{N-3} \\ x_{N-4} \end{bmatrix}, S_N = \begin{bmatrix} x_N \\ x_{N-1} \\ x_{N-2} \\ x_{N-3} \end{bmatrix} \quad (1)$$

Step 3: Compute i^{th} predicted vale from an output neuron by activating $(i\text{-}1)^{th}$ weights:

$$y_i = b + \boldsymbol{S}_i \cdot \boldsymbol{W}^{i-1} \tag{2}$$

, where $\cdot$ is a dot product operator and $\boldsymbol{W}^{i-1} = \left[w_1^{i-1}, w_2^{i-1}, w_3^{i-1}, w_4^{i-1}\right]^t$ (*t*: transpose) is a weight vector that was determined at $(i\text{-}1)^{th}$ training stage

Step 4: Update and weights, *i =1, 2, 3, 4,* with nullifying a bias effect (i.e., $b = 0$):

$$\boldsymbol{W}^i = \boldsymbol{W}^{i-1} + \alpha \cdot (t_i - y_i) \cdot x_i \text{ , } t_i\text{: the target ECG data} \tag{3}$$

During the process of training algorithm, the weights are updated adaptively for each feed forwards ECG segment $\boldsymbol{S}_i$ so as to minimize the residual error between the real ECG data X_i and the output vale y_i. In order to evaluate the performance in detecting R-peaks in ECG data, we use the MIT-BIH database [13] which recorded the long-term ECG readings with the sampling rate of 360 Hz (180 Hz in text file format) from the heart disease patients. For our experiments, the record titled with ECG100_1 (contains the normal sinus rhythms and one PVC beat) and ECG119_1 (includes a quite number of PVC abnormal heart beats) are considered. As a pre-processing step, a digital high-pass finite impulse response (FIR) filter is initially applied to the considered ECG data to eliminate the baseline wandering noise [14].

2. 2 ADALINE artificial neural network architecture

Fig.2-(a) shows that the part of residual error e_i, $i = 1, \ldots, N$, between the predictive value from an output neuron as an estimate of ECG data x_i (ECG100_1) and the real data. Fig.2-(b) describes the updating weights when i^{th} ECG segment $\boldsymbol{S}_i$ (ECG100_1) feeds into the input neurons. In this figure, the error signal is abruptly changed in the neighborhood area of R-peak. Thus we can utilize this fact that the initial R-peaks candidates can be selected with applying a moving average filter followed by low-pass FIR digital filter to smooth the residual peaks and to set the adaptive threshold line in order to delineate R-peak candidates.

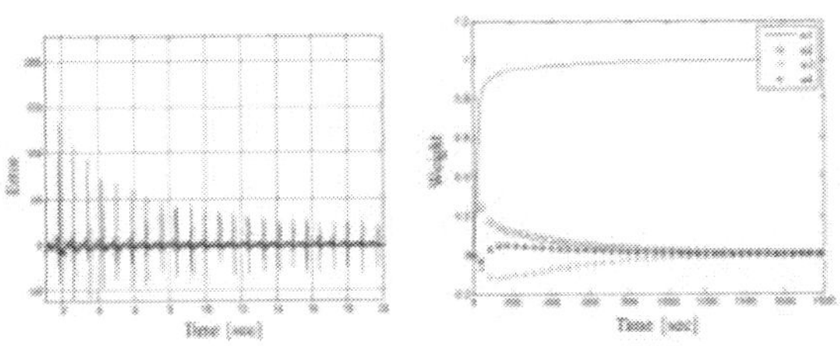

(a) (b)

Figure 2. (a) The part of residual error e^i for ECG100_1 dataset. (b) Updating weights of ADALINE neural net for ECG100_dataset.

3 Detection of R-peaks in ECG signal

In order to smooth the sharpness in the residual peaks, the moving average filter (window size, *L* is selected to be 20) is applied and *eps* (error in prediction signal from ADALINE) is defined as in equation (4).

$$eps^i = \left(\frac{1}{L}\right) \cdot \sum_{j=0}^{L-1} \left|e^j\right| \text{ , } i = 1, 2, \ldots, N \tag{4}$$

To set the threshold line that is adaptively tune to the trend of residual error signal, eps is low-pass filtered by a FIR digital filter (see Table 1 for filter design specification). As of a result, Fig.3 shows eps^i with low-pass filtered *eps* signal, $leps^i$, $i=1, \ldots, N$. Here $leps^i$ is used as a reference level to delineate the time intervals including R-peaks.

Table 1. Design specifications of low-pass FIR filter for extracting $leps^i$

Parameters	Value
Window function	Hanning
Sampling freq.	180 Hz
Filter order	599
Pass band cutoff freq.	0.5 Hz
Stop band cutoff freq.	1.5 Hz

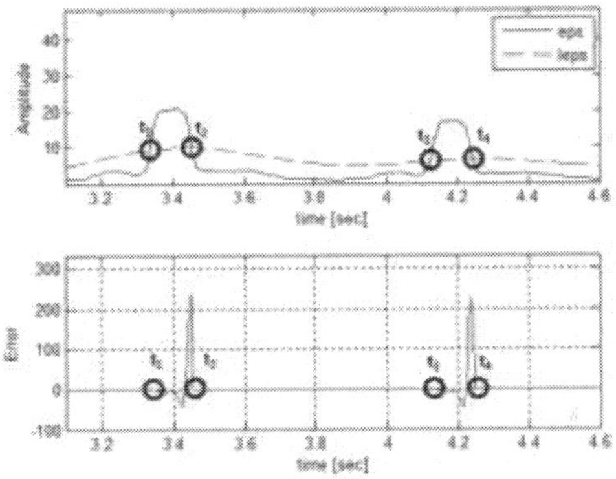

Figure 3. The illustration of residual error: eps^i (solid line), $leps^i$ (dotted line) in the top chart and the overall trend of $leps^i$ in the below one.

Finally, the time locations in R-peak can be resolved by computing the gradient value for each *leps*-segment for $t_i \le t \le t_j$. Fig.4 illustrates the result of R-peak detection as a symbol of 'o' for ECG100_1 and for ECG119_1. Note that two R-peaks in ECG119_1 are missed especially during the initial weights updating stage (i.e., the time less than 2 seconds) due to the inherent characteristics of neural network model. In other words, ADALINE needs some warming up time to trace the trend of input ECG data $\boldsymbol{S}_i$. The accuracy of estimating R-peaks by ADALINE can be slightly varied upon the size of moving average window *L*. Nonetheless, the overall performance of detecting R-peaks by ADALINE is superior to the conventional method that simply computes the gradient of ECG waveform. Table 2 and 3 summarize the performance from ADALINE and differential operator in terms of TP (True Positive), TN (True Negative), FP (False Positive) and FN (False Negative) in detecting R-peaks for ECG100_1 and

ECG119_1, respectively. In Table 2, Sensitivity and Accuracy are defined as follows [15];

$$\text{Sensitivity} = \frac{\text{TP}}{\text{TP+FN}} \times 100(\%),$$
$$\text{Accuracy} = \frac{\text{TP+TN}}{\text{TP+FP+FN+TN}} \times 100(\%) \qquad (5)$$

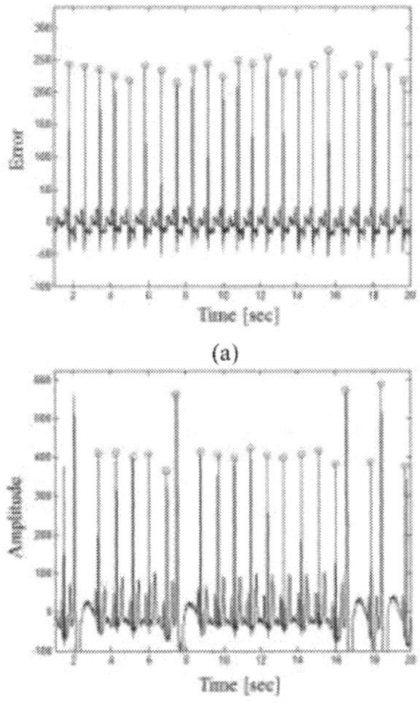

Figure 4. The results of R-peak detection as a symbol of 'o': (a) ECG100_1, $L = 20$ (b) ECG119_1, $L = 20$.

Table 2. Comparison of performance in detecting R-peaks by ADALINE using the different size of window L with applying differential operator to ECG100_1

	TP	TN	FP	FN	Sensitivity (%)	Accuracy (%)
Differential operator[5]	2,151	0	10	99	95.6	95.18
ADALINE (L=10)	2,258	0	1	0	100	99.96
ADALINE (L=15)	2,256	0	2	1	99.96	99.87
ADALINE (L=20)	2,256	0	1	2	99.91	99.87

Table 3. Comparison of performance in detecting R-peaks by ADALINE with applying the differential operator to ECG119_1 dataset

	TP	TN	FP	FN	Sensitivity (%)	Accuracy (%)
Differential operator[5]	1,343	0	365	220	85.92	69.67
ADALINE (L=10)	1,974	0	8	2	99.9	99.5
ADALINE (L=15)	1,975	0	0	2	99.9	99.9
ADALINE (L=20)	1,972	0	0	2	99.9	99.9

In summarizing our proposed algorithm, ADALINE neural net accompanied with the necessary digital filtering processes is described in Fig.5.

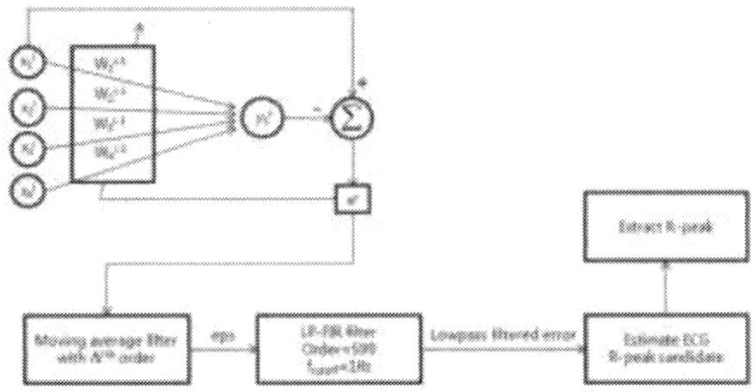

Figure 5. The proposed ADALINE neural net for detecting R-peaks in the ECG data.

4 Conclusion

In this research, we can find the fact that R-peaks in ECG can be automatically detected with the high accuracy of more than 99% by ADALINE neural net implemented with four input neurons and one output neuron. Concerning the clinical decisions for the diagnosis of arrhythmia, FN index is the most crucial feature. As shown in Table 2 and 3, the number of FN's are 2 ($L = 20$) for ECG 100_1 and ECG 119_1 dataset, respectively. The main cause of this misclassification is the limited capacity of ADALINE neural network model when it deals with updating weights especially during the initial training stage.

Acknowledgment

This research was supported by Basic Science Research Program through the National Research Foundation of Korea (NRF) funded by the Ministry of Education, Science and Technology (NRF-2013R1A1A2012393).

References

1. A. U. Rajendra, J. S. Suri, J. A. E. Spaan and S. M. Krishnan, Advances in Cardiac Signal Processing, Berlin Heidelberg: Springer, (2007).
2. M. L. Talb and A. Charef, "PVC discrimination using the QRS power spectrum and self-organizing meps," Computer Methods and Programs in Biomedicine, **94**, 223-231, (2009).
3. M. V. Kamath, M. A. Watanabe and A. R. M. Upton, Heart Rate Variability (HRV) Signal Analysis: Clinical Applications. CRC Press, (2013).
4. F. Chiarugi, V. Sakkalis, D. Emmanouilidou, T. Krontiris, M. Varanini and I. Tollis, "Adaptive

threshold QRS detector with best channel selection based on a noise rating system," IEEE Computers in Cardiology, 157-160, (2007).
5. J. Pan and W. J. Tompkins, "A real-time QRS detection algorithm," IEEE Transactions on Biomedical Engineering, **32**, 230-236, (1985).
6. C. Li, C. Zheng and C. Tai, "Detection of ECG characteristic points using wavelet transforms," IEEE Transactions on Biomedical Engineering, **42**, 21-28, (1995).
7. P. J. M. Fard, M. H. Moradi and M. R. Tajvidi, "A novel approach in R peak detection using Hybrid Complex Wavelet (HCW)," International Journal of Cardiology, **124**, 250-253, (2008).
8. C. Li, C. Zheng and C. Tai, "Detection of ECG characteristic points using wavelet transforms," IEEE Transactions on Biomedical Engineering, **22**, 21-28, (1995).
9. D. S. Benites, P. A. Gaydeck, A. Zaidi and A. P. Fitzpatrick, "A new QRS detection algorithm based on the Hilbert transform," IEEE Computers in Cardiology, 379-382, (2000).
10. L. Fausett, Fundamentals of Neural Networks, Upper Saddle River: Prentice Hall, (1994).
11. S. N. Sivanandam, S. Sumathi and S. N. Deepa, Introduction to neural networks using Matlab 6.0, McGraw-Hill Education, (2006).
12. Y. H. Hu and J. N. Hwang, Handbook of Neural Signal Processing, CRC Press, (2002).
13. A. L. Goldberger, "PhysioBank, PhysioToolkit, and Physionet: components of a new research resource for complex physiologic signals," Circulation, **23**, 215-220, (2000).
14. J. H. Kim, S. E. Park, J. W. Lee and K. S. Kim, "Design and implementation of digital filters for mobile healthcare applications," International Journal of Electronics and Electrical Engineering, **2**, 75-79, (2014).
15. W. L. Martinez, Computational Statistics Heandbook with Matlab, Chapman & Hall, CRC Press, (2008).

Rotating Machinery Vibration Signal Processing And Fault Diagnosis Based on LMD

Ruirui Bo and Ze Zhang

Department of Electronic and Information Engineering, Inner Mongolia University, Hohhot, 10126, China

Abstract. There are abundant of fault information in rotating machinery vibration signal. On account of the non-linearity and non-stationarity, the paper first does pre-process to the vibration signal using wavelet threshold de-noising method and this method can bring a smooth signal. Then it decomposes the vibration signal using local mean decomposition(LMD), which is effective to the vibration signal. The LMD decomposes the signal into many PFs as the frequency from high to low. These PFs are composed of the production of envelop signal and pure frequency modulated signal. Finally, it takes most use of the kurtosis which is sensitive to the fault impact. By calculating the kurtosis of PF, it can assess the distribution of fault impact signal in every frequency band, consequently distinguishing the operating state of bearing and recognizing the fault mode according to the growth of turtosis. The experiment of actual bearing vibration signal demonstrates that the methods this paper proposed can effectively diagnose the vibration fault and has good performance.

Keywords: LMD, wavelet threshold method, kurtosis, fault diagnosis.

1 Introduction

Giant rotating machinery like air compressor and pump and turbine are key production equipment in modern industry. With the substantial improvement of science and technology and automation, giant rotating machinery becomes high precision and high speed and serialization and automation[1]. The interrelation and coupling between different portions in device grows much closer. Once there occurs a fault in a certain portion, it not only has influence on the operation of machinery itself but also affects the subsequent production, causing chain reaction and breakdown of system, leading to great financial loss even disastrous consequence. The fault information is usually shown through vibration signal when there is fault in a certain giant rotating machinery device[2]. It is the primary approach to supervise and diagnose the operating state of device using vibration signal at present. Due to operating environment's complexity and variety of giant rotating machinery, its vibration signal usually has strong non-linearity and non-stationarity. Therefore, the traditional fault diagnosis method based on classical signal procession like FFT and wavelet analysis cannot meet the demand on fault feature extraction, and they cannot make ends meet of local detail feature and general view feature in time and frequency domain,and which makes it an urgent theoretical study and practical application requirement to develop modern self-adaptive signal analysis method based on time-frequency analysis aiming at non-linear and non-stationary signal[3].

2 Vibration signal pre-process

2.1 Vibration data acquisition

Bearing, as the significant component in rotating machinery, would receive key information of rotating machinery system's operation state by analyzing and processing vibration signal. The vibration data used in this paper are from America Case Western Reserve University bearing data center. The experimented bearing is 6205-2RS JEM SKF deep groove ball bearing. There are four states that are normal operating and outer raceway fault and inner raceway fault as well as ball fault,all these single point faults are introduced to the test bearing using electro-discharge machining with fault diameters of 0.18mm, 0.36mm, 0.53 mm, 0.71mm, and 1.02 mm, the paper chooses 0.53mm fault. Vibration data is collected using accelerometer and at 12,000 samples per second. Motor speeds is 1797. The driving end bears no load. The vibration signal wave of the four operating state are shown as fig.1.

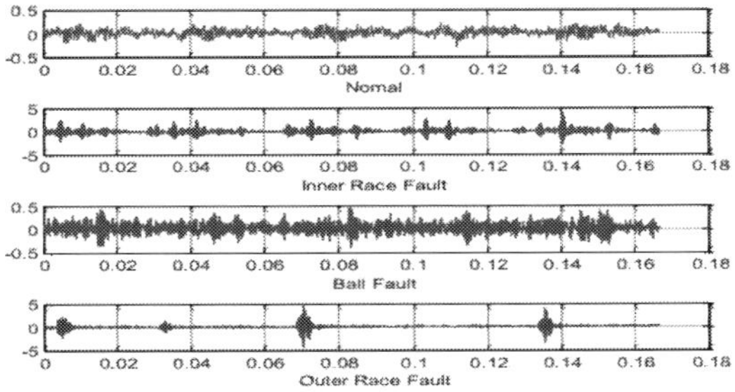

Fig.1: Vibration signal wave of the four operating state

2.2 Wavelet threshold De-noise

The vibration signal is corrupted by a great number of noise contribution besides fault information, which appears to be non-linear and non-stationary and with low signal noise ratio. When decomposing the signal with noise using LMD, the noise contribution may, on the one hand, affect the cyclic process and, on the other hand, bring down the timeliness but promote the decomposing time, even result in mode mixing phenomenon. It is the precondition and guarantee of signal fault diagnosis to reduce noise contribution effectively[4]. Wavelet threshold de-noise is a typical de-noising method, which has irreplaceable significance for its good time-frequency property in de-noise. The paper opts for SYM5 wavelet function and decomposes the signal to the fifth layer then uses minimaxi threshold chose to dispose coefficients for the sake of de-noise. The signal after de-noising are listed in fig. 2.

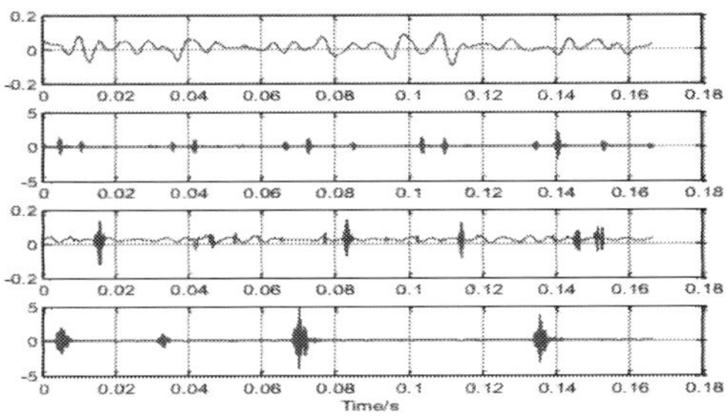

Fig.2: Denoised signal on four state

After wavelet de-noising, there is basically no noise contribution in the bearing vibration signal and the vibration is much smoother. From fig.2, it can see the real vibration signal. Analysis of the de-noised signal suggests that there are different levels of impact in the signal which is the main reflect of fault when operating.

3 Local Mean Decomposition

Essentially the LMD scheme involves progressively separating a frequency modulated signal from an amplitude modulated envelope signal. This separation is achieved by smoothing the original signal, subtracting the smoothed signal from the original signal, and then amplitude demodulating the result using an envelope estimate. The envelope estimate and the smoothed version of the original signal are both obtained by using moving averaging weighted by the time-lapse between the successive extrema of the original signal[5]. The instantaneous frequency can then be calculated from the frequency modulated signal. The corresponding envelope estimates are multiplied together to form a final envelope. This envelope is then multiplied by the frequency modulated signal to form a product function, which is subtracted from the original signal. The whole process is then repeated on the resulting signal, to produce a second product function, with an associated envelope and instantaneous frequency.This decomposition continues until the remaining signal contains no more oscillations.The resulting instantaneous frequency and envelope values can be plotted together in the form of a demodulated signal time-frequency representation[6]. For a certain non-stationary signal $x(t)$, the specific decomposing process is as follow:

Step 1: Searching for the local maximum(minimum)value point of the signal n_i ,then calculating the mean value of the adjacent extremum.

$$m_i = \frac{n_i + n_{i+1}}{2} \tag{1}$$

Connecting the adjacent mean value point m_i with straight line, smoothing the straight line using moving average method and obtaining the local mean function $m_{11}(t)$.

Step 2: Calculating the envelope estimate:

$$a_i = \frac{|n_i - n_{i+1}|}{2} \tag{2}$$

Connecting the adjacent envelope estimate point a_i with straight line, smoothing the straight line using moving average method and obtaining the envelope estimate function $a_{11}(t)$.

Step 3: Separating the local mean function $m_{11}(t)$ from the original signal $x(t)$.

$$h_{11}(t) = x(t) - m_{11}(t) \tag{3}$$

Step 4: Using the envelope estimate function $a_{11}(t)$ to demodulate $h_{11}(t)$.

$$s_{11}(t) = \frac{h_{11}(t)}{a_{11}(t)} \tag{4}$$

The ideal situation is that $s_{11}(t)$ is pure frequency-modulated signal now, i.e. envelope estimate function $a_{12}(t) = 1$. If not, taking $s_{11}(t)$ as original signal and repeating the step1-step4 until $s_{1n}(t)$'s envelope estimate function $a_{1(n+1)}(t) = 1$, i.e. $s_{1n}(t)$ is a pure frequency-modulated signal. But in general it cannot meet the ideal condition, so it sets a reference quantity Δ, when $1\text{-}\Delta \le a_{1n}(t) \le 1 + \Delta$, the iteration stops.

Step 5: Calculating the envelope signal (instantaneous amplitude function):

$$a_1(\mathrm{t}) = a_{11}(\mathrm{t})a_{12}(\mathrm{t})\ldots a_{1n}(\mathrm{t}) = \prod_{q=1}^{n} a_{1q}(\mathrm{t}) \tag{5}$$

Step 6: Obtaining the first PF component by multiplying envelope signal $a_1(\mathrm{t})$ and pure frequency-modulated signal $s_{1n}(\mathrm{t})$.

$$P_1(\mathrm{t}) = a_1(\mathrm{t})\mathrm{s}_{1n}(\mathrm{t}) \tag{6}$$

It contains the highest frequency component of original signal, $P_1(\mathrm{t})$ is a simple component amplitude-modulated and frequency-modulated signal, $a_1(\mathrm{t})$ is the instantaneous amplitude of PF, its instantaneous frequency $f_1(\mathrm{t})$ can be calculated from pure frequency-modulated signal.

$$f_1(\mathrm{t}) = \frac{1}{2\pi}\frac{d[\arccos(\mathrm{s}_{1n}(\mathrm{t}))]}{dt} \tag{7}$$

Step 7: After separating the first PF component from the original signal $x(\mathrm{t})$, it obtains the new signal $u_1(\mathrm{t})$. Taking $u_1(\mathrm{t})$ as the original signal and repeat step1-step6 for k times until $u_k(\mathrm{t})$ is the monotone function.

$$\begin{cases} u_1(\mathrm{t}) = x(\mathrm{t}) - P_1(\mathrm{t}) \\ u_2(\mathrm{t}) = u_1(\mathrm{t}) - P_2(\mathrm{t}) \\ \vdots \\ u_k(\mathrm{t}) = u_{k-1}(\mathrm{t}) - P_k(\mathrm{t}) \end{cases} \tag{8}$$

Ultimately, the original signal $x(\mathrm{t})$ can be expressed by all PF components and the residual $u_k(\mathrm{t})$.

$$x(\mathrm{t}) = \sum_{p=1}^{k} P_p(\mathrm{t}) + u_k(\mathrm{t}) \tag{9}$$

Step 8: Combining the instantaneous amplitude and instantaneous frequency of all PF components again according to time-frequency-amplitude 3-D distribution. Using the LMD to decompose the four operating state vibration signal, the PF components of each state are as fig.3-fig.6.

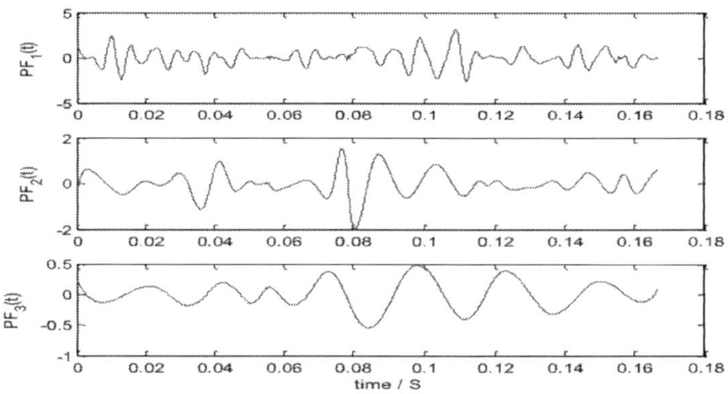

Fig.3: PF component of normal state after LMD

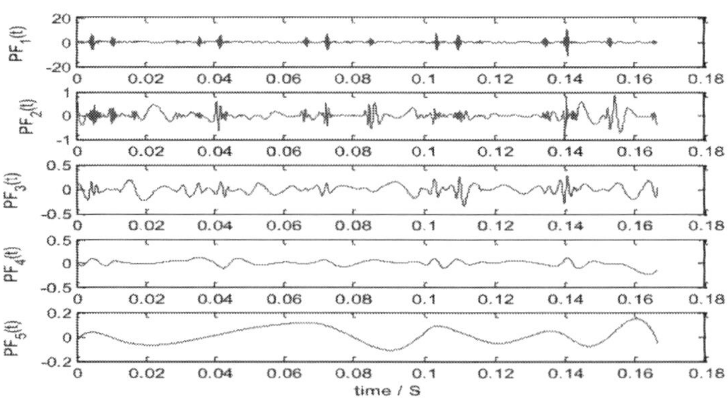

Fig.4: PF component of inner raceway fault after LMD

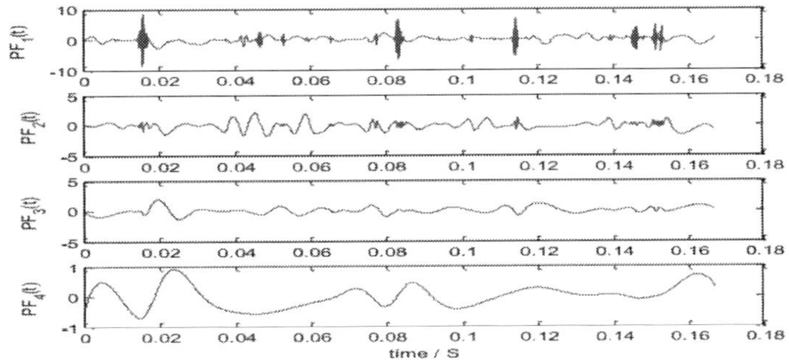

Fig.5: PF component of ball fault after LMD

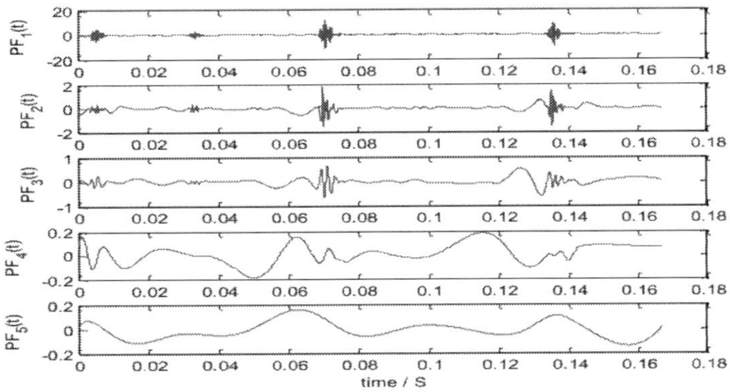

Fig.6: PF component of outer raceway fault after LMD

4 Rotating bearing fault diagnose based on PF kurtosis

Diagnosis information acquisition as well as fault feature extraction and state recognition are fundamental content of rotating bearing's state surveillance and fault diagnosis while fault feature extraction is the most significant and difficult problem. It directly affects the accuracy of diagnosis and reliability of prediction, in a way is the key factor to implement the rotating bearing fault diagnosis.

Kurtosis is a kind of measurement of degree of peak. A big kurtosis corresponds to a sharp wave peak and suggests that the signal has impact nature. As for the rotating bearing vibration signal, sharp peaks are usually caused by fault. For this reason, the paper combines kurtosis and LMD, thereby calculating kurtosis of signal in different frequency band after LMD. When retaining the advantage of kurtosis's sensibility to the fault impact, it studies the distribution of fault impact signal frequency in different frequency band.

Kurtusis is a statistic reflecting degree of peak, it describes the distribution character of vibration signal. The definition of kurtosis K is:

$$K = \frac{E(x-\mu)^4}{\sigma^4} \tag{10}$$

Here, μ is the mean of signal x, σ is the standard deviation of signal x, $E(t)$ is the value of expectation of variable t [7].

In general, fault information is usually included in high frequency signal, as a consequence, the paper opts for the first three PF components of the four state after LMD and calculates their kurtosis in the form of table I.

Table 1: Kurtosis of PF on four state

	PF1	PF2	PF3
Normal Data	4.092	5.400	2.727
Inner Race Fault	35.300	6.283	3.371
Ball Data	14.200	8.866	4.190
Outer Race Fault	54.350	15.350	7.070

From the table, it can be known that when the bearing operating in normal state, there is no impact caused by fault, its vibration signal is relatively steady and kurtosis is comparatively small. When fault occurs, period impact character appears in the vibration signal, the kurtosis of PF containing main fault information increase in large scale. Kurtosis of PF1 has the most great growth, on account that there are many impact components caused by fault in these PFs. Impact from different types of fault has different distribution in PF, thus the kurtosis are in different values just as shown in table I. The kurtosis of PF1 on inner raceway fault increases 9 times, the ball fault is 3.5 times and outer raceway fault 13 times.Therefore the paper determines the fault mode of rotating bearing from the scale of kurtosis increase.

DOI: 10.1051/matecconf/20165410002

5 Conclusions

Wavelet threshold de-noise can remove the noise contribution effectively from vibration signal and expose the real signal to light. LMD has well self-adaptability when processing non-linear and non-stationary signal, and it decomposes the vibration signal into PF components with variety frequency. By calculating the kurtosis of every PF component, it can directly assess the existence of impact in PF component signal with quantification and knows the operating state of bearing from kurtosis. Moreover, it can determines the mode of fault from the growth of kurtosis. Actual vibration signal experiment proves that the fault analysis based on LMD vibration signal process has good performance.

Acknowledgements

The authors of the paper would like to thank College of Electronic and Information to provide us with equipment to finish the experiment and the paper. We also appreciate our classmates in Control Engineer Laboratory for technical support and helpful discussions.

References

1. Y. Amirat, M. Benbouzid, T. Wang, & S. Turri, Performance analysis of an EEMD-based hilbert transform as a bearing failure detector in wind turbines. *2014 First International Conference on Green Energy ICGE*, 193-198, (2014).
2. Z. Junhong & L. Yu, Bearing fault diagnosis based on improved LMD. *2011 International Conference on Transportation, Mechanical, and Electrical Engineering (TMEE). Changchun, China*, 2544-2547, (2011).
3. X. Chen, C. Yin, & W. He, Feature extraction of gearbox vibration signals based on EEMD and sample entropy. *2013 10th International Conference on Fuzzy Systems and Knowledge Discovery (FSKD)*, 811-815, (2013).
4. J. Jenitta & A. Rajeswari, Denoising of ECG signal based on improved adaptive filter with EMD and EEMD. *Proceedings of 2013 IEEE Conference on Information and Communication Technologies (ICT 2013)*, 957-962, (2013).
5. Z. Li, Z. Wan, X. Xiong, & X. Liao, The application of LMD and BP neural network in gear fault diagnosis. *2013 25th Chinese Control and Decision Conference (CCDC)*, 4609-4612, (2013).
6. J. S Smith, The local mean decomposition and its application to EEG perception data. Journal of the Royal Society Interface, **2**, 5, 443-454, (2005).
7. Z. Zhang, W. Li, W. Gong & J. Zhong, An improved EEMD model for feature extraction and classification of gunshot in public places. *21st International Conference on Pattern Recognition (ICPR 2012) Tsukuba, Japan*, 1517-1520, (2012).

MATEC Web of Conferences **54**, 10003 (2016) DOI: 10.1051/matecconf/20165410003
MIMT 2016

Removal of electrocardiogram distortion in portable high-power medical device

Hyung-Sik Kim[1], Sung-Jun Park[1], Tae Kyung Ha[2], Jeong-Han Yi[1]

[1]*Department of Biomedical Engineering, BK21+ Research Institute of Biomedical Engineering, College of Biomedical & Health Science, Konkuk University, Chungju, Chungbuk 380-701, South Korea*

[2]*Department of Surgery, College of Medicine, Hanyang University, Seoul, 133-791, South Korea*

Abstract. In this study, we propose a method to remove bio-signal distortions generated when the power is at transient condition in a high-powered portable medical device. To verify the method, three units were designed that consists of a main control unit, bio-signal measurement unit, and power unit. Also, each units were separated with galvanic isolation. For stable operation at the surge condition, the power unit was composed using push-pull converter topology. The design has been examined in both transient and steady state of the operating conditions. The experiment result shows that galvanic isolation between each units was efficient in high-powered potable medical device.

1 INTRODUCTION

Recently, as medical emergency situations gain interest, the existing electronic medical devices used by connecting power-lines are gradually becoming portable [1, 2]. Especially, as the criteria on ambulance become strict, developments on first-aid medical devices that are powered by compact batteries and stored in vehicles are in progress [3]. In addition, due to the improvement in standard of living, interest in health is increasing, the demand on compact portable medical device constantly raising with easy storage and convenient use to utilize even in a home [4].

Among these medical devices, automated external defibrillator (AED) has become an essential medical device with the tendency of increase in cardiovascular diseases, and each country advises to necessarily provide AEDs within public places, trains, cinemas, planes, and etc., [5, 6]. To allow even the ordinary people to use AEDs at cardiac arrest, they are designed in simple and intuitive interfaces for automatic diagnosis of heart conditions, charging and delivery energies for defibrillations. Also, AEDs can be used in various environments because they are powered by batteries. However, since around 150~360 Joule of energy is needed for heart resuscitation, AED converts the low voltage of its inner battery to a high voltage up to about 1,500 ~ 2,000 volts for capacitor. At this point, to increase the success rate of defibrillation, the voltage of the capacitor has to reach its target voltage in a short time. Therefore it needs a sophisticated power converter design that the size and the efficiency of the overall circuit is considered. Thus, for voltage conversion and fast charging, most of the AEDs are made consuming a lot of current for initial stage of charge. At this time, because of a lot of current from the battery is instantly delivered to the main circuit, bio-signal distortion occurs due to the noise generated from the operation of power converter and drop of battery voltage in a moment. For this reason, the bio-signal diagnostic software generates a wrong signal leading to malfunctioning of an AED. To prevent this condition, various methods are carried out to ensure stable operations such as in aspects of circuits (analogue circuit, power factor correction, and firmware) [7, 8] and software (electrocardiogram detection algorithm) [9, 10], however, the underlying is not being solved.

In this study, each of circuit units such as the main control, power, and bio-signal measurement unit was designed and separated by galvanic isolation to eliminate distortion of bio-signal at the transient condition of power source.

2 Methods

The configuration of the proposed system is presented in Figure 1. The overall system is controlled by ARM family microcontroller, STM32F103D (ST Microsystems, USA), positioned at the main control unit. This microcontroller is composed of a 32-bit CortexTM-M3 CPU, 112 GPIOs (general purpose input output), 21 channels of 12-bit ADC (analog-to-digital converter), 12 channels of DMA (direct memory access) with a maximum operating frequency of 72 MHz, and etc.,

To prevent mutual interference between each circuit units, galvanic isolation which is the electrical separation, is done by having different ground points of over two circuits. In this study, all 3 units are designed in the primary and secondary side. The units are all composed of active circuits, each side needs a power source independently. For the isolation barrier, the clearance is set to 3.5 mm according to the 'Means of Patient Protection (MOPP)' in the IEC60601-3rd edition presented by the international medical device standards [11].

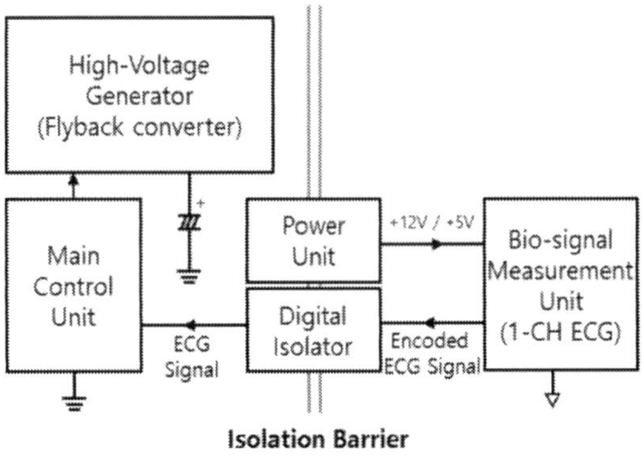

Figure 1. Overall configuration of proposed design.

2.1 Main control unit

The main control unit controls overall system operation, controls status, processing an external input, storage data and monitoring a system, etc. Also, it produces high voltage for heart defibrillation. The high voltage of 1,500 ~ 2,000 volts is generated from a battery voltage of 10.8 volts using fly-back power converter topology (Figure 2).

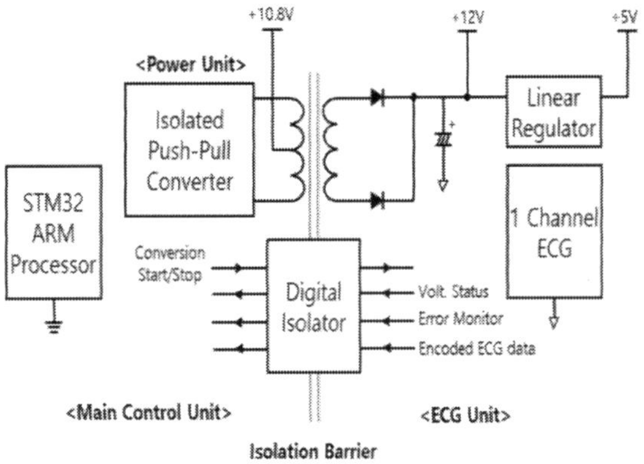

Figure 2. Implementation of main control, power, and bio-signal measurement unit.

2.2 Power unit

The isolated push-pull converter topology is used for the power unit. Since this topology can be simply composed using two transistors and a small size transformer for isolation, it is widely used for compact systems. Considering the degree of voltage drop of the battery voltage, 10.8 volts, at high-voltage charge, the voltage at full charge of the battery, and the voltage drop by a long-term use, the unit is configured to reliably generate a power source of +12 Volts / 0.5 Amperes to the secondary side which is isolated in the range of 9.5 ~ 12 volts. UC3845 current mode PWM converter (Texas Instruments, USA) and PMIC (power management integrated circuit) are used for the configuration. For voltage regulation, the voltage at secondary side is done by a voltage divider using two resistors and the voltage is converted to digital encoded signal through 555 timer. The encoded secondary side voltage is transmitted to the STM32 microcontroller positioned at the primary side using a digital isolator, ADuM2401 (Analog Device, USA) quad-channel digital isolator. This is decoded by STM32 which checks the generated voltage on secondary and controls the UC3845 with turn-on and -off for voltage regulation.

For the isolation transformer, EE1312 magnetic core (TDK, Japan) of ferrite material is used and the turn ratio of primary and secondary was 22 and 27, respectively. A 10 pins of vertical type EE1312 bobbin (Dae Heung Bobbin Co., Ltd, South Korea) was used. The winding is done using 0.25 mm diameter tri-insulation wire (TIW) for the primary side, and 0.3 mm diameter general purpose copper wire for the secondary side. A details of winding was shown in Figure 3. To comply with the standards presented by the international medical device regulations, tri-insulation wire was used. The inductance of primary side was 0.51 mH and leakage inductance was 0.28 μH. The inductance of secondary side was 0.93 mH and leakage inductance was 4.97 μH. The inductances was measured at 100 kHz using impedance analyser HP4192A (Hewlett Packard, USA).

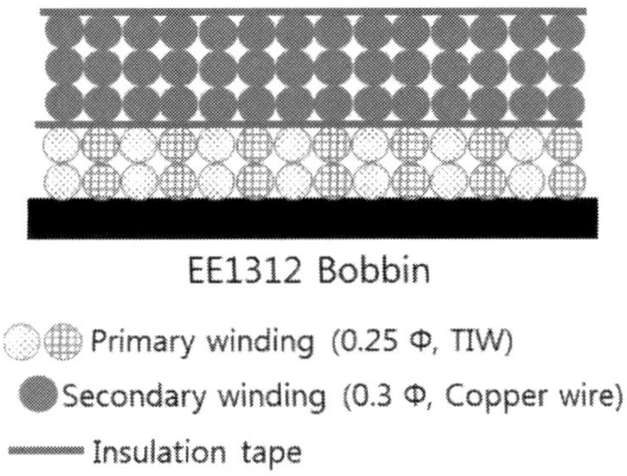

Figure 3. Winding details of transformer for push-pull converter.

The voltage of 12 volts generated at the secondary side was converted into 5 volts using LM7805 (Texas Instrument, USA) linear regulator. The voltage of 12 volts is used for the electrocardiogram amplifier, and 5 volts for electrocardiogram amplifier, 555 timer, digital isolator, and supervisor circuit to detect an invalid operation (Figure 2).

2.3 Bio-signal measurement unit

To measure the cardiac signal, 1-Channel electrocardiogram (ECG) amplifier is configured. Figure 4 shows the schematic of the electrocardiogram amplifier. For the preamp which directly contacts with patients, instrumentation operational amplifier INA128 (Texas Instruments, USA) was used. This amplifier has a 1010 Ωof input impedance, 50 μV offset voltage, 0.5 μV/℃ drift, 120 dB of common-mode rejection ratio, and excellent noise characteristics.

For the right-leg driver and an additional amplification, LMV324 (Fairchild Semiconductor, USA) precision dual Rail-to-Rail output operational amplifier were respectively used. A two first order passive low- and high-pass filter consisted of resistor and capacitor was connected in series for filtering of the cardiac signal. The bandwidth of the bandpass filter which consisted from two passive filter was in range of 0.75 ~ 300 Hz.

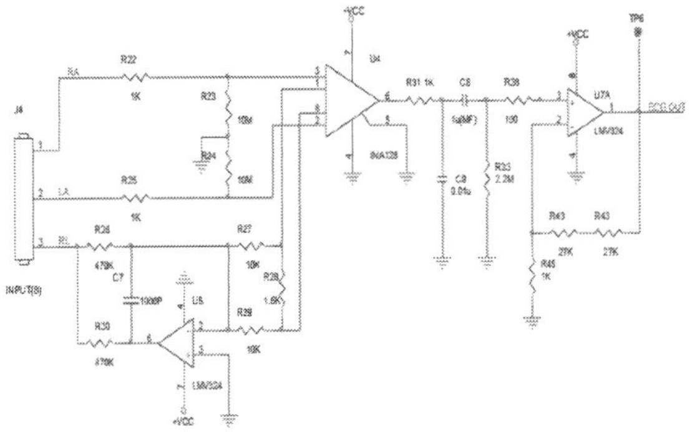

Figure 4. Schematic of 1-channel electrocardiogram amplifier.

The measured electrocardiogram signal is encoded into a digital signal using AD7741 (Analog Device, USA) voltage-to-frequency converter, and transmitted to the main control unit through ADuM2401 (Figure 2). The transmitted signal is input to STM32 which is then encoded and stored to the micro-SD card using DMA channel. The stored ECG data can be viewed in a personal computer (PC) using a custom made viewer.

2.4 Testing the implementation of isolation barrier

The implemented circuit was tested with patient simulator (Model# 217A, Dynatech Nevada Inc, USA) and on two power conditions such as 'Charging Start' and 'Re-Charging' status. The 'Charging Start' state is where the capacitor charged voltage is in 0 volt with no energy charged and the 'Re-Charging' state is a segment where the energy loss occurred by the natural discharge of inside of the circuit's time constant after full recharge is restored. The logged ECG data in the memory card was extracted using a custom maded ECG viewer to text file. Two datas that before and after application of isolation barrier were compared with MATLAB (Mathworks, USA) software.

3 Results

The implemented circuit is shown in Figure 5. The overall size which includes the 3 units is about 9 x 11 cm^2. The area of the proposed circuit was larger about 20 % than prior circuit. The output of the push-pull converter is confirmed to generate stable output voltage of 12 volts within the input voltage range of 9.5 ~ 12 volts. Load and line regulation of the push pull converter was 0.40 % and 0.87 %, respectively.

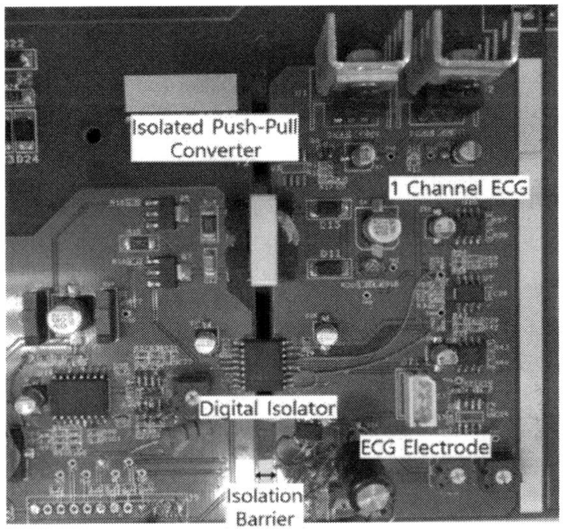

Figure 5. Picture of the prototype implemented units.

Figure 6 shows the custom made ECG viewer. This program could plot entire and enlarged waveform of the electrocardiogram, and record the status of the system and detection results according to the elapsed time; such as Power on/off, ECG lead, Analyzing, Shock advised, Press shock, Ventricular fibrillation (V-Fib), Asystole, Charging, Discharging, and etc.,

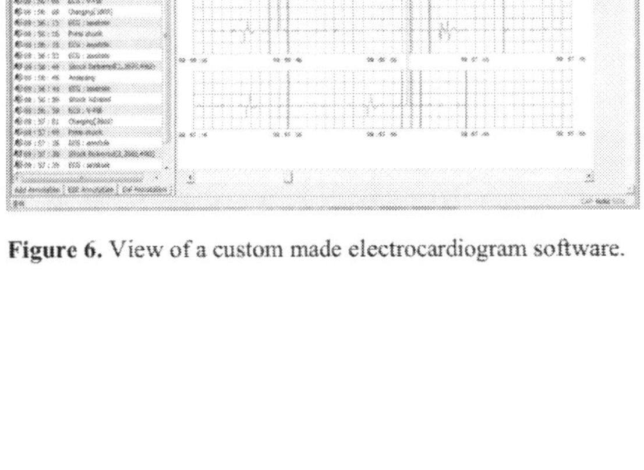

Figure 6. View of a custom made electrocardiogram software.

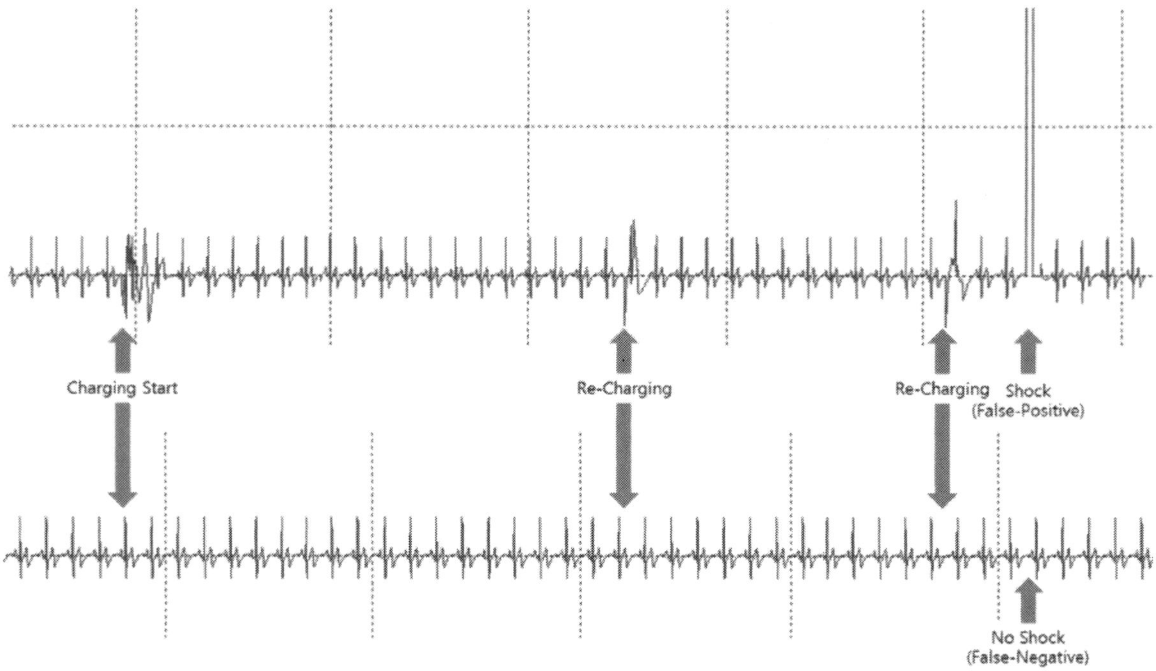

Figure 7. Comparison of the electrocardiogram signal between before (upper) and after (lower) application the isolation barrier.

Figure 7 shows the results of before and after application of the isolation barrier. Before applying the isolation barrier, distortion in EEG signal was occurred at both of the 'Charging Start' and in each of 'Re-Charging' status (Figure 7 upper). Also even in normal situation, an output signal for defibrillation is generated by the distortion (False-Positive condition) from the decision of integrated algorithm inside the STM32 microcontroller to detect and decision a shockable rhythms or not. This situation could be fetal to the normal heart because these energy delivery casuses fibrillation even to a normal person. Especially, the signal distortion is more severe in the 'Charging Start' state compared to the 'Re-Charging' state due to the voltage drop of the battery and high current flows into the uncharged capacitor for faster charge.

After applying the isolation, distortion in ECG waveform was almost disapeared and eliminated (Figure 7 lower). Also the false-positive condition did not occurred.

4 Conclusions

In this study, we proposes a method to attenuate bio-signal distortion generated in high power portable medical devices. By applying galvanic isolation between the main control unit, power unit, and bio-signal measurement unit, the signal distortion on bio-signal generated during energy charge is eliminated.

Although the size of the circuit is about 20% bigger than the size prior to applying the galvanic isolation, this is considered to be overcome by designing a sophisticated printed circuit board (PCB) and downsizing the area of other circuits. Also this approach can reduce the additional time and cost for developing a shockable rhythm detection algorithm to deal with signal distortions. Therefore it is considered that this can be widely applied to small and medium sized medical device companies.

Acknowledgement

This research was supported by the Ministry of Trade, Industry & Energy (MOTIE), Korea Institute for Advancement of Technology (KIAT) through the Encouragement Program for The Industries of Economic Cooperation Region (Grant number R0004491).

References

1. M.Y. Smith, J.D. Depue, C. Rini C, Pain Med. **8**, Suppl 3 (2007)
2. N.J. Lee, J. Starren, S.A. Bakken, AMIA Annu. Symp. Proc, **1021** (2005)
3. G. Pistoia, Batteries for Protable Devices (Elsevier, Rome, 2005)
4. K.M. Conrad, P.A. Reichelt, S.A. Lavenderm J. Gacki-Smith, S. Hattle, Appl. Ergon. **39**, 6 (2008)
5. A.L. Hansell, M. Blangiardo, L. Fortunator, S. Floud, K.D. Hoogh, D. Fecht, R.E. Ghosh, H.E. Laszlo, BMJ, **347**, f5432 (2013)

[a] Corresponding author: hskim98@kku.ac.kr

6. K.G. Kanz, M.V. Kay, P. Biberthaler, W. Russ, S. Wessel, C.K. Lackner, W. Mutschler, Resusci. **62**, 2 (2004)
7. L.A. DeLuca, A. Simpson, D. Beskind, K. Gral, L. Stoneking, U. Stolz, D.W. Spaite, A.R. Panchal, K.R. Denninghoff, Annls Emergen. Med. **59**, 2 (2012)
8. P.F. Meyer, P.D. Gadsby, D. Van Sickle, W.E. Schoenlein, K.S. Foster, G.P. Graber, Med. Biologi. Eng. Comput. **43**, 2 (2005)
9. D.M. González-Otero, J. Ruiz, S. Ruiz, U. Irusta, U. Ayala, E. Alonso, BioMed Res. Int., **2014**, 865967 (2014)
10. M.I. Sobhy, A.E.D. Shehata, Int. Bifurcation. Chaos, **10**, 12 (2000)
11. International Electrotechnical Commission, IEC60601 3rd Ed. (2013)

Computational Study of Fluid Flow Inside a Concentric Reducer

N.Subaschandar

Department of Mathematics and Computational Sciences, BIUST, Palapye, Botswana

Abstract. Scale growth in pipe fittings is a major problem for an alumina refinery. Investigation into the scale growth mechanism at an alumina refinery found almost 60% more scale growth in concentric reducer when compared with the connecting straight pipe sections for similar flow conditions. The current study is an attempt to characterise the fluid flow inside a concentric reducer, especially streamwise and crossflow fluctuations using computational fluid dynamics methodology. The present study shows that streamwise component of instantaneous velocity fluctuation decreased and cross component of the instantaneous velocity fluctuation increased inside the concentric reducer. Present Computational results have shown good agreement with the experimental data.

1 Introduction

In alumina industry, scale growth in pipe fittings is a major problem for an alumina refinery. Investigation into the scale growth mechanism at an alumina refinery found almost 60% more scale growth in concentric reducer when compared with the connecting straight pipe sections for similar flow conditions[1]. In the present research an attempt is made to study the behaviour of streamwise and crossstream components of instantaneous fluctuating velocity inside the concentric reducer. A fully developed turbulent flow through an axisymmetric contraction producing constant strain rate was studied in the work[2]. They found that the streamwise component of fluictuating velocity is decreased and there was a similar increase in cross flow components in such a way that total turbulent kinetic energy remains constant in the initial section of the contraction. The streamwise component almost vanishes in the later sectFion of the contraction[2]. Recently, extensive measurements, mainly focusing on the behaviour of fluctuating velocity components as the flow goes through the reducer, have been carried out and presented in[1]. It was reported a decrease in the streamwise component of fluctuating velocity and an increase in the crossflow components of fluctuating velocity as the flow goes through the reducer[1]. Flow through contractions have been studied in[3-6]. Also axisymmetric flow through an expansion-contraction duct was studied in[7]. The present computational work studies the fluid flow through the concentric reducer, mainly focusing on the behaviour of fluctuating velocity components.

2 Computational model

The concentric reducer studied here is shown in the Figure 1. Computational domain extends on both side of the reducer by a distance of 4 diameters lengths. The water is the flow medium and flow is turbulent. Fluid is treated as incompressible fluid and, because of symmetry, the flow is treated as axisymmetric thus reducing the grid size and computational time. A uniform speed of 0.27m/s is considered at the inlet plane. Reynolds number based on the uniform speed of water and upstream diameter of the reducer is about 27,000. Pressure outlet condition was used at the exit plane. K-ε turbulence model was used to model turbulence for all computations. Computations were carried out using the commercially available CFD software SOLIDWORKS. The computational domain had 112000 cells inside the reducer and this grid size was reached after carrying out grid independence study. Streamwise and crossflow fluctuating velocity components were computed at several streamwise stations inside the concentric reducer, as shown in Figure 2, to compare present computational results with experimental results[1].

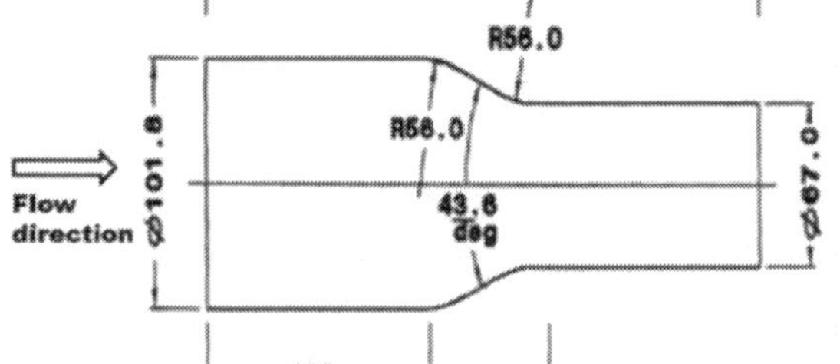

Figure 1 Sketch of the reducer geometry(All dimensions are in mm).

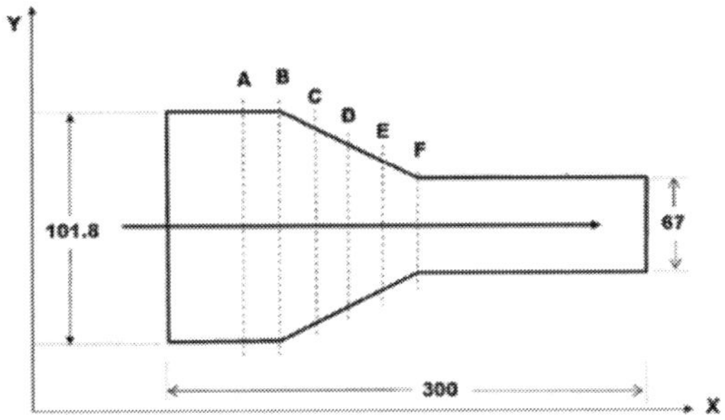

Figure 2 Streamwise locations at which the comparisons are made. AB=BC=CD=DE=EF=8. (All dimensions are in mm).

3 Results and Discussion

Figure 3 shows the comparison between computed streamwise component of velocity fluctuations and experimental data[1] at various streamwise locations inside the concentric reducer. It can be seen that the agreement between the computed and experimental results is reasonably good.

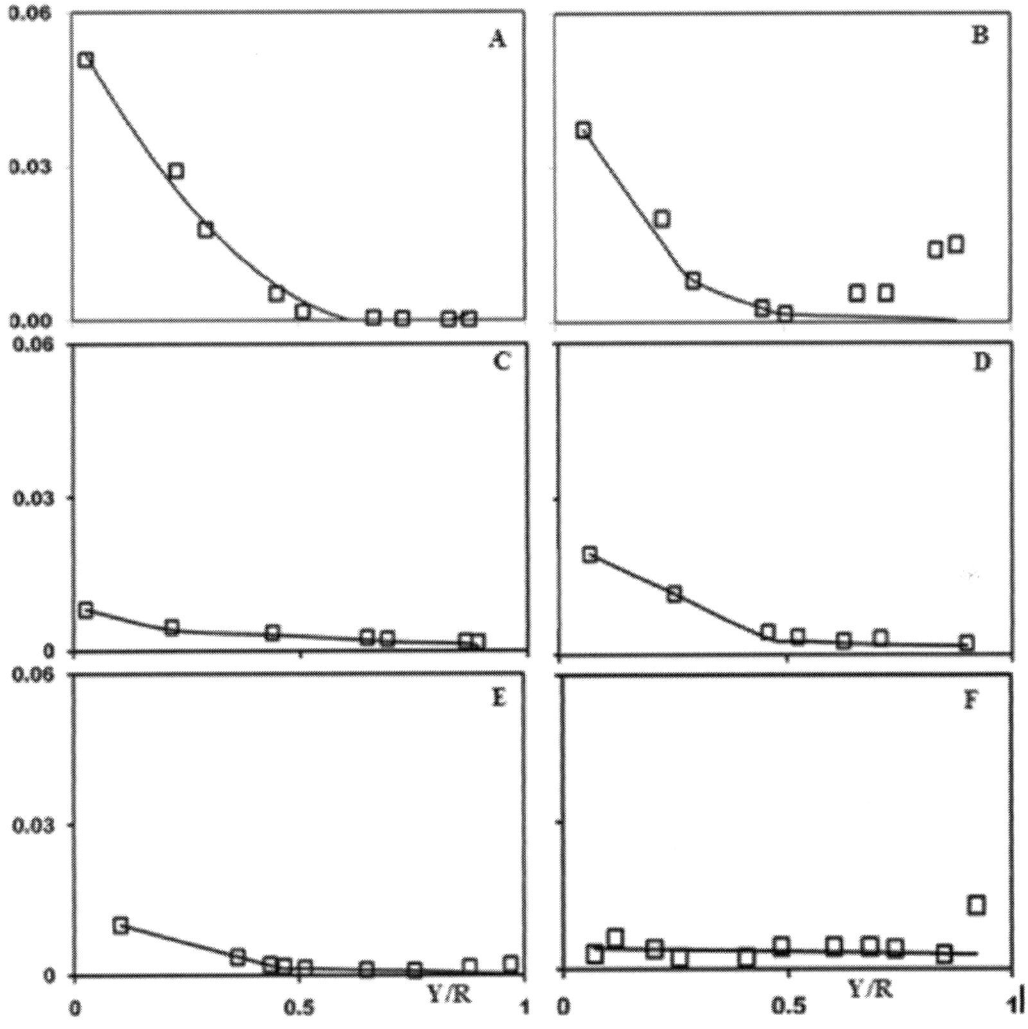

Figure 3 Variation of streamwise component of velocity fluctuation(u) through different sections.
Vertical axis –(u^2/U^2)

In the Figure 3B, the agreement between the computed result and experimental data is not good in later part of the graph; computations deviate from the experimental data as normal distance is increased. Figure 4 shows the comparison of computed results with that of the experimental data[1]. It can be seen from this figure that the agreement is not good, eventhough the computational results follow the trend of the experimental data. The experimental data has lot of scatter, which can be seen in the Figure 4, that makes comparison between computed results and experimental data very difficult. Figure 5 shows the comparison of computed results with that of the experimental data[1]. It can be seen from this figure, that even though there is a large scatter in the experimental data the computations follow the trend of the experimental results. We can see that the computations capture the small gradual decrease in the streamwise component of fluctuating velocity and the rise and fall of cross stream fluctuating velocity component along the streamwise direction at a location Y/R=0.2,which is near the wall. It is suspected that one reason for the reduction in the streamwise and crossstream components of fluctuating velocity could be the increase in the fluid velocity inside the reducer.

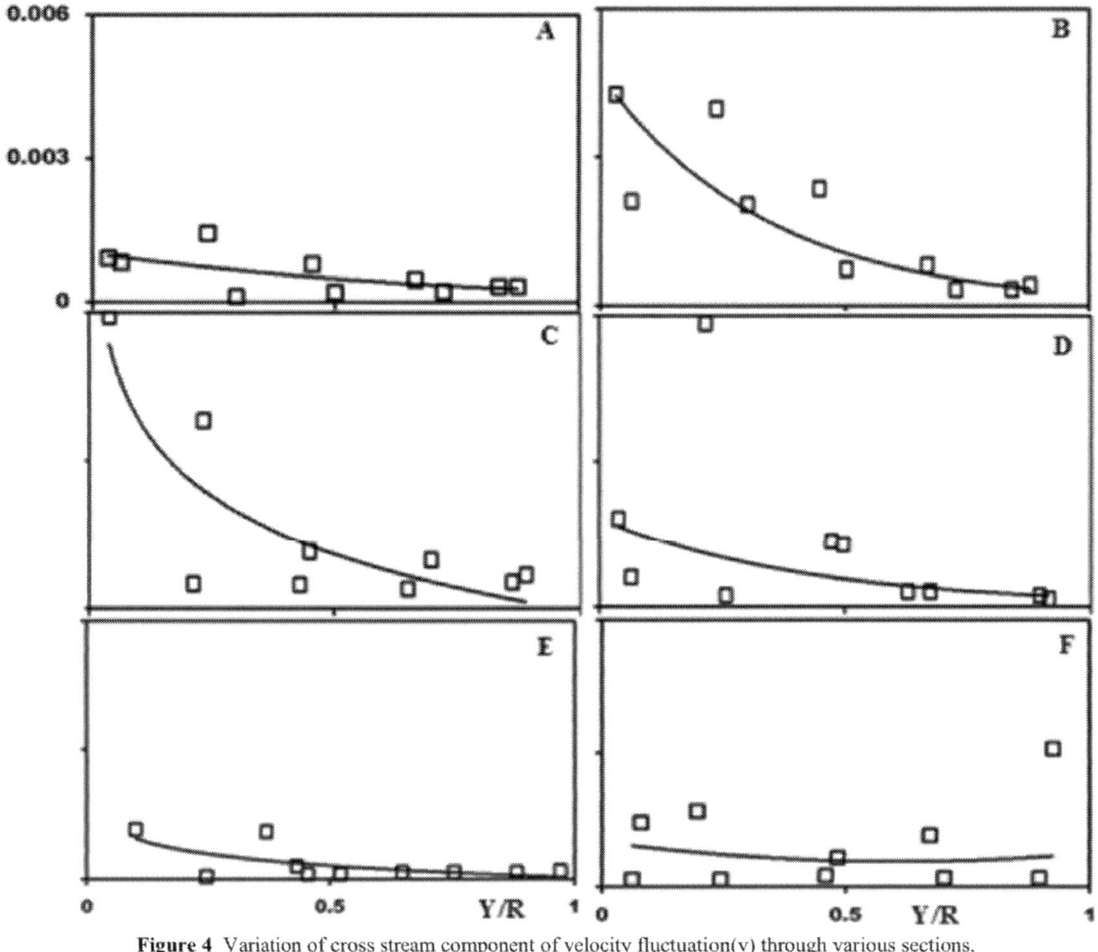

Figure 4 Variation of cross stream component of velocity fluctuation(v) through various sections.
Vertical axis – (v^2/U^2)

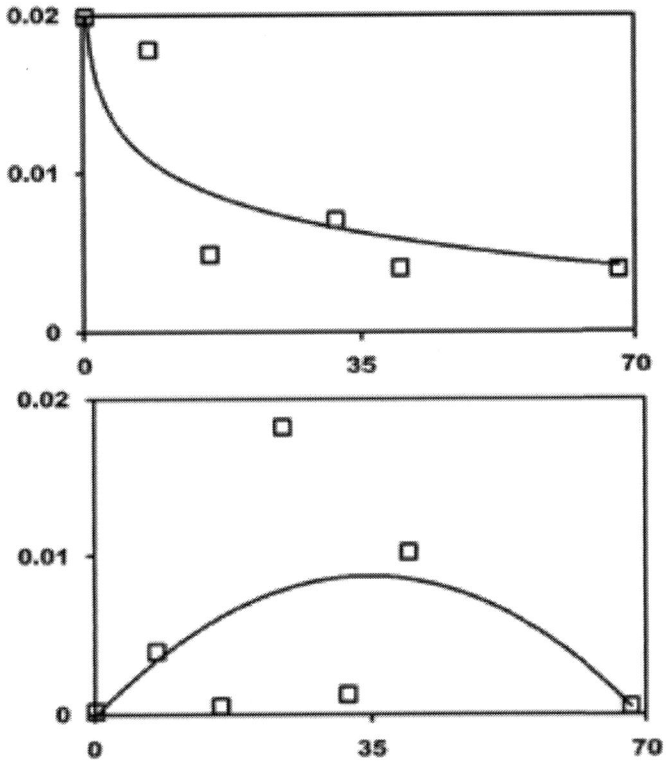

Figure 5 Variation streamwise and cross stream fluctuating velocity components near wall. Vertical axis for (A) : (u^2/U^2); For (B) : (v^2/U^2)

Fluid flow inside a concentric reducer is studied computationally. Computed results compare reasonably well with the experimental data. Computed results show that the streamwise component of fluctuating velocity decreases with normal distance. The computations do not predict the behaviour of crossstream fluctuating velocity component, but the trend is captured. The computations predict the variation of streamwise component of fluctuating velocity along the axial distance at a location close to the wall reasonably well. But the cross stream component of velocity fluctuation is not predicted satisfactorily. It is suspected that the increase in mean velocity in the reducer could be a reason for the decrease in the streamwise intensity.

Acknowledgements

Author acknowledges the support received from the Department of Mathematics and Computational Sciences and BIUST authorities in carrying out this research.

References

1. T. Rashid, Master's thesis, Dept. of Mech. Engg., Central Queensland University, QLD, (2005).
2. L. E. Tobergsen and P. Krogstad, Proceedings of the 13thAustralasian Fluid Mech. Conf., Monash University, Melbourne, **1**, 378 (1998).
3. E. M. Kim, R. A. Brown, and R. C. Armstrong, J. of Non-Newtonian Fluid Mech., **13**, 341, (1983).
4. J. P. Rothstein and G. H. McKinley, J. of Non-Newtonian Fluid Mech., **86**, 61 (1999).
5. J. P. Rothstein and G. H. McKinley, J. of Non-Newtonian Fluid Mech., **98**, 33 (2001)
6. P. Das, M. M. K Khan, M. G. Rasul, and S C. Saha, Proc. of 11th Int. Conf. on Heat transfer, Fluid Mechhermodynamics, **289** (2012).
7. M. Tachibana, N. Kawabata, and M. Sakamoto, Nihon Reororji Gakkaishi, **15**, 210 (1987)

MATEC Web of Conferences **54**, 11002 (2016) DOI: 10.1051/matecconf/20165411002
MIMT 2016

Mean Velocity in the Asymmetric near Wake Behind a GAW(2) Airfoil

N. Subaschandar

Department of Mathematics and Computational Sciences, BIUST, Palapye, Botswana

Abstract. The two-dimensional asymmetric near wake behind a GAW(2) airfoil section is studied at low speeds. Mean velocity measurements have been made in the near wake. Mean velocity profiles in the near wake are presented. Also integral parameters based on the mean velocity profiles have been presented here. In the near wake region mean velocity profiles are seen to exhibit symmetry within a distance of about 60 trailing edge momentum thicknesses downstream of the trailing edge.

1 Introduction

Turbulent wake region is the region of flow behind the trailing edge of a body where the upstream boundary layers merge into one single shear layer. The evaluation and understanding of flow in the in the near wake is important as the flow in this region is still influenced by the upstream trailing edge conditions. Most of the investigations in the near-wake, theoretical as well as experimental are confined to the two-dimensional wake flows [1–4]. At the edge of the trailing edge of a streamlined body, where the boundary layers transform into a wake, the skin friction suddenly drops to zero, and at the same time, the wall restriction on the normal-to-wall velocity component is removed. In symmetric wakes, this allows mixing across the wake centreline and initiates inner near wake[1]. In asymmetric wakes, both mean flow and turbulent transport across the wake centreline takes place. Therefore, the relative intensities and the relative size of the turbulence on both sides become important in the development of wake. In addition, there is longer-range influence due to equalisation of the pressure field, when there is a pressure variation in the trailing edge region, as in the case of airfoil experiencing lift. In general, these effects appear as combined and interact with each other. Leuchter[5] studied asymmetric generated downstream of a flat plate by contouring the ceiling and the floor of the tunnel differently. The asymmetry obtained in this way is somewhat closer to the airfoil wake, but is mild and makes the external flow itself asymmetric. Studies[6,7], on the other hand, used a resistance such as a screen to retard flow on one side of the body. The asymmetric wake generated in this way is more akin to a mixing layer between two different streams of different total pressures, and they show wake-like behaviour only in the region close to the trailing edge, and quickly start to show the mixing layer type of profiles. The aim of the present study is to analyse and understand near wake flow developing from asymmetric trailing edge conditions. The GAW(2) airfoil which is a 13% thickness General Aviation Wing was chosen for this study because of its strong relevance in many applications. GAW(2) airfoil section was not modified in anyway to make the flow less complicated at the trailing edge. GAW(2) is an asymmetric airfoil so that the asymmetry at the trailing edge is natural, unlike the experiments referred in[8,9]. Research study [8] modified the top surface of symmetric flat plate by gluing sandpaper for the full length of the model. Study carried out in[9] used a flexible flat plate to get asymmetric conditions at the trailing edge. Research work[10] studied extensively the three-dimensional asymmetric turbulent near wake behind a swept wing with GAW(2) airfoil cross section. Our focus was concentrated in the wake region close to the trailing edge until about 75 trailing edge momentum thicknesses distance downstream of the trailing edge.

2 Experimental Setup

The experiments were conducted in the 300x1500mm boundary layer wind tunnel of the National Aerospace Laboratories, Bangalore. The GAW(2) airfoil model, having a chord length(C) of 600mm and a span of 300mm, was mounted vertically in the test section. The experiments were carried out at a freestream velocity of 30 m/s, providing a chord Reynold number of $1\text{x}10^6$. GAW(2) airfoil is an asymmetric airfoil and it was not modified in any way to avoid flow complexities close to the trailing edge. Figure 1 shows the schematic of the experimental setup. Mean velocity measurements were carried out in the wake in the distance range of 1-75 trailing edge momentum thicknesses behind the trailing edge.

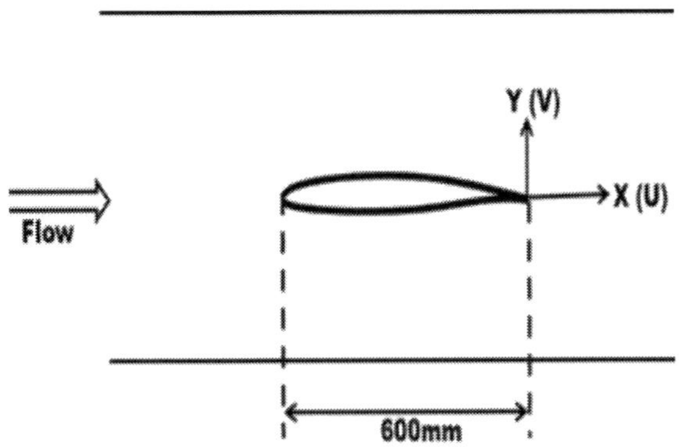

Figure 1 Schematic of the experimental setup

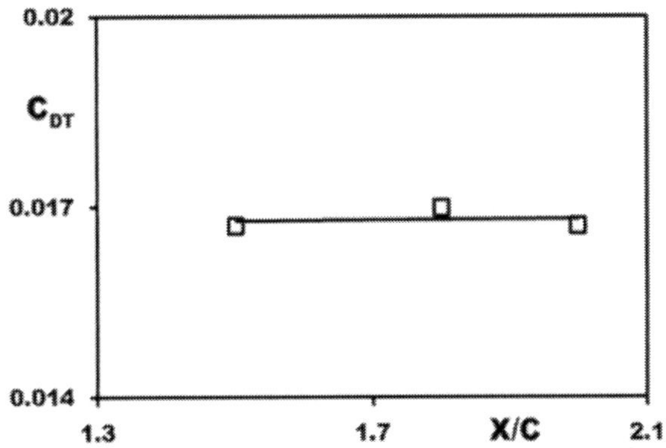

Figure 2 Total drag coefficient estimated using wake survey method at three stations.

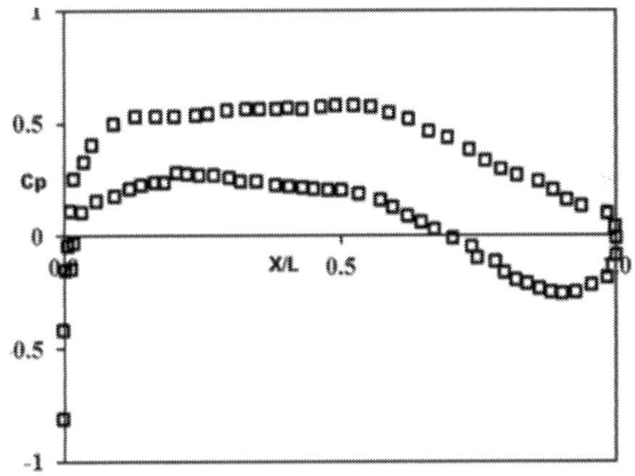

Figure 3 Pressure distribution on the airfoil model

3 Initial Conditions

The development of wake behind any streamlined body depends on the initial conditions of the flow at the trailing edge. The values of displacement thickness, momentum thickness and shape factor for both top and bottom surfaces of the airfoil at a location just upstream of the trailing edge are given in the Table 1. The values of gross parameters on the top and bottom surfaces indicate the existence of strong asymmetry at the trailing edge. Total drag was computed by measuring mean velocity profiles at three streamwise stations(namely X/C=1.5, 1.75 and 2.0) and using wake survey method. The total drag coefficient at these three streamwise stations are plotted and shown in Figure 2. The plot shows very little difference in the values of total drag at these streamwise stations confirming the existence of two-dimensionality in the flow.

Table 1 Paramaters at the trailing edge

Parameter →	δ*(mm)	θ(mm)	H
Top surface	3.6	2.1	1.534
Bottom surface	0.8	0.5	1.416

4 Results and Discussion

Figure 3 shows the static pressure distribution on the airfoil along the chord length. The pressure coefficient distribution indicates that the boundary layer is only under a moderate adverse streamwise pressure gradient and the flow is attached near the trailing edge. These observations are supported by the fact, as it will be seen later, that the shape factor has a value of about 1.5 at the trailing edge.

Figure 4 shows the development of streamwise component of mean velocity profiles at various locations in the near wake. The profiles are asymmetric with respect to x-axis and also the minimum velocity location starts shifting toward the lower side of the wake as the streamwise distance increases. Mean velocity profiles display a tendency towards symmetrisation by a distance of about 60 trailing edge momentum thicknesses. The location of Y_m where the minimum velocity occurs is laterally displaced signifying that there is a net lateral (i.e. downward) transfer of momentum within the wake.

Figures 5 and 6 show the development of minimum value of mean velocity and wake half-thickness with streamwise distance in the wake. Wake half-thickness b is the sum of the wake half-thicknesses on the upper and lower sides of the wake. Turbulent diffusion makes wake half-thickness and minimum velocity increase rapidly with streamwise distance.

Figures 7 and 8 show the development of momentum thickness and shape factor along the streamwise distance. One can observe a rapid decrease in the values of momentum thickness near the trailing edge and thereafter the momentum thickness is nearly constant.

 DOI: 10.1051/matecconf/20165411002

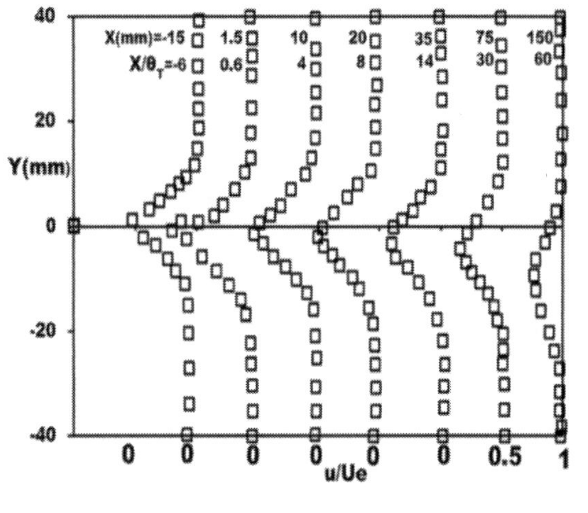

Figure 4 Profiles of streamwise component of mean velocity in the near wake

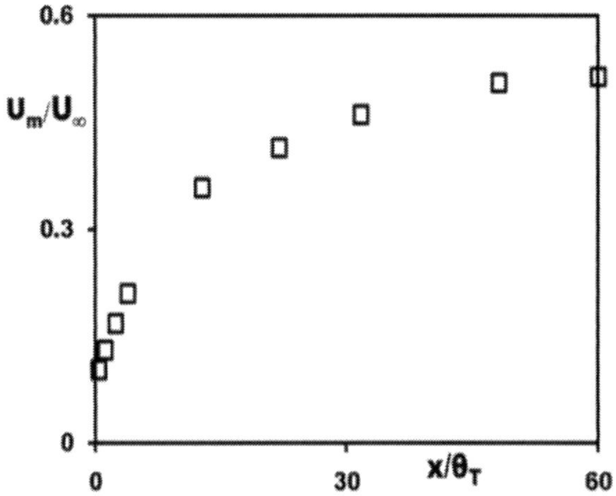

Figure 5 Variation of minimum velocity in the near wake

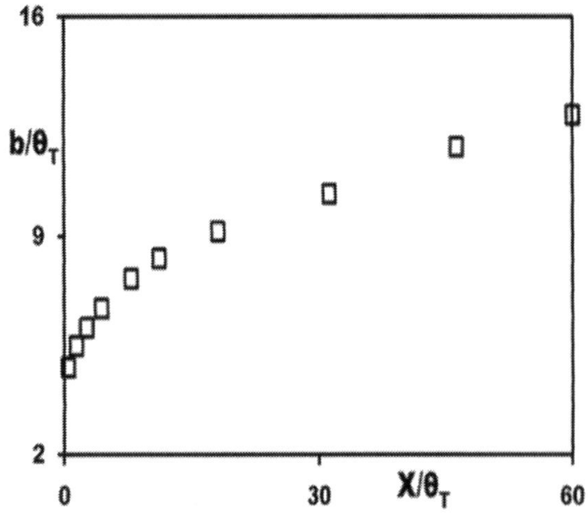

Figure 6 Variation of Wake half thickness in the near wake

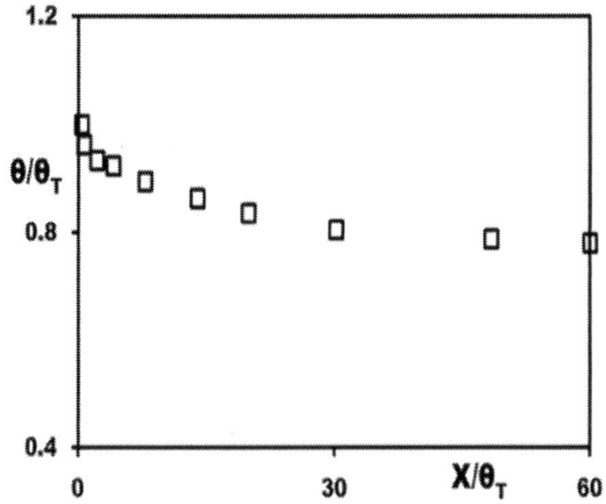

Figure 7 Momentum thickness variation in the near wake.

The shape factor evolves rapidly in the region upto $\frac{X}{\theta_T} = 30$, and thereafter the evolution is slow; by a distance of 60 trailing edge momentum thicknesses the shape factor only slightly greater than 1, its asymptotic value at very large streamwise distances.

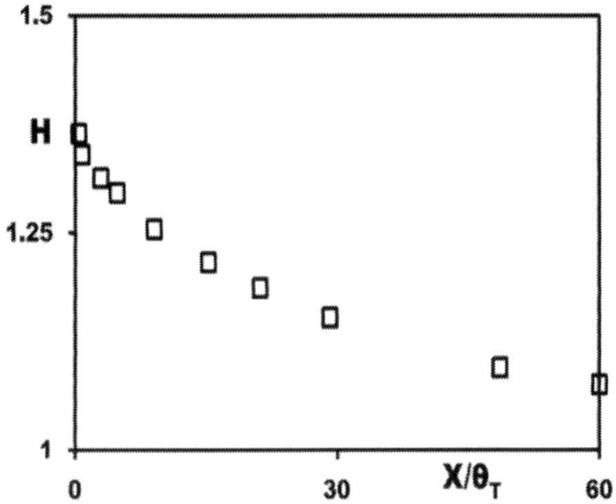

Figure 8 Variation of shape parameter in the near wake

5 Conclusions

Mean velocity measurements have been carried out in the near wake flow region behind a GAW(2) airfoil at low speeds. Profiles of mean velocity exhibit asymmetry close to the trailing edge and show a tendency toward symmetrisation around a distance of 60 trailing edge momentum thicknesses. The present experiments indicate that far-wake structure has not been attained within a distance of 60trailing edge momentum thicknesses. The data presented in this paper would be useful to validate various turbulence models, numerical schemes and computational codes.

Acknowledgements

Author acknowledges the support received from the Department of Mathematics and computational Sciences and BIUST authorities in carrying out this research

References

1. J. Andreopolous and P. Bradshaw, J. of Fluid Mech., **100**, 639 (1980).
2. R. Chevary and L. S. G. Kovaznay, AIAA J., **8**, 1641 (1969).
3. B. R. Ramaprian, V. C. Patel, and M. S. Sastry, AIAA J., **20,** 1228 (1982).
4. N. Subaschandar and A. Prabhu, J. of Aerospace Sci. and Tech., **1**, 61 (1999).
5. O. Leuchter, *Turbulence in Internal Flows*, Editor. S. N. B. Murthy, Hemisphere Co., Washington D.C., 371(1974).
6. B. E. Thomson, J. H. Whitelaw, J. of Fluid Mech., **157**, 305 (1985).
7. M. A. Badri Narayanan, S. Raghu, and E. G. Tulapurkara, AIAA J., **23**, 97 (1983).
8. B. R. Ramaprian, V. C. Patel, and M. S. Sastry, Iowa Institute of Hydraulic Research Report 231, University of Iowa, Iowa City, USA.(1981)
9. A. Nakayama, H. P. Kreplin, Phys. of Fluids, **6**, 2430 (1994).
10. N. Subaschandar and A. Prabhu, J. of Aircraft, **39**, 290 (2002).

MATEC Web of Conferences **54**, 11003 (2016) DOI: 10.1051/matecconf/20165411003
MIMT 2016

Mean Velocity Charateristics in the Far Wake behind a GAW(2) Airfoil

N. Subaschandar

Department of Mathematics and Computational Sciences, BIUST, Palapye, Botswana

Abstract. The two-dimensional asymmetric far wake behind a GAW(2) airfoil section is studied at low speeds. Mean velocity measurements have been made and analysed in the far wake, where initial asymmetric wake has become symmetric with respect to the minimum velocity line. Governing equations have been solved with appropriate boundary conditions in the far wake. Self-similarity solutions have been obtained for the governing equations for the mean velocity distributions. Results of self-similarity analysis have been compared with experimental data.

1 Introduction

Turbulent wake region is the region of flow behind the trailing edge of a body where the upstream boundary layers merge into one single shear layer. The far wake region is at a large distance from the trailing edge where the flow is less influenced by the initial conditions at the trailing edge. The evaluation and understanding of flow in the far wake is important as the flow in this region is solely dependent on the gross parameters like total drag of the body. The aim of the present study is to analyse and understand far wake flow developing from initial asymmetric trailing edge conditions. The GAW(2) airfoil which is a 13% thickness General Aviation Wing was chosen for this study because of its strong relevance in many applications. Many of the studies carried out in the past were concentrated in the symmetric far wake flows[1-5]. Self-similarity solutions have been obtained in the symmetric far wake in such flows. The far wake flow, which is developing from initial asymmetric conditions at the trailing edge, has not been studied well. The present research was started with an aim to describe far wake, with initial asymmetric conditions, in terms of self-similarity solutions for governing equations. The three-dimensional asymmetric turbulent near wake behind a swept wing with GAW(2) airfoil cross section has been studied extensively[6]. Their studies showed that the turbulent wake which was asymmetric near the trailing edge exhibited symmetry by about a distance of 60 trailing edge momentum thicknesses downstream of the trailing edge[6]. With this in mind, our focus was concentrated in the wake region beyond a distance of 200 trailing edge momentum thicknesses downstream of the trailing edge so that symmetry has been reached in the region of our interest.

2 Experimental Setup

The experiments were conducted in the 300x1500mm boundary layer wind tunnel of the National Aerospace Laboratories, Bangalore. The GAW(2) airfoil model, having a chord length of 600mm and a span of 300mm, was mounted vertically in the test section. The experiments were carried out at a freestream velocity of 30m/s, providing a chord Reynold number of $1\text{x}10^6$. Figure 1 shows the schematic of the experimental setup. Mean velocity measurements were carried out in the wake in the distance range of 1-800 momentum thicknesses behind the trailing edge. The mean velocity profiles are presented at three streamwise stations furthest from the trailing edge so as to avoid the uncertainty regarding the symmetry of the flow.

3 Similarity Analysis

In this section, an attempt is made to extend the self-similarity analysis which is valid for symmetric two-dimensional far wake to the far wake flow evolving from initial asymmetric trailing edge conditions. In this section far wake flow is analysed using continuity and momentum equations. Most of the analysis follows the steps of two-dimensional far wake analysis given in[4].

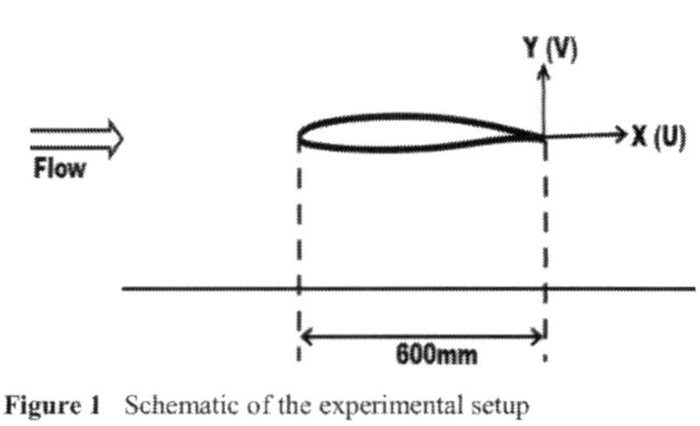

Figure 1 Schematic of the experimental setup

With the notations shown in Figure 1, the equations for the far wake flow are

$$\frac{\partial U}{\partial X}+\frac{\partial V}{\partial Y}=0 \tag{1}$$

$$U\frac{\partial U}{\partial X}+V\frac{\partial U}{\partial Y}=\frac{1}{\rho}\frac{\partial \tau}{\partial Y} \tag{2}$$

The freestream velocity outside the wake is assumed to be constant(U_∞) and $\tau=\mu\frac{\partial U}{\partial Y}-\rho\overline{uv}$ is the total shear stress. We seek a similarity solution of the form

$$\tau=\rho W_0^2 g(\eta),\ W=W_0 f(\eta) \tag{3}$$

where $W = U_\infty - U$ and $\eta = Y/B$ and W_0 and B are the local velocity and length scales, respectively. Eqs 1 and 2 reduce to

$$\left[\frac{U_\infty B'}{W_0}-\underline{B'f}\right]\eta f'-\left[\frac{U_\infty W_0'}{W_0^2}B-\underline{\frac{W_0'B}{W_0}f}\right]f-\underline{f'\int\left\{\frac{W_0'B}{W_0}f-B'\eta f'\right\}d\eta}-g'=0 \tag{4}$$

At large distances from the body, the velocity defect is small, i.e $W_0 << U_\infty$, and the underscored terms are of second order or smaller. Hence

$$\frac{(U_\infty B)'}{W_0}\eta f'-\frac{B(U_\infty W_0)'}{W_0^2}f-g'=0. \tag{5}$$

If this asymptotic wake is self-preserving, then Eq. 5 leads to the following well-known result for a far wake with a nonzero momentum thickness[7,8]

$$B\sim X^{1/2}\ \text{and}\ W_0\sim X^{-1/2}. \tag{6}$$

These half-power growth and decay laws can therefore be used to measure the approach of the mean flow to asymptotic conditions.

In order to evaluate the constants in Eq. 6 and determine the velocity profile f(η), it is necessary to invoke a turbulence closure model. Thus, for example, introduction of a constant eddy-viscosity , ν_T, viz

$$\frac{\tau}{\rho}=\nu_T\frac{\partial U}{\partial Y}=\nu_T\frac{W_0}{B}f'\ \text{or}\ g=\frac{-\nu_T}{W_0 B}f'.$$

reduces Eq.5 to

$$\frac{\nu_T}{W_0 B}f''+\left(\frac{U_\infty B'}{W_0}\right)\eta f'-\left(\frac{U_\infty B W_0'}{W_0^2}\right)f=0. \tag{7}$$

which is identical with that for laminar flow except that **v** has to be replaced by ν_T [7]. With the appropriate boundary conditions, the solution of Eq.7 is

$$f=\frac{W}{W_0}=\exp(-4\eta^2\ln 2). \tag{8}$$

$$B=4\sqrt{\ln 2}\left(\frac{\nu_T X}{U_\infty}\right)^{1/2}. \tag{9}$$

$$\frac{W_0}{U_\infty}=\frac{1}{\sqrt{\pi}}\left(\frac{U_\infty C_d d}{\nu_T}\right)^{1/2}\left(\frac{C_d d}{X}\right)^{1/2}. \tag{10}$$

where $C_d=\frac{D}{(\frac{1}{2})\rho U_\infty^2 d}$ is the drag coefficient and d is a characteristic length of the body. The overall momentum shows that $C_d d = 2\theta$ where

$$\theta=\int_{-\infty}^{\infty}\frac{U}{U_\infty}\left(1-\frac{U}{U_\infty}\right)dY.$$

is the constant, asymptotic momentum thickness of the wake. Using **θ** as a length scale, equations (9) and (10) become

$$\left(\frac{B}{\theta}\right)^2=16\left(\frac{\nu_T}{U_\infty\theta}\right)\left(\frac{X}{\theta}\right)\ln(2). \tag{11}$$

$$\left(\frac{U_\infty}{W_0}\right)^2=4\pi\left(\frac{\nu_T}{U_\infty\theta}\right)\left(\frac{X}{\theta}\right). \tag{12}$$

From an examination of some early results from the wake of cylinders, Schlichting deduced a value of 0.0444 for $\frac{\nu_T}{U_\infty\theta}$ [7]. However, Townsend quotes a value of 0.032[8], which is identical with the average value deduced from the asymptotic wake data[5]. A value of 0.032 was found to be suitable in the present analysis also. It should be noted that the constants in the above results are derived on the basis of a constant eddy-viscosity model, which is not entirely realistic in view of the marked intermittency of the flow over a large part of the wake.

4 Results and Discussion

In this section, results of the above analysis are compared with the data measured in the asymmetric wake behind a GAW(2) airfoil. The appropriate value of θ was determined from the data of the most downstream station in order to avoid any uncertainties occurring from the measurements near the trailing edge. Figure 2 shows measured mean velocity profiles in the far wake flow at several streamwise stations. Since it was observed that the velocity profiles had reached symmetry only one half of the profiles are shown in the figure. Also, shown in the figure is the self-similar exponential curve given in the Eq. 8. It can be seen from this figure that the measured data are in good agreement with the exponential asymptotic profile except for some small departures near the outer edge of the wake.

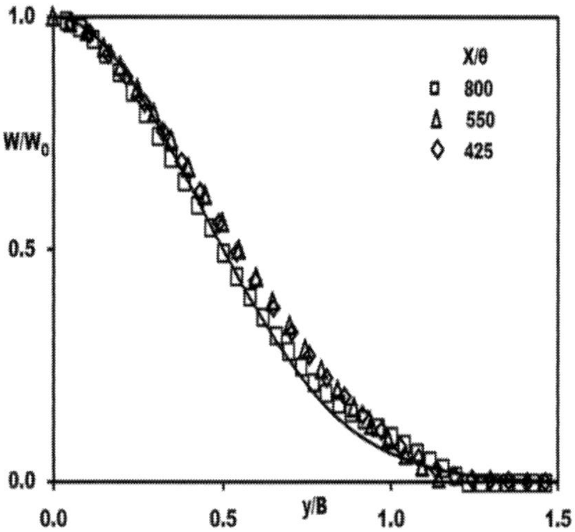

Figure 2 Mean velocity profiles in similarity variables.

Figure 3 shows the variation of wake half thickness in the far wake. This figure also shows the relationship for wake half thickness as given by the Eqs.11 with $\frac{\nu_T}{U_{\infty}\theta} = 0.032$. From this figure it is clearly seen that the wake half thickness does not follow the asymptotic power laws with the prescribed constant value of eddy-viscosity until about X/θ = 300 and after that the agreement between the measured data and the analytical results is reasonably good.

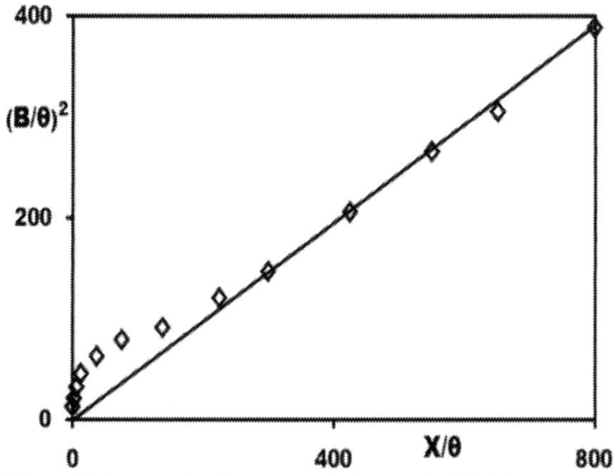

Figure 3 Streamwise development of wake half-thickness

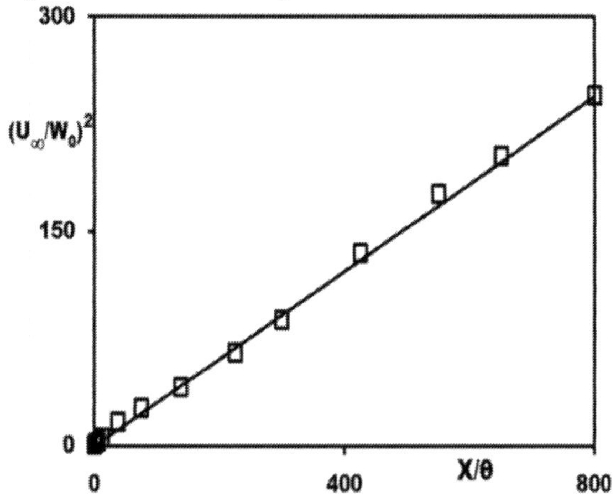

Figure 4 Streamwise development of inverse of maximum velocity defect

Figure 4 shows the variation of inverse of maximum velocity defect in the far wake. This figure also shows the relationship for inverse of maximum velocity defect as given by the Eq. 12 with $\frac{\nu_T}{U_{\infty}\theta} = 0.032$. From this figure it is clearly seen that the inverse of the maximum velocity defect does not follow the asymptotic power law with the prescribed constant value of eddy-viscosity until about X/θ = 300 and after that the agreement between the measured data and the analytical results is reasonably good.

In summary, the experimental data in the turbulent wake behind a GAW(2) airfoil indicate that the velocity profiles have become symmetric in the far wake. The experimental data in the turbulent far wake behind the GAW(2) airfoil indicate that the wake becomes independent of the initial conditions and approaches as asymptotic state in mean velocity at around X/θ = 300.

5 Conclusions

Mean velocity measurements have been carried out in the far wake flow region behind a GAW(2) airfoil. It is shown that the self-similarity analysis of a symmetric two-dimensional far wake flow can be extended to the far wake flow of an initially asymmetric wake flow. The results of self-similarity analysis agree well with experimentally measured mean velocity data in the far wake. The wake half-thickness and inverse of maximum velocity defect agree well with the experimental data in the far wake and shown to follow the half power laws derived in the self-similarity analysis.

Acknowledgements

Author acknowledges the support received from the Department of Mathematics and computational Sciences and BIUST authorities in carrying out this research.

References

1. R. Chevray and L. G. Kovasznay, AIAA J., 7(8), 1641, (1969).
2. P. J. Pot, NLR TR 79063, National Aerospace Laboratory NLR, Netherlands, (1979).
3. J. Andreopoulos and P. Bradshaw, J. of Fluid Mech., **100**, 639, (1980).
4. B. R. Ramaprian, V. C. Patel, and M. S. Sastry, Iowa Institute of Hydraulic Research Report 231, University of Iowa, Iowa City USA, (1981).
5. R. Narasimha and A. Prabhu,, J. of Fluid Mech., **54** , 1 (1972).
6. N. Subaschandar and A. Prabhu, J. of Aircraft, **39**, No 2, 290 (2002)
7. H. Schlichting, *Boundary Layer Theory*, McGraw-Hill Book Co., New York, (1968)
8. A. A. Townsend, Turbulent Shear Flow, Oxford University Press, G. B., (1956).

Performance Improvement of a Typical Manifold using Computational Fluid Dynamics

N. Subaschandar* , G. Sakthivel#

Department of Mathematics and Computational Sciences, BIUST, Palapye, Botswana
#*COMMAIR, Osborne Park , WA , Australia*

Abstract. This paper presents an analysis of fluid flow in a typical manifold. A commercially available Computational Fluid Dynamics (CFD) package has been used to analyze the flow pattern inside the manifold. Pressure drop in whole fluid domain and mass flow rate in all the outlets in the manifold have been computed. Based on the preliminary results, a modification to inside of the main pipe of the manifold has been incorporated. The performance of the modified design has been compared with the original design from the point of view of average temperature and mass flow rate at the outlets. A simple modification has been shown to improve the uniformity of temperature and mass flow at the outlets.

1 Introduction

Flow in manifold is of great importance in many industrial processes when it is necessary to distribute a large fluid stream into several smaller streams and then to collect them into one discharge stream. Flow inside a manifold in which two or more fluids with different velocities, temperatures and/or concentrations are mixed is encountered in various thermal and industrial equipment, e.g., chemical reactor, combustion chamber, piping system in power plants. The flow in manifolds is extensively encountered in many industrial processes when it is necessary to distribute a large fluid stream into several parallel streams and then to collect them into one discharge stream, such as fuel cells, plate heat exchanger, radial flow reactor, and irrigation. One of the typical mixing manifolds is found in the HVAC (Heating, Ventilating, Air-conditioning) systems. The flow field in the mixing manifold is accompanied by the flow separation and reattachment, secondary flow, and strong turbulence. Therefore, it is expected that the turbulent fluid, heat and mass transport processes between the hot and cold flows is quite complex. In order to develop effective methods of the mixing promotion and control, one has to fully understand the detailed flow characteristics and mechanism of the mixing of flows in the manifold. For most applications, the goal of manifold design is to achieve a uniform flow distribution through all of the lateral exit ports. A great number of experimental, analytical, and numerical studies deal with flow in manifold. The flow in distribution manifold has been studied by several investigators [1–9]. In general, all previous studies for manifolds with different applications had shown that typical manifold design does not give a uniform flow distribution at outlets. Therefore, the objective of the present study was to predict the flow distribution through each outlet and to develop an optimized design having a better flow distribution through outlets. The purpose of this analysis is to establish the influence of the inflow velocity (flow rate) along uniformity of the exit flows through the multiple holes. This work focuses on the prediction of pressure drop in the main stream which affects the mass flow rate of the gas flowing out through the outlet holes. Computational fluid dynamics methodology has been used to analyse the manifold geometry.

2 Computational fluid dynamics model

The model studied here is shown in Figure 1. It is a simple manifold with two inlet pipes. One inlet pipe is at the top and cold water is passed through this inlet. The second inlet pipe is in the horizontal orientation joining the main pipe at about half the height from the top. Hot water is passed through this inlet pipe. There are four outlet holes in the bottom horizontal pipe section through which the mixed water goes out. The diameter and length of the horizontal pipe at the bottom are 40mm and 300mm. The diameter and length of the vertical pipe is 30mm and 400mm respectively. The small horizontal pipe through which hot water is passed through has diameter of 30mm and a length of 40mm. The diameter of the four outlets at the bottom horizontal pipe are same(5mm). The models were built and CFD analysis was made using the commercially available software SOLIDWORKS. In the preliminary study, the original

 DOI: 10.1051/matecconf/20165411004

geometry of the manifold is analysed. The Table 1 gives the inlet and outlet conditions for all computations.

Table 1 : Conditions at the inlet pipe and outlet pipe.

Boundary	Mass flow rate(Kg/s)	Temp (^{0}C)	Pressure
Cold Water Inlet	1.0	20	
Hot water Inlet	0.4	80	
Outlets	-	-	Atmospheric Pressure

Figure 2 shows a typical grid used for computation for the original geometry. Original geometry had about 126,000 cells in the computational domain. In all subsequent computations, the grid size was kept around 126,000 so as to avoid the influence of grid size while comparing the results. K-ε turbulence model is used for all computations.

3 Results and Discussion

Figures 3a, 3b and 3c show the contours of static pressure, velocity and temperature inside the manifold model. From these figures one can see that non-uniformity exists in the exit planes.

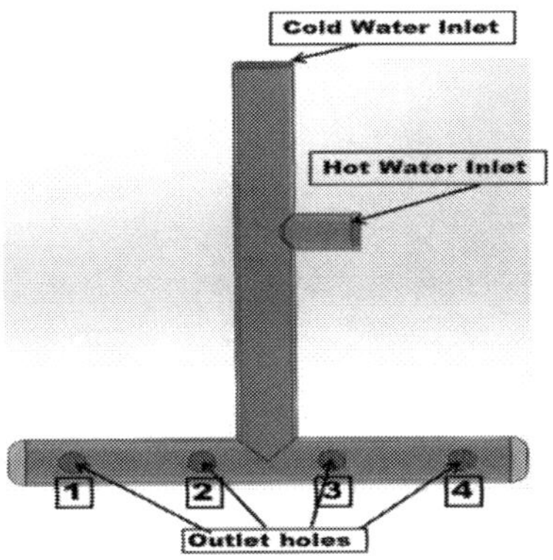

Figure 1 Original geometry with inlet pipes and outlets shown

Figure 2 Typical grid inside the computational domain

From Figure 3c we can see that hot water and cold water do not mix evenly and, almost, both the streams are separated. It was observed that the existence of non-uniformity at the outlets is mainly due to non-mixing of cold and hot water streams. Based on these preliminary results of the original geometry, a design modification was carried out. Two humps were placed in the vertical pipe below the hot water inlet pipe on both sides as shown in Figure 4. Both humps are in the flow direction, having a length of 35mm, giving minimum disturbance to the flow. Humps extend upto the centreline of the vertical pipe.

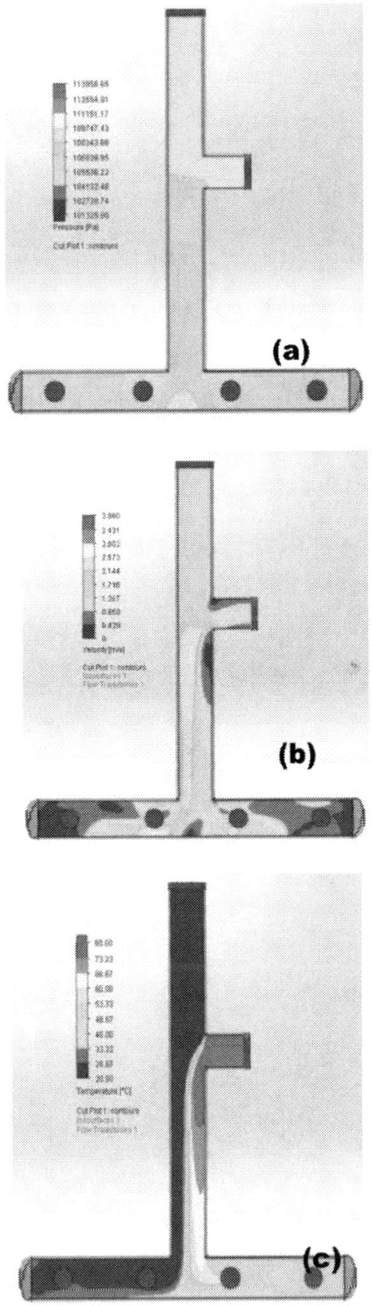

Figure 3 Contours of (a) Pressure, (b) Velocity and (c) Temperature inside the manifold for the original model

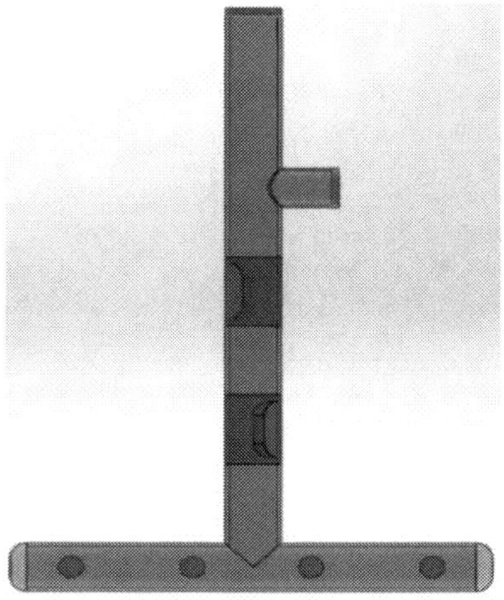

Figure 4 Modified geometry with two humps inside the vertical pipe.

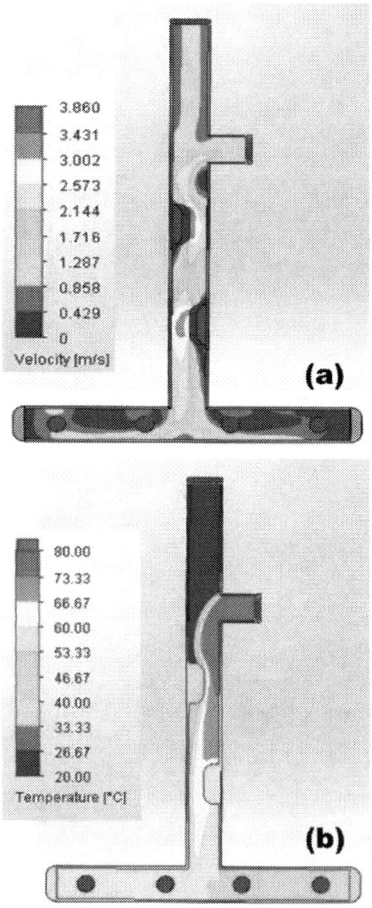

Figure 5 Velocity(a) and temperature(b) contours inside modified manifold geometry

The modified geometry was analysed using computational fluid dynamics method. Figures 5a and 5b show the contours of velocity and temperature inside the manifold. As we can see from these figures that the velocity and temperature are more evenly distributed and there is proper mixing of cold and hot water at the outlets. Table 2 gives the mass flow rates and temperature at the four outlets for both original and modified geometries. From this Table 2, also, we see that mass flow rate and temperature values are almost equal in all the four outlets. Here Mass flow rate has been expressed as a percentage of total mass flow rate of fluids passing through both inlet pipes. Figure 6 gives the contours of temperature distribution inside the manifold for both original and modified geometry for comparison. This figure clearly shows the effect of design modification carried out in improving the flow quality and flow uniformity at the outlets.

Table 2 : Mass flow rates and Temperatures at the four outlets for the modified geometry

	Original Model		**Modified Model**	
Outlet	Mass Flow Rate %	Temp(^{0}C)	Mass Flow Rate %	Temp(^{0}C)
1	28.82	27.24	25.61	51.39
2	20.48	23.03	24.91	50.55
3	21.76	51.36	23.77	53.24
4	28.94	46.54	25.71	56.09

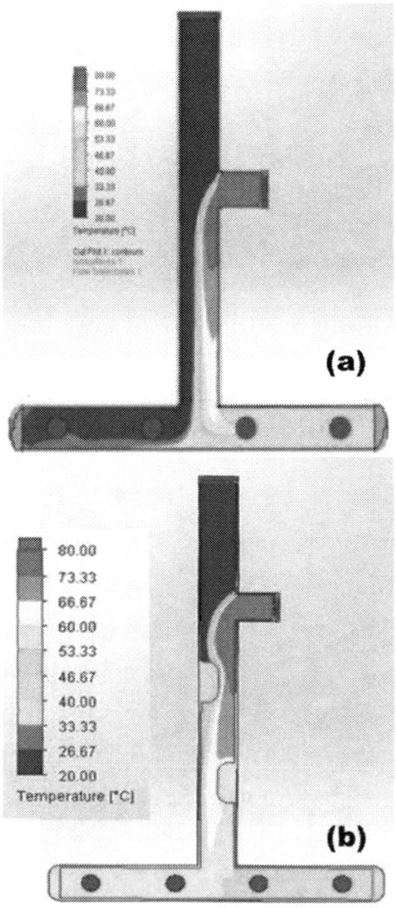

Figure 6 Contours of temperature inside the manifold for (a) Original model, (b) Modified model

4 Conclusions

Computational fluid dynamics methodology has been used to analyse the flow characteristics inside a manifold. The original geometry is found to have flow non uniformity in terms of mass flow rates and temperatures at the outlets. It is shown that hot water and cold water streams were, almost, not mixing inside the manifold, causing the maldistributions in mass flow rates and temperatures in the outlets. Based on the preliminary results a design modification has been implemented by adding two humps in the vertical pipe along the flow direction. Results show that the maldistribution at the outlets for mass flow rate and temperatures has decreased significantly. It is shown that this simple design modification has improved the flow quality by giving nearly equal mass flow rates and temperatures at the outlets. It has been shown that the computational fluid dynamics methodology can be used effectively and successfully to analyse and improve flow quality inside a manifold.

Acknowledgements

Author acknowledges the support rendered by Department of Mathematics and Computational Sciences, BIUST, Botswana in carrying out this research.

References

1. M. S. Gandhi, A. A. Ganguli, J. B. Joshi, and P. K. Vijayan, Chemical Engineering Research and Design, **90**, 487 (2012).
2. R. A. Bajura, J. of Engng for Gas Turbines and Power, **93**, 7, (1971).
3. R. A. Bajura and E. H. Jones Jr., J. of Fluids Engng, Transactions of the ASME, **98**, 4, 654 (1976).
4. A. K. Majumdar, Applied Mathematical Modelling, **4**, no. 6, 424 (1980).
5. M. K. Bassiouny and H. Martin, Chemical Engineering Science, **39**, no. 4, 693 (1984).
6. M. K. Bassiouny, H. Martin, Chemical Engineering Science, **39**, 4, 701(1984).
7. J. S. McNown, Transactions of ASCE, **119**, 1103 (1954).
8. A. Acrivos, B. D. Babcock, and R. L. Pigford, Chemical Engineering Science, **10**, 112 (1959).
9. R. L. Pigford, M. Ashraf, and Y. D. Mlron, Ind. Eng. Chem. Fundam. **22**, 463 (1983).

Analysis of Turbulent Axisymmetric Near Wake

T. Yane and N. Subaschandar

Department.of Mathematics and Computational Sciences, BIUST, P.B.16, Palapye, Botswana

Abstract. Axisymmetric near wake flow behind a body of revolution has been analysed. It is shown that the logarithmic layer of the upstream turbulent boundary layer continues to be valid for some more distance in the near wake and as the streamwise distance is increased the logarithmic layer is slowly getting destroyed. It is also shown that the central line velocity exhibits a logarithmic behaviour with streamwise distance.

1 Introduction

Turbulent flows are ubiquitous in natural phenomena and engineering applications [1] therefore a mathematically tractable description of them is desirable for their prediction and control. Flow region near an afterbody of an airborne vehicle has technological as well as scientific interest on account of its contribution to aerodynamic drag. The turbulent boundary layer approaching the base is subjected to large pressure gradient and to longitudinal and transverse curvature. It converts vorticity into the recirculating region near the base. On the other hand, the latter has an upstream influence on the boundary layer. This interaction is weak in supersonic case and is relatively well understood, but the problem is more serious in subsonic flow cases.

The near wake is the region of the wake flow just behind the trailing edge of the body where flow development takes place without the constraints imposed by the body. Turbulent wakes are one of the least studied flows, mostly due to their long development length and the difficulty in measuring low-level turbulent intensities in the far wake. Most of the previous works have focused on the behaviour of planewakes, such as those produced by circular cylinders, flat plates and airfoils[2-5]. Most of the analytical work available in the literature are related to two-dimensional turbulent near wake flows[6-8]. Alber[6] studied the development of the turbulent near wake of a thin flat plate using method of asymptotic expansions. His analysis showed that the velocity profile varied logarithmically with normal distance in the inner near wake. Subaschandar and Prahbu[7] carried out an asymptotic analysis of a two-dimensional turbulent near wake behind a flat plate. Walker et al[8] provided an analytical solution to the two-dimensional turbulent near wake problem. These reports indicated that a logarithmic variation in the normal direction exists in the overlapping region between the inner near wake and the outer near wake regions.

The axisymmetric turbulent wake is a flow that has puzzled researchers for more than a half-century, since measured results have been either inconclusive or contradictive. The axisymmetric wake is a challenging flow to measure because of the small velocity deficit, the slow decay of the velocity deficit downstream and a turbulence intensity of the same order as the deficit. In fact, the far axisymmetric wake still is at the threshold of what is possible to measure today using even the best wind tunnels and the most stable low-noise anemometer equipment. Also, unlike many other free shear flows for which the local Reynolds number remains constant or increases downstream, in the axisymmetric wake it drops slowly. Thus viscous effects continuously become more important until eventually they may dominate (if the flow extends far enough downstream). These varying viscous effects, together with the many very different possibilities for the structure of the near wake from different generators, considerably complicates interpretation of the data. Axisymmetric turbulent near wake flow has not been studied very well. Available research reports on axisymmetric near wake are pertaining to experimental works[9-13]. These studies essentially highlight special features of thick axisymmetric turbulent boundary layers and axisymmetric turbulent near wake regions. There are no analytical works that describe the axisymmetric inner near wake behaviour.

The present work makes an attempt to describe the behaviour of mean velocity profiles in the inner near wake behind behind a long slender cylinder. The turbulent flows, because of memory, take long time and distance before adjusting and coming to equilibrium with a new environment[14]. In view of this memory, it is reasonable to assume the velocity and length scales determining the turbulent boundary layer near training edge should also be relevant velocity and length scales for the near-wake.

2 Model for near wake:

Figure 1 gives the schematic of the problem along with the coordinate system and the corresponding velocity components. In this study, we focus our attention in the inner near wake region. The governing equations for inner near wake can be obtained by making a local asymptotic analysis, similar to that made for a fully developed turbulent boundary layer and two-dimensional turbulent near wake[7,15]. The limiting conditions are $\left(\frac{U_\tau}{U_\infty}\right) \to 0$ and $\left(\frac{\delta U_\tau}{\nu}\right) \to \infty$, where U_τ is the skin friction velocity at the trailing edge. Under these limiting conditions the non-dimensionalised governing equations for inner and outer near wake regions are

$$\frac{\partial(y_+U_+)}{\partial x_+} + \frac{\partial(y_+V_+)}{\partial y_+} = 0 \qquad (1)$$

$$U_+\frac{\partial U_+}{\partial x_+} + V_+\frac{\partial U_+}{\partial y_+} + \frac{1}{y_+}\frac{\partial}{\partial y_+}\left\{ky_+\frac{\partial U_+}{\partial y_+}\right\} = 0. \quad (2)$$

where $U_+ = \frac{U}{U_\tau}, V_+ = \frac{V}{U_\tau}, y_+ = \frac{YU_\tau}{\nu}, x_+ = \frac{XU_\tau}{\nu}, U_d = \frac{U_\infty - U}{U_\tau}, \tau = \mu_\tau \frac{\partial U_+}{\partial y_+}$.

For simplicity of the analysis, an assumption, similar to the one made in reference[6], for the eddy viscosity is given by $\mu_\tau = kU_\tau y_+$. The matching between the inner and outer near wake leads to the boundary conditions in the overlapping region as $\frac{U}{U_\tau} = \frac{1}{k}\ln\frac{yU_\tau}{\nu} + B_1$, where B_1 is a constant. B_1 depends on the radius of the cylindrical body[16].

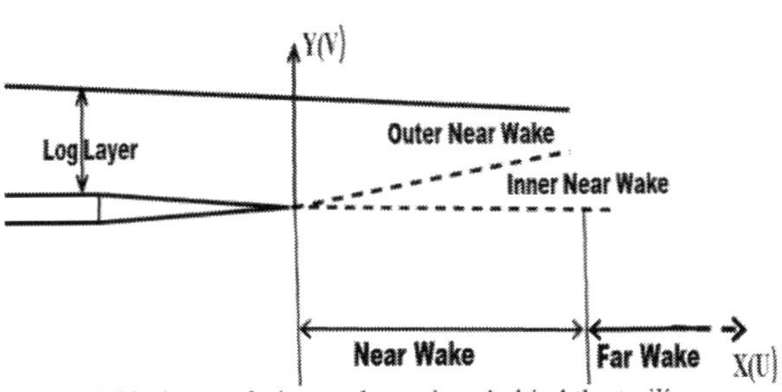

Figure 1 Various turbulent wake regions behind the trailing edge of an axisymmetric body

3 Analysis

We seek for a solution of Eq. 2 and the continuity Eq. 1 which is valid for a large y_+ such that for a fixed value of X_+.

$$\lim_{y_+\to\infty} U_+ \sim \frac{1}{k}\ln y_+ + B_1. \qquad (3)$$

This suggests a solution of the form,

$$U_+ = S(X_+) + \frac{1}{k}F'(\eta). \qquad (4)$$

Where $\eta = \frac{Y_+}{g(X_+)}$ and prime denotes differentiation with respect to η or X_+. We can get an equation in terms of V_+ from the continuity Eq. 1 and substituting into the momentum Eq. 2 to get an equation as follows;

$$\frac{1}{\eta}[\eta^2F'']' - \frac{gS'}{k}[F''\eta] + gSS' = \frac{gS'}{k}\left[\frac{\eta F''}{2} - F'\right] + \frac{2g'}{k^2}\left[FF'' - \frac{F}{\eta}\int_0^\eta F d\eta\right]. \qquad (5)$$

To obtain the similarity equation for F we set the two functions of X_+ appearing in the equation to constants. Without loss of generality we may set the constants to 1, that is,

$$\frac{gS'}{k} = 1 \text{ and } gSS' = 1 \qquad (6)$$

On solving to get equations of S and g yields,

$$S(X_+) = \frac{1}{k}\ln(g) \text{ and } g[\ln(g) - 1] = k^2X_+ \qquad (7)$$

Using these equations in Eq. 5, we get

$$\frac{1}{\eta}[\eta^2F'']' - [F''\eta] + 1 = \frac{1}{\ln g}\left[\eta\frac{F''}{2} - F' + FF'' - \frac{F''}{\eta}\int_0^\eta F d\eta\right]. \qquad (8)$$

For a relatively large $\ln(g)$, we get

$$[\eta^2F'']' - [\eta^2F''] + \eta = 0. \qquad (9)$$

The boundary conditions for $F(\eta)$ at the axis are given by the requirement that $V_+ = 0$ on $y_+ = 0$ and hence the centreline conditions are;

$$F'(0) = 0 \text{ and } \lim_{\eta\to\infty}\eta F'' = 0. \qquad (10)$$

To satisfy the boundary conditions imposed by Eq. 3 for a large Y_+ at a fixed X_+ we require that

$$\lim_{\eta\to\infty} F' \to \ln\eta + Bk. \quad (11)$$

4 Similarity solution for F'_0

The solution to the differential Eq. 9 subject to the previously stated boundary condition is,

$$F'' = \frac{\eta - 1}{\eta^2}. \quad (12)$$

Integrating to obtain a solution for F' yields,

$$F' = \ln\eta + \frac{1}{\eta} \quad (13)$$

Using the asymptotic boundary conditions given by Eq. 11, Eq. 13 reduces to

$$\lim_{\eta\to\infty} F'(\eta) \sim F'(0) + B_1 + \frac{1}{\eta} + \ln\eta \ \sim \ln\eta + Bk. \quad (14)$$

and hence,

$$F'(0) = Bk - B_1. \quad (15)$$

Rewriting Eq. 14 using Eq.15 we have;

$$F' = \ln\eta + \varphi(\eta) + Bk - B_1. \quad (16)$$

5 Experimetal data for comparison

Patel and Lee [8] carried out experimental studies in the axisymmetric near wake behind a body of revolution. Reynolds number in their experiments, based on the body length and freestream velocity was 1.2×10^6. On analysing the pressure distribution data given in report[8], it was observed that the pressure gradients in the freestream and normal direction were small and hence their dataset was considered for comparing the analytical results of the present study. Jimeniz at al [9] performed measurements in the axisymmetric wake of a long slender cylinder and the Reynolds number range was 1.6×10^6 to 67×10^6. The results of the present analysis were compared with the experimental data of Jimenez at al [9], but not presented here because of page limit.

6 Results and discussion

Figure 2 presents the variation of mean velocity profiles in the normal direction at different streamwise stations, comparing with the experimental data[8]. This figure shows the logarithmic lines and shows the existence of logarithmic layer in the inner near wake. The logarithmic layer slowly gets destroyed as the streamwise distance increases.

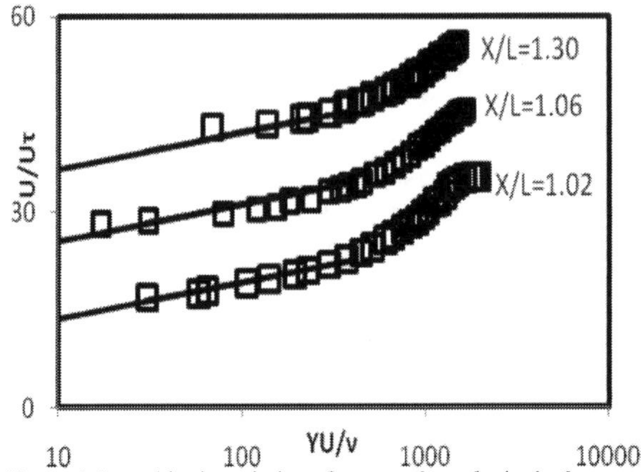

Figure 2 Logarithmic variation of streamwise velocity in the normal direction. □ experimental data[8], ——— log line.

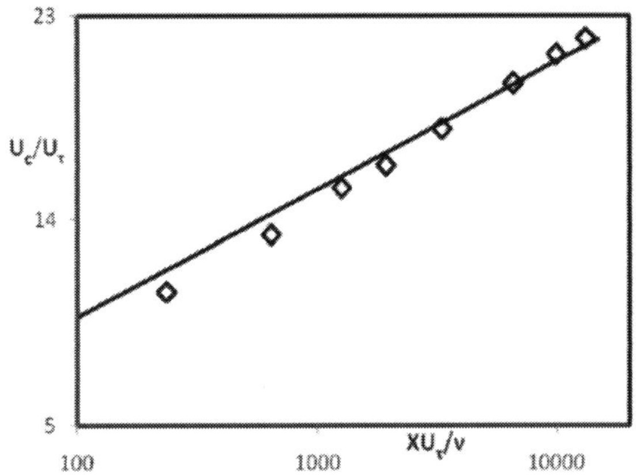

Figure 3 Variation of the centerline velocity with streamwise distance. ◇ Experimental data[8], ——— log line.

The behaviour of the solution for large X_+ in the inner near wake can be seen from the variations of the centerline velocity given by $S(X_+)$ and the length scale of the inner wake given by g from the Eq. 7. Taking logarithm on both sides of the second equation of Eq. 7 the relation between g and X_+ becomes more explicit in the form

$$\ln(g) + \ln[\ln(g) - 1] = 2\ln k + \ln(X_+). \quad (17)$$

This shows that, to the first order, ln(g) varies like $\ln(X_+)$. Substituting for ln(g) in the Eqs. 7, we get a relation for $S(X_+)$ as

$$S(X_+) = \frac{1}{k}[\ln(X_+) - \ln[\ln(g) - 1] + 2\ln k + C. \quad (18)$$

This shows that the centerline velocity variation for large X_+ is dominated by the term $(1/k)\ \ln(X_+)$. The other terms are very much small compared to this term. The variation of In(g), can also be considered small over the region of interest in X_+, which is of couple of decades' extent, and for all practical purposes S can be considered to vary like

$$S = \frac{1}{k} \ln ((X_+) + C_1 \qquad (19)$$

over the range of interest in the downstream direction for large X_+. C_1 is a constant which can be evaluated from experimental data. Figure 3 shows the variation of the centreline velocity with streamwise distance for the experimental data[8]. Figure 3 shows the logarithmic line and we can see the logarithmic variation of the centreline velocity for large streamwise distance, as given by Eq. 19. Logarithic behaviour of mean velocity profiles has been reported in the experimental data given in[12] also.

In summary, mean velocity profiles, in the inner near wake, vary logarithmically with distance in the normal as well as streamwise directions.

7 Conclusions

An axisymmetric turbulent inner near wake development has been studied analytically. It is shown that the same limiting condition $\frac{U_\tau}{U_\infty} \to 0$ as used for an asymptotic treatment of a fully developed axisymmetric turbulent boundary layer for a large Reynolds number is applicable to the analysis of the axisymmetric turbulent near wake. The streamwise component of the mean velocity has been shown to vary logarithmically with normal distance in the normal direction in the inner near wake. This logarithmic layer is getting destroyed slowly as the streamwise distance increases. The centreline velocity is shown to vary logarithmically in the streamwise direction for large streamwise distances. Available experimental data validate the results of the analysis.

Acknowledgements

Authors acknowledge the support received from the Department of Mathematics and computational Sciences and BIUST authorities in carrying out this research.

References

1. S. B. Pope, Turbulent flows, Cambridge University Press (2000).
2. S. J. Goldstein, J. Aeronaut. Sci. **15**, 189 (1948).
3. A. A. Townsend, *The Structure of Turbulent Shear Flow*. Cambridge University Press. (1956).
4. I. Wygnanski, F. Champagne, & B. Marasli, J. Fluid Mech. **168**, 31 (1986).
5. H. Oertel, Annual Rev. Fluid Mech. **22**, 539 (1990).
6. I. E. Alber, AIAA J., **18**, 1044 (1980).
7. N. Subaschandar and A. Prabhu, J. Aerospace Science and Technology., **1**, 1089 (1999)
8. J. D. A.Walker and D. Bocucz, J. of Fluid Mech., **112**, 100 (1988).
9. V. C. Patel and Y. T. Lee, IIHR Report 210, IIHR, The Univ. of Iowa, USA. (1977).
10. J. M. Jimenez, M. Hultmark, and A. J. Smits, J. of Fluid Mech., **659**, 516 (2010).
11. V. C. Patel, Y. T. Lee, and O. Guven, In Turbulent Shear Flows – I , Springer-Verlag, Berlin, 137 (1979).
12. R. A. Merz, R. H. Page, and C. E. G. Prziregel, AIAA J., **16**, 656 (1978).
13. V. Krishnan and K. S. Yajnik, Proceedings of 10th Australasian Fluid Mech. Conf., University of Melbourne, DEC 1989, 9.26 (1989).
14. R. Narasimha and A. Prabhu, J. of Fluid Mech., **54**, 1 (1972)
15. N. Afzal and R. Narasimha, J. of Fluid Mech., **74**, 113 (1976).
16. G. N. V. Rao and N. R. Keshavan, J. of Applied Mech., **39**, 1, 25 (1972).

MATEC Web of Conferences **54**, 11006 (2016)
MIMT 2016
DOI: 10.1051/matecconf/20165411006

The Dynamic Responses of the Submersible Vehicle Mast with Different Cross-Sectional Shapes subjected to Underwater Explosion

Ching-Yu Hsu [1], Cho-Chung Liang [2], Tso-Liang Teng [3] and Hai-Anh Nguyen [4]

[1]*Department of Marine Mechanical Engineering ,ROC Naval Academy, Taiwan*
[2]*Department of Mechanical and Automation Engineering, Da-Yeh University, Changhu , Taiwan*
[3]*Hsiuping University of Science and Technology, Taichung, Taiwan*
[4]*Department of Mechanical Engineering, National Chiao Tung University, Taiwan*

Abstract. A submersible vehicle mast is a device that extended above the surface of the water while the vehicle remains hidden below. Masts mounted above a submersible vehicle are used to support navigational instruments and various electronic devices, such as radar antennas and sensors. Submersible vehicle mast must be designed to survive extreme loading conditions, such as underwater explosions (UNDEX). The present study applied the finite element method, Cole's empirical formulation and acoustic-structure coupling method to simulate an UNDEX and investigate the survival capability of a damaged submersible vehicle mast. A shape optimization problem was solved for the submersible vehicle mast models subjected to UNDEX with different cross-sectional shapes are circle, ellipse and streamline shape, respectively. The submersible vehicle mast was modelled with cylindrical shapes. First, the cylindrical and rectangular plate in an UNDEX models were conducted and simulation results are close to the failure modes shown in experiments of Kwon and Ramajeyathilagam. Second, Three prototypes of submersible vehicle masts subjected to UNDEX were presented using the same method. The results were show the model with a circular cross-section is stronger than other designs when subjected to an UNDEX. The analytical results could offer a valuable reference for submersible vehicle mast design.

1 Introduction

During underwater explosion (UNDEX), the sudden release of energy from a conventional high-explosive or nuclear weapon generates a shockwave and forms a superheated, highly compressed gas bubble in the surrounding water [1].

The primary concern in naval engineering and offshore structure research is predicting how submerged structures are damaged by UNDEX. In particular, cylindrical shell structures are crucial components of submersible vehicles and marine structures. Numerical methods for analysing submerged structures exposed to UNDEX have been successfully implemented. For example, Kwon [2] simulated the nonlinear dynamic response of a cylinder subjected to a side-on shock wave by using both numerical simulation and experimental (1993). Ramajeyathilagam [3] used a box model set-up under a shock tank and explosive charges of PEK-I to perform numerical investigations on thin rectangular plates subjected to UNDEX loading with the CSA/GENSA software (2004). Shin [4] presented ship shock modeling and simulation for far-field underwater explosion by applying the LS-DYNA code coupled with USA code (2004). Liang [5] studied a preliminary study of the transient responses of a 2000-ton patrol boat with shock loading using the finite element method (FEM) coupled with the second Doubly Asymptotic Approximation (DAA2) (2006). Rajendran [6] focused on the phenomenological evolution of blast damage of plates, arises from the primary concern for the design of plated structures against blast load. Zhang [7] analysed dynamic response of non-contact UNDEX on the warship and the supercharging boiler under the combination load of the shock wave and the bubble by using ABAQUS (2011). Li [8] reviewed the propagation of the shock wave and bubble pulse. The dynamic response of a cylindrical shell was examined in a water pool by using the MSC.DYTRAN software. A fluid-structure interaction method was provided to define the interaction between the water and the cylindrical shell (2012). Zhang [9] developed a procedure with the doubly asymptotic approximation method to study the problem of transient responses in a ship hull structure subjected to an UNDEX bubble (2014). Wang [10] examined the dynamic response of ship structures with the combined effect of shock wave load and bubble pulsation subjected to close-in non-contact UNDEX (2014). Chen [11] conducted a numerical study of protective effect of polymer coating on the circular steel plate response to near-field underwater explosion (2015).

In the aforementioned literature the UNDEX response was primarily studied for ships and submerged structures; however, no studies exist on dynamic responses affecting the shapes of submersible vehicle masts. The mast is extended above the surface of the water while the submersible vehicle remains hidden below. Figure 1 shows the mast and the other systems in a submarine [12].

This study developed a procedure to investigate the dynamic responses of submersible vehicle mast exposed to UNDEX. Simulation results of the three samples were compared to determine the influence of the cross-section of the dynamic response of a mast in explosion shock wave conditions to identify the most effective cross-sectional shape in three prototypes. This simulation problem was based on an UNDEX experiment in which a submerged test cylinder modelling a submarine mast was exposed to a pressure shock wave produced by a 9-kg TNT explosive charge.

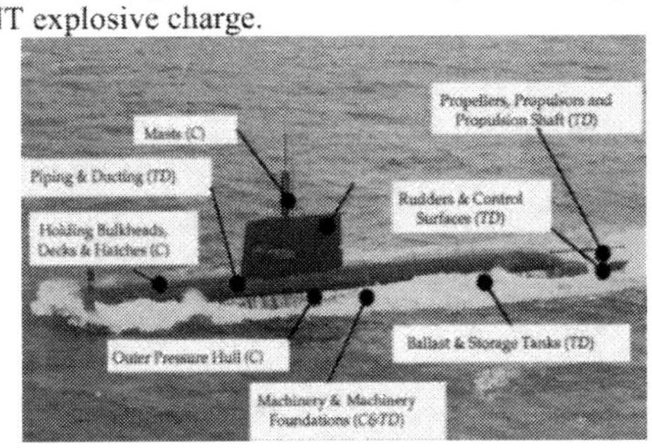

Figure 1. Submarine masts system [12]

2 Theoretical background

Generally the load caused by the shock wave is very high but the corresponding duration is very short (several milliseconds) showed as Figure 2.

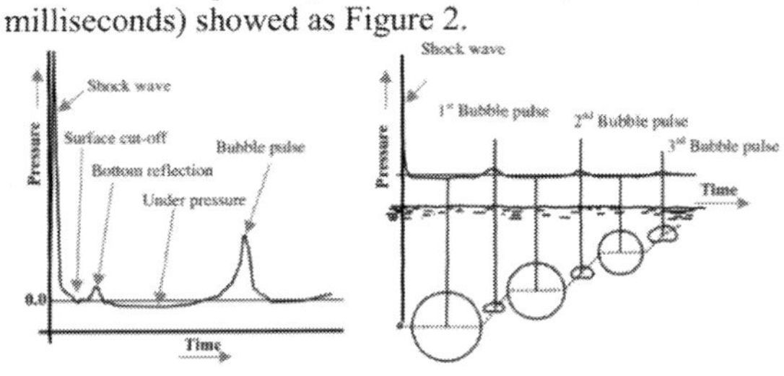

Figure 2. Phenomenon of the underwater explosion: shock wave and high-pressure bubble [1]

Empirical equations were determined to define the profile of the shock wave [1] and can be expressed as follows:

$$P(t) = P_{\max} e^{-t/\lambda} \quad (1)$$

$$P_{\max} = K_1 \left(\frac{W^{1/3}}{R} \right)^{A_1} \quad (2)$$

$$\lambda = K_2 W^{1/3} \left(\frac{W^{1/3}}{R} \right)^{A_2} \quad (3)$$

P(t): the pressure profile of the shock wave (MPa);
W: the weight of the explosive charge (Kg)
R: the distance between explosive charge and target (m)
P_{max}: the peak pressure of the wave. (MPa)
λ :the shock wave decay constant. (ms)

K_1, A_1, K_2, A_2 are constants which depend on explosive charge type, listed in Table1.

Table 1. Shock wave parameter values

Parameter \ Charge type	**HBX-1** [13]	**TNT** [1]	**PETN** [13]
K1	53.51	52.12	56.21
A1	1.144	1.180	1.194
K2	0.092	0.0895	0.086
A2	-0.247	-0.185	-0.257

3 Validation of the numerical method

This simulation was based on an UNDEX experiment in which a submerged test cylinder was exposed to a pressure shock wave produced by a 60-lb (27.3-kg) HBX-1 explosive charge. Kwon and Fox [2] originally described the experiment with a set of selected experimental results.

The test cylinder was made of T6061-T6 aluminium, the linear elastic material model is adopted to describe the mechanical property of the material has shown in Table 2.

Table 2. Engineering properties of the T6061-T6 aluminum

Property	Value
Young's modulus (MPa)	75.6×10^3
Poisson's ratio	0.33
Mass density (Kg/m^3)	2784
Static yield stress (MPa)	300

It had an overall length of 1.067m, an outside diameter of 0.305m, a wall thickness of 6.35 mm, and 24.5-mm-thick welded end plates. The cylinder was suspended horizontally in a 40-m-deep fresh water test quarry. The 27.3-kg HBX-I explosive charge and the cylinder were placed at a depth of 3.66 m. The charge was centred at the side of the cylinder and located 7.62 m from the cylinder surface. The suspension depths, charge offset, and duration of the test were selected such that cavitation of the fluid was not substantial and no bubble pulse occurred. Strain gauges were placed in multiple locations on the outer surface of the test cylinder, as shown in Figure 3.

Two pressure sensors were positioned 7.62 m from the charge; they were away from the cylinder, but were at the same depth as it. These sensors provided an experimental determination for the pressure versus time history of the spherical incident shock wave as it traveled by the point on the cylinder closest to the charge (strain gauge location B1). Figure 4. shows a time history curve of the incident pressure wave recorded by the sensors (experiment data [2]) and numerical simulation with incident pressure calculated by equations (1), (2) and (3).

The Figure 5. (a)-(b) shows the cylinder and field mesh and the test cylinder was meshed with 2,400 S4R finite strain shell elements. The external fluid was consisted of 45,586 elements AC3D4 acoustic tetrahedral elements. The definition of S4R element (Shell, 4-node, Reduced integration) and AC3D4 (Acoustic Continuum, 3D, 4-node) in detail are described in literature [14] and as shown in Figure 5.(c)-(d).The outer boundary of the

DOI: 10.1051/matecconf/20165411006

external fluid is represented by a cylindrical surface with spherical ends.

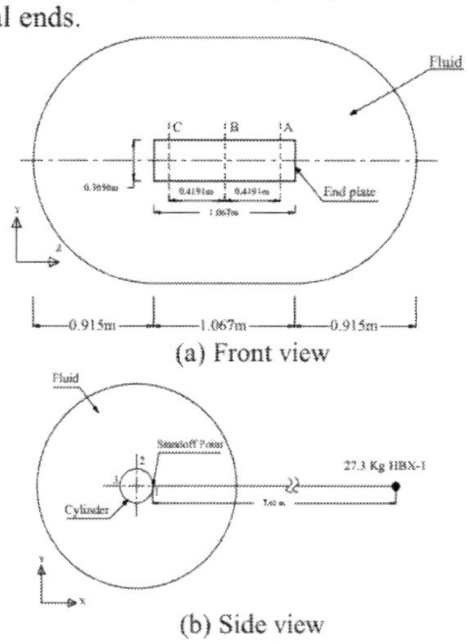

(a) Front view

(b) Side view

Figure 3. The arrangement of the model and strain gage locations on test cylinder

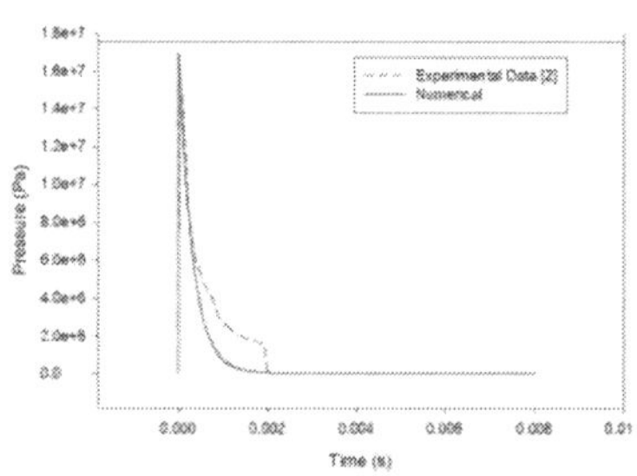

Figure 4. Incident pressure wave transient time history

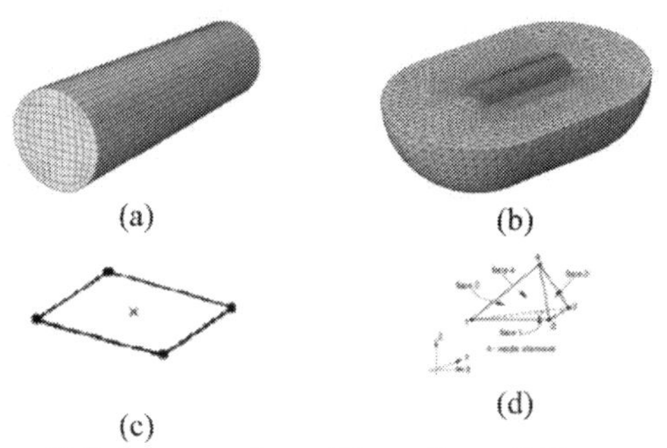

(a) Cylinder mesh, (b)acoustic fluid mesh, (c) S4R element and (d) AC3D4 element

Figure 5. Finite element model [14]

Figure. 6 to Fig. 11 contain time history of the test cylinder strains obtained from the analysis, with experimental shock loading data for locations of the strain measurements provided in Figure 3. The axial and hoop strains obtained from experimental and numerical analyses were consistent, as shown in the Figure 6-Figure 11. The numerical response had the same shape as the experimental results (the numerical response according to Cole's empirical formula and the numerical response according to the experimental data [2] inputted into ABAQUS were identical). In addition, the results used for Cole's formula as shock loading input improved slightly compared with those used for the experimental value (Figure 11). However, several variations existed when comparing data between the numerical results and experimental data [2]. The data matched the experimental results closely at B1 and B2, but less closely at position A, C (except for axial strain of C1). Numerical and experimental results matched more closely in areas with lower values of strain.

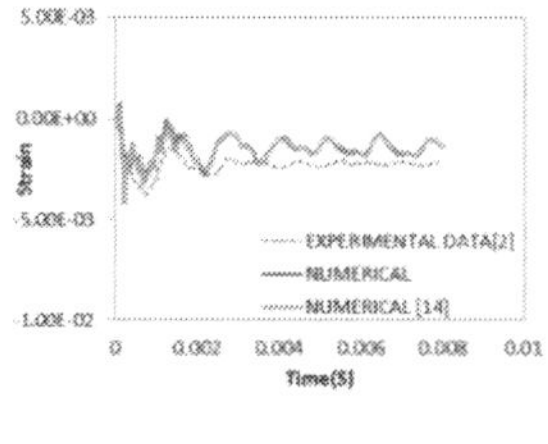

Figure 6. Experimental and numerical comparison for position A1(axial strain)

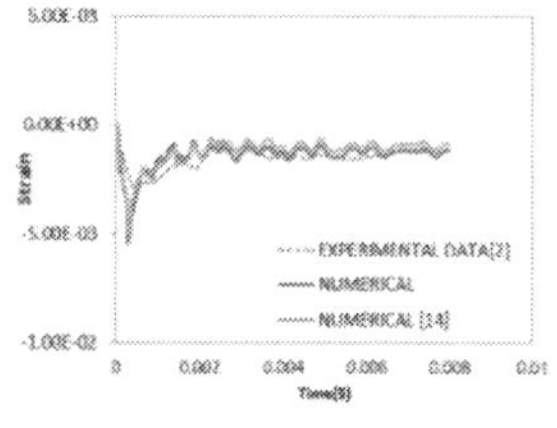

Figure 7. Experimental and numerical comparison for position A2(hoop strain)

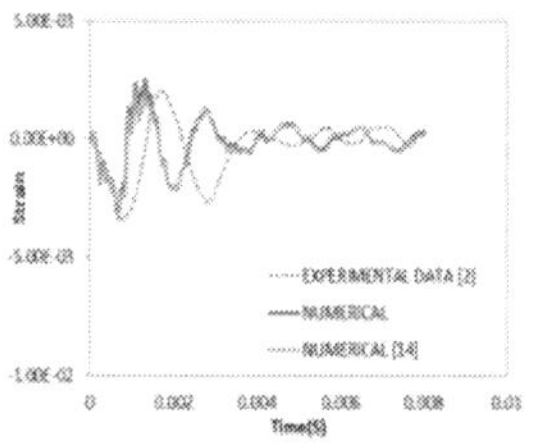

Figure 8. Experimental and numerical comparison for position B1(axial strain)

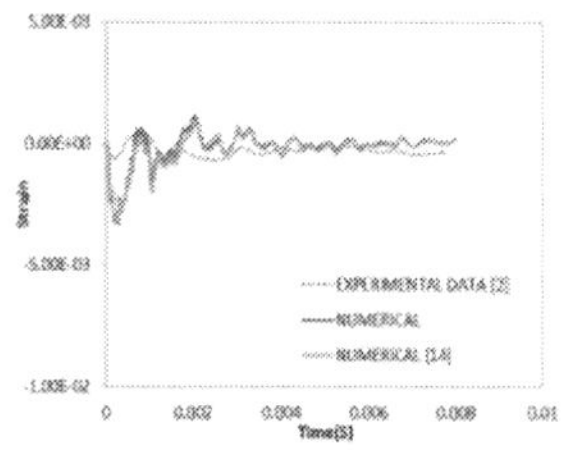

Figure 9. Experimental and numerical comparison for position B2(hoop strain)

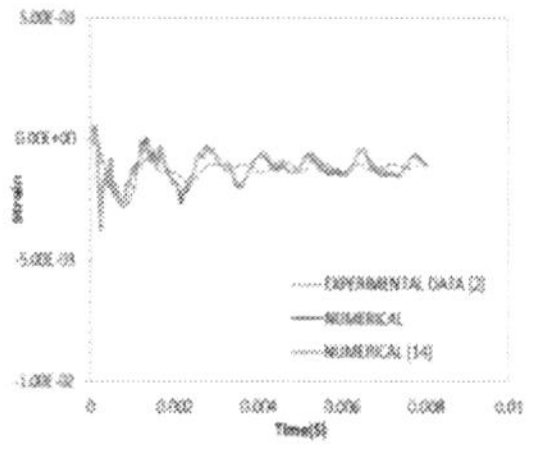

Figure 10. Experimental and numerical comparison for position C1(axial strain)

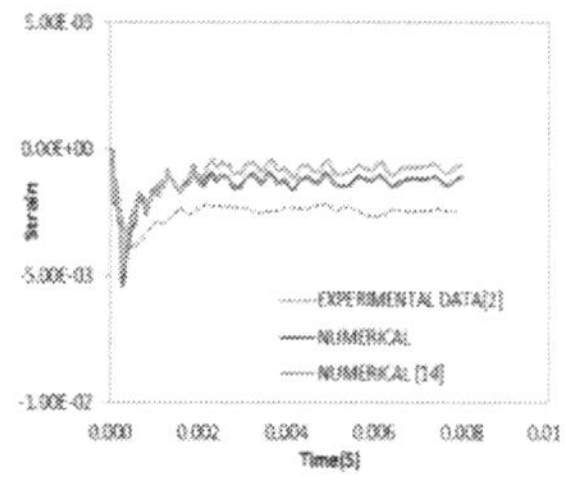

Figure 11. Experimental and numerical comparison for position C2(hoop strain)

4 Numerical simulation of submersible vehicle mast with variable cross-sectional types

A simplified submersible mast as a cylinder was suspended vertically in a 1-m-deep fresh water test quarry. A spherical charge of 9-kg TNT was centred at the bottom of the cylinder and located 10m from the cylinder surface. The location and the simulation of the experiment are shown in Figure 12 with the source point and standoff point as they relate to an incident acoustic wave loading. The source point represented the actual physical location of the explosive charge relative to the structure; the standoff point represented the location of the incident wave (shock front) at the start of the analysis. In this simulation, the standoff point was placed at the point on the fluid-structure interface that was closest to the source point. A non-reflecting boundary condition was set on the surfaces of the fluid model. The boundary condition at the bottom surface of the cylinder was assumed to be fixed.

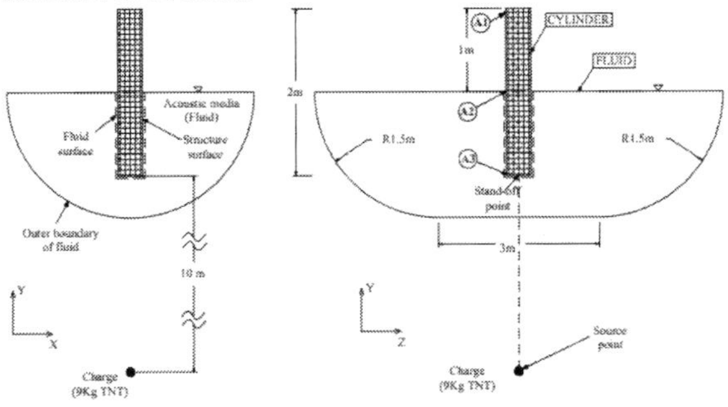

Figure 12. Schematic diagram of simulation and locations of A1, A2, A3

The test cylinder was made of steel. It had length of 2m and thickness and thick welded end caps of 2mm. To effectively compare the effect of the cross-section of the model, three samples were created with no substantial differences. The surrounding area, weight, and height were considered factors when creating three models .The shape and circumference of cylinders are shown in Figure 13. The variable cross-sectional model was generated at the equivalent circumferential length. Three different perimeters of the cross-sectional models (0.4398 m, 0.4488 m, and 0.4739 m) were used to represent the three models.

The finite element model is shown in Figure 14. The mesh of Model-1 consisted of 1,067 elements; Model-2 consisted of 1,185 elements and Model-3 consisted of 1,115 elements. The element size of 0.03m was set for all three models. The element connectivity was such that each shell normal was directed into the external fluid. A total of 105,000 four-node acoustic tetrahedral elements (AC3D4) were used in the model of the flow field. The outer boundary of the external fluid was represented by a cylindrical surface with spherical ends. The characteristic radius of the outer boundary was 1.5 m. The properties of three models materials are shown in Table 3 (The elastic/plastic materials model is adopted to describe the mechanical property of Model-1; Model-2 and Model-3).

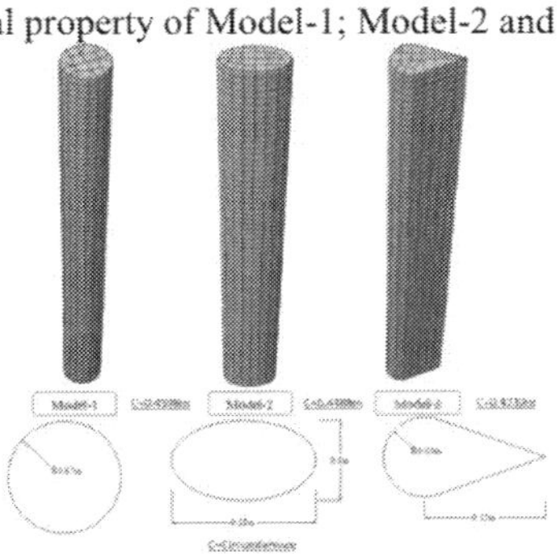

Figure 13. Three models of simplified submersible mast as a cylinder with different cross-sectional types

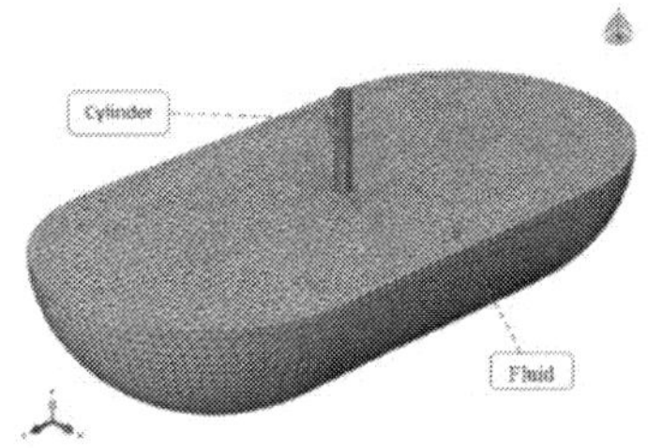

Figure 14. The finite model of the cylinder and external fluid

Table 3. Properties of cylindrical model and external fluid

Material	Parameter	Value	Unit
Steel	Mass density	7800.0	kg/m^3
	Young's modulus	210×10^9	Pa
	Poisson's ratio	0.3	-
	Yield's stress	400×10^6	Pa
Water	Mass density	1000.0	kg/m^3
	Speed of sound in water	1463	m/s
	Bulk modulus	2.1404×10^9	Pa

5 Results and discussion

Using the validated method in Section 3, the numerical time histories of each measuring point are shown in Fig.15. Three measuring points A1, A2, A3 were chosen on each model, as shown in Figure 12. Points A1 and A3 were chosen on the top and bottom of the cylinder. Point A2 was selected in the middle of the cylinder.

Figure 15 shows the response results of velocity. The greatest velocity occurred in Model-2 and Model-3, followed by Model-1. At position A1, the peak velocity value of Model-3 was approximately 14.7m/s, compared with approximately 2.3m/s for Model-2 and 1.4m/s for Model-1. In Model-3, the velocity value at A1 (top) was larger than that at A2 and A3 (middle and bottom, respectively). At A2 and A3, the velocity values of Model-2 were higher than those at the other two models. However, at approximately 0.001s, an upward trend appeared in Model-2 at A2 at approximately 10 m/s. In general, Model-1 showed smaller velocity values compared with Model-2 and Model-3.

Figure 16 – Figure 18 shows response of the submarine mast models from 0 to 0.006sec. The von Mises stress of Model-2 and Model-3 rose quickly to around 5.0×108 Pa after the 0.001s. By contrast, the figures for Mode-1 remained below 3.5×108Pa throughout 0.006s. A quick glance at the graph reveals that the value of von Mises stress of Model-2, Model-3 is highest. The filled contours figures illustrate the displacement responses of the whole model .These analyses showed that the response of Model-1 subjected to an UNDEX shock with a circular cross-section was a more effective shape for the submarine mast.

6 Conclusions

Numerical simulations of a simplified submersivble vehicle mast as a cylinder model subjected to underwater shock wave loading were conducted in this study, and a procedure to analyze the shock response was developed. The vlocity, displacement, von Mises stress, and time histories were obtained when a structure was subjected to underwater shock.

This study showed that the FEM, Cole's formulas, and acoustic-structure coupling method were effective tools for identifying the response of the structure subjected to UNDEX analysis. The shock wave caused by the detonation of a 9-kg TNT charge was detonated 10 m away from the center of the bottom of the submarine mast model.

The results indicated that with the same UNDEX conditions, the simulation results of the submarine-mast model were compared. The mast model with a circular cross-sectional form was stronger than a streamlined and elliptical shape, and can be applied quantitatively to real structural design.

Location	Results
A1	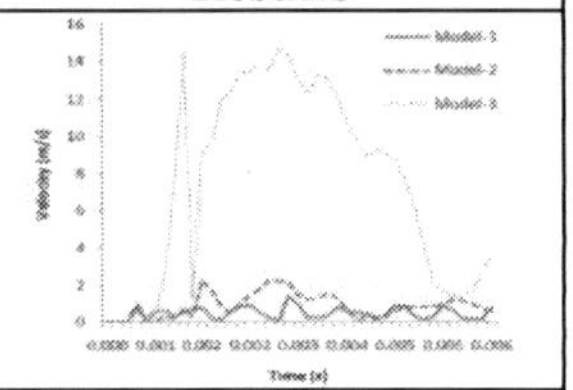
A2	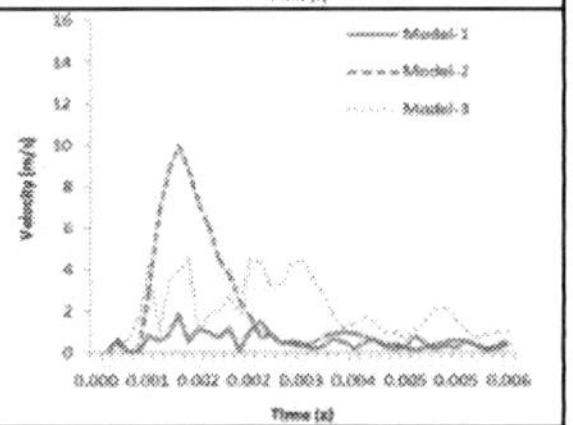
A3	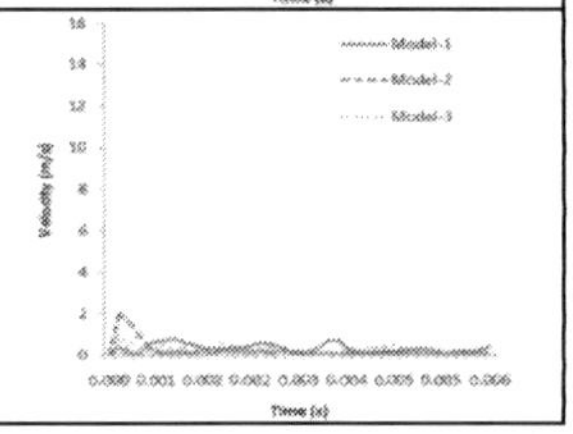

Figure 15 Time histories of velocity of A1, A2, A3

Figure 16. von-Mises and deformation response for Model-1

Figure 17. von-Mises and deformation response for Model-2

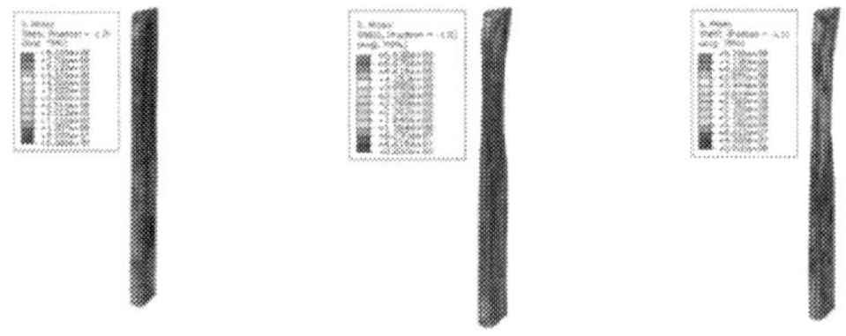

Figure 18. von-Mises and deformation response for Model-3

Acknowledgements

The authors would like to acknowledge the financially supporting this work under contract MOST-104-2221-E-012-004.

References

1. R. H. Cole, Underwater explosion. New Jersey. Princeton University Press (1948)
2. Y. W. Kwon and P. K. Fox, Underwater shock responses of a cylinder subjected to aside-on explosion. Comput. Struct. **48**, 4, 646–737 (1993)
3. K. Ramajeyathilagam and C. P. Vendhan, Deformation and rupture of thin rectangular plates subjected to underwater shock. I. Int. J. Impact. Eng. **30**, 6, 699-719 (2004)
4. Y. S. Shin, Ship shock modeling and simulation for far-field underwater explosion. Comput. Struct. **82**, 2211-2219 (2004)
5. C. C. Liang and Y. S. Tai, Shock response of a surface ship subjected to noncontact underwater explosions. Ocean Engineering. **33**, 748-772 (2006)
6. R. Rajendran and JM. Lee, Blast loaded plates. Marine Structures. **22**, 2, 99-127 (2009)
7. A. Zhang, W. Zhou, S. Wang, and L. Feng, Dynamic response of the non-contact underwater explosions on naval equipment. Marine Structures. **24**, 4, 396-411 (2011)
8. J. Li and J. Rong, Experimental and numerical investigation of the dynamical response of structures subjected to underwater explosion. Eur. J. Mech. B Fluids. **32**, 59-69 (2012)
9. N. Zhang, Z. Zong, and W. P. Zhang, Dynamic response of a surface ship structure subjected to an underwater explosion bubble. Marine Structures, **35**, 26-44 (2014)
10. H. Wang, X. Zhu, Y. S. Cheng, and J. Lin, Experimental and numerical investigation of ship structure subjected to close-in underwater shock wave and following gas bubble pulse. Marine Structures, **39**, 90–117 (2014)

 DOI: 10.1051/matecconf/20165411006

11. Y. Chen, F. Chen, Z. P. Du, Y. Wang, P. D. Zhao, and H. X. Hua, Protective effect of polymer coating on the circular steel plate response to near-field underwater explosion. Marine Structures. **40**, 247-266 (2015).
12. A. P. Mouritz, E. Gellert, P. Burchill, and K. Challis, Review of advanced composite structures for naval ships and submarines. Compos. Struct. **35**, 21-24 (2001)
13. W. D. Reid, The response of structure of surface ships to underwater explosions. Ship Structures and Materials Division, DSTO Aeronautical and Maritime Research Laboratory, DSTO-GD-0109, (1996)
14. ABAQUS user's and theory manuals, version 6.11.1 Dassault Systèmes, RI, USA (2011)

MATEC Web of Conferences **54**, 11007 (2016)
MIMT 2016
DOI: 10.1051/matecconf/20165411007

Comments on Variational Method and Energy Method in Computational Mechanics

Xingbo WANG Biao LU and Ying TIAN

Department of Mechatronics,Foshan University,Foshan City, PRC,528000[1-3]

Abstract. The article demonstrates that the energy method utilized in computational mechanics is a special form of the variational method through mathematical deduction and comparison to the two methods. The result discloses a basic fact that mechanical computations performed by different methods are intrinsically related in some way. The article also points out that a deep study on this intrinsic relation is necessary and helpful to guarantee loyalty of the computational results to the computed objectives.

1 Introduction

Computational mechanics is a subject that combines various skills and techniques in both mechanics and computer science. Ever since the being of the subject, people have developed tens of methods to perform kinds of mechanical computations, e.g., boundary element method (BEM)[1], the finite element method (FEM)[2], the finite volume method (FVM)[3], the collocation method (CM)[4], Numerical manifold method (NMM)[5], the meshless method or element free method (EFM)[6] and many other methods as are introduced in reference [7]. Literatures[8]-[10] also show that the Galerkin method(GM) , Ritz method (or Rayleigh–Ritz method) (RM) are frequently applied to derive all the BEM, FEM, NMM and so on. Also it is can see that, the energy methods (E_M)[11][12,13], the variational method (VariM)[14] and the weighted residual method (WRM) [15] are considered fundamental principles in computational mechanics. Since the ultimate purpose of computational mechanics is to solve numerically a partial differential equation (PDE) that is derived from a mechanical problem, I present in reference [16] a hierarchy diagram, as shown in Fig 1, to classify the classes of the methods describe the relationship of the methods mentioned above.

In the diagram, one can see that, either of the GM and the RM can deduces FEM, BEM, EFGM, or FDM while the GM and the RM can have different sources: the GM is deduced only from the WRM while he RM can be derived from every one of the WRM, the VariM and the E_M.

Now come at least three questions in front of the engineers:
(1). Are the GM and the RM identical ones?
(2). Are the WRM, the VariM and the E_M identical ones?
(3). If the GM and the RM are not identical or the WRM, the VariM and the E_M identical are not identical, which one is sure?

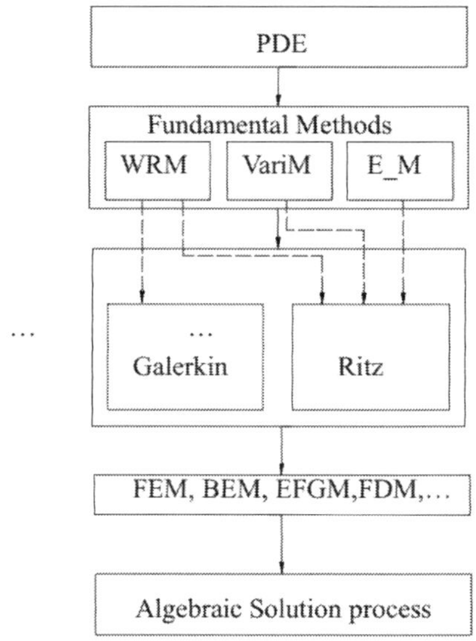

Fig 1. Hierarchy diagram of numerical methods to solve PDEs

Weile DS and Wildman RA might answer the question (1) in their article [8]. They say " the Rayleigh–Ritz and Galerkin methods are in principle very different, they often lead to the same set of discretized equations, are closely related, and are often discussed as if they were identical." But what are the answers to the reminders? Not finding an explicit answer, this paper proves theoretically that the E_M is identical to VariM under some special conditions. That is to say, the E_M is a special VariM.

2 From Variational Method to Energy Method

The VariM are methods based on the variational principle. In order to show that the E_M is a special VariM, we begin with the general principle of variational method, and then finish the proofs.

2.1 Equivalent variational form of a PDE

Given a PDE

$$L\phi - f = 0 \quad in \ \Omega \tag{1}$$

where $L = -(\sum_{i,j=1}^{2} \partial_j (a_{ij}(x_1, x_2)\partial_i) + a_{00}(x_1, x_2))$ is an operator of 2nd order PDE , and Ω is the domain of the function ϕ and f.

Multiplying (1) by a variation $\delta\phi$, and integrate over the domain Ω, it yields

$$\int_{\Omega} d\Omega \underbrace{(L\phi - f)}_{PDE} \delta\phi = 0 \tag{2}$$

Then integrate by part is used to transfer the derivatives in the operator L to $\delta\phi$, and the δ symbol is brought outside the integral and obtain an equation in the form

$$\delta(\int_{\Omega} \Phi(\phi) d\Omega) = 0 \tag{3}$$

Denote (3) by

$$\delta I(\phi) = 0 \tag{4}$$

and the equation (4) is the equivalent variational form of the PDE (1) in which $I(\phi)$ is the variational form of the PDE.

The equation (4) is a general form of general meaning. When applied in concrete problem, it may have definite form. For example, if L be a symmetric, positive definite, real operator, and a PDE is with Dirichlet boundary condition, as shown by

$$\begin{cases} L\phi = f & in \ \Omega \\ \phi = g & on \ \partial\Omega \end{cases} \tag{5}$$

the variational form of the PDE is

$$I(\phi) = \frac{1}{2} < L\phi, \phi > - < \phi, f > \tag{6}$$

If the PDE is with Neumann boundary condition

$$\begin{cases} L\phi = f & in \ \Omega \\ \sum_{i,j}^{n} a_{i,j} \partial_n \phi = h & on \ \partial\Omega \end{cases} \tag{7}$$

then its variational form is

$$I(\phi) = \frac{1}{2} < L\phi, \phi > - < \phi, f > - < h, \phi >_{\partial\Omega} \tag{8}$$

where

$$< Lu, v > = \int_{\Omega} (Lu) v d\Omega, < f, v > = \int_{\Omega} f v d\Omega, < h, v >_{\partial\Omega} = \int_{\partial\Omega} h v d(\partial\Omega)$$

The equations (6) and (9) are fundamental equations to deduce Rayleigh-Ritz Method, for which we omit details here.

2.2 Energy Methods

In mechanics, the energy methods refer to either the virtual work (or virtual displacement) method, or the complementary virtual work method. Now we prove that both methods are special forms of VariM.

2.1.1 Virtual Work Method (VWM)

Virtual work method comes from is also called principle of virtual work (PVW) or principle of virtual displacement (PVD). Virtual work δW is work done by a force F along a virtual displacement $\delta \boldsymbol{d}$

$$\delta W = \boldsymbol{F} \cdot \delta \boldsymbol{d} \tag{9}$$

In a generalized coordinate system of N generalized coordinates $(q_1, q_2, ..., q_N)$, the PVM says

$$\delta W = \sum_{i=1}^{N} Q_i \delta q_i = 0 \tag{10}$$

where $\{Q_i\}_{i=1}^N$ are the generalized forces acting on the system, $\{\delta q_i\}_{i=1}^N$ are virtual displacements and δW are the virtual work done by $\{Q_i\}_{i=1}^N$ along $\{\delta q_i\}_{i=1}^N$.

Now consider in equation (9), F(r) is a conservative force along a path r, then work done by F(r) from r1 to r2 on the path is

$$W = \int_{r_1}^{r_2} \boldsymbol{F}(\boldsymbol{r}) \cdot d\boldsymbol{r} = W(r_2) - W(r_1) = W(r_1, r_2) = \Delta W \tag{11}$$

Note that the change of potential energy Π is just the work done by conservative forces from r1 to r2, namely

$$W = -(\Pi_2 - \Pi_1) = -\Delta\Pi \tag{12}$$

By (11), one can see

$$\boldsymbol{F} = \nabla W = -\nabla\Pi \tag{13}$$

where ∇ is gradient operator.

Thus the component Fd of F in direction $\boldsymbol{d}$ is

$$F_d = \boldsymbol{F} \cdot \boldsymbol{d} = -\boldsymbol{d} \cdot \nabla\Pi = -\frac{\partial\Pi}{\partial \boldsymbol{d}} = \frac{\partial W}{\partial \boldsymbol{d}} \tag{14}$$

Substituting (14) into (1) yields

$$\delta W = -(\frac{\partial \Pi}{\partial q_1}\delta q_1 + \frac{\partial \Pi}{\partial q_2}\delta q_2 + ... + \frac{\partial \Pi}{\partial q_N}\delta q_N) = -\delta\Pi \quad (15)$$

If $\boldsymbol{F}$ can be decomposed into an internal force $\boldsymbol{F}_i$ and external forces $\boldsymbol{F}_e$, then the work can also be set into internal work W_i and external work W_e. And then the total potential energy Π (TPE) is defined as

$$\Pi = U + V \quad (16)$$

where $U = -W_i$ is internal potential and $V = -W_e$ is external potential energy.

And

$$\delta\Pi = (\frac{\partial \Pi}{\partial q_1}\delta q_1 + \frac{\partial \Pi}{\partial q_2}\delta q_2 + ... + \frac{\partial \Pi}{\partial q_N}\delta q_N) \quad (17)$$

is also defined the first variation of total potential energy Π.

By the minimum potential energy theorem (MPET), a conservative mechanical system is in a stable state of equilibrium if and only if the total potential energy is a relative minimum. By variational principle of stationary point, the MPET is equivalent to the statement: a system is in equilibrium if and only $\delta\Pi = 0$ for each and every virtual displacement, which is called mechanical variational theorem (MVT).

Comparing to (4), it is known that the MVT is actually a variational form in TPE. Hence we finish demonstrating that the VWM is a special form of VariM.

2.1.2 Complementary Virtual Work Method(CVWM)

Complementary virtual work is virtual work δW done by virtual force $\delta \boldsymbol{F}$ along a real displacement $\boldsymbol{r}$.

$$\delta W = \delta \boldsymbol{F} \cdot \boldsymbol{r} \quad (18)$$

With respect to PVW, principle of complementary virtual work (PCVM) says that, the necessary and sufficient condition for a system to be in static equilibrium is that zero complementary virtual work is done for all virtual forces acting on the system, namely

$$\delta W = \sum_i \delta \boldsymbol{F}_i \cdot \boldsymbol{r}_i = \boldsymbol{0} \quad (19)$$

where $\delta \boldsymbol{F}_i, \boldsymbol{r}_i$ are virtual forces and displacements respectively.

In a conservative system, total complementary potential energy (TCPE) is defined by

$$\Pi^* = U^* + V^* \quad (20)$$

where $U^* = -W_i$ is complementary internal potential and $V^* = -W_e$ is complementary external potential energy.

And by (18), the first variation of total complementary potential energy is

$$\delta\Pi^* = \sum_i \frac{\partial \Pi^*}{\partial \boldsymbol{F}_i}\delta \boldsymbol{F}_i \quad (21)$$

Like the MVT, it can derive that $\delta\Pi^* = 0$ is the sufficient and necessary condition for a conservative mechanical system to be in a stable state of equilibrium. Obviously, $\delta\Pi^* = 0$ is a variational form of the total complementary potential energy Π^* and is of course a special variational form of VariM..

3 Comments and Expectation

3.1 Conclusions About E_M and VariM

Deductions in section 2 show that the E_M, which is essentially PVW, is an instance of the VariM. The Total potential energy or Total complementary energy is the functional for the variation, which is demonstrated by (17) and (21). Therefore we say that, the E_M is a special form of the VariM while the inverse is false. This can be illustrated with Fig 2.

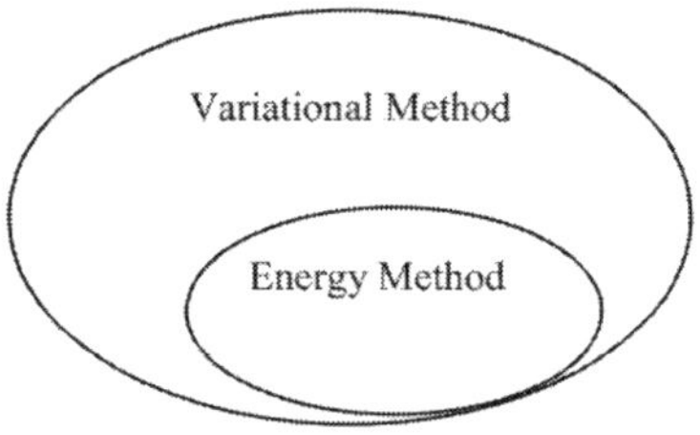

Fig 2. E_M is a subset of VariM

The relationship between the energy method and the variational method discloses a basic fact that foundation of mechanical computations is intrinsically related, and they are just "all roads lead to home". The most importance of the intrinsic relation is that it is in some way to ensure the loyalty of the computational results to the computed objectives.

3.2 About Weighted Residual Method (WRM)

Fig 1 shows that the WRM is more general a method than the VariM and the E_M. In fact, anyone who has gotten an excellent background of computational mechanics knows that, the WRM can deduce more concrete computational methods than the VariM. For example, the WRM has deduced collocation method, sub-domain method, least square method, momentum method and even the Galerkin method while the VariM has up to now deduced the Ritz method. Nevertheless, the WRM still intersects with the VariM owning to the fact that the Galerkin method of the WRM can derive the same form of governing equation as that of the Ritz method of the VariM if the variation of a functional exists. Namely, the WRM intersects with the VariM via the Galerkin-Ritz method. Fig 3 demonstrates the relationship.

Another fact existing in computational mechanics is that the VariM has strong mathematical foundations and the E_M can control the lower and upper bound of computations in an appreciate range while people up to now have no way to control the bound of the residual. This may be the reason that people prefer using the VariM and the E_M to the WRM, particularly in computational solid mechanics.

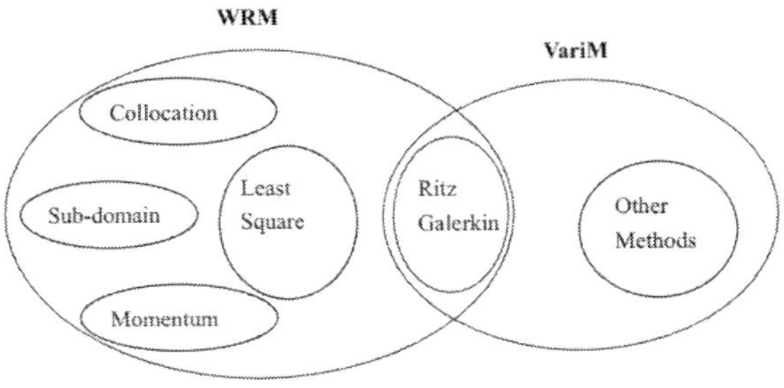

Fig 3. Ritz-Galerkin mtheod bridegs WRM and VariM

3.3 Expectation

Up to now it has been known that the energy method, the variational method and the weighted residual method are three fundamental methods in computational mechanics and the energy method is essential a special form of the variational method. It is also known that the weighted residual method is more general and incorporates with the variational method in some way, e.g., the Ritz-Galerkin method. This situation leaves a lot of imagination and work for us to do. On the one hand, the energy method is a subset of the variational method, and on the other hand the variational method has gotten an intersection with the weighted residual method. Now we expect to know if this intersection can be enlarged under some universal principle. If it can, what is the universal principle? If it cannot, what are the barriers? The answers to these questions need more and deeper researches in foundation of computational mechanics. It may concerns more intrinsic essence of mechanics and might bring more ingenious results for mechanics, which are our hopes.

Acknowledgment

The research work is supported by the national Ministry of science and technology under project 2013GA780052, Department of Guangdong Science and Technology under projects 2015A030401105 and 2015A010104011, Foshan Bureau of Science and Technology under projects 2013AG10007, Special Innovative Projects 2014KTSCX156, 2014SFKC30 and 2014QTLXXM42 from Guangdong Education Department. The authors sincerely present thanks to them all.

References

1. Q. H. Qin, Boundary Element Method, Springer New York, 2009
2. S. Cen, Ch. F. Li, S. Rajendran, and Zh. Q. Hu, Advances in finite element method. Mathematical Problems in Engineering, 2014,volume 2014
3. L. M. Surhone, M. T.\ Tennoe, S. F. Henssonow, Finite Volume Method.Betascript Publishing, **6**, 12-17, (2010).
4. Arithmetic A. Collocation Method, Betascript Publishing, 2013
5. G. W. Ma, X. M. AN, and L. He. The numerical manifold method: A review. International Journal of Computational Methods, **7**, 1, 1-32, 2010.
6. G. R. Liu, *Meshfree Methods: Moving Beyond the Finite Element Method*, CRC Press,2002
7. E. Stein, R. D. Borst, and T. J. R. Hughes, *Encyclopedia of Computational Mechanics*, John Wiley and Sons,2014
8. D. S. Weile, R. A. Wildman, G. Pisharody and A. Mohan, *Galerkin Method (Rayleigh -Ritz Method)*, John Wiley & Sons, Inc., 2005
9. V. Thomee, *The Standard Galerkin Method*, Springer, 2006, 1-24
10. S. Ilanko. Rayleign-Ritz method for structural analysis, ISTE Willy, 2014.
11. Eisley J. G., A. M. Waas, *Work and Energy Methods*, John Wiley & Sons, Ltd, 2011
12. T. H. G. Megson, *Introduction to Aircraft Structural Analysis*, Elsevier,2010
13. T. H. G., Megson Structural and stress analysis, Virtual Work and Energy Methods, Elsevier Ltd, 2014
14. M. Struwe, Variational methods: Applications to nonlinear PDE and Hamilton systems(2nd Edition). Springer-Verlag, 1996
15. B. A. Finlayson, The method of weighted residuals and variational principles, Philadelphia, 1972.
16. X. B. WANG, Mathematical and mechanical Method in Mechanical Engineering. Press of National University of Defense Technology, 2013.

Solving of the contradictory problem of the precision- Stability by using the extenics theory

OLARU Adrian[1] and OLARU Serban [2]

[1]University Politehnica of Bucharest, Splaiul Independentei nr.313, Bucharest, Romania
[2] S.C.ROMSYS S.R.L. Bucharest, Romania

Abstract. Assisted analyze of the elements and systems with Extenics theory is necessary to choose the optimal value and the optimal minim field called the attraction field of the working point. With this study easily can be determined all influences to the current working point. The Extenics theory was defined by prof.Cai Wen in 1983 and contents the mathematical, philosophy and logical models with the goal to solve the contradictory problem. This paper show the applying the Extenics theory to the optimal choose of the working area of the hydraulic cylinder in the precision-stability field and define some classes of this precision-stability by using the Extenics transformations and dependent functions. The research was made by using the proper virtual LabVIEW instrumentation with the possibility to work on-line and to see all transformation of the dynamic behavior. We established the affiliation of one precision-stability classes by see the changes of the proper form of the dependent function values in the used 2D dimensions. The paper shown the assisted study case of the contraction transformation defining finally the minimal attraction field like the optimized place of the working point.

1 Introduction

Precision and stability of all dynamic systems is one of the more important contradictory problem what must be solved. This problem is contradictory because if we try to increase the precision, imposed from the application, decrease the stability and will be possible to touch the limit of stability, when the element or system can't be controlling more, otherwise if we try to increase the stability, the precision will decrease and the element or system is very slow and will don't respect the minimal promptitude desired limit. In this situation the Extenics theory must be applied to solve the contradictory problem.

The Extenics theory was created by prof. Cai Wen in 1983 by publishing the paper "Extension Set and Non-Compatible Problems"[1]. Its goal was to solve contradictory problem and also nonconventional ideas in many fields like technical, social, philosophical, architectural, *etal.* Usually the Extenics theory operate with some extension transformation what change the problem from the contradictory field in to non-contradictory by extend the universe of discourse and respect the same value of the dependent function. Some used mining of the transformation [1-3] are substitution, increasing/decreasing, expansion/contraction, decomposition, duplication and some complex transformation like expansion-duplication, approximation-duplication, substitution- increasing, *etal.* Many of the contradictory problem can't be solved by using only one transformation and must be applied many successive or simultaneously transformations.

In the proper paper [4] was defined the minimal attraction field instead of the attraction point established in the paper [5], because in the dynamic behavior all elements and systems don't work only in one singular point, but in one closer field around this point called *working minimal attraction field.* Attraction points like the optimal working point, defined in [4] was changed in this paper with *the working minimal attraction field* like to be optimal solution solving the contradictory problem, because in the real work the element and system change his working point in each moment, or by applying the proper dependent formula for 2D dimensions will be not possible to find only one approved situation because will be choose for most of them like full- desirable, partial desirable, partial acceptable, *etal.* The solution can't be stable, but must be controlled to be inside of one know area established,(desirable, acceptable or compound of them) defined by Extenics theory [1-3] and [6].

In the last proper paper was determined, by applying the basic element and logical equation of the Extenics theory, that the optimal point to solve the contradictory problem precision-stability is the point defined by the *x coordinate of the extreme precision point* and *y coordinate of the extreme stability point.* This solution was determined in [5] by applying only the basic

Extenics logic element was considered that not to be the optimal solution and for that the research was continued with the Extenics extended transformation and with proper defined dependent 2D functions.

In this paper will be applied the extension/contraction Extenics transformation and calculus of the dependent function from Extenics theory [Cai Wen 1983, Yang Chunyan, Cai Wen 2002] with some proper considerations.

The method consist in to determine the new attraction field and calculate the dependent function. If the dependent function is zero, the point is on the boundary field. If the dependent function have other values, the working point can be inside or outside of the desirable or acceptable field. After applying more increasing/decreasing transformation to obtain minimum of the desirable and acceptable field, will be calculated again the dependent function. If the dependent function have the same value, the functional point is in the same class of the precision-stability, in other conditions must be changed his value and will be obtained a higher class of the precision-stability.

With this paper we introduce new term in the study of the optimal solving the precision-stability contradictory problem: *the classes of the precision-stability*. This operation will be resume more and more before obtained the same class of the working point. The operation will be continue with increasing/decreasing transformation of the current field and with verifying the affiliation at the precision-stability same class.

2 Study case of the hydraulic cylinder's precision-stability

2.1 Modeling and simulation.

In all study cases the modeling and simulation offer some on-line results what can be used to choose some constructive and functional parameters to obtain one dynamic behavior without vibration, one short acceleration time, minimal stationary error, one bigger Bode frequency, one higher cutting Bode frequency, one higher proper and natural frequencies, *etal*. In the proper research paper [5] was shown that the Extenics obtained working point is formed by x coordinate of the precision limit point and y coordinate of the stability limit point, because the y coordinate determine the increasing of the proper frequency with effect to the increasing the precision and decreasing the stability and x coordinate determine the increasing of the damper factor with effect to the increasing of stability and decreasing the precision. In this case (partial solving case) the solution of the contradictory problem can be written in the following form $S = \begin{pmatrix} x(precision\_\lim itpo\text{int}) \\ y(stability\_\lim itpo\text{int}) \end{pmatrix}$ and the compatible degree functions $k(x)$, $k(y)$ and compatible degree of the contradictory problem $K(P_x)$ and $K(P_y)$.

$$k(x) = \frac{x - \inf\_\lim\_goal\_x}{\sup\_\lim\_goal\_x - \inf\_\lim\_goal\_x};$$
$$k(y) = \frac{y - \inf\_\lim\_goal\_y}{\sup\_\lim\_goal\_y - \inf\_\lim\_goal\_y}; \quad (1)$$

The root's locus characteristic with the desirable field can be seen in Fig. 1.

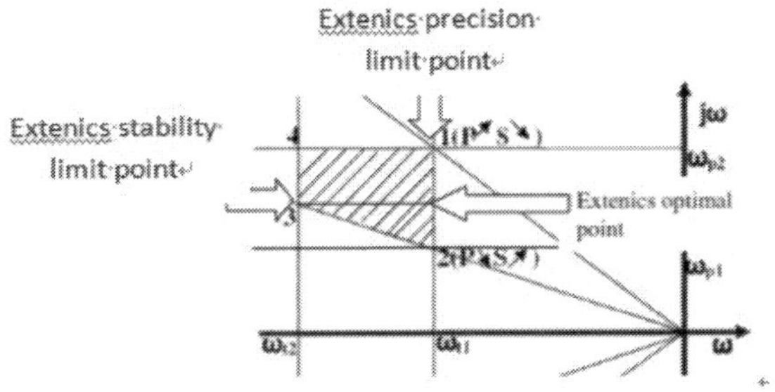

Figure 1. Solution after apply the basic Extenics logical method (partial solution).

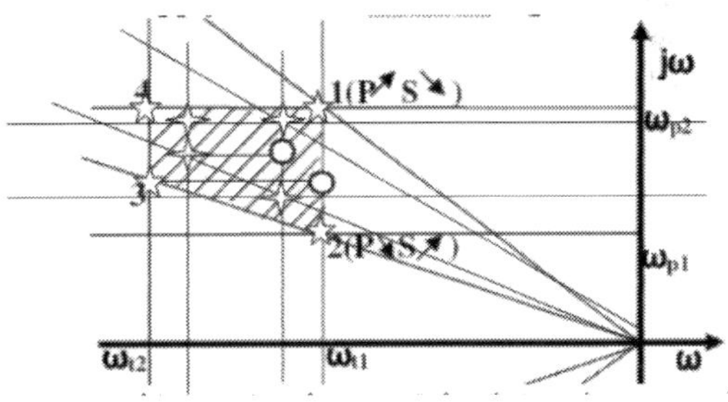

Figure 2. The Extenics working point that partial solving the contradictory problem after applying the Extenics extended/contraction transformation.

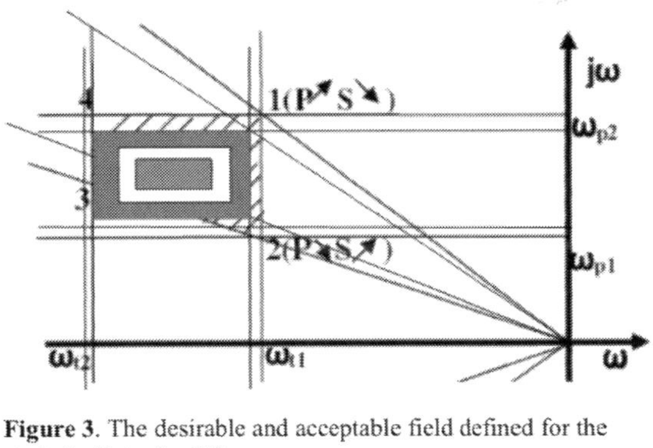

Figure 3. The desirable and acceptable field defined for the study by Extenics *2D*.

In the modeling step will be used the dependent formula defined in proper paper [6] in the following form:

$$k(x,y)=\begin{Bmatrix}k(x)\\k(y)\end{Bmatrix}=\begin{Bmatrix}\dfrac{\rho(x,X_0)}{\rho(x,X)-\rho(x,X_0)}\\\dfrac{\rho(y,Y_0)}{\rho(y,Y)-\rho(y,Y_0)}\end{Bmatrix} \tag{2}$$

where: ρ (x,X_0) is the distance between the working point x and the desirable interval X_0, ρ (x, X) is the distance between the working point x to the acceptable interval X, the same means for the oy axes. The dependent complex formula will be analyzed by logical Extenics theory to choose the optimal point between all analyzed.

2.2 Extenics transformations and dependent formula.

Extenics theory use the extended transformation method to change the contradictory problem in one non contradictory. The dependent formula show when the working point is inside or in outside of the imposed, in this case, 2*D* intervals (desirable and acceptable) what assures optimal solution of the contradictory problem. For this study was used the Extenics contraction transformation [3] of the desirable intervals for x and y axes. Contraction transformation was applied with the reason that working point (x,y) to be farther from the constraint precision-stability area imposed by the researcher.

Extension/Contraction transformation was applied with α factor for each of axes. About this factor will adapted the α' factor that will be applied for all cases studied parameters of the hydraulic cylinder. This factor will be calculated versus limits values for each of the parameters and will be different for all parameters.

$$\begin{aligned}&TX_0=\alpha X_0;TX=\alpha X\\&TY_0=\alpha_0 Y_0;TY=\alpha Y\\&\alpha_i'=\frac{\inf(\lim param.value)_i}{\sup(\lim.param.value)_i}\\&\alpha_i''=\frac{\alpha_i'}{nr.steps}\end{aligned} \tag{3}$$

where $\alpha>1$ for the extension and $0<\alpha<1$ for the contraction what will be applied to the interval X and Y, α' is the *total adapted increasing/decreasing factor* and α'' is the *adapted increasing/decreasing factor for the first step* value. Some of the data must be increasing and other decreasing, finally defined the *minimal working attraction field* [6]. The other way of the research was to define different values of the factor. When the $\alpha''<<1$ will be calculated other value with the condition to respect the step value, all values for increasing or decreasing will be multiplier or de-multiplier with this new value, other will be calculate with $1-\alpha''$ or $1+\alpha''$.

With the first one way was determined the point $\boldsymbol{P}_1$ and with the second one way was determined the point $\boldsymbol{P}_2$. The other points were the maximal point of the second research way $\boldsymbol{P}_3$, and $\boldsymbol{P}_{4,5}$ the points of the middle of intervals. For each step were applied different values to determine the convergence process. The decreasing factor was determined by de multiply the current factor by around 2. After applying the contraction factor will be obtained the data what are shown in the Table 1 for the first way of the research when the factor was the same for all parameters, like the extension/contraction factor of the field and in the Table 2 are shown the data obtained after applying the different factor for the parameters what is in the condition $\alpha''<<1$. The Extenics transformation will be applied to determine the desirable field for each of the Stability and Precision parameter's values. Each of the current determined field will be the new desirable field and the last one will be the acceptable field in each iteration process. The novelty of the Extenics applied method was to use in the research the different factor for increasing and decreasing values for each of the studied constructive and functional parameters limits and alternating them or each iteration to assure the convergence process and finally to obtain the *minimal working attraction field*. The value of the transformation factor α'' was de multiplied by 2 for each steps, to assure the asimptoticaly neighbour to the minimal attraction field.

Table 1. Data calculated after applied Extenics transformation for the value of α'<1

α=0.1		α=0.05		α=0.025		α=0.012	
S	P	S	P	S	P	S	P
91.3	85.5	86.7	89.77	88.9	87.52	87.83	88.57
31.5	2.2	29.9	2.31	29.17	2.36	28.81	2.38
188.1	201.3	197.5	191.2	192.5	196	194.8	193.6
45	44	42.75	41.8	41.68	42.84	42.18	42.32

α=0.006		α=0.003		α=0.001	
S	P	S	P	S	P
88.35	88.03	88.08	88.29	88.16	88.2
28.63	2.39	28.54	2.39	28.51	2.39
193.7	194.8	194.2	194.2	194	194
42.43	42.06	42.3	42.18	42.25	42.22

Table 2. Data calculated after applied Extenics transformation for the value of α'<<1

α'=2.5		α'=1.2		α'=0.6		α'=0.3		α'=0.15		α'=0.07	
S	P	S	P	S	P	S	P	S	P	S	P
14	5	11.6	6	4.64	9.6	3.24	6.72	3.72	5.71	3.98	5.31

2.3 LabVIEW simulation results and discussions of the solving contradictory problem hydraulic cylinder's precision-stability.

Contradictory problem is one of the more important think to solve in the technical field, because not be possible to design one hydraulic cylinder that can be precise and stable without Extenics solving the contradictory problem.

LabVIEW simulation contents the simulation of the root's locus and dependent formula. After applying the mathematical model of the Extenics dependent function for 2D dimension and increasing/decreasing transformations were obtained some results, Figs. 4-7. The Extenics theory was accomplished with the proper remarks and concepts. We remark that the partial solution is not singular and inside of the minimal attraction field and also is not single. For that was applied the dependent formula for more other working points P_1-P_5 what there are very close to this determined area to establish if all these points are in the same precision-stability class or represent the new one precision-stability class. Analyzing the obtained simulation data for the dependent formula, Fig.7 we can remarks the following: all points is in the acceptable- desirable 2D field, exception, $\boldsymbol{P_4}$; $\boldsymbol{P_1}$ is the optimal working point determined by Extenics theory. $\boldsymbol{P_{1,2,3,5}}$ there are in the same precision-stability class by the established desirable- acceptable 2*D* fields. $\boldsymbol{P_4}$ is not acceptable point in the precision-stability class. The research can be more precise and for that can be used one new contraction transformation to reduce the acceptable field and will remain inside of the acceptable field only $\boldsymbol{P_{1,2,3}}$. These remarks are very important for the researchers to establish very easily and correctly the precision-stability class of the hydraulic cylinder, like was remarked in this study case.

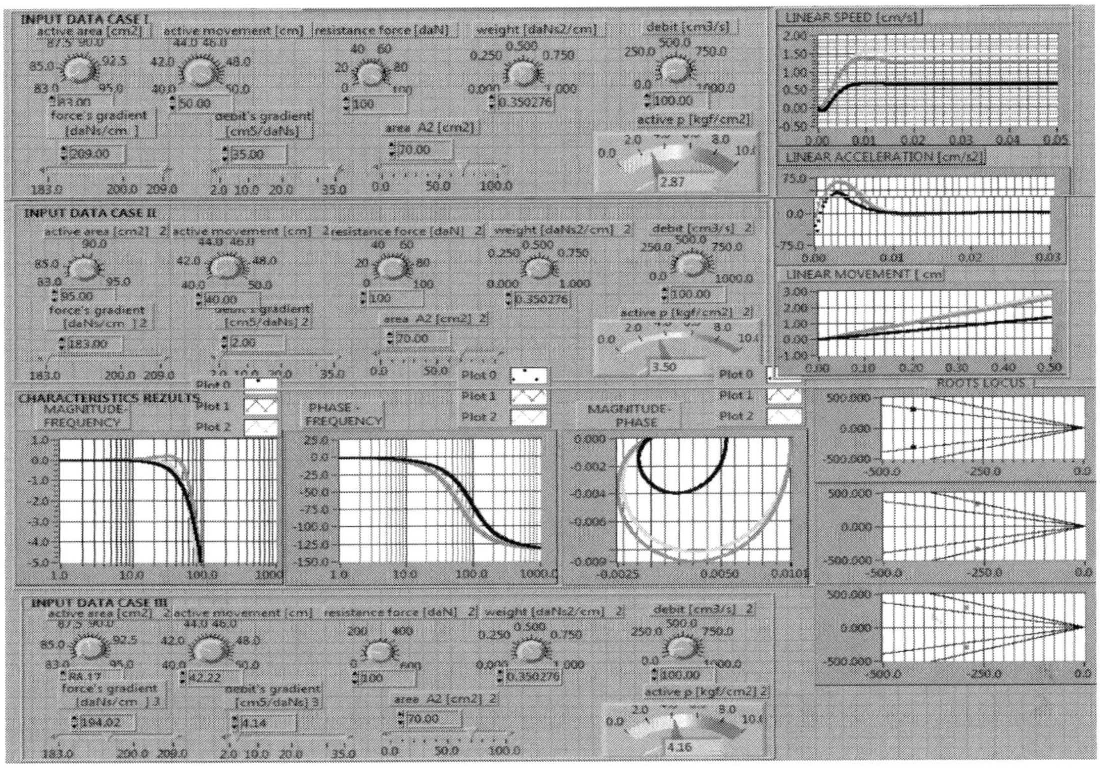

Figure 4. The comparative results with the limits values interval of the precision and stability and the optimal value established after applied the Extenics complex transformation.

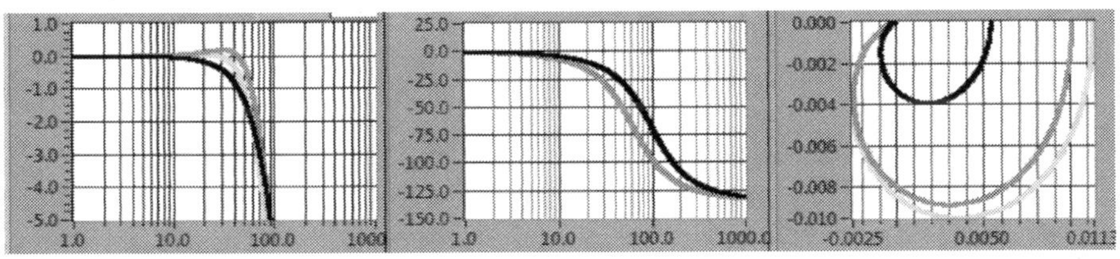

a

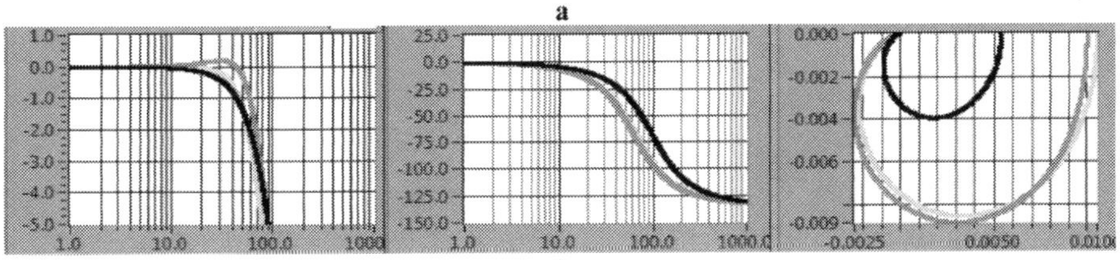

b

DOI: 10.1051/matecconf/20165411008

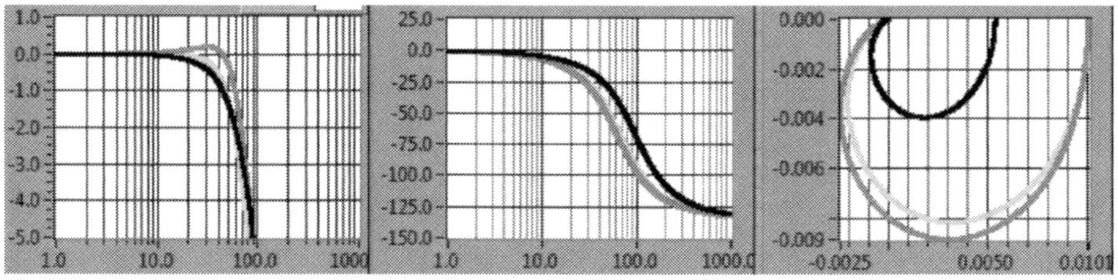

c

Figure 5. Comparative frequency characteristics Bode and Nyquist with the intermediate and with the final minimal attraction field dynamic behavior established by applying the Extenics theory.

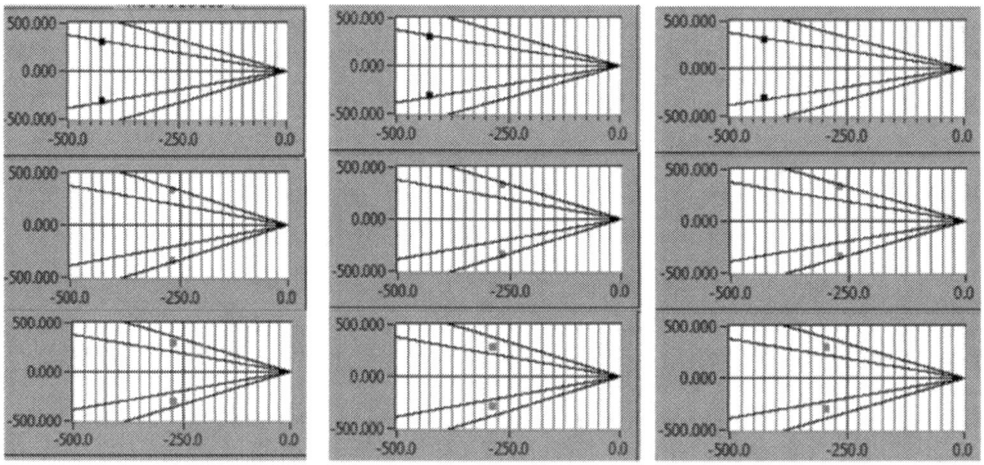

Figure 6. Comparative root's locus for the intermediate and for partial optimal solution after applying the Extenics theory.

The studied points $P_{1\text{-}5}$ have the following simulation parameters:

$$P_1=\begin{Bmatrix}88.16\\2.39\\194\\42.22\end{Bmatrix};P_2=\begin{Bmatrix}88.16\\3.98\\194\\42.22\end{Bmatrix};P_3=\begin{Bmatrix}88.16\\5.31\\194\\42.22\end{Bmatrix};P_4=\begin{Bmatrix}88.16\\20\\194\\42.22\end{Bmatrix};P_5=\begin{Bmatrix}88.16\\15\\194\\42.22\end{Bmatrix}. \tag{4}$$

where the data means: aria A=88.16[cm^2]; flow gradient a_m=2.39;3.98;5.31;20;15[cm^5/daNs]; force versus velocity gradient b_m= 194[daNs/cm]; active movement c=42.22[cm]. After applying the Extenics theory and the assisted determination of the dependent *2D* formula for each of these points, we were obtained the results from (5). The applied values were determined by apply the logical Extenics theory with AND function between the goals and conditions matter-elements[5].

The values of the point's parameters were obtained by convergence process of the Extenics transformation and after applying *min (precision_value, stability_value)*function.

$$\begin{aligned}
&P_1=\begin{Bmatrix}286{,}556\\286{,}979\end{Bmatrix}\begin{Bmatrix}k_1(x)=-0.025;k<0\Rightarrow x\in X\\k_1(y)=0.03;0<k<1\Rightarrow y\in Y_0\end{Bmatrix}\Rightarrow acceptable_desirable_po\text{int}\\
&P_2=\begin{Bmatrix}292{,}964\\286{,}979\end{Bmatrix}\begin{Bmatrix}k_2(x)=0.145;0<k<1\Rightarrow x\in X_0\\k_2(y)=-0.100;-1<k<0\Rightarrow y\in Y\end{Bmatrix}\Rightarrow desirable_acceptable_po\text{int}\\
&P_3=\begin{Bmatrix}298{,}324\\297{,}505\end{Bmatrix}\begin{Bmatrix}k_3(x)=0.288;0<k<1\Rightarrow x\in X_0\\k_3(y)=-0.208;-1<k<0\Rightarrow y\in Y\end{Bmatrix}\Rightarrow desirable_acceptable_po\text{int}\\
&P_4=\begin{Bmatrix}358.773\\340.217\end{Bmatrix}\begin{Bmatrix}k_4(x)=-0.21;-1<k<0\Rightarrow x\in X\\k_4(y)=-1.195;k<-1\Rightarrow y\notin Y_0,Y\end{Bmatrix}\Rightarrow acceptable_nonacceptable_po\text{int}\\
&P_5=\begin{Bmatrix}338{,}623\\327{,}842\end{Bmatrix}\begin{Bmatrix}k_5(x)=0.175;0<k<1\Rightarrow x\in X_0\\k_5(y)=-0.909;-1<k<0\Rightarrow y\in Y\end{Bmatrix}\Rightarrow desirable_acceptable_po\text{int}
\end{aligned} \tag{5}$$

DOI: 10.1051/matecconf/20165411008

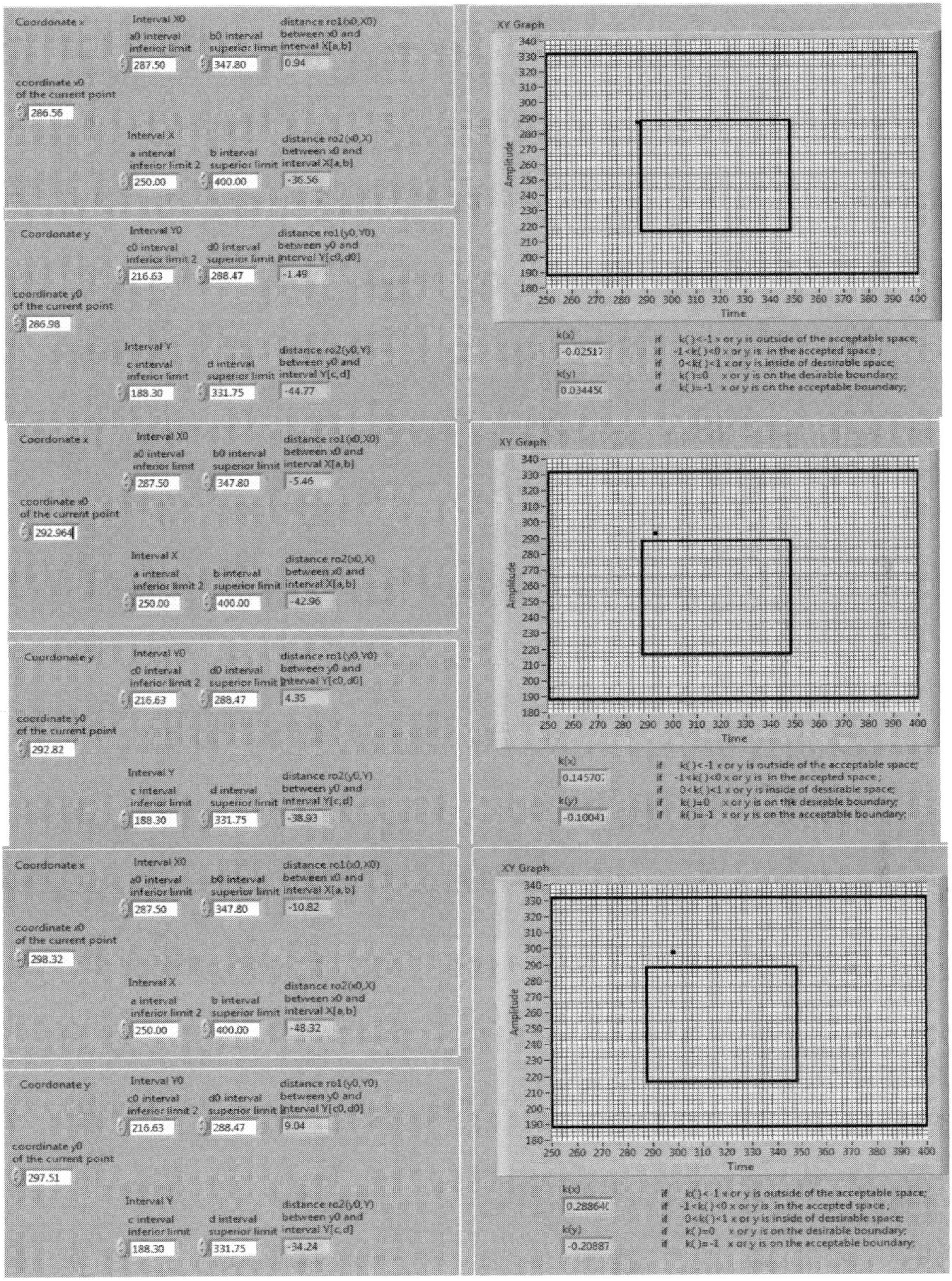

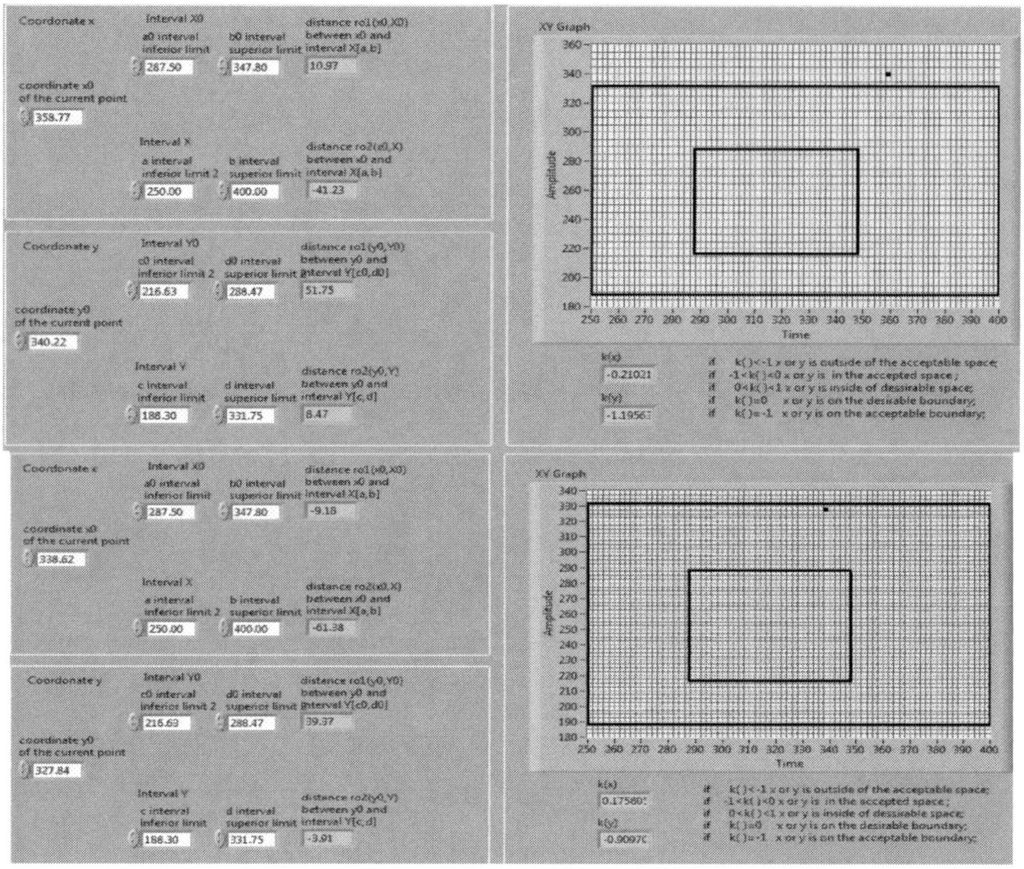

Figure 7. The desirable and acceptable fields and determined depend *2D* formula for the all working researched points P_1-P_5.

We can define 9 classes of the acceptable, desirable and complex acceptable-desirable and non-acceptable fields of the Extenics theory in 2D dimensions modeling of the dependent function.

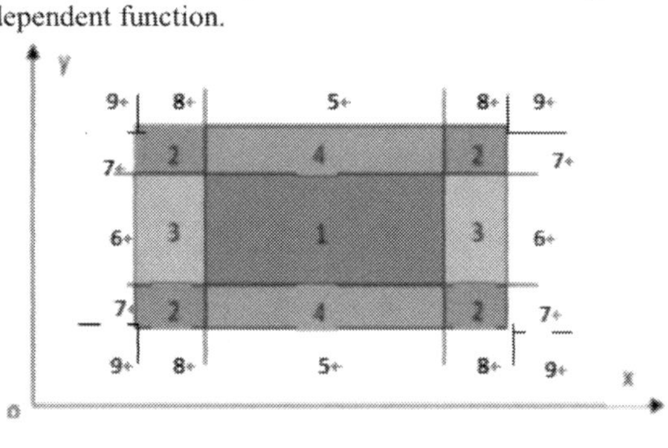

Figure 8. The desirable and acceptable fields in the 2D dimensions of Extenics.

1-Full desirable $\begin{Bmatrix} x \Rightarrow desirable \\ y \Rightarrow desirable \end{Bmatrix};$

2- Full acceptable $\begin{Bmatrix} x \Rightarrow acceptable \\ y \Rightarrow acceptable \end{Bmatrix};$

3- Partial acceptable $\begin{Bmatrix} x \Rightarrow acceptable \\ y \Rightarrow desirable \end{Bmatrix};$

4- Partial desirable $\begin{Bmatrix} x \Rightarrow desirable \\ y \Rightarrow acceptable \end{Bmatrix};$

5-Partial acceptable $\begin{Bmatrix} x \Rightarrow desirable \\ y \Rightarrow nonacceptable \end{Bmatrix};$

6-Partial acceptable $\begin{Bmatrix} x \Rightarrow nonacceptable \\ y \Rightarrow desirable \end{Bmatrix};$

7- Partial acceptable $\begin{Bmatrix} x \Rightarrow nonacceptable \\ y \Rightarrow acceptable \end{Bmatrix};$

8- Partial acceptable $\begin{Bmatrix} x \Rightarrow acceptable \\ y \Rightarrow nonacceptable \end{Bmatrix};$

9- Full nonacceptable $\begin{Bmatrix} x \Rightarrow nonacceptable \\ y \Rightarrow nonacceptable \end{Bmatrix}$

3 Conclusions and future work

By applying the extended transformation and dependent formula for 2D were obtained the optimal results what solve the contradictory problem. The applied method LW instrumentation assures the precise results and validate some of Extenics relations. The obtained results, the applied method, all LabVIEW instrumentation and all remarks about 2D classes assures for the research the way for solving better the contradictory problems. For the future will be applied the complex extension transformation method and extension strategy generating system.

References

1. W. Cai, *The Extension Set and Non- compatible Problems*, Editor W. Z. Chien, AMM in China, International Academic Publisher, **2**, 1-21, (1990).
2. W. Cai, *Matter-element Models and Their Applications*, Beijing: Science and Technology Document, Publishing House, (1994).
3. C. Y. Yang, and W. Cai, *Extenics: Theory, Method and Application*, Science Press, China, (2002).
4. A. Olaru, *Optimal Solving of the Contradictory Problem* between *Hydraulic Cylinder's Precision and Stability with Extenics Theory*, ICMAE2013, Proceedings, Moskow, 2013.
5. Florentin Smarandache, *Generalization of the Distance and Dependent Function in Extenics to 2D, 3D and nD,* Science and Technology Document, Publishing House, (2012).
6. A. Olaru, *Assisted Analyze of the Extenics Dependent Functions in 1D,2D,3D and nD Dimensions with LabVIEW Instrumentation*, ICNCT2013, Proceedings, Seul, 2013.

A GaAs MMIC-based X-band Dual Channel Microwave Phase Detector based on MEMS Microwave Power Sensors

Di Hua , Junyan Tan and Chunhua Cai

HoHai Univerity College of Internet of Things Engineering, Changzhou, Jiangsu 213022 China

Abstract. By sensing the output powers of two channels with MEMS microwave power sensors, a wideband 8-12GHz dual channel microwave phase detector is presented. In order to detect phase difference between reference signal and testing signal in entire 0-360° at X-band, artificial 45° phase lead and 45° phase lag are introduced in two channels, additionally. The proposed phase detector is composed of power dividers, power combiners, 45° (at 10GHz) coplanar waveguide (CPW) transmission delay lines, MEMS capacitive power sensors and thermoelectric power sensors. The two CPW transmission lines result in 45° phase lead and 45° phase lag at 10GHz, and the two different kinds of power sensors are used to detect the combined powers. The fabrication of the phase detector is compatible with GaAs Microwave Monolithic Integrated Circuit (MMIC) technology. The phase detection measurement is accomplished at 10GHz with 0-180° phase shift between testing signals generated by a tunable analog phase shifter. Phase detection results in two channels show that the normalization of the measured results fit the calculated results well, though the results of phase lead have a little of deviation.

1 Introduction

Phase detectors are widely used in phase demodulators, phase-locked loops, and phase-measuring equipment. In these applications, phase detection is usually performed at relatively low frequencies, and frequently after down conversion of the original high-frequency signal [1]. Phase detection directly performed at microwave frequencies would, in many cases, lead to a reduction in hardware complexity. The existing technologies for microwave phase detection are based on diodes, CMOS and vector combination. Diode phase detectors take use of the square-law detection characteristic of diodes, but have problems with impedance matching [1-4]. CMOS phase detectors employ FET multiplier to realize phase detection, and will consume extra power, since DC bias is needed [5]. Meanwhile, diode phase detectors and CMOS phase detectors are sensitive to temperature variations. Based on vector combination and microwave power meters, vector combination phase detectors cannot detect phase difference between -180° to +180°, since the output is symmetrical in -180° to +180° [6]. Also, the phase detector in [6] had not been realized on one chip.

Composed of a power combiner and a thermoelectric power sensor, the single channel microwave phase detector had been reported [7, 8]. The proposed phase detector has simple structure and does not consume any DC power, but the measured result shows that phase shift cannot be measured or deduced in 0 to 360°. In this paper, the dual channel microwave phase detector divides reference signal and testing signal into two channels, and introduces additional 45° phase lead and 45° phase lag between the reference signal and the testing signal°. The phase difference can be finally calculated through the outputs of the two kinds of power sensors. In channel I, the divided testing signal and the divided reference signal with additional 45° phase delay are combined. The divided testing signal with additional phase delay is combined with the divided reference signal in channel II. The output powers of the two channels are detected by the MEMS power sensor. Two kinds of microwave power sensors are adopted because of the high power-handling capability of the capacitive power sensor and the high sensitivity of the thermoelectric power [9-12]. The membrane of the MEMS capacitive power sensor will be slightly pulled down with low power applied. The power will be absorbed by two load resistors, and be converted to the thermo-voltage by the thermopiles. When the output powers are incidentally large, the output powers result a detectable displacement of the MEMS membrane. The movement can be measured capacitively. Finally, the phase difference between testing signal and reference signal can be deduced according to the detected output power of the two channels.

2 Analysis and Design

The proposed dual channel can be divided into two parts. Dual channel structure is a four-port signal processing structure, and outputs the signals containing the phase difference between the reference signal and the testing signal in two channels. The other parts are microwave power sensors, including MEMS thermoelectric power sensors and MEMS capacitive power sensors.

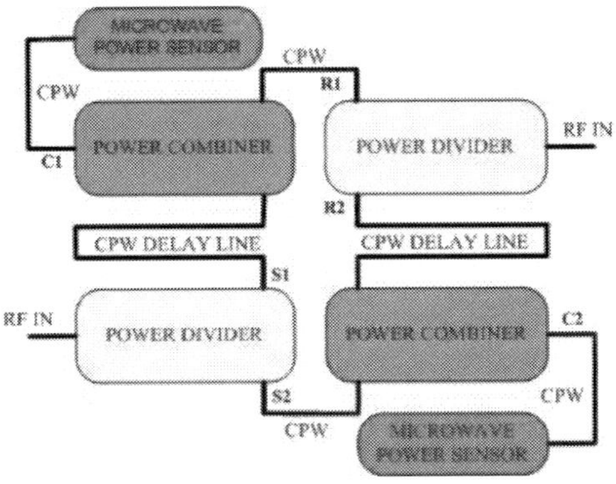

Figure 1. Schematic overview of the dual channel phase detector

2.1 Dual Channel Structure

The dual channel structure is composed of two power dividers), two power combiners and two phase shifters. Fig. 1 shows the schematic circuit of the dual channel microwave phase detector. Testing signal Us and reference signal Ur are applied to input ports, and the two signals should have the same power in order to simplify the calculation. Then, testing signal is divided into Us1 and Us2 by one power divider, which have the same magnitudes and the phase shifts, and reference signal is divided into Ur1 and Ur2 by the other power divider, respectively. Ur1 transmits through a 45° phase shifter, and be combined with Us1 by one power combiner. Symmetrically, Ur2 transmits through another 45° phase shifter, and be combined with Us2 by the other power combiner.

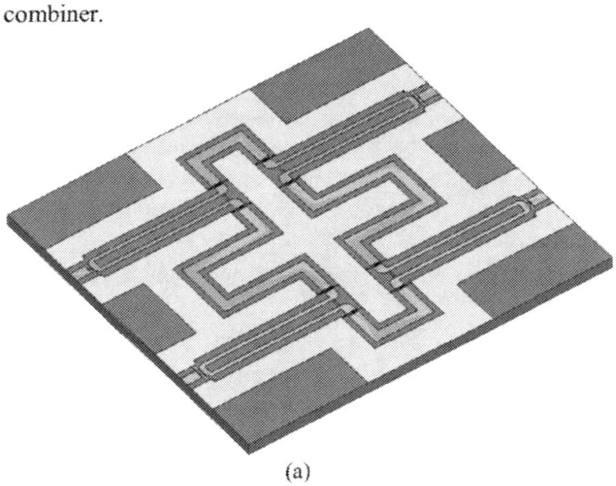

(a)

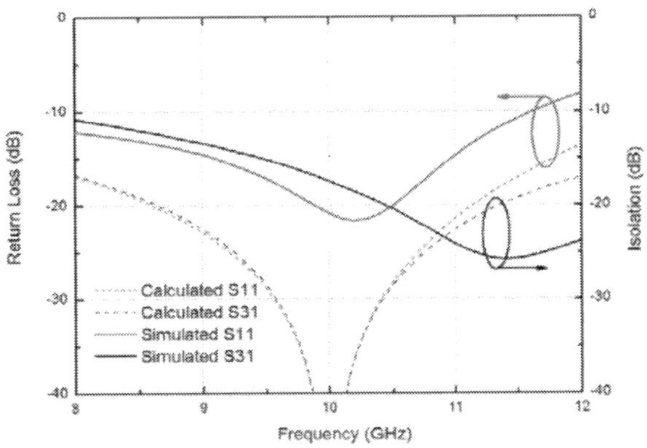

(b)

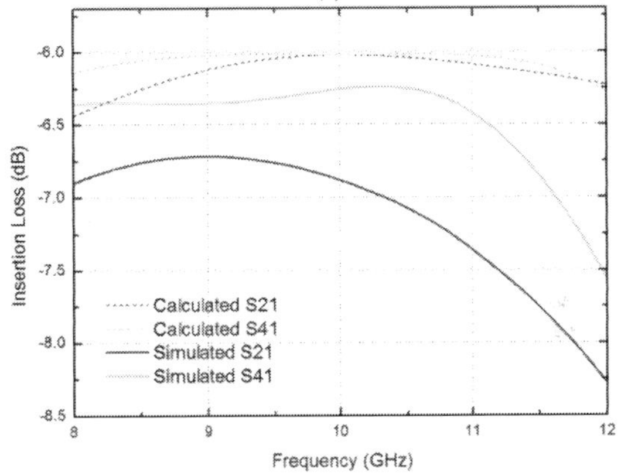

(c)

Figure 2. (a) The simulating dual channel structure and (b) (c) its simulated S-parameters

Fig. 2(a) shows the dual channel structure. As can be seen in Fig. 2, over a 75% bandwidth centered at 10GHz, the simulated return loss of input ports is less than -13.15dB and the insertion losses between port 1, 2 and 1, 4 are -7.75dB and -6.85dB, respectively. The isolation between input ports 1, 3 is less than -12.02dB. Since the insertion loss of ports 1, 2 and ports 1, 4 are almost equal, the insertion loss S12 and S14 are defined to S in order to simplify the research. When the testing signal and reference signal are applied to Ports 1, 3, the out put powers of Ports 2, 4 can be expressed as

$$P_{out2} = Z_0 S^2 U^2 (1 + \cos(\varphi + 45^\circ)) \quad (1)$$

$$P_{out4} = Z_0 S^2 U^2 (1 + \cos(\varphi - 45^\circ)) \quad (2)$$

2.2 MEMS capacitive power sensor and thermoelectric power sensor

The output powers of the dual channel structure are detected by two kinds of microwave power sensors, the MEMS capacitive power sensor and the thermo-electric power sensor. The two kinds of power sensors are adopted to increase the detective range. When relative high level signal is applied, the membrane will be attracted. While relative low level signal is inputted, the signal will be detected by the thermo-electric power sensor.

DOI: 10.1051/matecconf/20165412001

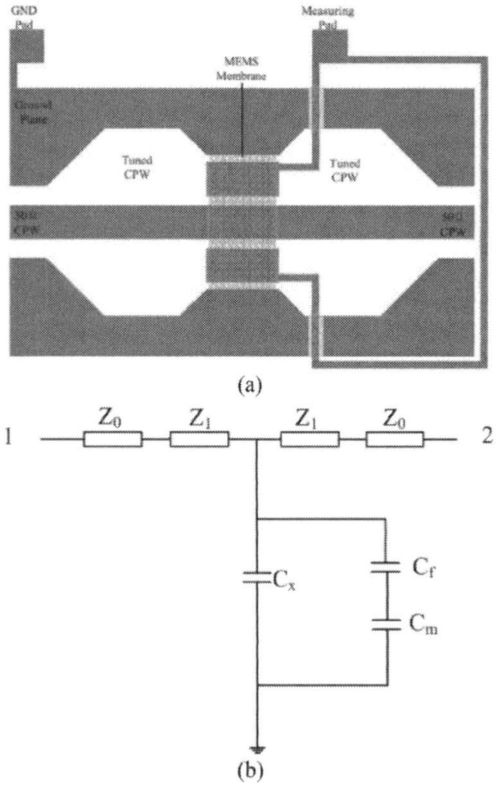

Figure 3. (a)Schematic view of the MEMS capacitive power sensor (b) the corresponding circuit

The MEMS capacitive power sensor is composed of 50Ω CPW, tuned CPW and a fixed-fixed MEMS membrane suspended on the signal line in Fig. 3(a). The anchors of the membrane are connected to the ground plane of the CPW, while the CPW is terminated by two parallel 100Ω TaN resistors. Due to the discontinuity of the shunt MEMS membrane, the CPW near the membrane are tuned to realize well impedance matching. Fig. 3 (b) shows the schematic view and the schematic circuit of the capacitive power sensor, where Z1 is the characteristic impedance of the tuned CPW, Cx is the capacitance between the center signal line and the membrane, Cf is the fringing capacitance between the center signal line and the measuring pad, Cm is the capacitance between the measuring pad and the membrane. Since Cf is relatively small, Cf and Cm can be removed for simplification.

When the processed microwave signals are applied to the two capacitive power sensors, the membrane is attracted. The thermoelectric power sensor is composed of 50 Ω terminating resistor and thermo-couples, whose principle is based on Seebeck effect. The thermo-couple is two jointed different materials. When relative low level combined signals are applied to the MEMS capacitive power sensor, the signals will not attract the membrane. Almost all the combined signals are absorbed by the terminating resistors and converted to DC thermovoltage output by the thermopiles [13].

3 Fabrication

The fabrication of the microwave phase detector is compatible with GaAs MMIC process. Surface micromachining technology is used to fabricate the thermopiles, and bulk micromachining technology is employed to reduce thermal loss and increase the sensitivity of the power sensor. In this detector, the CPW is designed to have 50Ω characteristic impedance. In [14], the process steps of the detector are briefly described as follows

(1) The thermopiles are made of Au and n+ GaAs with a doping concentration of 1.0 × 1018cm-3. The Au is made by sputtering a 500/2000Å AuGeNi/Au layer and using a lift-off process, and the n+ GaAs is made of an ion implantation layer. The thermopile has length of 80μm.

(2) A TaN layer is sputtered and patterned to form the load resistors and isolation resistors with the square resistance of 25.

(3) The lower plates of the MIM capacitors are made by evaporating a 0.3-μm-thick Au layer, and AuGeNi/Au layer is used as adhesion layer.

(4) A 1000 Å Si3N4 is deposited to form the dielectric layers of the MIM capacitors by PECVD.

(5) A 500/1500/300Å Ti/Au/Ti seed layer is evaporated and patterned. After removing the top Ti layer, the CPW transmission line and the upper plates of the MIM capacitors are formed through electroplating a 2-μm-thick Au layer.

(6) The substrate is thinned to about 100μm by a wafer grinding process, and the substrate underneath the thermopiles and the load resistors is etched to 20μm in order to reduce thermal losses using dry etching technology.

(7) The sacrificial layer of polyimide below the membrane and the air bridge is removed using a developer and the alcohol is utilized to get rid of the residual water in the MEMS membrane.

The fabrication of the microwave phase detector is compatible with GaAs MMIC process. Surface micromachining technology is used to make the thermopiles, and bulk micromachining technology is employed to reduce thermal loss and increase the sensitivity of the power sensor in Fig. 4. In this detector, the CPW is designed to have 50Ω characteristic impedance. In [8], the process steps of the detector are briefly reported.

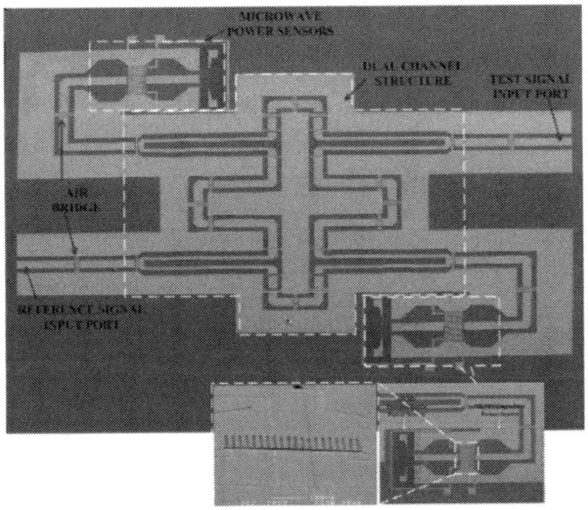

Figure 4. SEM picture of the whole dual channel microwave phase detectora and the MEMS membrane

4 Measurement Result

Phase detection environment is described as follow. The testing signal and the reference signal are generated by an Agilent power splitter and an analog phase shifter, which can make the phase shift change between 0 to 240°. A nut is used to control the phase delay of the phase shifter. The initial phase delay is about 180° instead of 0 because of the phase delay of the adapters and the connecting cables. According to the instruction book of the phase shifter, the phase delay step is about 9°at 10 GHz per round. The MEMS capacitive power sensor has the advantage of high power-handling capability, while its disadvantage is that relatively low test signal and reference signal can't attract the membrane. Since the phase detection environment is restricted by the maximum output power of the signal generator, the phase detection measurement is accomplished by the thermoelectric power sensor only, while the MEMS capacitive power sensor is measured, separately.

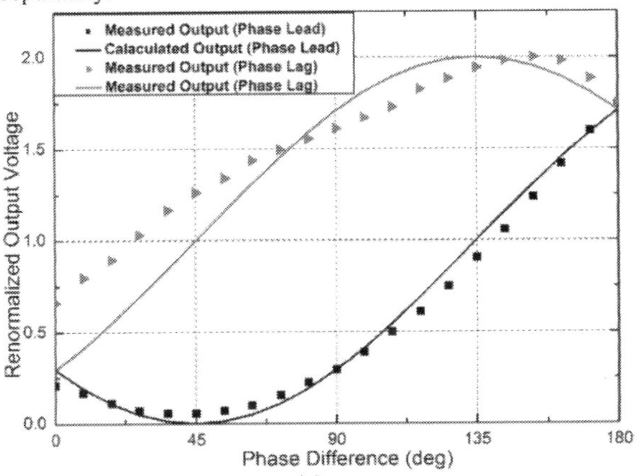

(a)

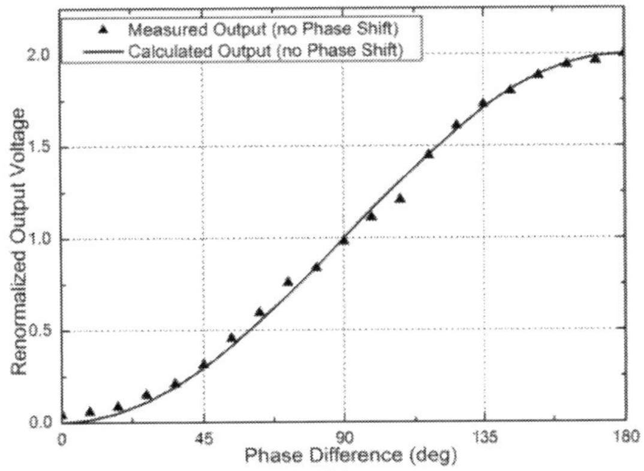

(b)

Figure 5. Normalizations of measured results and calculated results with 45° phase lead, 45° phase lag and no phase shift

Fig. 5 shows the normalized phase detection results in three different situations. Fig. 5 (a) show additional 45° phase lead and 45° phase lag between testing signal and reference signal, respectively, and Fig. 5 (b) shows no phase shift is added. As can be seen in Fig. 5, the measured results and the calculated results are matched well, which means that the phase detection principle is validated, while the results in Fig. 5 (a) have a little of deviation. The reason of deviation may be that the discreteness of the fabrication process makes the detector asymmetrical.

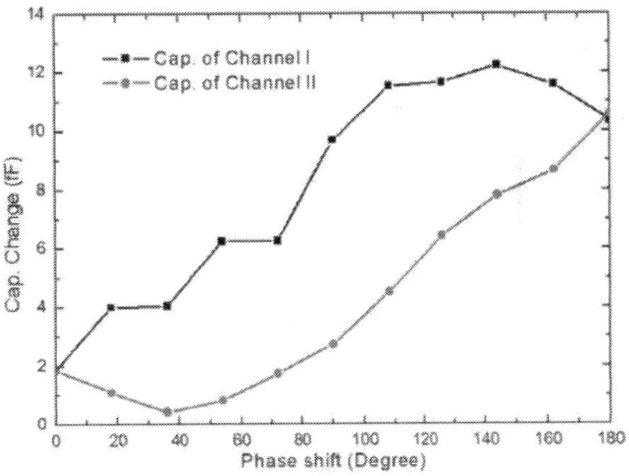

Figure 6. Measured capacitance change versus phase shift with 25dBm signal applied

Fig.6 shows the measured capacitance change versus the phase shift in two channels. Two methods are involved to make the capacitance change can be detected: 1. the maximum power level of the signal generator are applied. 2. the insertion loss of the measurement environment is optimized to reduce the power loss. The calculated phase differences from Fig. 5 and 6 include the effective phase difference and the initial phase difference. The simulation result of the dual channel structure shows that the insertion loss of channel I and II are 6.4dB and7.0dB, respectively. The maximum power levels of the testing signal and the reference signal should be less than 20dBm, which means that the power level of the signal generator should be less than 23dBm (3dB insertion loss of the power divider, used to generate the test and reference signals). For the thermoelectric power sensor, the high power limitation of the input testing signal and reference signal is estimated to be 20dBm.

The dimension of the MEMS membrane is 400μm*200μm*2μm with 20μm*10μm holes formed. The goal of the small holes is to release the sacrificial layer under the membrane. The designed height of the initial air gap is 1.6μm. A pair of 100μm*200μm*0.3μm measuring electrodes are fabricated between the anchors and the center signal line with a 0.1μm thick Si_3N_4 layer deposited. The calculated initial capacitance between the electrodes and the membrane is 0.110pF, and the fringing capacitance is neglected. Figure 7 shows the simulated displacement and the corresponding capacitance versus different voltages using ConventorWare v2006. The capacitive microwave power sensor is measured at 8, 9, 10, 11, and 12GHz, and the average sensitivity is 7.2 fF W^{-1} [15].

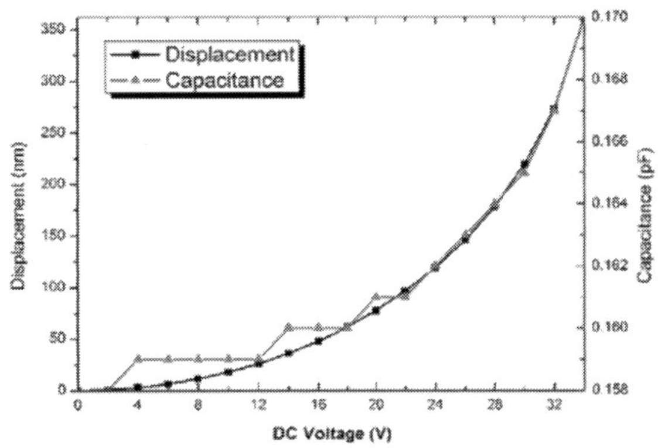

Figure 7 Simulated displacement and the corresponding capacitance versus different voltages

5 Conclusion

In order to extend the detectable phase range, a wideband 8-12GHz symmetrical dual channel microwave phase detector is presented in this paper, and the phase detector is accomplished with GaAs MMIC technology. The detector takes use of 45° phase lead and 45° phase lag in two channels, and two thermoelectric microwave power sensors are used to detect the two output powers. The MEMS capacitive power sensor provides wide power-handling capability, which means that the proposed phase detector has higher power-handling capability than diode-based phase detectors. Phase shift between reference signal and testing signal can be calculated, determinately and uniquely. The measured S-parameters show that the detector are well impedance matched, and the normalizations of measured output voltages agree well with the calculated results in different situations. The dual channel microwave phase detector can be widely applied to phase-measuring equipment and radar systems.

Acknowledgement

This work is supported by the Fundamental Research Funds for the Central Universities (2014B01914).

References

1. Ohm G and Alberty M. "Microwave Phase Detectors for PSK Demodulators," IEEE Microwave Theory and Techniques 1981, 29(7), 724-731.
2. Mohammad M. "A Wide Dynamic-range Phase Detector for Advanced Communication Systems Application," WESCANEX 93. 'Communications, Computers and Power in the Modern Environment.' IEEE Conference Proceedings, May, 1993, pp. 192-195.
3. Setc B. and Reckko C. "Broadband Microwave Phase Detector with Diode Ring," IEEE Microwaves, Radar and Wireless Communications, MIKON-2000, 13th International Conference, May, 2000, 1, pp. 65-68.
4. Slomski R and Stec B. "Broadband microwave phase discriminator with ring detectors," IEEE Microwaves, Radar and Wireless Communications, MIKON-2004, 15th International Conference, May, 2004, 3, pp: 936-939.
5. Shahriary I, Lin S.T. and Weller K. "A Practical Wide Band GaAs Phase Detector," Microwave and Millimeter-Wave Monolithic Circuits 1982, 82(1): 47-49
6. Lehto A, Tuovinen J, Borić O and Räisänen A. "Accurate Millimeter Wave Antenna Phase Pattern Measurements Using the Differential Phase Methode with Three Power Meters," IEEE Antennas and Propagation, 1992, 40(7), 851-853.
7. D. Hua, X.P. Liao and Y.C. Jiao, "A MEMS phase detector at X-band based on MMIC technology," in Sensors, 2009 IEEE, 2009, pp. 506-508.
8. D. Hua, X.P. Liao and Y.C. Jiao, "X-band microwave phase detector manufactured using GaAs micromachining technologies," Journal of Micromechanics and Microengineering, 2011, 21(3): 035019.
9. Y. Cui, X.P. Liao and Z. Zhu, "A novel microwave power sensor using MEMS fixed-fixed beam," in Sensors, 2011 IEEE, 2011, pp. 1305-1308.
10. Z.X. Yi, X.P. Liao, Z. Zhu; , "An 8-12GHz capacitive power sensor based on MEMS cantilever beam," Sensors, 2011 IEEE , vol., no., pp.1958-1961, 28-31 Oct. 2011
11. D.B. Wang, X.P. Liao, T. Liu, "A Novel Thermoelectric and Capacitive Power Sensor With Improved Dynamic Range Based on GaAs MMIC Technology," Electron Device Letters, IEEE , vol.33, no.2, pp.269-271, Feb. 2012
12. De-Bo Wang, Xiao-ping Liao, "A terminating-type MEMS microwave power sensor and its amplification system," J. Micromech. Microeng. 2010, 20(7):075021
13. Milanovic V, Hopcroft M, Zincke C.A., et al. "Optimization of CMOS MEMS microwave power sensors," Circuits and Systems, Proc. IEEE ISCAS '99, May, 1999, 5:144 -147.
14. D. Hua, X.P. Liao and J.Q. Huang, "A GaAs MMIC-based dual channel microwave phase detector at X-band," in Sensors, 2012 IEEE, 2012, pp. 1-4.
15. Y. Cui, X.P. Liao and Z. Zhu, "A novel microwave power sensor using MEMS fixed-fixed beam," in Sensors, 2011 IEEE, 2011, pp. 1305-1308.

MATEC Web of Conferences **54**, 12002 (2016) DOI: 10.1051/matecconf/20165412002
MIMT 2016

Rapid prototyping of Networks-on-Chip on multi-FPGA platforms

Junyan TAN[1], Virginie FRESSE[2], Frédéric ROUSSEAU[3]

[1]*College of IoT, Hohai University, 21300 Changzhou, CHINA*
[2]*Hubert Curien Laboratory UMR CNRS 5516, 42000 Saint-Etienne, France*
[3]*TIMA Laboratory, UJF/CNRS/Grenoble INP, 38000 Grenoble, FRANCE*

Abstract. Experimental approaches used for architecture exploration and validation are often based on configurable logic device such as FPGA. NoC architectures require multi-FPGA platforms as the resources of a single FPGA are not big enough. Partitionning a NoC on multi-FPGA requires special techniques for allocating communication channels, physical links and suitable resource allocation scheme. We present a scalable emulation platform and its associated design flow based on a multi FPGA approach that allows quick exploration, evaluation and comparison of NoC solutions. The efficiency of our approach is illustrated through the deployment of the Hermes NoC and its exploration on several FPGA platforms.

1 INTRODUCTION

Networks on Chip (NoCs) have emerged as a viable option for designing scalable communication architectures for SoC and MPSoCs for signal and image processing applications [1]. However, the design of NoC means making several architectural choices, just like buffer sizing, flow control policies, topology selection. These choices must be made at the design time keeping in mind that the final NoC must stratify a set of critical constraints which depend on the target application such as: latency, energy consumption, design time. The design space being very wide, automation of the design flow and automatic architecture exploration must be considered to ensure a rapid evaluation and test of each solution. Mathematical and experimental approaches are used to accelerate the architecture exploration. Experimental approaches use simulation or emulation with different abstraction levels. FPGA devices are commonly used for emulation and test. Today, several NoC architectures have successfully been implemented on mono FPGA platforms [1][2][3]. One single FPGA device usually does not offer enough resources any more to support a complete large Network/ System on Chip architecture. Such large systems need to be partitioned over several reconfigurable devices, most of time on multi-FPGA platforms for emulation. Existing tool target the mono FPGA platform and there is no any development tool proposed for the multi-FPGA implementation. In [4], a NoC specific multi-FPGA technique is proposed. The major drawback of this work is that the platform was tested with synthetic traffic generators, and authors have not studied the impact of time accuracy loss on real world applications.

Porting a NoC on multi-FPGA platforms means that designers have to manually adapt the architecture with a manual partitioning and to manually integrate inter-FPGA communication blocks. Such adaptations lead to a significantly increased design time.

The aim of this work is to propose a generic design flow for the emulation of large NoC-based MPSoCs on multi FPGA platforms. This design flow integrates a tool that automatically builds the NoC architecture that is partitioned and ready for the multi FPGA implementation. The tool also includes logic blocks to handle the inter-FPGA communications, and all necessary logic required for emulation.

2 DESIGN FLOW FOR THE GENERATION OF THE MULTI-FPGA EMULATION PLATFORM

A new design flow must be proposed to generate the emulation architecture to multi-FPGA platforms. The design flow takes as inputs 1) the NoC architecture, 2) parameterized adaptation blocks: a description of the multi-FPGA platform: the number of available FPGA chips, the type (resources) of FPGA and the physical links used for inter-board communications, 3) emulation blocks: traffic generator and traffic receptors.

2.1 NoC architecture

Network on Chip (NoC) are communication architectures with high scalability, high performance and energy efficient customized solution. NoC architecture is composed of Network Interface (NI), Switch, Links and

Resources. These basic elements are connected using a topology to constitute the NoC architecture.

Data transmitted in NoC architectures are sent through messages. Several data can be sent with one message and one data can be sent with several messages. One message is a set of packets and a packet is a set of flits (Flow Control Unit). The flit is the basic element transferred by a NoC. Packets are sent with idle times specified with data injection rate. Data injection rate is defined as the ratio of the amount of receiving data on its ability to carry data. A 50% data injection rate indicates that the packets use 50% of the bandwidth.

2.2 Adaptation blocks for inter-FPGA communication

Both parallel links and serial links can be used for the intercommunication between FPGA. The high-speed serial link is used as inter-FPGA communication medium in this paper but any other serial or parallel communication can be inserted with the identical methodology. FPGA giants Xilinx and Altera integrate high-speed serialiser/deserialiser (SerDes)[3][5]. High-speed SerDes are made of two functional blocks: the Parallel in Serial out (PISO) block and the Serial In Parallel Out (SIPO) block.

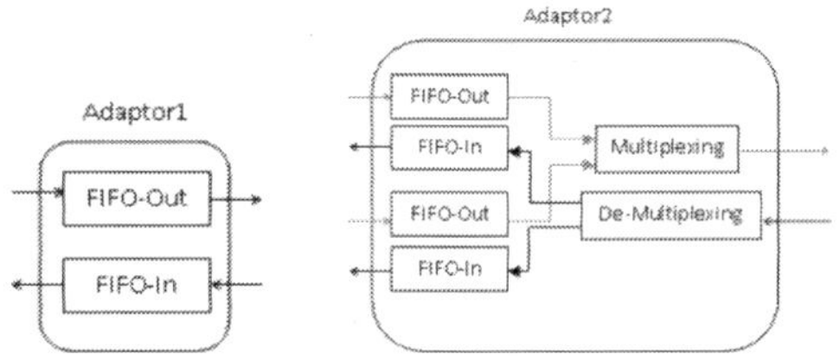

Figure 1. Parameterized IP Block for inter-FPGA communications

Parameterized adaptive blocks are designed for connecting links of the NoC inside the FPGA to IP blocks dedicated to external high-speed serial links. The partitioning depends on the number of physical links (NPL) and the number of inter-FPGA (NIF) links specified. Two scenarios may exist:

1. $N_{PL} >= N_{IF}$: one inter-FPGA link uses one physical link. Adaptations consist in inserting FIFO to adapt frequencies. FIFOs are designed to store one packet.
2. $N_{PL} < N_{IF}$: one inter-FPGA link uses several physical links. In this case, multiplexers, de-multiplexers and FIFOs blocks are inserted between the serial link and the NoC architecture.

Two VHDL parameterized adaptive IP blocks are designed (Figure 1) and inserted in library of the design flow. Adaptor1 contains parameterized FIFOs and is used for scenario 1). Adaptor2 contains parameterized FIFO, multiplexer and de-multiplexer blocks required for scenario 2). These blocks are automatically parameterized according to external and internal frequencies, size of data, number of serial-links and size of NoC.

2.3 Traffic generators

For the emulation of the NoC, IP blocks connected to the NoC are replaced by deterministic traffic generators (TG). These traffic generators simulate the traffic flow between IP blocks inside the NoC with a stochastic traffic distribution to reproduce the behavior of a real IP block. Several traffic generator models have been proposed but none of them are suitable for image and signal processing applications. For example the source address is not given, the number, position and type of data to send either. The proposed TG contains respectively the address of the initiation core, address of the destination cores, the size of transmitted packet, the number of packets to be sent to the destination core, the number of inter-FPGA link crossed.

Several emulations are proposed:

- First emulation explores timing performance according to the data injection rate. The data injection rate is automatically and dynamically generated from a 0% to a 100% load.
- Second emulation is based on a given data injection rate (specified by the designer as a constant value).

2.4 Traffic Receptors

The traffic flow generated by traffic generators is sent through the NoC and then received by traffic receptors. Traffic receptors analyze received packets and extract transmission performances of the NoC. Two types of traffic receptor exist: the statistics from traffic receptor and the trace traffic receptor. Both type of traffic receptors are parameterized VHDL blocks inserted in the design flow.

2.5 Design Flow

The design flow depicted in Figure 2 automatically generates the emulation architecture for mono-FPGA or multi-FPGA platform.

The design flow is based on the existing NoC architecture implemented on a mono-FPGA platform. The designer can insert at the input of the design flow any existing NoC as long as the HDL description is available. From the **classical NoC architecture**, the designer manually selects the number of FPGAs, the number of external serial links and the partitioning of the NoC with a text file containing several fields. The designer associates switches to the target FPGA in the ***partitioning field.*** He indicates how to connect the switches together in the ***connection field***. Then the inter-FPGA communication using aurora blocks are associated to switches in the ***specifying field***. In the example depicted in Figure 3, switches (0,0)(0,1) are implemented on FPGA_0 and other nodes on another FPGA. Switch (0,0) from the first FPGA is connected to switch(0,2) with one aurora communication block. Based on Lex and Yacc, this text file is analyzed, and then, the tools remove the internal connections (links between switches) of the NoC description, inserts parameterized adaptation blocks detailed in 2.2 and serial blocks generated by the development tool associated to the FPGA from the libraries of the design flow. Adaptation blocks integrate mux-demux blocks and FIFOs that are automatically parameterized according to the size and number of flits to transmit. The partitioning selected implies that the node is associated to its switch. Both cannot be implemented on two devices. It is the reason why the text file contains only information about the switches, not about nodes anymore.

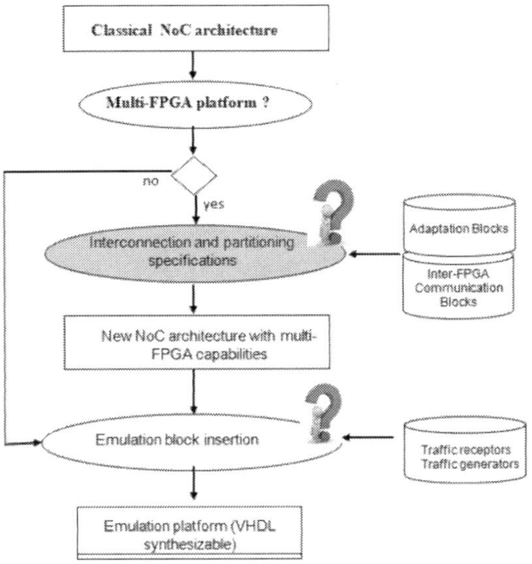

Figure 2. Design flow for the generation of the emulation platform.

The design flow generates a **new NoC architecture with multi-FPGA capabilities** according to the designer requirements. At this step only the communication architecture is generated. Without any text file, the NoC is not partitioned and is implemented onto one FPGA only. In this case, the design flow directly goes to step **Emulation block insertion** presented later.

```
** Partitionning
Switch (0,0) = FPGA_0
Switch(0,1)= FPGA_0
Switch (0,2) = FPGA_1
Switch(0,3)= FPGA_1
**Connections
Switch(0,1)=Switch(0,2)
** Specifying aurora
Aurora_0_in=switch(0,1)
Aurora_0_out=switch(0,2)
```

Figure 3. Example of partitioning specified in the design flow

Then for mono-FPGA or multi-FPGA platforms, traffic generators and traffic receptors are connected to all nodes of the NoC in the **Emulation block insertion step** according to the type of emulation detailed in section 3. This step inserts to all switches TG and TR (by instantiating components declared in the package) whatever the traffic required. The complete emulation architecture on multi-FPGA platform is then generated in HDL description language. This system is synthesized and implemented using existing commercial tools for Synthesis and Place and Route tools inserted in the design flow. The designer implements the complete emulation platform without any hardware requirements.

3 DESIGN FLOW FOR THE HERMES NOC ON XILINX FPGA PLATFORMS

The design flow previously presented is adapted to the Hermes NoC on Xilinx Virtex 5 platforms. Xilinx ISE 14.1 with all integrated tools is used.

3.1 Hermes NoC and ATLAS tool

Hermes is a NoC created by the Catholic University of Rio Grande do Sul (Porto Alegre, Brazil) [1]. This NoC is a 2D packet switched Mesh. The main components of this infrastructure are the Hermes switch and IP cores (Figure 4). The Hermes switch has routing control logic and five bi-directional ports. The local port builds the connection between the switch and its local IP core. All ports possess input buffers for provisional storage of information.

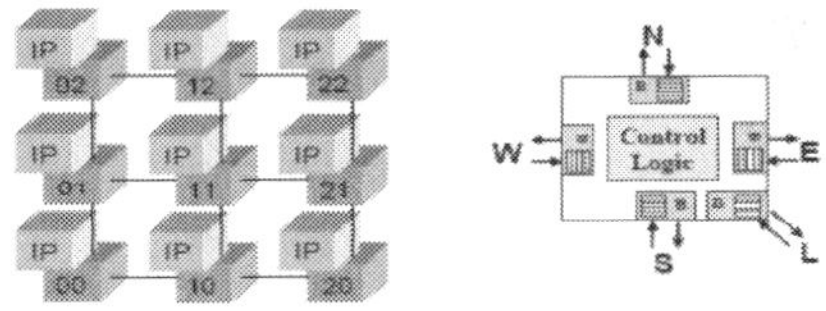

Figure 4. The Hermes NoC and switch architecture

Hermes uses the wormhole flow control. There are several routing algorithms usable in Hermes. The XY algorithm is often used in 2D networks because of its determinism and minimization.

ATLAS is an open source environment designed to automate the generation of the Hermes VHDL structure [6]. Several features can be parameterized in the ATLAS environment: size of flit, buffer depth, number of virtual channels, flow control strategies. These parameters are easily set by the designer to match the specifications of the algorithm.

3.2 Reconfigurable FPGA board for emulation

The multi-FPGA platform used for the NoC emulation is the ML506 evaluation board. Each platform contains a Virtex 5 XC5VSX50 FPGA. The ML506 offers the ability to create high speed serial designs utilizing the RocketIO™ GTP transceivers [7]. RocketIO transceiver with Aurora protocol is used in our emulation platform. It requires a limited design effort to integrate it into an existing design as Xilinx development tools automatically generates the IP blocks dedicated to the serial interface.

Table 1. Utilization of resource on FPGA for different NoC implementations

		Slice Registers	Gain of Slice Registers	Slice LUTs	Gain of Slice LUTs
Mono- FPGA		16% (5260/32640)	X	27% (8831/32640)	X
Multi-FPGA 1 Version1	On one FPGA	10% (3441/32640)	-34%	16% (5356/32640)	-39%
	Total	10% (6882/65280)	30%	16% (10712/65280)	21%
Multi-FPGA 2 Version2	On one FPGA	8% (2898/32640)	-45%	15% (5094/32640)	-42%
	Total	8% (5796/65280)	10%	15% (10098/65280)	14%

3.3 Criteria for emulation on multi-FPGA

The experimental study is mainly based on the average latency, the number of LUTs and registers.

3.4 Emulation platforms of Hermes NoC

The emulations platforms generated by the design flow are based on a 4*3 Mesh. Previous exploration showed that the maximum number of nodes on a single Virtex5 FPGA is 64 nodes. We deliberately use a small NoC to compare results with the mono-FPGA platform. For the multi-FPGA platforms, the design flow splits the NoC, inserts the adaptors blocks and the Xilinx RocketsIO IP blocks. The time for the multi-FPGA emulation platform is several minutes only (depending on the PC and tools used).

Figure 5 and Figure 6 presents the emulation platform on a multi-FPGA architecture with respectively $N_{PL} = N_{IF}=3$ and ($N_{PL\ (=1)} < N_{IF\ (=3)}$). The resources used for all emulation platforms (including the mono-FPGA platform) are depicted in Table 1.

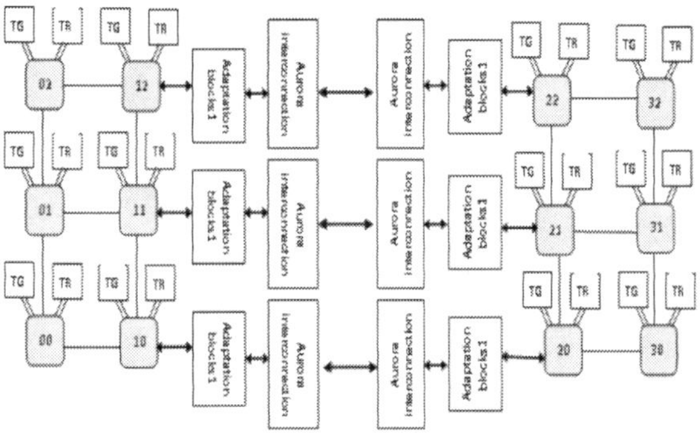

Figure 5. Emulation platform generated by the proposed design flow for the multi-FPGA implementation (scenario 1: $N_{PL} = N_{IF}$)

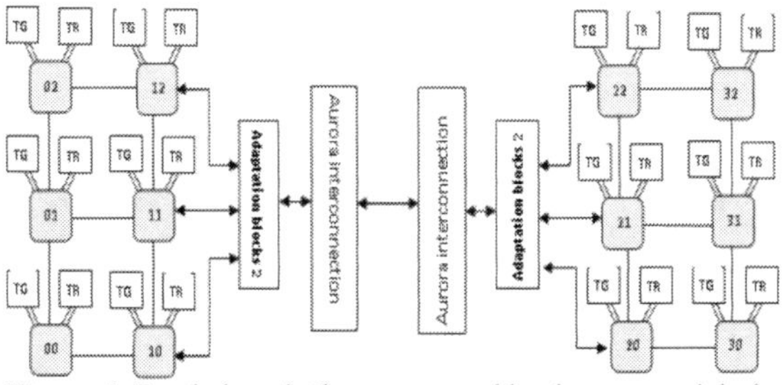

Figure 6. Emulation platform generated by the proposed design flow for the multi-FPGA implementation (case 2: $N_{PL} < N_{IF}$).

The multi-FPGA version1 emulation platform requires 30% more registers and 21% more LUTs than the mono-FPGA emulation platform. The number of registers increases with a factor of 10% and the number of LUTs 14% for version 2 compared to the mono-FPGA emulation platform. With both emulation platforms, the size (number of nodes) of the NoC can be easily extended as the number of resource required for the inter-FPGA communication is low. We assume that the number of nodes depends on the number of FPGA *64 nodes. Moreover, the number of resources for a reconfigurable logic device and for an application specific device can be predicted from multi-FPGA platforms. The most precise evaluation (with 10-15% of resources removed) is from the $N_{PL} = 1$ scenario.

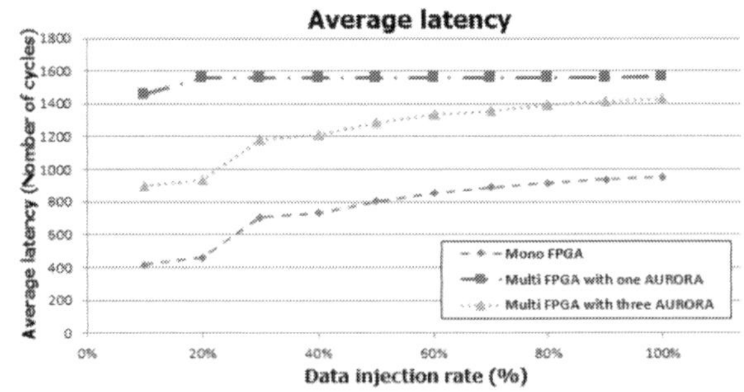

Figure 7. The average latency according to 3 implementation solutions.

The timing evaluation is then based on the sending of 50 packets. Figure 7 presents the average latency for all implementations. For the mono-FPGA scenario, the traffic receptor receives 15 flits with the average latency of 92 cycles below the saturation point.

The saturation points can be extracted. This point corresponds to the saturation of the architecture. The saturation points are identical for the mono-FPGA and the platform in version1. Multiplexing data for the inter-FPGA communication (version2) changes the position of the saturation point. As the number of inter-FPGA crossed is extracted and specified for each packet, the timing performance can be extracted from case 1. Timing for transmitting packet through the aurora protocol (t_{aurora}) takes 24 cycles for initialization and 1 cycle/flit. FIFOs at the input (t_{FIFO_in}) and output (t_{FIFO_out}) take 2 cycles for controlling and 1 cycle/flit. Therefore it is possible to obtain the timing performance on the mono FPGA by removing these three latencies. For the experiment, the added latency for the inter-FPGA communication is 478 cycles. By removing this time, the end to end latency obtained from case 1is very close to the mono-FPGA implementation by few cycles. The precise timing exploration and the saturation point extraction can be made when $N_{PL} = N_{IF}$. Timing evaluations are more difficult when $N_{PL} <> N_{IF}$.

4. CONCLUSION

A fast and automatic design flow for the implementation of NoC on multi-FPGA platform is proposed. The emulation platform generated is based on a synthesizable code of an existing NoC architecture, high speed serial links and multi-FPGA platforms. This design flow is an "open" design flow: all inputs (NoC, inter-FPGA links or multi-FPGA platform) can be changed or updated. From a partitioning given by the user, the design flow automatically splits the communication architecture, inserts adaptation blocks and emulation blocks and generates the code of the multi-FPGA architecture. Several implementations can be used for resource and timing evaluations.

The design flow generated two types of emulation platform used for timing and resource evaluation. The platform using one link for one inter-FPGA communication will be used for timing estimation (including the saturation point). The platform using one link for all inter-FPGA communication will be used for resource estimation (i.e. to predict the number of resource required for the NoC) and to extend the NoC on multi-FPGA platform in an optimal architecture choice.

ACKNOWLEDGEMENT

This paper is supported by "the Fundamental Research Funds for the Central Universities" (2014B01814).

References

[1] F. Moraes, N. Calazans, A. Mello: "HERMES: an Infrastructure for Low Area Overhead Packet-switching Networks on Chip", Integration, the VLSI Journal, vol. 38, no. 1. pp 69–93. Oct. 2004.

[2] C. A. Zeferino, A. A. Susin: "SoCIN: A Parametric and Scalable Network-on-Chip", Proc. 16th Symposium On Integrated Circuits and System Designs, pp 169-174. 2003.

[3] C. Hilton, B. Nelson: "PNoC: a flexible circuit-switched NoC for FPGA-based Systems", in Field Programmable Logic, Aug. 2005. pp 24–26.

[4] Kouadri A., Senouci B., Petrot F., "Scalable Multi-FPGA Platform for Networks-On- Chip Emulation", Application - specific Systems, Architectures and Processors, 2007. ASAP. IEEE International Conf. on , pp.54-60, 9-11 July 2007.

[5] L. Dave: "SerDes Architectures and Applications". Technical Report of National Semiconductor Corporation.

[6] http://www.inf.pucrs.br/~gaph/AtlasHtml/AtlasIndex_us.html

[7] http://www.xilinx.com/

Non-initial overlap bulk silicon comb-drive actuator with low-voltage and large-displacement

Chunhua Cai[1], Di Hua[1] and Ming Qin[2]

[1]College of internet of things engineering, Hohai University, 213022 Changzhou, China
[2]Key Laboratory of MEMS of the Ministry of Education, Southeast University, 210096 Nanjing, China

Abstract. Two different kinds of bulk-Si comb-drive actuators are designed in this work: initial overlap comb-drive actuator and non-initial overlap comb-drive actuator. They are fabricated by a simple post-CMOS bulk micromachining process. A cascade folded beam is designed to achieve large displacement at low driving voltages. And non-initial overlap and unequal wide comb fingers are used to reduce the side instability and improve the pull-in voltage. The measurement results show that the non-initial overlap comb-drive actuator improves the pull-in voltage by 73.3% than the initial overlap comb-drive actuator, and the maximum displacement of the non-initial overlap actuator is larger by 78.9% than that of the initial overlap actuator.

1 Introduction

Comb-drive actuators based on the principle of electrostatic forces have become a typical component in MEMS devices due to its low power consumption, high speed and good compatibility with microfabrication process [1]. Since the first time developed by Tang et al. in 1989 [2], comb–drive actuators have be used in a variety of applications including tweezer [3], optical switch [4], micro-tribotester [5] and Micro XY–stage [6-8], etc.Generally, comb-drive actuators usually need a high drive voltage to achieve a large displacement. In addition, the desired displacement of the comb-drive actuators is constrained by the side instability. In order to overcome this drawback, some effective approaches have been developed for a large displacement by alleviating side instability, such as optimal suspension designs [9-11], linearly engaging comb fingers [12-13], and appropriate first and last comb fingers [14]. However, there is scarcely comb-drive actuator that can achieve 50μm within 25V driving voltages [9-17].

In this work, the non-initial overlap and unequal wide comb fingers are utilized to reduce the side instability and improve the pull-in voltage. A cascade folded beam is used to achieve to achieve the displacement of more than 50μm at 24V driving voltages within a relatively small size. The whole size of the comb-drive actuator is 1.14mm^2. In addition, the comb-drive actuators are fabricated using a post-CMOS bulk micromachining process.

Furthermore, the fabrication tolerances together with their effects on the performances of the comb-drive actuator are also discussed.

2 The post-CMOS process

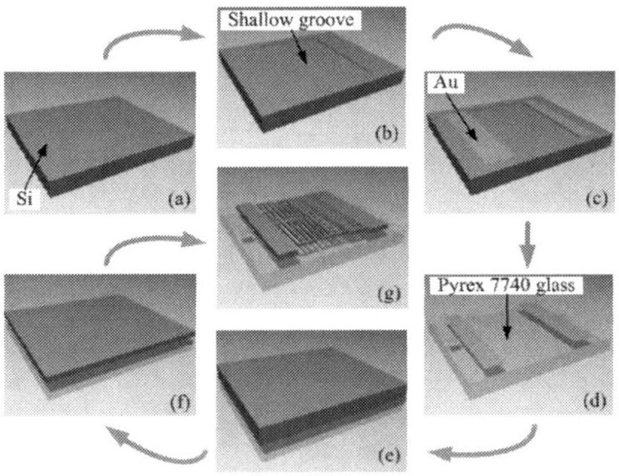

Figure 1. Fabrication process of the comb-drive actuator based on post-CMOS technology.

This research demonstrates a bulk silicon post-CMOS MEMS process based on the Au–Au bonding technology to fabricate the electrostatic type comb-drive actuators. In Figure 1 (b), the 4μm shallow grooves with precise geometry are fabricated on the silicon substrate (Figure 1 (a)) by RIE (reactive ion etch) process, leaving a small clearance where the movable comb fingers of the comb-drive actuators can extend on the glass chip. As shown in Figure 1 (c) and Figure 1 (d), 500Å/1000Å Ti/Au is sputtered onto the Pyrex7740 glass wafer and the silicon substrate. As indicated in Figure 1 (e), the silicon substrate is flip-chipped onto the Pyrex7740 glass wafer by using the Au–Au thermal bonding process. The Au-Au bonding technique is practicable, reliable, and capable of providing pads for electrical interconnection and hermetic packing of MEMS devices [18].

Then the Si wafer is thinned to about 34μm from the backside by the CMP (chemical mechanical planarization) process at the temperature of around 30 ℃ with the alkaline SiO_2 slurry (Figure 1 (f)). At last, the Bulk-Si micro-structures are released through the ICP (inductively coupled plasma) process by the ionized gas of C_4F_8 (Figure 1 (g)).

3 Design of the bulk silicon comb-drive actuators

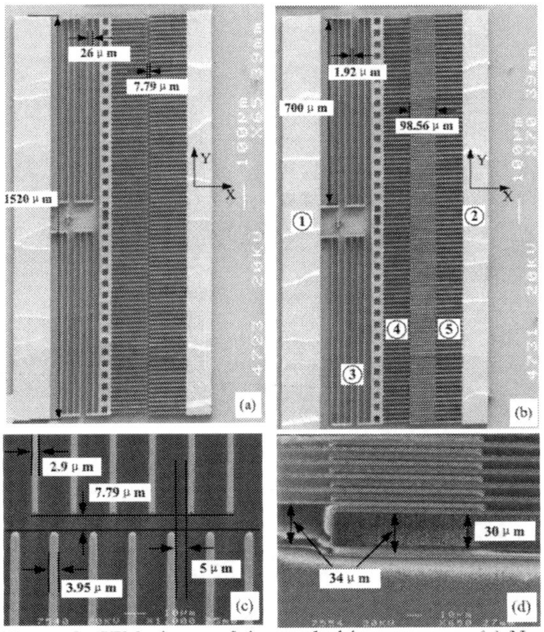

Figure 2. SEM photos of the comb-drive actuators: (a) Non-initial overlap comb-drive actuator, (b) Initial overlap comb-drive actuator, (c) Non-initial overlap comb fingers, (d) Profiles of the initial overlap comb fingers. ①~②-Input pads of driving voltages and output pads of capacitance, ③- Folded beams, ④-Movable comb fingers, ⑤-Fixed comb fingers.

With the development of the comb–drive actuators techniques, bigger stroke, higher speed and higher accuracy are required in various applications. However, some obstacles have been present accordingly, such as side instability, high applied voltage and large device footprint. To overcome these shortcomings, a novel non-initial overlap bulk silicon comb–drive actuator is designed and fabricated in this work to achieve a large displacement at a low driving voltage within a relatively small size, as shown in Figure 2 (a). As a comparison, an initial overlap comb-drive actuator and a non-initial overlap comb-drive actuator are described as shown in Figure 2 (b). The sizes of the two actuators are the same except overlap segments.

As shown in Figure 2(b), the electrostatic force F_x generated between the movable comb fingers and the fixed comb fingers along the moving direction (x–direction) is given as:

$$F_x = \frac{N\varepsilon_0 t U^2}{g} \tag{1}$$

where N is the number of movable comb fingers, ε_0 is the permittivity constant of air, t is the thickness of the comb-drive fingers, g is the fingers gap spacing between the movable comb finger and the fixed comb finger and U is the applied voltage.

As shown in Figure 2(a), the movable comb fingers and the fixed comb fingers of the comb-drive actuator are not overlap. With a driving voltage U and the comb fingers non overlap, the electrostatic force $F_{x\text{-}non\text{-}overlap}$ generated between the movable comb drives and the fixed comb drives along the moving direction (x–direction) is given as

$$F_{x-non-overlap} = \frac{1}{2}\frac{N\varepsilon_0 (w_m + w_f) t U^2}{(L + L_1 - x)^2} \tag{2}$$

Where w_m is the width of movable fingers, w_f is the width of fixed comb fingers, L is the length of the comb finger, L_1 is the length of the non-initial overlap segment, and x is the displacement of the movable comb fingers in the moving direction.

Based on equation (1) and equation (2), the electrostatic force ratio of non-overlap comb fingers and overlap comb fingers can be expressed as:

$$\frac{F_{x-non-overlap}}{F_x} = \frac{1}{2}\frac{(w_m + w_f) g}{(L + L_1 - x)^2} \tag{3}$$

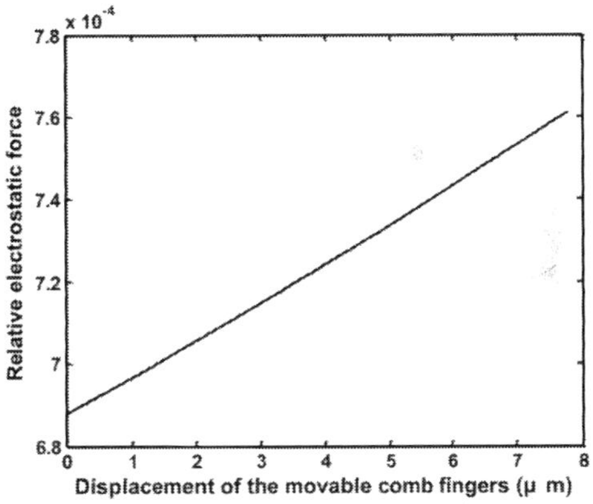

Figure 3. The relative electrostatic force versus the displacement of the movable comb fingers.

Figure 3 shows the electrostatic force ratio as a function of the displacement of the movable comb fingers. Under the conditions of thickness t = 30μm, gap between comb fingers g = 5μm, width of the movable comb fingers w_m = 2.9μm, width of the fixed comb fingers w_f= 3.95μm, length of comb fingers L = 150μm, length of the non-initial overlap segment L_1 = 7.79μm. When the displacement of the movable comb fingers increases from 0 to 7.79μm, the relative electrostatic force of the comb-drive actuator increases from $6.88*10^{-4}$ to $7.61*10^{-4}$. The curve derives that the displacement change rate of the non-initial overlap actuator with the comb fingers non-overlap are lower than that of the non-initial overlap actuator with the comb fingers overlap.

4 Tolerance of DRIE effects on the performance of the actuators

Due to the etching tolerance by deep reactive ion etching (DRIE), the measured results are different from the initial design. Generally, wider trenches tend to become more reentrant. Consequently, the fabrication tolerance generated by DRIE affects the electrical and mechanical performance of the comb-drive actuators [19].

4.1 Tolerance of DRIE effects on the capacitance of the initial overlap comb-drive actuator

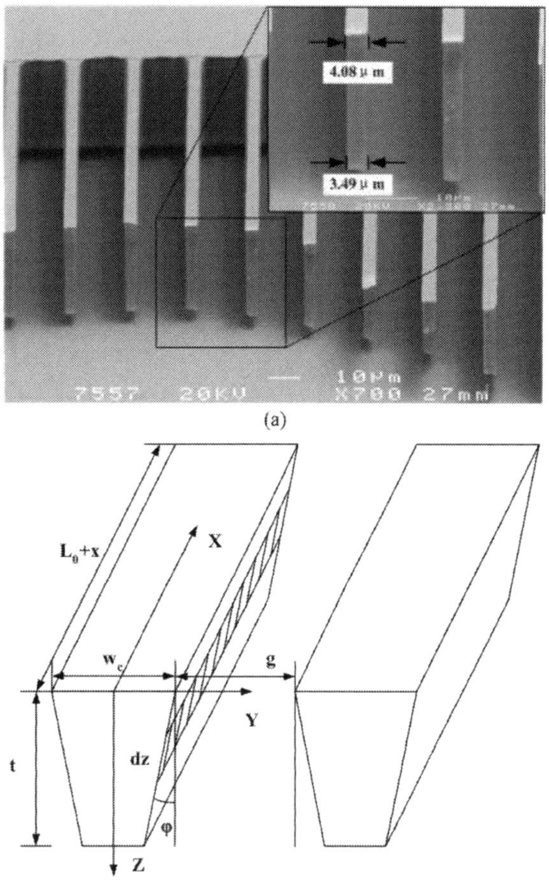

(a)

(b)

Figure 4. The fixed comb fingers with negative etching profile: (a) SEM images of the broken fixed comb fingers; (b) Scheme of the comb fingers with negative etching profile.

The profile tolerance of comb fingers and flexure beams caused by the DRIE process were studied [19]. Figure 4 (a) shows the SEM images of the comb fingers with the negative profile which are measured from the intentionally broken pieces of the comb-drive actuators. The thickness of the comb fingers and the cascade folded beams is 30μm. The capacitance between the movable comb fingers and the fixed comb fingers, which excludes the fringing capacitance, could be expressed as:

$$C_{tot\text{-}profiles} = -N \cdot \frac{\varepsilon_0 (L_0 + x)}{\tan\varphi} \cdot \ln(1 - \frac{t}{g} \cdot 2\tan\varphi) \qquad (4)$$

4.2 Tolerance of DRIE effects on Stiffness of the Folded Beams

Due to the etching tolerance by deep reactive ion etching (RIE), the cascade folded beams also have negative profiles. And wider trenches tend to become more reentrant [19]. As a result, the slope angle θ of the cascade folded beams is around -1.2°. The negative profiles effect on the stiffness of the folded beams. Consequently, they effect on the performances of the comb-drive actuator.

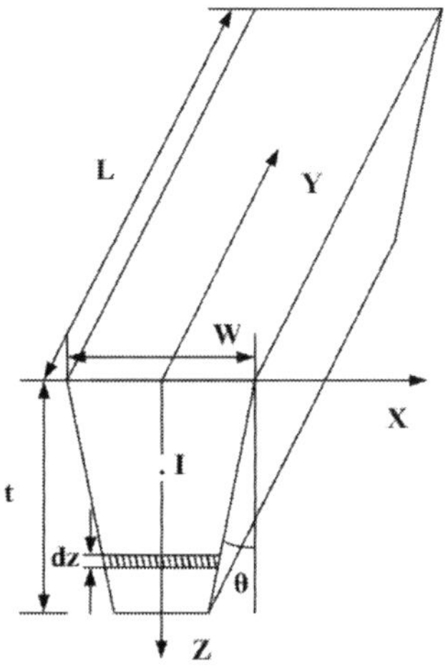

Figure 5. Scheme of the negative etching profiles of the cascade folded beam with fabrication tolerance caused by the DRIE process.

Figure 5 shows a scheme of the folded beam with the negative slope profile. Assuming that the width of the flexure is W and slope angle is θ. The width of the folded beam becomes the function of the depth as expressed as:

$$w(z) = W + 2z \cdot \tan\theta \qquad (5)$$

The inertial momentum $I_{x\text{-}slope}$ and the stiffness in the x-direction of the folded beams K_x is calculated as:

$$I_{x-slope} = \int_0^t \frac{(W + 2z \cdot \tan\theta)^3}{12} dz = \frac{(2t \cdot \tan\theta + W)^4 - W^4}{96 \cdot \tan\theta} \qquad (6)$$

$$K_{x-slope} = \frac{12EI}{L^3} \cdot \frac{4}{3} = E \cdot \frac{(2t \cdot \tan\theta + W)^4 - W^4}{6 \cdot \tan\theta \cdot L^3} \qquad (7)$$

4.3 Tolerance of DRIE effects on deflection of the comb-drive actuator

With a driving voltage U, the electrostatic force F_x generated between the movable comb fingers and the fixed comb fingers along the moving direction (x-direction) is given as

$$F_{x-profiles} = \frac{\partial}{\partial x}(\frac{1}{2} C_{tot-profiles} U^2) \qquad (8)$$

For the overlap comb fingers structure with consideration of edge field, the electrostatic force F_x

along the moving direction(x–direction) and the displacement can be expressed as:

$$F_{x-profiles} = -\frac{N}{2} \cdot \frac{\varepsilon_0}{\tan\varphi} \cdot \ln(1 - \frac{t}{g} \cdot 2\tan\varphi) \cdot U^2 \quad (9)$$

Based on equation (7) and (9), the displacement of the comb-drive actuator with negative etching profiles can be calculated as:

$$x_{profiles} = \frac{-\frac{N}{2} \cdot \frac{\varepsilon_0}{\tan\varphi} \cdot \ln(1 - \frac{t}{g} \cdot 2\tan\varphi) \cdot U^2}{E \cdot \frac{(2t \cdot \tan\theta + W)^4 - W^4}{6 \cdot \tan\theta \cdot L^3}} \quad (10)$$

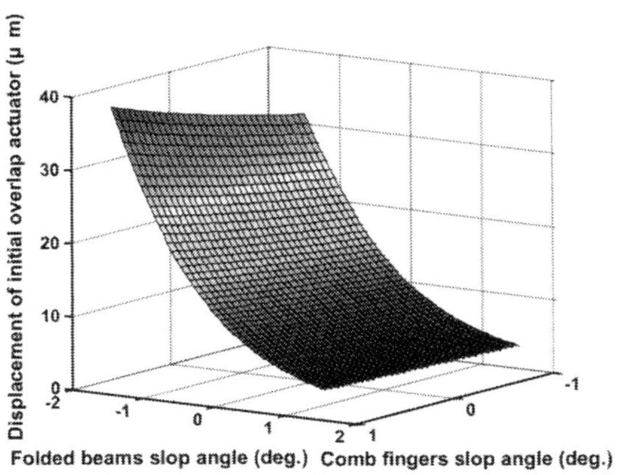

Figure 6. Displacement of comb-drive actuator versus slope angle of folded beams and comb fingers.

Figure 6 shows that the displacement of the actuator increases rapidly when the slope angle of the folded beams θ decreases. On the contrary, the displacement of the actuator increases slowly when the slope angle of the comb fingers φ increases. Excluding the fringing capacitance of the actuator, the displacement of the comb-drive actuator is 24.612μm with the slope angle of comb fingers φ = -0.5634°, the slope angle of the fold beam θ = -1.2° and the driving voltage U = 15V.

5 Measurements

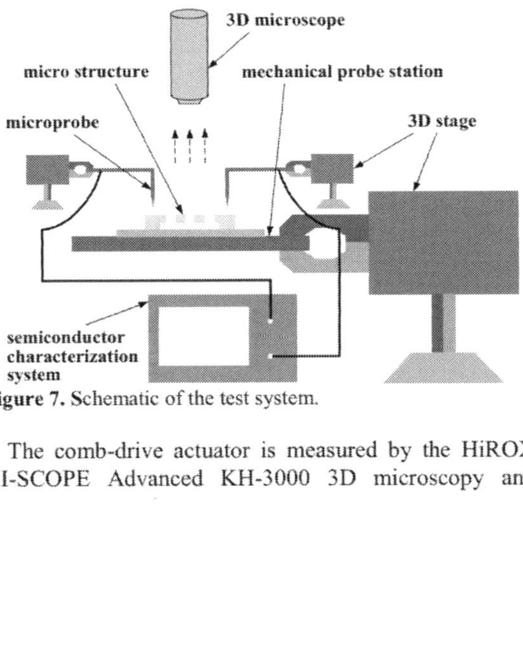

Figure 7. Schematic of the test system.

The comb-drive actuator is measured by the HiROX HI-SCOPE Advanced KH-3000 3D microscopy and KEITHLEY 4200-SCS semiconductor characterization system with the applied voltage in the range from 0V to 15V for the initial overlap comb-drive actuator and 0V to 24V for the non-initial overlap comb-drive actuator.

As shown in Figure 8, the displacement change rate of the initial overlap comb-drive actuator is 0.125μm/V^2. The initial overlap actuator is pulled in at 15V driving voltages. The maximum displacement of the initial overlap actuator is 28.5μm. The measured displacement change rate of the initial overlap actuator is larger by 186.7% than the initial design value. That is the reason that the folded beams have negative etching profiles. The negative etching profiles decrease the stiffness of the folded beams by 62.35%. Consequently, that will increase the deflection of the comb-drive actuator by 165.5%. Consistency between the experimental results of the initial overlap comb-drive actuator and the modified designed curve that considered the fabrication tolerance is observed. For the fringing capacitance between the movable comb fingers and the fixed comb fingers of the initial overlap comb-drive actuator is excluded, the electrostatic force of the initial overlap comb-drive actuator decreases. As a result, the measured displacement of the initial overlap comb-drive actuator is larger than the modified displacement $x_{profiles}$ of the actuator that considered the fabrication tolerance.

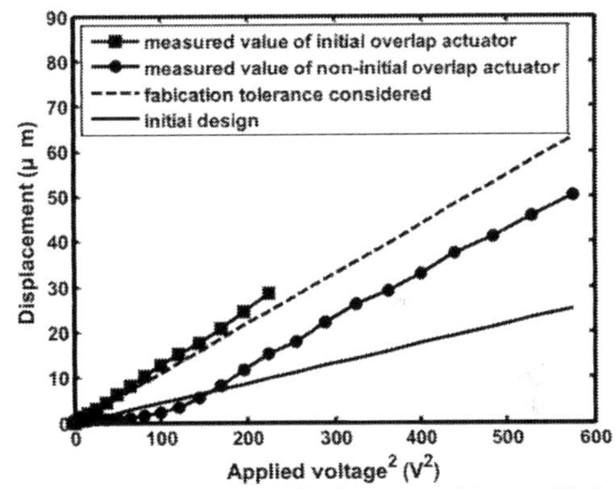

Figure 8. Measured results of displacement of the two kinds of comb-drive actuators.

The displacement change rate of the non-initial overlap comb-drive actuator is 0.03μm/V^2 and the comb fingers of the non-initial overlap actuator are non-overlap with the driving voltage in the range from 0V to 14V. And the displacement change rate of the non-initial overlap comb-drive actuator is 0.114μm/V^2 and the comb fingers of the non-initial overlap comb-drive actuator are overlap with the driving voltage in the range from 15V to 24V. The pull-in voltage of the non-initial overlap actuator is 24V. And the maximum displacement of the non-initial overlap actuator is 50.127μm.

6 Conclusions

Two kinds of bulk silicon comb-drive actuators with low voltage and large displacement has been designed and fabricated by a novel and simple bulk micromachining post-CMOS process. A cascade folded beam is designed to achieve large displacement at low driving voltages. And non-initial overlap and unequal wide comb fingers are used to reduce the side instability and improve the pull-in voltage. The measurement results show that the non-initial overlap comb-drive actuator improves the pull-in voltage by 73.3% than the initial overlap comb-drive actuator, and the maximum displacement of the non-initial overlap actuator is larger by 78.9% than that of the initial overlap actuator. With the optimized cascade folded beams and the unequal wide and non-initial overlap comb fingers, the displacement of the non-initial overlap actuator can reach 50.127μm at 24V driving voltages. The fabrication tolerances together with their effects on the performances of the comb-drive actuator are also discussed. The experimental results of the comb-drive actuator are in good agreement with the theoretical predictions.

Acknowledgment

This work was supported by the Fundamental Research Funds for the Central Universities (2014B02014) and the National High Technology Research and Development Program of China (2013AA041106).

References

1. J. Li, Q. X. Zhang, and A. Q. Liu, "Advanced fiber optical switches using deep RIE (DRIE) fabrication," Sensors and Actuators A(physical), vol. 102, no. 3, pp.286-295, Jan. 2003.
2. W. C. Tang, T. C. H. Nguyen, and R. T. Howe, "Laterally driven polysilicon resonant microstructures," Sensors and Actuators A(physical), vol. 20, no. 1-2, pp.25-32, Nov. 1989.
3. X. Liu, J. Tong and Y. Sun, "A millimeter-sized nanomanipulator with sub-nanometer positioning resolution and large force output," Smart Mater. Struct, vol. 16, no. 5, pp.1742-1750, Oct. 2007.
4. K. Wang, M. Sinclair, G. K. Starkweather, and K. F. Böhringer, "An Electrostatic Zigzag Transmissive Microoptical Switch for MEMS Displays," Journal of Microelectromechanical Systems, vol. 16, no. 1, pp.140-154, Feb. 2007.
5. Z. Guo, Y. Meng, H. Wu, C. Su, and S. Wenb, "Measurement of static and dynamic friction coefficients of sidewalls of bulk-microfabricated MEMS devices with an on-chip micro-tribotester," Sensors and Actuators A(physical), vol. 135, no. 2, pp.863-869, Apr. 2007.
6. T. Harness, R. R. A. Syms, "Characteristic modes of electrostatic comb-drive X-Y microactuators," Journal of Micromechanics and Microengineering, vol. 10, no. 1, pp.7-14, Mar. 2000.
7. C. H. Kim, H. M. Jeong, J. U. Jeon, and Y. K. Kim, "Silicon Micro XY-Stage With a Large Area Shuttle and No-Etching Holes for SPM-Based Data Storage," Journal of Microelectromechanical Systems, vol. 12, no. 4,pp.470-478, Aug. 2003.
8. L. Gu, X. Li, H. Bao, B. Liu, Y. Wang, M. Liu, Z. Yang and B. Cheng, "Single–wafer-processed nano-positioning XY-stages with trench-sidewall micromachining technology," Journal of Micromechanics and Microengineering, vol. 16, no. 7, pp.1349-1357, Jul. 2006.
9. G. Zhou, and P. Dowd, "Tilted folded-beam suspension for extending the stable travel range of comb-drive actuators," Journal of Micromechanics and Microengineering, vol. 13, no. 2, pp.178-183, Mar. 2003.
10. S. Krylov, and Y. Bernstein, "Large displacement parallel plate electrostatic actuator with saturation type characteristic," Sensors and Actuators A(physical), vol. 130-131, no. 14, pp.497-512, Aug. 2006.
11. M. A. Erismis, H. P. Neves, R. Puers, and C. V. Hoof, "A low-voltage large-displacement large-force inchworm actuator," Journal of Microelectromechanical Systems, vol. 17, no. 6, pp.1294-1301, Dec. 2008.
12. A. Ming, T. Li, P. Zhou, and Y. Wang, "An Electrostatic MEMS Actuator with Large Displacement Under Low Driving Voltage," Journal of semiconductors, vol. 29, no. 9, pp.1703-1707, Sep. 2008
13. J. D. Grade, H. Jerman, and T. W. Kenny, "Design of Large Deflection Electrostatic Actuators," Journal of Microelectromechanical Systems, vol. 12, no. 3, pp.335-343, Jun. 2003.
14. Y. C. Chen, I. C. M. Chang, R. Chen, and M. T. K. Houc, "On the side instability of comb-fingers in MEMS electrostatic devices," Sensors and Actuators A(physical), vol. 148, no. 1, pp.201-210, Nov. 2008.
15. A. E. Franke, J. M. Heck, T.-J. King, and R. T. Howe, "Polycrystalline silicon-germanium films for integrated microsystems," Journal of Microelectromechanical Systems, vol. 12, no. 2, pp.160-171, Apr. 2003.
16. J. C. Chiou and C. F. Kuo, "Development of Vertical Electrostatic Comb-Drive Actuator Using Magnified Cascade Configuration," Japanese Journal of Applied Physics, vol. 46, no. 10A, pp.6546-6549, Oct. 2007.
17. Y. Gerson, S. Krylov, B. Ilic, and D. Schreiber, "Large displacement low voltage multistable micro actuator," Proceedings of the IEEE 21st International Conference on Micro Electro Mechanical Systems, pp.463-466, Jan. 2008.
18. R. F. Wolffenbuttel, and K. D. Wise, "Low-temperature silicon wafer-to-wafer bonding using gold at eutectic temperature," Sensors and Actuators A(physical), vol. 43, no. 1-3, pp.223-229, May. 1994.
19. E. Jing, B. Xiong, and Y. L. Wang, "Low temperature Au-Si wafer bonding," Journal of Micromechanics and Microengineering, vol. 20, no. 9, 095014(6pp), Sep. 2010.

MATEC Web of Conferences **54**, 12004 (2016) DOI: 10.1051/matecconf/20165412004
MIMT 2016

A sub 1V CMOS bandgap reference with two diodes

Hyung-Jin Choi[1], Jeong Cho[2] and Soo-Won Kim[2]

[1]*Department of Nano-Semiconductor Engineering, Korea University, Seoul, Republic of Korea*
[2]*Department of Electrical Engineering, Korea University, Seoul, Republic of Korea*

Abstract. A small size, sub 1V CMOS bandgap voltage reference with the temperature coefficient (TC) of 12ppm/°C from 0°C to 100°C is presented in this paper. The proposed circuit offers several performance advantages over previous bandgap voltage reference circuits. Compared to a conventional structure, it achieves a smaller area by using only two diodes. Also, it can fulfil sub 1V reference voltage by using temperature coefficient of threshold voltage to generate a current proportional to absolute temperature. This is because TC of this current on proposed circuit is bigger than that of conventional structure. Detailed analysis of the proposed bandgap reference is provided, which is simulated using 0.35μm CMOS process. The proposed bandgap reference occupies an active area of 0.0067mm^2. The simulation results are the voltage reference output voltage of 0.843V and an accuracy of ±0.08%.

1 Introduction

Voltage reference is at most essential part on mixed-signal applications, such as analog-to-digital (AD) and digital-to-analog (D/A) converters, regulators. Generally, precision of voltage reference over supply voltage and operating temperature is important factor, because it limits the resolution of these applications. So bandgap reference which makes voltage reference less sensitive to temperature is important. Recently, the demand for small size and low power is increasing. As a result, the demand for bandgap references which have small size and sub 1V voltage reference is also increasing. For size of bandgap reference, the number of diodes is the critical factor, because diodes are implemented by bipolar transistor transistors whose size is very huge. Thus, the number of diodes must be decreased to reduce active area. The conventional bandgap references generally use more than 9 diodes, leading to large die area. And they have voltage reference range from 1.1V to 1.3V [1, 2]. This is because of low positive temperature coefficient of current. Sub 1-V bandgap voltage references are described in [3-8]. However, since these bandgap references have complex structure or have high current, they occupy large area or consume high power.

A sub-1V bandgap reference (BGR) with a small area is presented in this paper. The following chapters show conventional BGR structure and proposed BGR structure including mathematical analysis and the simulation results.

2 Conventional bandgap reference

Generally, BGR uses negative temperature coefficient (TC) and positive TC. These two TCs counterbalance each other. So, it can make voltage reference independent to temperature. Figure 1 shows the schematic of conventional BGR. It uses base-emitter voltage of bipolar transistors or, forward bias voltage of a diode to make a negative (TC). That negative TC value is approximatively 1.5mV/ °C, is explained in [9, 10]. Conventional BGR also uses difference between forward voltages of diodes to make positive TC, when current flowing diodes is same. Because their difference is proportional to the absolute temperature (PTAT). Figure 2 shows concept of difference between forward voltages of diodes, and equation of PTAT voltage is given by (1).

$$\begin{aligned}\Delta V_D &= V_{D1} - V_{D2}.\\ &= V_T \ln\frac{I}{I_O} - V_T \ln\frac{I}{nI_O}.\\ &= V_T \ln n. \end{aligned} \tag{1}$$

(I_O is saturation current of diode)
(n is number of diode D2 in Fig. 2)

V_T is proportional to temperature, and is presented in (2).

$$V_T = kT/q. \tag{2}$$

(k is Boltzmann constant)
(T is absolute temperature)
(q is quantity of electric charge)

So, difference between forward voltages of diodes has positive TC. Conventional BGR uses both negative TC and positive TC to make voltage reference independent to temperature. Amplifier in Fig. 1 makes node voltage of A equal to V_{D2}. Also, amplifier makes current flowing D2 equal to current flowing D1. This is because output of amplifier is input of p type mosfets, which have same width and length. As a result, voltage drop across R1 is difference between forward voltages of diodes and has positive TC. Current flowing R1 is given by

$$\begin{aligned} I &= \frac{V_{D1} - V_{D2}}{R1}. \\ &= \frac{1}{R1}(V_T \ln\frac{nI}{I_O} - V_T \ln\frac{I}{I_O}). \\ &= \frac{V_T}{R1}\ln n. \end{aligned} \tag{3}$$

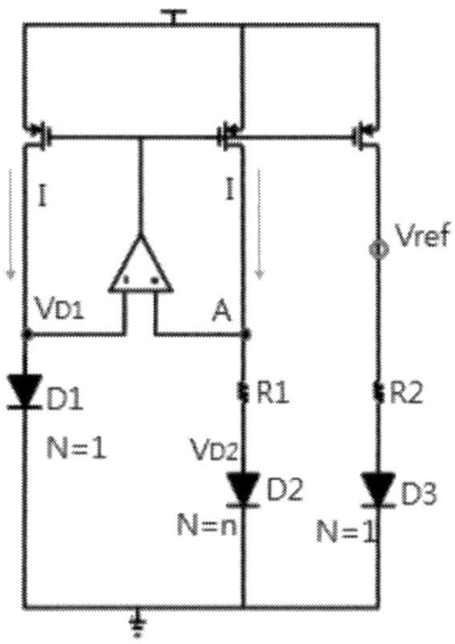

Figure 1. Schematic of the conventional bandgap reference.

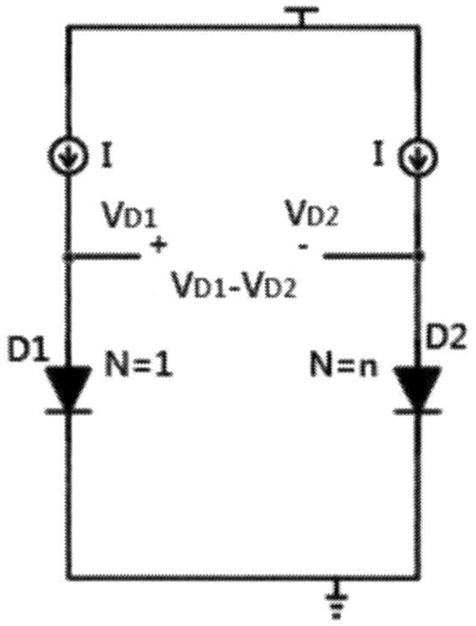

Figure 2. Conceptual schematic of PTAT voltage.

Eq. (3) shows that the current flowing R1 (I) is growing up by increasing temperature, because difference between forward voltages of diodes is PTAT. P type mosfets copy the current flowing R2 from that of R1. So voltage drop across R2 is also PTAT. As a result, V_{ref} is given by,

$$\begin{aligned} V_{ref} &= V_{D3} + R_2 I. \\ &= V_{D3} + \frac{R_2}{R_1} V_T \ln n. \end{aligned} \tag{4}$$

V_{D3} is turn on voltage of D3, which has negative TC, while voltage drop across R2 is PTAT. So both negative TC and positive TC counterbalance each other. TC of V_{ref} is represent by

$$\begin{aligned} \frac{\partial V_{ref}}{\partial T} &= \frac{\partial V_{D3}}{\partial T} + R_2 \frac{\partial V_T}{\partial T} \ln n. \\ &= \frac{\partial V_{D3}}{\partial T} + \frac{R_2}{R_1}\frac{k}{q} \ln n = 0. \end{aligned} \tag{5}$$

(n is number of D2/number of D1)

Therefore, V_{ref} can be independent on temperature.

3 Proposed bandgap reference

Figure 3 shows the schematic of proposed BGR. It only uses two diode (D1, D2). Proposed circuit uses forward bias diode, which has negative TC when the current flowing diode D1 is fixed. And, it also uses current flowing R1, which has positive TC. It is possible to analyse positive TC of current flowing R1 by using property that currents flowing M1 and R1 in Fig. 3 are same, Since R1 and M1 are connected in series. The amount of current flowing them is same as

$$\frac{1}{2}u_p C_{ox}\frac{W}{L}(V_{DD} - A - V_{th})^2 = \frac{V_A - V_B}{R_1}. \tag{6}$$

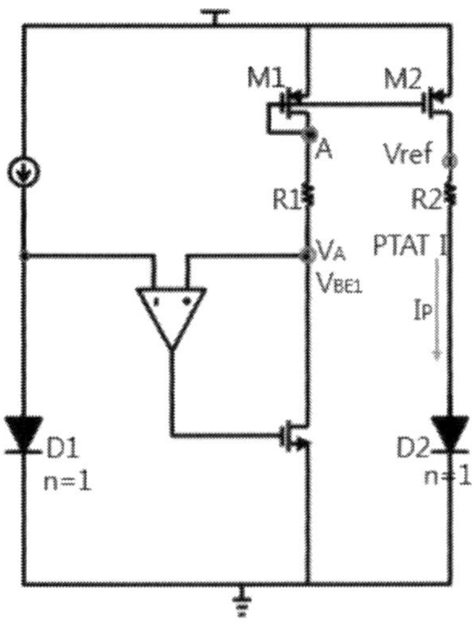

Figure 3. Schematic of the proposed bandgap reference.

If the current flowing R1 in Fig. 3 is analysed about temperature, temperature variation of current on R1 is given by (7).

$$u_p C_{ox}\frac{W}{L}(V_{DD}-A-V_{th})(-\frac{\partial A}{\partial T}-\frac{\partial V_{th}}{\partial T})=\frac{1}{R_1}(\frac{\partial A}{\partial T}-\frac{\partial V_A}{\partial T}).$$

$$\frac{\partial A}{\partial T}=\frac{\frac{1}{R_1}}{\frac{1}{R_1}+g_{mp}}\frac{\partial V_A}{\partial T}-\frac{g_{mp}}{\frac{1}{R_1}+g_{mp}}\frac{\partial V_{th}}{\partial T}.$$

$$(u_p C_{ox}\frac{W}{L}(V_{DD}-A-V_{th})=g_{mp})$$

$$\frac{\partial I_P}{\partial T}=\frac{1}{R_1}(\frac{\partial A}{\partial T}-\frac{\partial V_A}{\partial T})$$

$$=\frac{1}{R_1}(\frac{\frac{1}{R}}{\frac{1}{R}+g_m}\frac{\partial V_A}{\partial T}-\frac{g_m}{\frac{1}{R}+g_m}\frac{\partial V_{th}}{\partial T}-\frac{\partial V_A}{\partial T}). \quad (7).$$

$$=-\frac{1}{R_1}(\frac{g_m}{\frac{1}{R_1}+g_m})(\frac{\partial V_A}{\partial T}+\frac{\partial V_{th}}{\partial T}).$$

Eq. (7) shows that the current flowing R1 (I_p) is growing up by increasing temperature, because V_A and V_{th} have negative TC. As a result, I_P is PTAT. P type mosfets copy the current flowing R2 from that of R1. Reference voltage V_{ref} is represented by

$$V_{ref}=V_{D2}+R_2 I_P. \quad (8)$$

which V_{D2} is turn-on voltage of D2. When the temperature increases, V_{D2} decreases due to the negative TC of V_{D2}, while the voltage drop across R2 increases due to the increase of I_P. TC of V_{ref} is given by

$$\frac{\partial V_{ref}}{\partial T}=\frac{\partial V_{D2}}{\partial T}+R_2\frac{\partial I_P}{\partial T}.$$

$$=\frac{\partial V_{D2}}{\partial T}-\frac{R_2}{R_1}(\frac{g_m}{\frac{1}{R_1}+g_m})(\frac{\partial V_A}{\partial T}+\frac{\partial V_{th}}{\partial T}). \quad (9)$$

$$=\frac{\partial V_{D2}}{\partial T}-\frac{R_2}{R_1}(\frac{g_m}{\frac{1}{R_1}+g_m})(\frac{\partial V_{D1}}{\partial T}+\frac{\partial V_{th}}{\partial T})=0.$$

As the temperature goes up, increasing of voltage drop across R2 counterbalances decreasing at V_{D2}. Therefore, V_{ref} can be independent on temperature.

The PTAT current TC of proposed BGR is bigger than that of conventional structure [1], when amount of PTAT currents is same. Thus, when PTAT current counterbalances negative TC of V_{D2}, amount of PTAT current in proposed BGR is smaller than that of conventional structure. As a result, voltage drop across R2 is also smaller than that of conventional art, and V_{ref} is low voltage, which is under the 1V.

4 Use BMR for current source

A beta-multiplier reference (BMR) insensitive to supply voltage variation, is used as the current source in bandgap reference as shown in Figure 4. The BMR current is given by (10).

$$I_C=\frac{2}{u_n C_{ox}(W/L)_N}\cdot\frac{1}{R_S^2}(1-\frac{1}{\sqrt{K}})^2.$$

$$=\frac{2}{u_n C_{ox}(W/L)_N}\cdot\frac{1}{R_S^2}(1-\frac{1}{\sqrt{K}})^2=\frac{A}{u_n}. \quad (10)$$

$$(A=\frac{2}{C_{ox}(W/L)_N}\cdot\frac{1}{R_S^2}(1-\frac{1}{\sqrt{K}})^2)$$

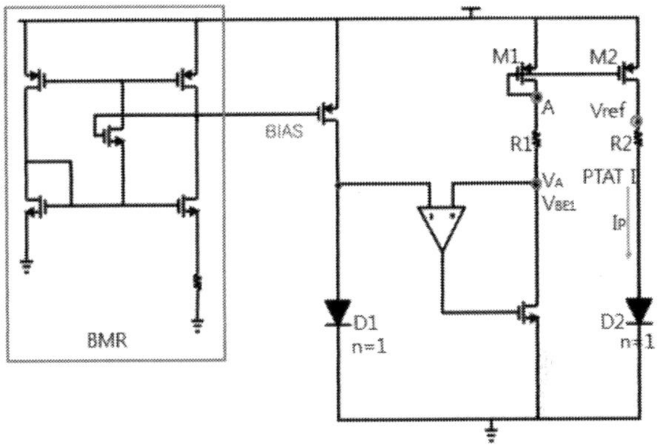

Figure 4. Proposed BGR with BMR.

However The BMR current is affected according to temperature variation. Since this effect can vary TC of turn-on voltage of diode (V_D) , TC of BMR current must be considered when TC of V_D is analysed. TC of BMR current is given by

$$\frac{\partial I}{\partial T}=-A\frac{1}{u_n^2}\frac{\partial u_n}{\partial T}. \quad (11)$$

TC of V_D considering TC of BMR current is given by

$$\frac{\partial V_{BE}}{\partial T}=\frac{\partial V_T}{\partial T}\ln\frac{I_C}{I_O}+V_T(\frac{1}{I_C}\frac{\partial I_C}{\partial T}-\frac{1}{I_O}\frac{\partial I_O}{\partial T}).$$

$$=\frac{\partial V_T}{\partial T}\ln\frac{I_C}{I_O}-\frac{V_T}{u_n}\frac{\partial u_n}{\partial T}-\frac{V_T}{I_O}\frac{\partial I_O}{\partial T}.$$

$$=(\frac{\partial V_T}{\partial T}\ln\frac{I_C}{I_O}-\frac{V_T}{I_O}\frac{\partial I_O}{\partial T})-\frac{V_T}{u_n}\frac{\partial u_n}{\partial T}. \quad (12)$$

$$=\frac{V_{BE}-(4+m)V_T-E_g/q}{T}-\frac{V_T}{u_n}\frac{\partial u_n}{\partial T}.$$

$$\approx\frac{V_{BE}-(4+m)V_T-E_g/q}{T}.$$

(I_O is saturation current of diode)

Eq. (12) shows that temperature variation of BMR current has no impact on temperature coefficient of turn on voltage of diode at all.

5 Simulation Results

The proposed BGR and conventional BGR are designed and simulated with 0.35μm CMOS process. Layout view of Conventional BGR is presented in Figure 5. It has large active area, which is $0.0132mm^2$. Figure 6 shows a layout view of the proposed BGR. It occupies an active area of $0.0067mm^2$, which is smaller than of [1], [2] and [3]. Conventional BGR approximatively has twice area compared to area of proposed BGR.

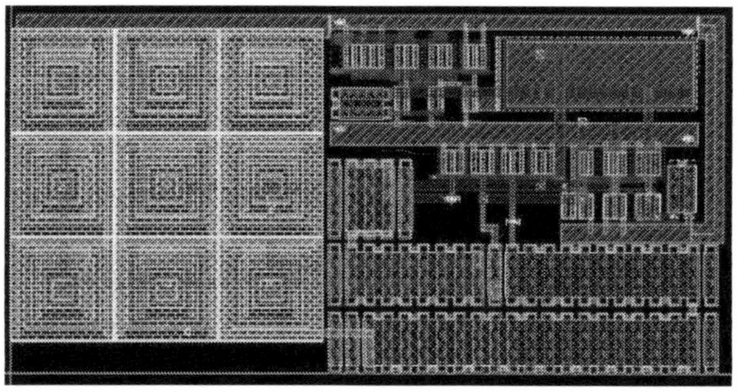

Figure 5. Layout view of the conventional BGR..

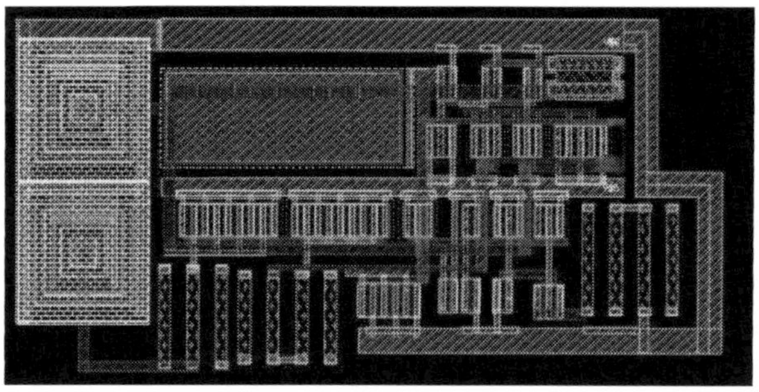

Figure 6. Layout view of the proposed BGR.

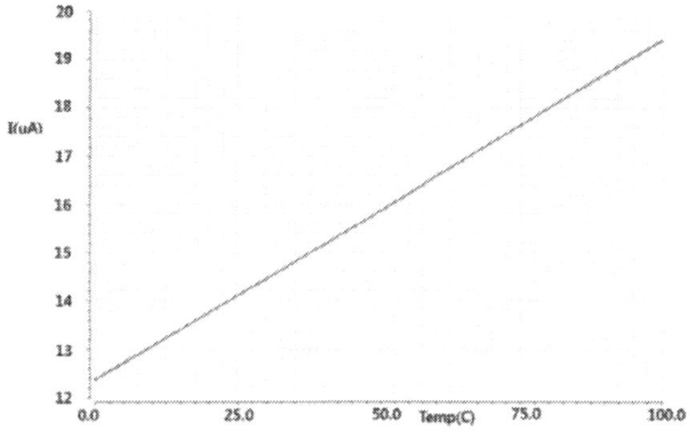

Figure 7. Simulation result of *I*

TC of PTAT current *I* in conventional bandgap at Fig 1 is 0.7nA/°C, which is described in Figure 7. It is measured when *I* at 0°C is overall 12.4uA. Whereas, TC of PTAT current I_{PTAT} in proposed BGR is 1.65nA/°C, as shown in Figure 8. This TC value is also measured when I_{PTAT} at 0°C is overall 12.4uA. Proposed BGR has bigger TC value of PTAT current than that of conventional BGR. TC of PTAP current in proposed BGR is 2.35 times larger than that of conventional BGR, which is the same result expected in the previous chapter. A slope of I_{PTAT} can be changed by adjusting W/L ratio of M1 and M2. Conventional BGR output voltage of bandgap reference is represented in Figure 9. Output voltage V_{REF} is 1.168V, which is larger than 1V. Peak to peak error for voltage is 1.2mV, and TC is 12ppm/°C in the temperature range from 0°C to 100°C. Figure 10 shows output voltage of proposed BGR. Output voltage of proposed BGR V_{REF} is 0.843V. The maximum voltage variation and the TC are 1.2mV and 12ppm/°C in the temperature range from 0°C to 100°C, respectively. Proposed BGR has almost same precision for temperature variation compared to conventional BGR, but it has small reference voltage that is sub 1V

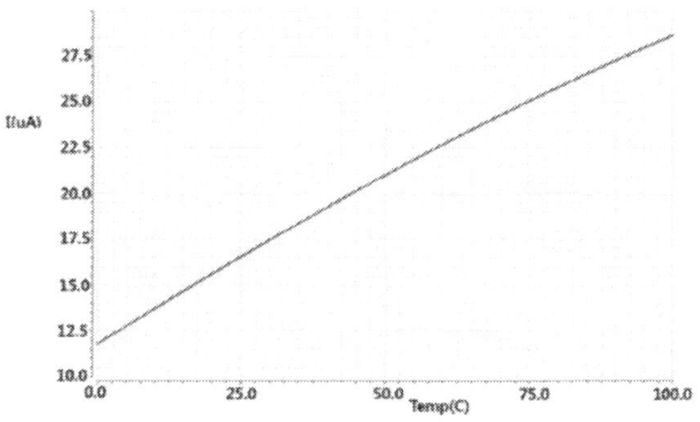

Figure 8. Simulation result of I_{PTAT}.

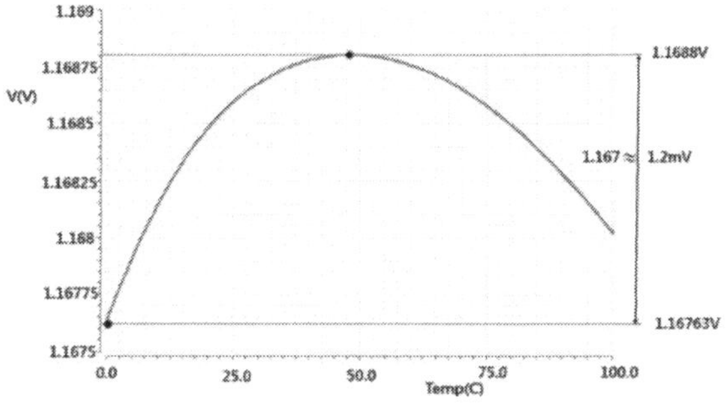

Figure 9. Simulation result of conventional BGR V_{REF}.

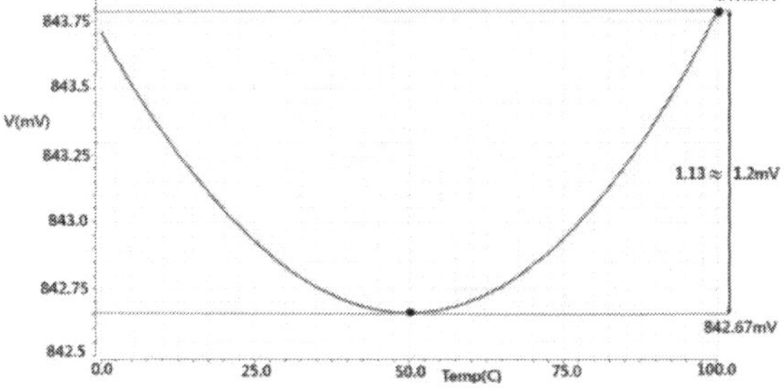

Figure 10. Simulation result of proposed BGR V_{REF}.

6 Conclusion

This paper presents the CMOS bandgap reference with small size and sub 1-V operation by using only two diode and TC of Vth to make PTAT current, respectively. It occupies an active area of $0.0067mm^2$. The simulation results are the voltage reference output of 0.843V, the accuracy of ±0.08%, and TC of 12ppm/°C from 0 °C to 100°C.

Acknowledgement

This work was partly supported by the IT R&D program of MSIP/IITP [2014-050-005-038, Development of Fast Wireless Charge Technology for 2A Mobile Device Supporting Dual Standard].

References

1. P. R. Gray and R. G. Meyer, *Analysis and Design of Analog Integrated Circuits*. New York :Wiley pp. 289-296 (1977),
2. R. J. Widlar, "New developments in IC voltage regulators," *IEEE J.Solid-state Circuits*, vol. **6**, pp. 2-7, Feb. (1971)
3. Banba, Hironori, et al. "A CMOS bandgap reference circuit with sub-1-V operation." *Solid-State Circuits, IEEE Journal of* **34.5**: 670-674 (1999)
4. Sanborn, Keith, Dongsheng Ma, and Vadim Ivanov. "A sub-1-V low-noise bandgap voltage reference." *Solid-State Circuits, IEEE Journal of* **42.11**: 2466-2481 (2007):
5. Jiang, Yueming, and Edward KF Lee. "Design of low-voltage bandgap reference using transimpedance amplifier." *Circuits and Systems II: Analog and Digital Signal Processing, IEEE Transactions on* **47.6**: 552-555 (2000).
6. Chen, Suheng, and Benjamin J. Blalock. "Switched capacitor bandgap voltage reference for sub-1-V operation." *IEICE Electronics Express* **3.24**: 529-533 (2006).
7. Wiessflecker, Martin, et al. "A sub 1V self clocked switched capacitor bandgap reference with a current consumption of 180nA." *Circuits and Systems (ISCAS), 2012 IEEE International Symposium on*. IEEE, 2012 : pp 2841-2844(2012).
8. Shrivastava, Aatmesh, et al. "5.4 A 32nW bandgap reference voltage operational from 0.5 V supply for ultra-low power systems." *Solid-State Circuits Conference-(ISSCC), 2015 IEEE International*. IEEE, 2015.
9. Ohte, Akira, and Michiaki Yamagata. "A precision silicon transistor thermometer." *Instrumentation and Measurement, IEEE Transactions on* **26.4**: 335-341 (1977).
10. Freire, Raimundo Carlos Silverio, Sergio Daher, and Gurdip Singh Deep. "A highly linear single pn junction temperature sensor." *Instrumentation and Measurement, IEEE Transactions on* **43.2**: 127-132 (1994)

DLL-based 4-phase duty-cycle and phase correction circuit for high frequency clock tree

Jaewoo Song[1], Hee-Am Shin[2] and Soo-Won Kim[2]

[1]*Department of Nano-Semiconductor Engineering, Korea University, Seoul, Republic of Korea*
[2]*Department of Electrical Engineering, Korea University, Seoul, Republic of Korea*

Abstract. In high-speed data transmission applications such as double data rate memory and double sampling ADCs, clock generation and distribution circuits must provide the clocks with precise duty cycle of 50% and sufficient timing margin. The proposed DLL-based 4-phase duty-cycle and phase correction circuit, consisting of delay-locked loop (DLL) and 45 phase clock generator (SR latch) corrects distorted duty-cycle clock to 50% duty-cycle. The distorted duty-cycle input clock passes through the DLL. After the DLL is locked, the delay of delay line is identical to the period of input clock. Lastly, 4-phase, 50% duty-cycle clocks is generated from the combination of rising edges of signals at each 1/4 points of delay line. The proposed circuit is implemented in 65nm CMOS. The simulation results shows that the frequency range of the proposed circuits is 550-1600MHz, the maximum duty cycle error of the output clock can be less than 1% with the input duty cycle correction ranging from 25% to 80%. The phase difference with the 4-phase output clock is 250±3ps at a frequency of 1GHz. The measured power dissipation is 4.3mW.

1 Introduction

Recently, game consoles or high-performance TVs are need to high frequency bandwidth DRAM for improving the processing speed. However, an increasing in power consumption and a reducing of timing margin according to improvement of frequency have become a problem.
One of the solutions is using multi-phase signals. But there are some unwanted factors such as jitter, skew, and PVT variations that make distortion of duty-cycle and/or a phase-skew between multi-phase clocks. For this reason, the duty-cycle correction (DCC) and phase correction (PC) are essential blocks for the circuits using multi-phase clock signals.

Many studies about DCCs have been progressed and classified into analog type and digital type. Analog DCCs usually employ a pulse-width control loop (PWCL) [1]-[3]. Because of its analog and feedback loop, this type of DCCs achieves higher correcting accuracy, but suffers from relatively slow settling behaviour, loop instability, and charge pump mismatch.

All-digital duty cycle corrector (ADDCC) have two major types of architecture: synchronous-mirror-delay (SMD) and time-to-digital converter (TDC). And some applications use DCC with delay-locked loop (DLL) to eliminate the phase error.

The SMD-based ADDCCs [4]-[5] use a half-cycle delay line (HCDL) composed of a matched delay line and SR latch to generate a 50% duty-cycle clock. However, the pulse width generated by two delay-line is easily affected by process corners making mismatch problem.

The TDC-based DCCs [6]-[8] quantize a period of input clock into a digital code, and then generate half-cycle delay time using this digital code. The TDC-based DCCs can reduce the lock-in time, however duty-cycle error of the output is dependent on TDC resolution. Furthermore, it is difficult to design a high-resolution and a wide range TDC with low power.

Although the DCCs can make a 50% duty-cycle output clock, phases of the input and output clock are misaligned. To solve this problem, the DCC with DLL is proposed [9]. However, the DCC with DLL has a design challenge and circuit complexity of the DLL has to be minimized to reduce the power consumption and area.

Previously introduced circuits are single phase DCC architectures and some circuits are difficult to implement for multi-phase clock scheme. In this paper, a DLL-based 4-phase DCC and PC circuit for multi-phase interface is proposed. This scheme uses DLL to detecting the input clock period. The DLL guarantees precise detecting operation because of feedback loop. And then 50% duty-cycle signals with 90° phase difference are made from the combinations of rising edge of four signals from delay line. Due to the using of analog type DLL, proposed design can achieve relatively precise duty-cycle correction and phase correction. Furthermore, errors caused by a mismatch are eliminated by using current mismatch reduced charge pump and clock generation structure.

The rest of this paper is organized as follows. The architecture of proposed 4-phase DCC and PC and detailed circuit description are presented in Section 2.

Section 3 discusses the simulation results. Section 4 gives the conclusion with summary.

2 Proposed DCC and PC architecture and circuit description

2.1 Architecture

2.1.1 Architecture

Figure 1 shows the block diagram of the proposed 4-phase DCC and PC. There are two major parts of this circuits: a DLL and a 4-phase clock generator (CG). The DLL is comprised of a phase and frequency detector (PFD), a voltage controlled delay line (VCDL), a charge pump (CP) and a loop filter (LF). The CG consists of a pulse generator (PG), a single to differential converter (SDC), 2 types of SR latch, and a latched comparator.

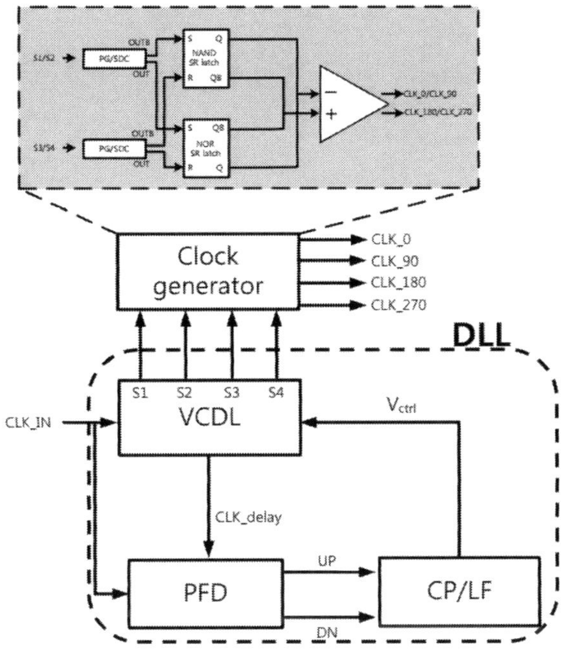

Figure 1. Block diagram of the proposed 4-phase DCC and PC.

2.1.2 Circuit operation

The input clock passes through the VCDL and the input clock is compared with the delayed clock to align the positive edges of the two signal at PFD. The PFD of the DLL detects the phase and frequency error between the input clock and the delayed clock, and then the PFD outputs "UP/DOWN"-signal control the CP to convert a phase error into a control voltage. The control voltage controls the delay of VCDL to eliminate the phase error between the input clock and the delayed clock. When the phase error is removed, the DLL is locked.

After the DLL is locked, the delay between the input clock and the delayed clock are equal to the period of the input clock. And the output signals at every 1/4 point of VCDL apply to the input signal of the clock generator. The rising edges of the input signals (S1 – S4) trigger pulse generator that makes a pulse. The SDC makes the pulse into differential short pulses. The differential signal passes through each types of SR latch. To get 50% duty cycle clock, the SR latches sense the rising edge of the input signals (S1/S3 or S2/S4) which rising edges have 180° phase difference. And its output signals enter into the latched comparator. Finally, the output signals of latched comparator are 4-phase clock signal with 50% duty cycle.

2.2 Circuit description

In this section, detailed DCC and PC circuits are introduced. The DLL block is examined first, followed by an in-depth description of the clock generator.

2.2.1 Voltage controlled delay line (VCDL)

The VCDL which is consist of current starved delay cell is illustrated in Figure 2. Both the N and the P side of mosfets control the rising and falling time of inverter. Total delay(T_{del}) is number of delay cell(N) • delay of unit cell(T_{ck}) and delay difference between each point (S1-S4) is identical to T_{del} /4. So if the DLL is locked, T_{del} is equal to the input clock period and the difference of delay between each points is 1/4 of the input clock period(=90 ° phase difference). Each points have a dummy cell to balance the capacitance loading.(this dummy cell is not described in Figure 2)

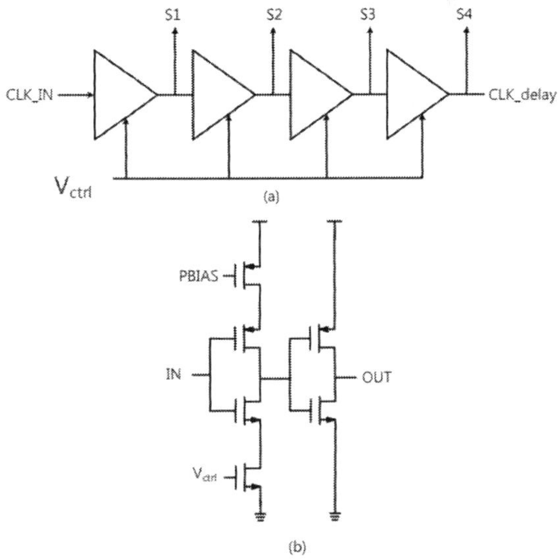

Figure 2. (a) Voltage controlled delay line (b) Delay cell.

2.2.2 Phase frequency detector (PFD)

The conventional phase frequency detector is used as shown in Figure 3. This circuit detects a phase error between the input clock and the delayed clock. It is also insensitive to duty cycle because of using a D-flip flop. When the input clock is high, output of the positive edge triggered D-flip flop goes high and generates UP signal. In a similar way, the delayed clock becomes high, output of the D-flip flop is high and generates DOWN signal. Both UP and DOWN signal are high, in that case, both D-flip flops are reset and UP and DOWN signal goes low. When the input phase error is eliminated, UP and DOWN signals make small and same pulse width signal at the same time when the both input clocks become high.

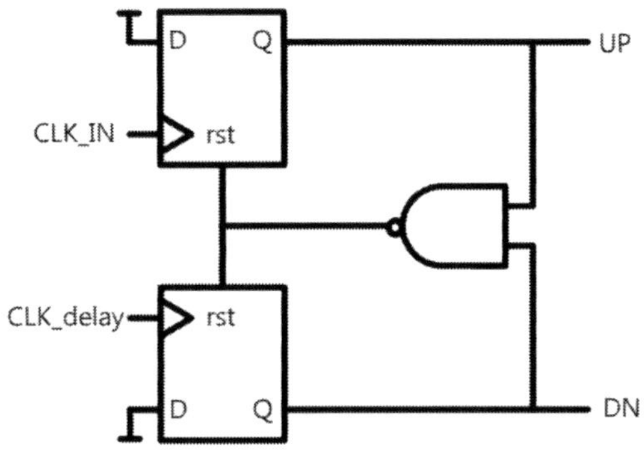

Figure 3. Conventional phase frequency detector.

2.2.3 Charge pump (CP)

Figure 4 shows the current mismatch reduced charge pump. This circuit converts phase error detected by the PFD into delay-line control voltage. The UPB and DN signals turn-on mosfet and provide current path onto node V_{ctrl}, and charge or discharge to move V_{ctrl} UP or DOWN. For ideal operation, I_{up} and I_{dn} need to be equal. But there are some non-ideality causing current mismatch. To solve these problems, 2 OP-AMPs are used. One problem is charge sharing issue. When UPB and DN are open, node X and Y are biased by the unity-gain amplifier and this

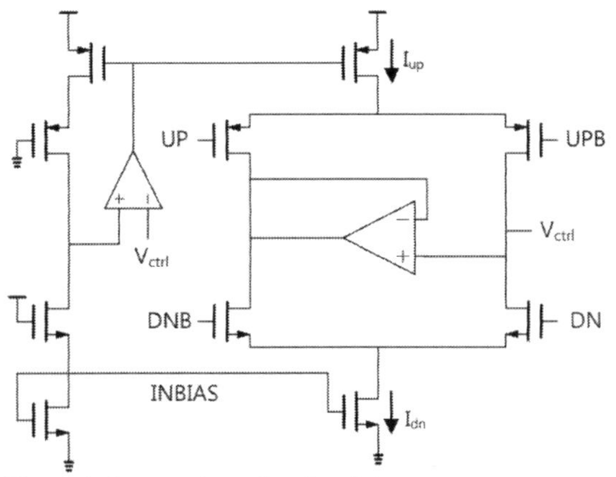

Figure 4. Current mismatch reduced charge pump.

structure suppresses charge sharing problem. Another mismatch is due to V_{ctrl} variation. This problem can be decreased by controlling the bias voltage of current source. The OP-AMP senses V_{ctrl}, modulates the bias voltage of I_{up} and therefore current mismatch is reduced.

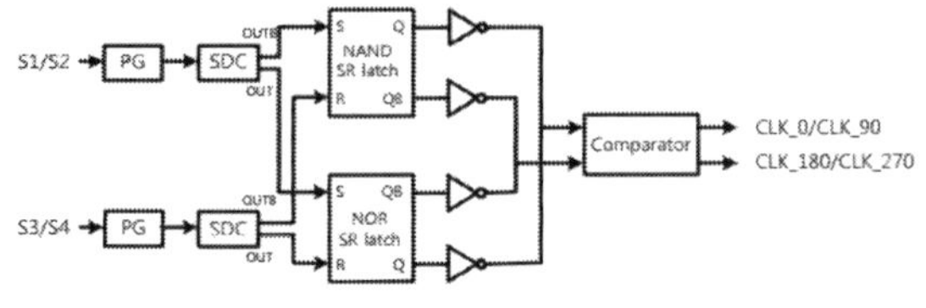

Figure 5. 4-phase Clock generator.

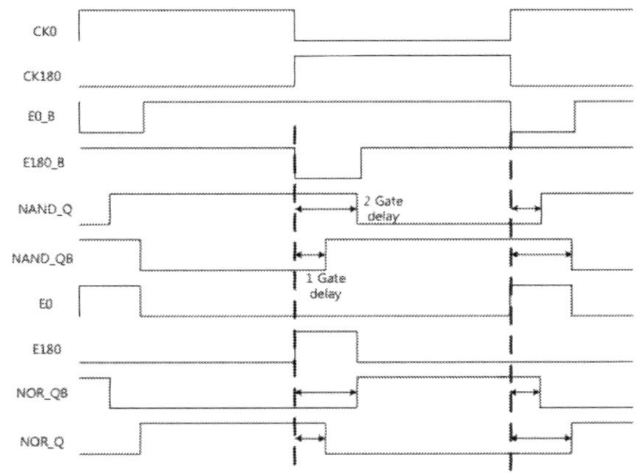

Figure 6. Timing diagram of SR latches.

2.2.4 Clock generator (CG)

The proposed clock generator design is used as shown in Figure 5. The CG has the two input signals with 180° phase difference, and senses the rising edges of the two signals and generates the 50% duty-cycle with 180°phase difference output clocks utilizing these rising edges.

To explain in detail, the input signal becomes high, the pulse generator makes a short pulse only to detect the rising edge regardless of the duty of the input signals. The SDC converts these short pulses to differential pulses. And the differential pulses through the SDC is applied to the different types of SR latch: NAND and NOR type. Table 1 shows the truth tables of SR latches. Since it has the structural difference, a pulse width of the output clocks is different: Figure 6 shows the timing diagram of SR latches. To decrease the difference of the output, the outputs of SR latches are interpolated. As a result, the difference between the rising edge and falling edge is about 1.5 gate delay, and the pulse width of the output clocks is approximately 50% of the input period. Moreover, by using latched comparator lowering the propagation delay, the error is further reduced.

DOI: 10.1051/matecconf/20165412005

Table 1. Truth tables of NAND and NOR SR latch

NAND SR latch			
S	**R**	**Q**	**QB**
0	0	1	1
0	1	1	0
1	0	0	1
1	1	No change	

NOR SR latch			
S	**R**	**Q**	**QB**
0	0	No change	
0	1	0	1
1	0	1	0
1	1	0	0

3 Simulation results

The proposed 4-phase DCC and PC is fabricated using a standard performance 65-nm CMOS process with a 1.2V power supply voltage. Figure 7 shows the layout of the circuits, and the active area is 165 μm x 79 μm. Figure 8 shows the output clock with the 1GHz and 40% duty-cycle input clock. The duty-cycle error is less than 5ps and the phase error is less than 3ps. Figure 9 shows the maximum duty cycle error and phase error of the output clock at typical process corner with the different input frequency and different input duty cycles. The maximum duty-cycle error is always smaller than 1% and the maximum phase error is always smaller than 1%. The input frequency of the proposed 4-phase DCC and PC ranges from 550MHz to 1.6GHz, and the input duty-cycle ranges from 25% to 80%.

Table 2 shows a performance comparison with prior works. The proposed circuit achieves the 4-phase 50% duty cycle output signals with a wide operation range, a wide duty cycle correction range with low output duty error and phase error and takes less chip area as well as power consumption.

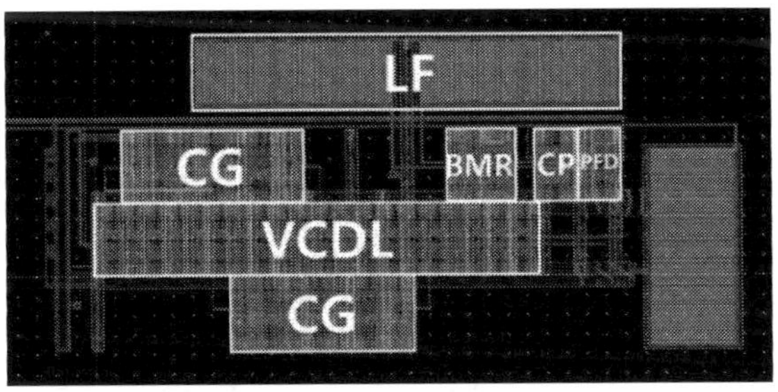

Figure 7. Layout of the proposed DCC and PC

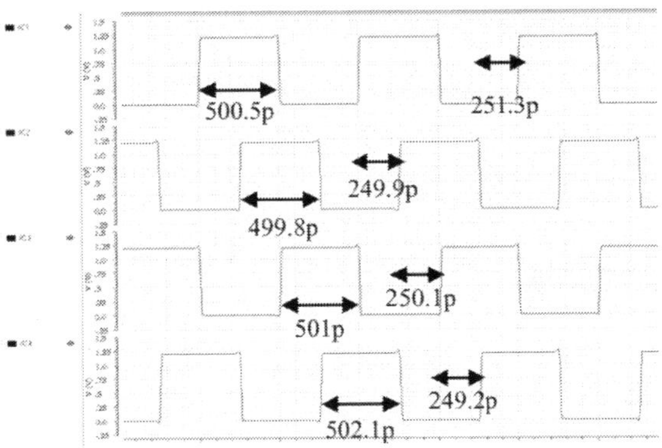

Figure 8. 4-phase output clock with the 1GHz, 25% duty cycle input clock.

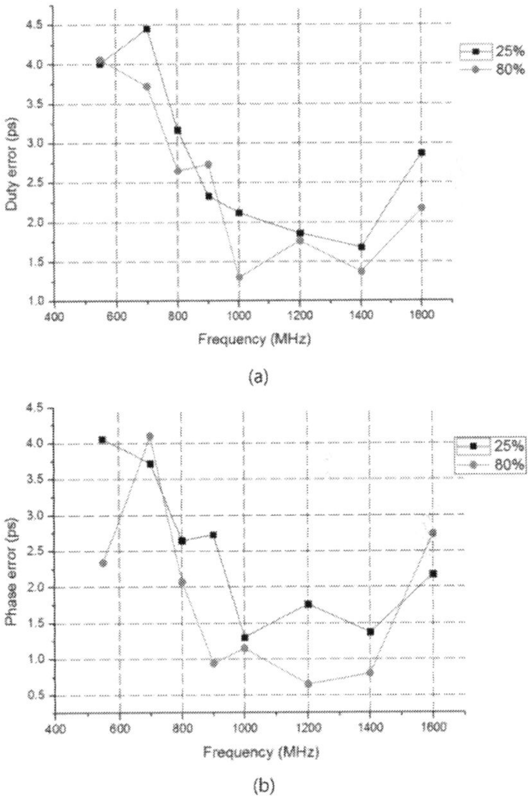

Figure 9. Maximum (a) duty cycle error and (b) phase error of the output clock at typical process.

4 Conclusion

In this paper, proposed 4-phase DCC and PC with a DLL is presented. The proposed DCC and PC circuit not only corrects the duty cycle of the output clock to 50% but also generates 4-phase clock with 90° phase difference. In addition, the proposed scheme can operate across a wide-frequency range and accomplish the DCC and PC despite a wide range of input duty cycle. Thus, it can be used for duty-cycle correction and phase correction applications, such as the DDR memory, multi-phase I/O interface.

Acknowledgments

This work was supported by the IT R&D Program of MKE/KEIT (10045400, HF-Band Magnetic Resonant Wireless Charging SoC for Mobile Device).

References

1. K. H. Cheng, C. W. Su, & K. F. Chang. A high linearity, fast-locking pulsewidth control loop with digitally programmable duty cycle correction for wide range operation. *Solid-State Circuits, IEEE Journal of*, **43**(2), 399-413. (2008).
2. F. Mu, & C. Svensson. Pulsewidth control loop in high-speed CMOS clock buffers. *Solid-State Circuits, IEEE Journal of*, **35**(2), 134-141. (2000)
3. H. Y. Huang, W. M. Chiu, & W. M. Lin. Pulsewidth control loop circuit using combined charge pumps and miller scheme. In *Solid-State and Integrated Circuits Technology, 2004. Proceedings. 7th International Conference on* (**Vol. 2**, pp. 1539-1542). IEEE. (2004, October)
4. Y. M. Wang, & J. S. Wang. An all-digital 50% duty-cycle corrector. In *Circuits and Systems, 2004. ISCAS'04. Proceedings of the 2004 International Symposium on* (**Vol. 2**, pp. II-925). IEEE. (2004)
5. Y. M. Wang, C. F. Hu, Y. J. Chen, & J. S. Wang. An all-digital pulsewidth control loop. In *Circuits and Systems, 2005. ISCAS 2005. IEEE International Symposium on* (pp. 1258-1261). IEEE. (2005, May)
6. J. Gu, J. Wu, D. Gu, M. Zhang, & L. Shi. All-digital wide range precharge logic 50% duty cycle corrector. *Very Large Scale Integration (VLSI) Systems, IEEE Transactions on*, **20**(4), 760-764. (2012)
7. J. R. Su, T. W. Liao & C. C. Hung. Delay-line based fast-locking all-digital pulsewidth-control circuit with programmable duty cycle. In*Solid State Circuits Conference (A-SSCC), 2012 IEEE Asian* (pp. 305-308). IEEE. (2012, November)
8. C. C. Chung & C. J. Li. A low-power delay-recycled all-digital duty-cycle corrector with unbalanced process variations tolerance. In *VLSI Design, Automation, and Test (VLSI-DAT), 2013 International Symposium on*(pp. 1-4). IEEE. (2013, April)
9. C. C. Chung, D. Sheng, & S. E. Shen. High-resolution all-digital duty-cycle corrector in 65-nm CMOS technology. *Very Large Scale Integration (VLSI) Systems, IEEE Transactions on*, **22**(5), 1096-1105. (2014)
10. J. C. Ha, J. H. Lim, Y. J. Kim, W. Y. Jung & J. K. Wee. Unified all-digital duty-cycle and phase correction circuit for QDR I/O interface.*Electronics Letters*, **44**(22), 1300-1301. (2008)

Table 2. Comparison with previous works

	[6] TVLSI '12	[7] ASSCC '12	[9] VLSI '14	[10] E-letter '08	This work
Supply(V)	1.8	1.8	1.0	1.8	1.2
Process	0.18 um	0.18 um	65nm	0.18 um	65 nm
The number of clock	1	1	1	4	4
Input duty cycle range	20 – 80 (%)	30 – 70 (%)	14 – 86 (%)	40 – 60 (%)	25 – 80 (%)
Output duty error	± 3.5% @1GHz	± 2.5%	± 1.6%	± 0.4%	± 1%
Output Phase error	-	-	-	± 3ps	± 4ps
Operating frequency	400Mhz – 2GHz	200 – 600 MHz	262MHz– 1.02GHz	800Mhz – 1.2GHz	550Mhz – 1.6GHz
Total power	3.6mW @2GHz	5.49mW @ 600MHz	6.5mW @ 1.02GHz	16mW @1GHz	4.3mW @ 1GHz
Area(mm^2)	0.025	0.0252	0.01	0.0896	0.013

AUTHOR INDEX

AUTHOR INDEX